AF805141

Mosquitoes

Omkar
Editor

Mosquitoes

Biology, Pathogenicity and Management

Springer

Editor
Omkar
Department of Zoology
University of Lucknow
Lucknow, Uttar Pradesh, India

ISBN 978-981-97-4162-5 ISBN 978-981-97-4163-2 (eBook)
https://doi.org/10.1007/978-981-97-4163-2

© The Editor(s) (if applicable) and The Author(s), under exclusive license to Springer Nature Singapore Pte Ltd. 2024
This work is subject to copyright. All rights are solely and exclusively licensed by the Publisher, whether the whole or part of the material is concerned, specifically the rights of translation, reprinting, reuse of illustrations, recitation, broadcasting, reproduction on microfilms or in any other physical way, and transmission or information storage and retrieval, electronic adaptation, computer software, or by similar or dissimilar methodology now known or hereafter developed.
The use of general descriptive names, registered names, trademarks, service marks, etc. in this publication does not imply, even in the absence of a specific statement, that such names are exempt from the relevant protective laws and regulations and therefore free for general use.
The publisher, the authors and the editors are safe to assume that the advice and information in this book are believed to be true and accurate at the date of publication. Neither the publisher nor the authors or the editors give a warranty, expressed or implied, with respect to the material contained herein or for any errors or omissions that may have been made. The publisher remains neutral with regard to jurisdictional claims in published maps and institutional affiliations.

This Springer imprint is published by the registered company Springer Nature Singapore Pte Ltd.
The registered company address is: 152 Beach Road, #21-01/04 Gateway East, Singapore 189721, Singapore

If disposing of this product, please recycle the paper.

Preface

The word mosquitoes in itself is quite frightening to human beings. They are tiny and common flying insects that live in most parts of the world. There are approximately more than 3600 species of mosquitoes found worldwide. They live both indoors and outdoors and can bite day and night. They are often active in the dark but can also be active and effective in the light.

Mosquitoes belong to the insect order Diptera. They possess a pair of compound eyes, a pair of pilose or plumose antennae (believed to detect carbon dioxide from a person's breath), piercing and sucking mouthparts, and two functional wings, the forewings. The hindwings are reduced into halters, which maintain balance during their flight. When they fly close to human being, a buzzing sound can be heard. In addition, they possess three pairs of jointed legs. Not all mosquitoes bite people and animals. Both the male and female mosquitoes feed on nectar and plant juices to obtain energy for flight. But the female mosquitoes feed on blood to get nutrients required to stimulate the egg ripening and layings. When they bite infected persons, they get infected with certain protozoans, nematode parasites, and viruses. Further, it takes just a few infected mosquitoes to start the outbreak of dreaded diseases in a community. They are considered prime human enemy as vectors.

The mosquito bite symptoms of human beings include a hard, puffy, and reddish brown bump soon after the bite or multiple bites in a day. Sometimes, small blisters or dark spots like bruises can be found instead of hard bump. The first bite of a mosquito may cause immune disorders in a person. Mosquito bite on inside of forearm of a person may cause more severe reactions with a large area swelling and redness, low grade fever, and swollen lymph nodes. We could not escape from mosquito bites despite closing the doors and windows well in advance of sunset. While fixing steel wire mesh on doors and windows is quite effective, it cannot completely protect us from mosquito bites. A mosquito bite to humans causes various diseases, like malaria, filaria, dengue, chikungunya, Japanese encephalitis, yellow fever, etc.

In view of the above issues, it has been planned to publish a book dedicated to mosquitoes. This book starts with a chapter on the silent architects, followed by origin and evolution, morphology, mosquito diversity and important species, biology, host finding, recognition, specificity and host finding, behavior and rhythmicity, their abiotic and biotic interactions, disease vectors and mechanism of disease transmission, mosquito-borne human diseases, mosquito-borne veterinary diseases, and management of mosquitoes. Mosquito management in itself is a tricky and difficult issue as adults are free living and terrestrial, while immature stages, especially the larvae are aquatic in habitat. One of the most characteristic features relatively different from other insects is their fast development of adaptation and biotransformation of insecticides applied to reduce/repel their population. Thus, a chapter on resistance to insecticides has also been taken. Mostly, we have been aware of the negative aspects of mosquitoes, but this book will cover the positive aspects of mosquitoes as well, like their use in research as a model besides other positive roles played by them, like in development of microneedles, their role in crop pollination, etc.

There are several existing books on mosquitoes, which emphasize either their taxonomy and distributional records or their role as vectors of humans and the livestock. This book has been planned with a holistic approach to bring to the fore both the negative and positive aspects of mosquitoes. Further, each chapter has been designed to include pointwise learning objectives at the beginning and conclusions and points to remember at the end. This approach makes the book students' and readers' friendly.

This book will not only be quite useful to the students of entomology, public health, medical and veterinary sciences, and applied life sciences but also to those involved in policy planning and vector management. In addition, the students preparing for competitive examinations and general public at large will also be benefitted.

I am grateful to various specialists and academicians for sparing time from their routine and agreeing to contribute chapters for this book. Most of the contributors, being senior specialists are likely to contribute their respective chapters with special emphasis on the topics covered by them.

Lucknow, Uttar Pradesh, India Omkar

Contents

Editor and Contributors

About the Editor

Omkar is a former Professor and Head, Department of Zoology, University of Lucknow. He has 36 years of teaching and 44 years of research experience in pesticide toxicity, as well as in the study of ladybird and parthenium beetles dealing with their ecology, behavior, sexual selection, and evolutionary biology. He has approximately 300 publications, including 24 written and edited books. He is a fellow of the National Academy of Sciences, India, and a recipient of Prof. T. N. Ananthakrishnan Foundation award and Rescholar award of Excellence in Agricultural Entomology. He has also been honored with Prof. G. S. Shukla Gold Medal, Saraswati Samman in Higher Education by the State Government, and ACCLAIM and Udeepan Awards from the University of Lucknow.

Contributors

Gautam Aditya Department of Zoology, University of Calcutta, Kolkata, West Bengal, India

Nasreen Akhtar ICMR-National Institute of Malaria Research, New Delhi, India

Paramita Banerjee Entomology Research Unit, Department of Zoology, The University of Burdwan, Burdwan, West Bengal, India

Sampa Banerjee Department of Zoology, University of Calcutta, Kolkata, West Bengal, India

Indranil Bhattacharjee Mosquito, Microbiology and Nanotechnology Research Units, Parasitology Laboratory, Department of Zoology, The University of Burdwan, Burdwan, West Bengal, India

Department of Zoology, Dr. Bhupendra Nath Dutta Smriti Mahavidyalaya, Hatgobindapur, West Bengal, India

Shreya Brahma Department of Zoology, University of Calcutta, Kolkata, West Bengal, India

Sunanda Burman Department of Zoology, The University of Burdwan, Burdwan, West Bengal, India

Goutam Chandra Department of Zoology, The University of Burdwan, Burdwan, West Bengal, India

Amlan Das Entomology Laboratory, Department of Zoology, University of Calcutta, Kolkata, West Bengal, India

Dipanwita Das Department of Zoology, Bagnan College, Bagnan, West Bengal, India

Susmita Das Entomology Laboratory, Department of Zoology, University of Calcutta, Kolkata, West Bengal, India

R. C. Dhiman ICMR-National Institute of Malaria Research, New Delhi, India

Rajnikant Dixit Host Parasite Interaction Biology Group, ICMR-National Institute of Malaria Research, New Delhi, India

Alex Eapen ICMR-National Institute of Malaria Research Field Unit, Chennai, Tamil Nadu, India

Sanjeev Kumar Gupta ICMR-National Institute of Malaria Research, New Delhi, India

Tameka Hazra Entomology Laboratory, Department of Zoology, University of Calcutta, Kolkata, West Bengal, India

Neera Kapoor School of Sciences, IGNOU, New Delhi, India

Jaspreet Kaur ICMR-National Institute of Malaria Research, New Delhi, India

Devinder Kaur Kocher Department of Zoology, Punjab Agricultural University, Ludhiana, Punjab, India

Gaurav Kumar ICMR-National Institute of Malaria Research, New Delhi, India

Sarita Kumar Department of Zoology, Acharya Narendra Dev College, University of Delhi, New Delhi, India

Abhijit Mazumdar Entomology Research Unit, Department of Zoology, The University of Burdwan, Burdwan, West Bengal, India

Geetanjali Mishra Ladybird Research Laboratory, Department of Zoology, University of Lucknow, Lucknow, Uttar Pradesh, India

Susobhan Datta Mudi Department of Zoology, University of Calcutta, Kolkata, West Bengal, India

Ida Anandi Mukhiya Entomology Laboratory, Department of Zoology, University of Calcutta, Kolkata, West Bengal, India

Omkar Ladybird Research Laboratory, Department of Zoology, University of Lucknow, Lucknow, Uttar Pradesh, India

Shweta Pasi ICMR-National Institute of Occupational Health, Ahmedabad, Gujrat, India

Pranesh Paul Department of Zoology, University of Calcutta, Kolkata, West Bengal, India

Soujita Pramanik Department of Zoology, University of Calcutta, Kolkata, West Bengal, India

Shriza Rai Ladybird Research Laboratory, Department of Zoology, University of Lucknow, Lucknow, Uttar Pradesh, India

John Ravindran Department of Vector Genomics, ICMR-National Institute of Malaria Research, New Delhi, India

S. Renuka ICMR-National Institute of Malaria Research Field Unit, Bengaluru, Karnataka, India

Pooja Rohilla Academy of Scientific and Innovative Research, Ghaziabad, Uttar Pradesh, India

Milita Roy Department of Zoology, Bejoy Narayan Mahavidyalaya, Itachuna, West Bengal, India

Sampurna Roy Entomology Laboratory, Department of Zoology, University of Calcutta, Kolkata, West Bengal, India

Goutam K. Saha Department of Zoology, University of Calcutta, Kolkata, West Bengal, India

Vaishali Saini Academy of Scientific and Innovative Research, Ghaziabad, Uttar Pradesh, India

Ankita Sarkar Entomology Research Unit, Department of Zoology, The University of Burdwan, Burdwan, West Bengal, India

Himmat Singh ICMR-National Institute of Malaria Research, New Delhi, India

Poonam Singh ICMR-National Institute of Malaria Research, New Delhi, India

Tanvi Singh Academy of Scientific and Innovative Research, Ghaziabad, Uttar Pradesh, India

Sanjay Tevatiya Department of Vector Genomics, ICMR-National Institute of Malaria Research, New Delhi, India

B. K. Tyagi Department of Biosciences, University Institute of Biotechnology, Chandigarh University, Gharuan, Mohali, Punjab, India

H. C. Vani ICMR-National Institute of Malaria Research Field Unit, Bengaluru, Karnataka, India

Tripti Yadav Ladybird Research Laboratory, Department of Zoology, University of Lucknow, Lucknow, Uttar Pradesh, India

Silent Architects: The Mosquitoes

1

Shriza Rai, Omkar, and Geetanjali Mishra

Abstract

From altering the outcomes of conflicts to steering scientific advancements, the impact of mosquitoes extends far beyond their small stature, leaving an undeniable imprint on the story of humanity. This chapter uncovers the lesser known stories of empires and their tryst with mosquitoes, exploring the impact of mosquito-borne diseases on the rise and fall of civilizations throughout history. From ancient civilizations to modern superpowers, the narrative navigates through the corridors of time, revealing how the persistent hum of mosquitoes shaped the destinies of empires. From Greeks to Romans to Mongols, the persistent presence of mosquitoes has played a crucial role in shaping the fates of empires. Furthermore, the immunity against mosquito-borne diseases of the Africans played a decisive role in determining their importance in slavery. With the slave trade and Columbian Exchange, the mosquitoes expanded their territory. Despite the concerted efforts to defend against her unrelenting assaults, the mosquito's deadly offensives and crimes against humanity persist unabated. Although the counterattacks have managed to decrease the annual casualties inflicted by mosquitoes, they remain the planet's deadliest hunter of human beings. The narrative delves into the mosquito's ability to outlast and overcome the effectiveness of control methods, particularly in the 1960s. The mosquito's response to extermination efforts involves strategic adaptations, with malaria parasites developing resistance to chloroquine and mosquitoes building immunity during exposure to DDT. The intricate interplay of survival strategies and the unintended consequences of their adversary have made them number one enemy of humans. The chapter unfolds, revealing a profound story where the

S. Rai · Omkar · G. Mishra (✉)
Ladybird Research Laboratory, Department of Zoology, University of Lucknow, Lucknow, Uttar Pradesh, India

© The Author(s), under exclusive license to Springer Nature Singapore Pte Ltd. 2024
Omkar (ed.), *Mosquitoes*, https://doi.org/10.1007/978-981-97-4163-2_1

tiniest entities exert a disproportionately large influence. It underscores the idea that history, with all its intricacies, is frequently moulded by the most unforeseen actors: The Mosquitoes.

Keywords

Vectors · Empire · Humanity · Mosquito-borne diseases · Rise and fall of civilizations · Ancient civilizations

Learning Objectives

1. Chronicles the ways mosquitoes shaped history and impacted human existence.
2. The mosquito, which was once dismissed as a mere annoyance, has emerged as a villainous force that left an indelible mark on the trajectory of human existence. This chapter very well demonstrates this.
3. The chapter explains that while bringing some of the greatest mighty kingdoms to peril, mosquitoes have not only changed the course of history but has emerged as the deadly invincible army.
4. It will increase our understanding of how the mosquito shaped course of time.
5. How the immunity against mosquito-borne diseases described the course of Columbian exchange and slave trade has been explained.
6. The menace of mosquitoes is not limited to civilians but has played crucial in military operations. And also, how they have affected military personnels.
7. From unnoticed creature to the deadliest predator of all time has been discussed, describing the deadly role mosquito has played.

They slaughtered our ancestors and derailed our history. And they're not finished with us yet.
–Brooke Jarvis

1.1 Introduction

Originating around 200 million years ago, mosquitoes began their evolutionary transformation into the blood-sucking insects we encounter today. The examination of fossils, resembling contemporary mosquitoes revealed a series of adaptive changes over time. As we emerged from the era of hunting and gathering, mosquitoes embarked on a captivating odyssey, forging connections with humans that unveiled a tapestry of consequences and discoveries. With the transformative rhythm of development in cities, the clearing of forests and cultivation of fields altered the landscapes, unintentionally creating an ideal habitat for mosquitoes, and fostering their flourishing expansion (Gillett 1971). These tiny creatures thrived in various ecosystems, inadvertently becoming vectors of deadly diseases that influenced the rise and fall of civilizations.

As the mosquito population continues to surge, they have evolved into a formidable army, waging an unyielding war against humanity. When female mosquitoes feed on our blood, they do not just leave angry red welts on the skin but also transmit evolved microbes acting as carriers for advanced pathogens while transmitting deadly diseases, like malaria, dengue, filariasis, etc. The United States Centres for Disease Control and Prevention has labelled these teeny tiny three-legged creatures as "the world's deadliest creature" (Global Health 2023). Weighing merely 2.5 mg (http://www.dupagehealth.org/ftb), mosquitoes cause over half a million deaths worldwide annually (DuPage County Health Department 2023; Gates 2018) (Fig. 1.1).

Vector-borne illnesses, such as malaria, dengue, Zika virus, etc., spread by these tiny tormentors, have inflicted significant suffering and have altered the trajectories of societies (Winegard 2019; Norrie 2016; Clark 2010; Sherman 2007). Malaria alone has been a formidable adversary, affecting millions and influencing the outcomes of wars and conquests (Slater 2014). The conquest of new territories was not only a clash of civilizations but also a struggle against the unseen enemy carried by mosquitoes.

Beyond the immediate health and economic impacts, mosquitoes have influenced cultural practices and societal norms. Regions prone to mosquito-borne diseases have developed various rituals, and their daily life has adapted to mitigate the risks. The omnipresence of mosquitoes in literature, art, and folklore reflects the enduring struggle against these subtle but formidable foes.

Despite efforts, the struggle against mosquitoes has serious implications beyond the annoyance of itching. Several challenges have been highlighted in the fight against mosquitoes. These include the high cost and limited effectiveness of biological control. Environmental concerns also hinder traditional methods, like swamp draining. Moreover, mosquitoes are developing resistance to new insecticides at a

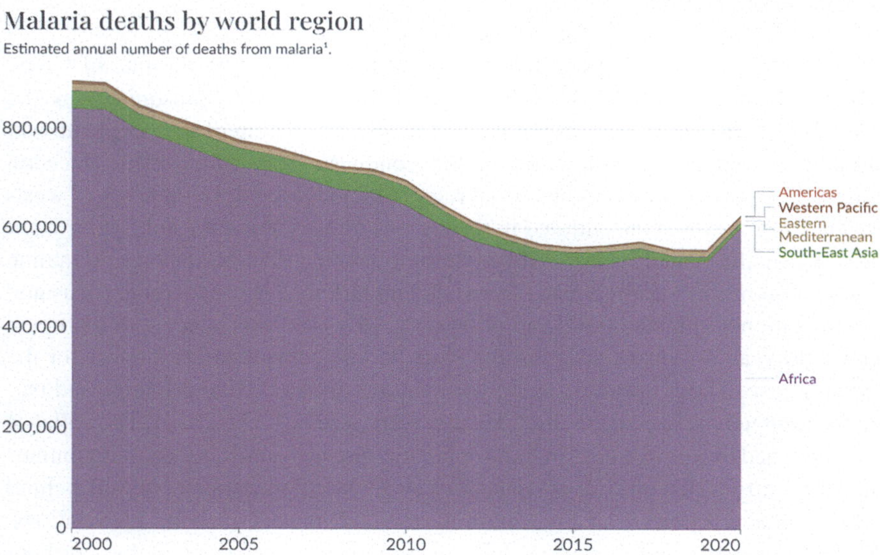

Fig. 1.1 Number of deaths caused by malaria worldwide from 2000 to 2020. Data source: Global Malaria Programme, WHO (2021), https://OurWorldInData.org/malaria

rapid pace. Additionally, the efficacy of drugs, like chloroquine, for malaria treatment is diminishing. The continuous quest for the perfect repellent remains a formidable challenge for researchers, underscoring the complexities involved in finding an optimal solution. Consequently, we find ourselves in a perpetual war against mosquitoes, from global health campaigns and pesticide use to backyard spraying. Lot of stories and events have been taken from T. C. Winegrard's *The Mosquito: A Human History of Our Deadliest Predator*, and it is highly recommended to read the book for detailed understanding. This chapter is a journey through time, where the unassuming mosquito takes the centre stage in a drama of conquests, resilience, and adaptations. A tale where the smallest among us wield an outsized influence, proving that history, in all its complexity, is often shaped by the most unexpected players: The Mosquitoes.

1.2 Mosquitoes and Empire Odyssey

Mosquitoes are often overlooked in the tapestry of history. At first, they appear as a blip in the vast scope of historical events. However, when observed closely, these diminutive creatures have played an outsized role in shaping the course of human affairs (Winegard 2019). Despite its miniscule size and featherweight, this unassuming insect has left an indelible mark on the pages of history, often steering the narrative in unexpected ways.

1.2.1 Battles of Marshes and Mosquitoes: Greece's Triumph Over Persia

Mosquitoes have played subtle but significant role in deciding the fate of Greek empire that went beyond battles. Through health challenges, mosquito took the role of the Third army, influencing the health and well-being of the people at war, ultimately changing the direction of the conquest. At a time when Macedon remained a rugged, mountainous tribal region, the ongoing developments of western civilization were concentrated in the regions of Mesopotamia and Egypt. Until 1200 BCE, the Middle East enjoyed stability with a balanced power dynamic among empires, like Babylonian, Assyrian, and Hittite. Their economic focus and specialization stimulated trade and prosperity. However, this peace was disrupted within 50 years by invasions from the "Sea People", mercenaries famous for the Trojan Horse. These invaders, along with natural disasters, like droughts and tsunamis, plunged the region into the "Ancient Dark Ages" (Cline 2014). The collapse was worsened by a devastating malaria epidemic spread by mosquitoes. Babylonians attributed it to a demon, Nergal (Black et al. 1998). This catastrophe left behind only ruins and a power vacuum, marking the end of once-thriving civilizations. While much of the Middle East sank into a cultural and evolutionary abyss following the devastation wreaked by marauding sea people, a new power emerged from the shadows in the eastern Persian Empire of Cyrus the Great, quietly rose to fame.

Under the leadership of Cyrus and his successors, Darius I and Xerxes, the Persian Empire continued to expand to include Egypt, Sudan, and eastern Libya. This foreshadowed a historic clash with Greece, a pivotal moment in ancient history. Cyrus brought together every nation under his regime except Greece. At that time, there was no unified "Greece". Instead, it comprised competing city-states, with Sparta and Athens as the top contenders for power. In 499 BCE, the Ionian Greeks, supported by Athens, revolted against Persian Emperor Darius I. Darius sought revenge and strategic control, leading to a full-scale invasion of Greece 7 years later. The Persian army crossed the Dardanelles Straits, marching through Thrace and Macedonia. However, Darius's avenging campaign turned disastrous. Approaching Athens, the Persian naval fleet was destroyed by a storm, and the land forces suffered setbacks, possibly due to a deadly combination of dysentery, typhoid, and malaria. Two years later, in 490 BCE, Darius launched a second campaign. This time, he opted for a different approach, avoiding the challenging northern overland route. Instead, he initiated an impressive amphibious landing, bringing a formidable force of 26,000 soldiers to Marathon, approximately 26 miles north of Athens. Despite being outnumbered, the amateur Athenian warriors were well-armed and heavily bronze-armoured. They cleverly contained the Persians in low-lying marshy camp sites. Within a week, the same deadly combination of disease mentioned earlier began to thin the Persian ranks. The geographical constraints of the Persian fleet's position, the landing grounds, and the Athenian defence setup made it impossible for the Persians to camp away from or navigate around the swamps. The battlefield's terrain and the Athenian strategy dictated the course of the battle. After a decisive victory for Athens, the disease-stricken Persians withdrew and set sail to launch an attack on Athens itself. Herodotus, the historian, documented that 6400 Persian bodies lay scattered on the battlefield, with an unknown number succumbing to the perils of the surrounding swamps. The Athenian army hurried to Athens and managed to thwart any Persian attempts to land. Sensing a missed opportunity, the Persians retreated, only to meet the surviving army's disillusionment with malaria ravages. However, this was not the end of the threat, for they would return under the succession and coming of a new leader, Xerxes, the late Emperor Darius.

Fuelled by a determination to avenge his father, Xerxes took personal command of an extraordinary and formidable force, combining both naval and land elements, reaching nearly 400,000 men. Against overwhelming odds, the 1500 remaining Greeks, with 300 Spartans commanded by King Leonidas, bravely confronted the Persian advance, fighting resolutely until their last breaths. This final stand gave rise to the legendary tale of the 300, a story that has since been embellished to the point of being unrecognizable, as exemplified by the historically distorted 300 movie franchise. Upon entering the coveted city, Xerxes was met with an unexpected sight—Athens stood deserted. The citizens and the retreating army had been evacuated by the fleeing fleet to the island of Salamis, leaving Athens deserted. Outraged by this, Xerxes commanded his navy to obliterate the coalition fleet stationed at Salamis, seeking retribution for what he considered an act of insolence. However, the Persian navy fell into a cleverly devised trap orchestrated by the Athenian general Themistocles. The Greeks skilfully lured the numerically superior Persian vessels into a narrow strait and swiftly blocked both entries with their more robust trireme ships. The Persian ground forces struggled in swampy terrain and faced resistance from Greek towns surrounded by marshes (Hughes 2014). Malaria and dysentery hit the Persian ranks hard, causing losses of over 40%. The weakened Persian army was defeated at the Battle of Plataea in August 479 BCE, putting an end to any Persian threat to Greece. The victories at Salamis and Plataea marked a turning point in the Greco-Persian Wars. Guided by General Anopheles, these wins shifted power to Greece, leading to its "Golden Age" and influencing the development of modern western society.

1.2.2 A Tragic Tale of Epidemics and War: Fall of Greece

Malaria had a detrimental impact on ancient Greece, contributing to its decline. The poet Homer mentioned malaria in the Iliad as "burning breath taints the red air with fevers, plagues, and death". Plato expressed regret that people had become so unhealthy that doctors had to create names for their illnesses and said "And we have made ourselves living cesspools, and driven doctors to invent names for our diseases". The prominent figures, like Sophocles, Aristophanes, and Hippocrates, documented its effects. The renowned Greek physician and the father of western medicine, Hippocrates (460–370 BCE) compared the deadly malaria season in summer and early fall to the arrival of Sirius, the Dog Star, terming it the "dog days of summer". He distinguished malaria from other fevers, emphasizing prevention over cure. Despite linking standing water to malaria, the role of mosquitoes remained unknown. Zoology professors Cloudsley-Thompson (1976) take a step further, acknowledging, "Hippocrates had a profound understanding of malaria; this subtle disease would go on to undermine and deteriorate the civilizations of ancient Greece and Rome". In the historical context of the ascension and decline of these superpowers, the mosquito, possessing skills comparable to earthly soldiers, played a crucial role in shaping the results of battles and campaigns (Karlen 1996; Jones 1907). During the Peloponnesian War, malaria worsened the burdens, influencing its outcomes. The rivalry between Athens and Sparta escalated, but disease intervened. The Plague of Athens in 430 BCE devastated the city, killing thousands and altering historical events. The subsequent Sicilian expedition in 415 BCE saw the Athenian army crippled by malaria, leading to a catastrophic defeat. The repercussions included the overthrow of Athens' democratic government and Sparta's economic decline. Endemic malaria further drained Greek resources and vitality, ultimately marking the end of Greece's golden age (Hughes 2014).

1.2.3 The Malarial End of Alexander the Great

As Alexander sought to conquer the world, the unassuming mosquito emerged as an unforeseen challenger, weaving its influence into the narrative of the demise of the greatest warrior of all time. Alexander the Great, groomed by his father Philip's innovative Macedonian army, emerged as a formidable leader (Freeman 2011; O'Brien 1992). With Greece in turmoil after the Peloponnesian wars, Philip united northern and central Greece, creating a powerful Macedonian force (Green 1991). Alexander's military prowess flourished as he quelled revolts and defeated a defensive coalition at the Battle of Chaeronea. After Philip's murder in 336 BCE, 20-year-old Alexander ascended to the throne, inheriting his father's vision of conquering Persia. Despite being outnumbered, Alexander achieved victories at Granicus, Issus, and Gaugamela (331 BCE), gaining control of the Persian Empire. He continued eastward, liberating Egypt and marching through unknown territories, including modern-day Turkmenistan, Uzbekistan, Tajikistan, Afghanistan, Pakistan, and India. Alexander's quest for world conquest faced a turning point at the Beas River in India (Bose 2003; Hammond 1997). Exhausted troops, supply challenges, and formidable enemies led to a halt in his eastward expansion. Although often sensationalized as a mutiny, the decision to return west was a pragmatic response to the army's fatigue and challenges. Alexander's legacy lies in his military achievements, cultural reciprocation, and empire-building strategies (O'Brien 1992). His conquests left an indelible mark on ancient history, but the Beas River marked the eastern limit of his vast empire.

As Alexander advanced into the Indus River valley, his army faced the deadly menace of mosquitoes carrying diseases, notably malaria (Atkinson et al. 2009), identified centuries earlier by Indian physician Sushruta. The mosquito-infested regions, combined with swampy terrains and monsoons, decimated his once-mighty army, reducing it to a fraction of its former strength. The unhealthy conditions and the relentless onslaught of diseases, along with other military complexities, led to the abandonment of Alexander's India campaign. There are different theories regarding this retreat. Some believe it was a strategic move by Alexander to avoid personal humiliation and preserve his perfect battle record. Others argue it was a sensible decision given the overwhelming challenges. Regardless, it marked the eastern limit of Alexander's conquests (Winegard 2019).

After returning to Persia, Alexander planned further campaigns, including an invasion of Arabia and North Africa, with aspirations of reaching the western Mediterranean. However, his grand plans were cut short. In Babylon, Alexander fell ill with a debilitating fever, likely caused by falciparum malaria transmitted by mosquitoes (Romm 2012; Atkinson et al. 2009). Despite his formidable conquests, Alexander succumbed to this inconspicuous enemy on June 11, 323 BCE, at the age of 32 (Doherty 2004). The unexpected death of Alexander the Great had profound consequences. It halted his ambitious plans for further conquests, preventing him from reaching the "ends of the world". His demise marked the end of his empire, as infighting among his generals led to its fragmentation. The enlightening legacy of Alexander's Hellenistic empire endured, spreading Greek influence across regions and fostering academic prosperity and progress. While Alexander's territorial gains faded, the cultural and intellectual impact of his conquests persisted for centuries (Martin and Blackwell 2012). The great conqueror Alexander fell not by the sword or the might of armies but at the merciless hands of a villainous mosquito. This tiny, blood sucking assailant, carrying the insidious weapon of malaria, brought down the undefeated warrior, revealing the sinister potency of nature's tiniest adversaries against even the most formidable of human foes. The mosquito, once a mere nuisance, emerged as an unexpected and malevolent force, casting a dark shadow over the illustrious legacy of Alexander the Great.

1.2.4 Apex to Abyss: Roman Empire's Dynamic Story

Nature, including bacteria and mosquitoes, played a role in shaping history of Rome. The centre of western power shifted to Rome after stops in Persia and Greece, as described in Harper (2017) "The Fate of Rome". Despite an initial victory in the First Punic War, Rome faced a tough challenge in the Second Punic War against the brilliant Carthaginian general Hannibal Barca (Goldsworthy 2001). The mosquito, previously involved in Greek and Persian conflicts, continued to impact history by influencing both the rise and fall of the Roman Empire.

Legend has it that Rome, founded in 753 BCE by Romulus and Remus, saw the brothers nurtured by a she-wolf (Fig. 1.2) and eventually led by Romulus after a fatal dispute (Spencer 2010). Unlike Greek city-states, Rome expanded by assimilating outsiders into its unified legal body, a unique approach that played a crucial role in its growth. Transitioning from a despotic monarchy to a democratic republic in 506 BCE, Rome gradually expanded its territory to include much of Europe, England, Egypt, North Africa, Turkey, the southern Caucasus, and the Mediterranean

Fig. 1.2 Romulus and Remus nurturing from their wolf surrogate mother, depicted in a bronze sculpture dating back to around 500–480 BCE, currently housed at the Capitoline Museums in Rome

region. The mosquito, alongside trade caravans and migrating columns, was inadvertently involved in Rome's expansion (Winegard 2019). The city of Rome was surrounded by 310 square miles of wetland, Pontine Marshes that were the absolute hotbed of malarial mosquitoes (Micallef 2016). Whenever invading armies sought to attack the city, Rome's formidable malarial shield proved undefeatable, preventing them from successfully capturing or maintaining control.

The vast geographic and ethnic diversity of the Roman Empire, combined with intersecting trade routes, facilitated the spread of malaria throughout Europe, reaching as far north as Scotland. As Rome aimed for dominance, it inevitably clashed with Carthage, another imperial power in the region (Goldsworthy 2016). Before Rome's founding, seafaring Phoenician traders had established outposts across the Mediterranean, including Carthage in Tunisia. This led to a contest between Carthage and Greek city-states for Mediterranean control. Carthage, after a failed Sicilian campaign, faced a mosquito-borne disaster similar to that of the Athenians. Despite this setback, Carthage expanded its empire in other regions. Rome and Carthage's territorial and economic interests clashed, leading to the First Punic War (264–241 BCE) over Sicily (Goldsworthy 2001). Although primarily a naval conflict, Rome's inexperienced navy triumphed, gaining control of Sicily, Sardinia, Corsica, and the Dalmatian Coast (D'Este 2008). The victory fuelled Rome's appetite for further expansion, leaving Carthage weakened at sea but still formidable on land.

In 218 BCE, Carthage, led by Hannibal, the son of Carthaginian leader Hamilcar Barca, wanted to avenge his father's loss to the Romans in the First Punic War (O'Connell 2011; Hoyos 2008; Lancel 1999). His plan was to infiltrate Rome by leading Carthaginian forces through the Alps, starting the Second Punic War. Not well known to Rome, the Pontine Marshes, a 310-square-mile area around the city, served as an unexpected defence due to lethal mosquitoes. These marshes were a challenging obstacle for invaders from various wars, including the Punic Wars and World War II. As Rome and Carthage evolved from humble origins to major players, the Pontine Marshes played a crucial role in their rivalry, influencing economic and territorial supremacy. The Pontine Marshes, also known as the Campagna, bordering Rome, housed swarms of deadly mosquitoes, acting as a natural defence equivalent to armies (Hughes 2014; McNeill 1998). An early Roman scholar vividly described the marshes as a frightening place where bloodsucking insects thrived in the summer heat, creating an unsettling environment with putrid vegetation (Geissler and Guillemin 2010). Invading armies, from the Punic Wars to World War II, faced fatal swarming in these marshes. Rome and Carthage, originating as small enclaves, eventually clashed for Mediterranean dominance, with the mosquitoes of the Pontine Marshes acting as referees.

The Greco-Roman culture that emerged after the Punic Wars dominated Europe, North Africa, and the Middle East for 700 years, profoundly shaping human civilization (Spencer 2010). The legacy of the Roman Empire still influences the world today, seen in Latin-based languages, legal systems, and the spread of Christianity. The victory in the Punic Wars also sparked the rise of Roman literature, with influential authors documenting the historical impact of mosquitoes. Roman scholar Varro, in the first century BCE, warned about the dangers of swamps and invisible creatures causing diseases. The Roman elite adopted the practice of building houses on high ground, influencing real estate trends for centuries. Roman physicians, following the Hippocratic tradition, reinforced the concept of miasma or "malaria", associating bad air with diseases, like malaria (Jones 1907). Classical Roman writers, including Pliny, Seneca, Cicero, Horace, Ovid, and Celsus, all referenced mosquito-borne diseases. The physician Galen, in the second century CE, provided detailed accounts of malaria fever and described filariasis or elephantiasis. Roman remedies included bloodletting, amulets with incantations, and prayers to the fever Roman goddess Febris.

As Roman influence spread across Europe, so did malaria (Harper 2017; McNeill 1998; Karlen 1996). The vast empire facilitated the exchange of ideas and innovations but also brought the persistent challenge of malaria, reaching as far north as Denmark and Scotland. Malaria became a chronic partner to Roman expansion. While the mosquito played a crucial role in ensuring Roman ascendancy over the Carthaginians, a century and a half later, it contributed to the downfall of the democratic Roman Republic and the rise of the emperor era under Julius Caesar. Julius Caesar, afflicted by malaria, crossed the Rubicon River in 50 BCE, setting the stage for the fall of the Republic (Retief and Cilliers 2004). Shakespeare noted Caesar's lifelong battle with malaria, highlighting his struggles during the campaign (Winegard 2019). If Pompey had confronted Caesar on the battlefield instead of fleeing, Caesar's risky move at the Rubicon could have ended in malarial disaster. Pompey was eventually defeated, seeking asylum in Egypt, where he was assassinated. Caesar and Cleopatra then deposed the Egyptian pharaoh. After Caesar's assassination in 44 BCE, subsequent Roman leaders, including Vespasian, Titus, and Hadrian, experienced bouts of malaria, with some succumbing to the disease. Caesar's heir, Octavian, and successor, Tiberius, faced recurrent malaria episodes from the Pontine Marshes' mosquitoes. Caesar's ironic plan to drain the marshes was never realized, altering the course of Roman history. Ironically, before his assassination, Caesar had planned a project to drain the Pontine Marshes, aiming to increase agricultural productivity. This initiative, if successful, would have significantly reduced mosquito populations, altering the course of Roman history. However, this alternative history was cut short with Caesar's demise. Caesar's ironic plan to drain the marshes was never realized, altering the course of Roman history.

The malaria-infested Campagna, protecting Rome from external threats, also weakened its armies abroad. Roman soldiers and administrators, unfamiliar with foreign malaria strains, faced reduced combat effectiveness during Germanic campaigns. Germans strategically forced Romans into marshes, taking advantage of the swamp-bred miasma, perhaps indicating a deliberate biological warfare. Malaria hampered Roman victories in Scotland and the Middle East at the Battle of the Teutoburg Forest (9 CE), while at home, it turned from a defender to a killer. The mosquito, a bigoted ally, contributed to Rome's decline and eventual fall.

The Romans, with their advancements in engineering and agriculture, unintentionally turned the mosquito from an ally to a foe. Ironically, their love for urban beautification elements, like gardens, cisterns, and fountains, combined with aqueducts and natural floods, inadvertently facilitated mosquito propagation, contributing to their own downfall. As Rome's population surged from 200,000 to over a million in the last two centuries BCE, aggressive deforestation and cultivation expanded mosquito habitats on the city's outskirts, including the Pontine Marshes (Winegard 2019). The Romans, in their forceful alteration of landscapes, faced nature's grim retaliation, primarily through malaria spread by mosquito bites (Hughes 2014). Malaria became a severe threat, earning the region the derogatory label "Roman Fever" due to Italy's well-documented malarial reputation (Sallares 2002).

Recurring malaria epidemics, including those triggered by a hurricane in 65 CE and the aftermath of Mount Vesuvius's eruption in 79 CE, plagued Rome, claiming thousands of lives (McNeill 1998). Tacitus documented the grim scenes of lifeless forms in houses and funeral-filled streets (Retief and Cilliers 2004). The Campagna, once fertile farmland, became fallow due to these disasters, facilitating marsh expansion and worsening malaria. Mussolini's efforts before World War II reclaimed the area. The resultant agricultural scarcity, combined with a population rise, fuelled a detrimental cycle of malaria-induced decline, impacting Roman society's health, agriculture, and politics. The decline in life expectancy, infant survival, and overall well-being marked a significant factor in the unravelling of the Roman Empire. The diminutive, multitudinous, and maligned mosquito, the real "Mistress of Rome", forced the collapse of the greatest empire in history

As Rome fell, the mosquito continued her unrelenting reign of terror, causing devastation and death throughout the medieval period. With the collapse of centralized Roman rule, the landscape changed, but the mosquito's deadly influence persisted. The migration of people and the disruption of established societies led to the spread of malaria in new territories. Towns and cities that once thrived fell into decline as malaria exacted its toll on populations and economies.

1.2.5 Infectious Epoch

The mosquito's impact was not limited to Europe; it played a significant role in shaping events across the globe. For instance, the ancient Egyptian pharaoh, Tutankhamun's fateful encounter with a mosquito bite might have contributed in his untimely demise. He was 9 years old when he became the king of Egypt during 18th dynasty of the new kingdom (c. 1332–1323 B.C.E.). His story would have been lost if the tomb was not discovered by the archaeologists (Brier 2023). The pharaoh was merely 19 years old when died. DNA analysis showed the boy king suffered from malaria. Examination of tuts mummy showed evidence of *Plasmodium falciparum* (Hawass 2013; Hawass and Saleem 2018). However, it is believed that malaria was not the cause of death rather a severe leg fracture was. Interestingly, mosquito wrath prevails beyond, and the excavator of the tomb of Tutankhamun, Carnarvon was found dead. A mosquito bite on his face had become infected, leading to deadly poisoning.

As civilizations rose and fell, the mosquito continued to be a formidable force. The Crusades, with their movement of armies and people across vast regions, facilitated the spread of malaria. The marshy lands surrounding the Holy Land were fertile breeding grounds for the *Anopheles* mosquito, and many Crusaders succumbed to the disease (Crosby and Alfred 1986).

In the following centuries, as exploration and trade connected distant corners of the world, the mosquito emerged as a silent but deadly companion. European powers, expanding their empires to the Americas, Africa, and Asia, unknowingly transported malaria to new lands (Drexler 2003). The introduction of malaria to the Americas had a profound impact on indigenous populations, who had no previous exposure to the disease and thus no immunity.

The influence of mosquitoes reached a tragic crescendo during the era of transatlantic slavery. The movement of enslaved people from malaria-endemic regions of Africa to the Americas introduced new malaria strains to the western hemisphere (Dubois and Garrigus 2017; Earle 1996). Malaria became a cruel and efficient tool of oppression, affecting both enslaved individuals and European colonizers. The disease shaped the demographics of the Americas and influenced the outcome of conflicts (Crosby 2006).

1.2.6 Mongols and Their Unexpected Adversary

The Mongols fierce army was also not spared by brutal mosquitoes. Genghis Khan was a leader who founded the Mongol Empire in the thirteenth century (Fig. 1.3). Born around 1162 as Temüjin, he united Mongol tribes through military conquests and alliances, becoming "the Great Khan" (Sverdrup 2017; McLynn 2016). Known for his military skills and organizational abilities, Genghis Khan expanded the empire across Asia and Europe. In 1220, Genghis Khan split his army into two. The main force moved east through Afghanistan and northern India, while a second group went north into Russia, sacking the Italian port of Kaffa in Crimea (Frankopan 2017). The Mongols devastated European Russia, the Baltic states, and Poland, causing widespread destruction and enslavement. As they probed Poland and Hungary, they quickly retreated east in 1223, raising questions about why they chose not to continue conquering Europe. Historians suggest that the decision to delay an invasion was influenced by the Mongol army's weakened state from malaria contracted in the Caucasus and along the Black Sea River systems. Genghis Khan, the leader, was also suffering from habitual bouts of malaria, and his death at 65 was likely a result of severe weakening from chronic malarial infection (Dowdeswell 2022; McLynn 2016).

The persistent thirst of mosquito for Mongol blood continued, impacting their empire. In 1241, unfavourable weather conditions in Hungary created a marshy environment with an abundance of malarial mosquitoes, negatively affecting the Mongol military. High humidity affected their bows, and the malaria parasite began its invasion. The mosquito, like a strategist, steered the Mongols away from Europe (Winegard 2019; Jackson 2005).

Fig. 1.3 Genghis Khan engaged in a battlefield confrontation, portrayed in an illustration found in a fourteenth-century chronicle by Rashid al-Din

While the Mongols had some limited success in the Holy Land, including brief stopovers in Aleppo and Damascus, malaria, other diseases, and strong defensive coalitions forced repeated retreats. The guardian mosquito, General *Anophele*s, played a role in thwarting the Mongol menace in the Levant, just as it had during earlier Christian campaigns (Winegard 2019). Despite some successes, the Holy Land, including Jerusalem, remained under Muslim control.

1.3 The Mosquito's Role in the African Slavery and the Columbian Exchange

The resistance of Africans to malaria, aided by traits, such as Duffy negativity, thalassemia, and sickle cell, coupled with their prior exposure to yellow fever in Africa, conferred hereditary shields and immunity against these diseases (Liu et al. 2014). Owing to lower mortality of Africans in comparison to non-Africans, European owners of mining operations and plantations found African slaves to be advantageous. Thus, genetic immunities and prior seasoning of Africans became crucial in the Columbian Exchange and the development of New World mercantilist economic markets (Horwitz 2008).

During his final voyage, Columbus and his crews experienced severe mosquito and malaria infestations along the northern shoreline of central America that they named the region "the Mosquito Coast". Established in 1510 following this voyage, Darien was the first European colony on the mainland of the Americas. The Darien settlement was ruled by blood thirsty mosquitoes. Spanish residents at Darien suffered greatly from the relentless attacks of mosquitoes, falling ill and dying (Earle 1996). Thus, Columbus's voyage marked the commencement of a new world order, introducing deadly mosquitoes and diseases to the Americas through the Columbian Exchange.

The mosquito, acting as both a killer and a saviour, played a pivotal role, in preventing Africa from foreign powers. However, despite this protection, the collaboration of Africans with European objectives allowed the slave trade in Africa to thrive. Africans facilitated the delivery of their compatriots into European subjugation, a task the mosquito made challenging for Europeans alone. Without African slavery, New World mercantilist plantation economics would have faltered (Hong 2011), quinine might not have been discovered, and Africa would have retained its original character. The entire course of the Columbian Exchange might have been vastly different or potentially non-existent.

African genetic immunities and resistance to mosquito-borne diseases, including malaria, yellow fever, and dengue, rendered them more resilient and productive on mosquito-infested plantations. Their ability to survive where others perished made Africans valuable for profit generation.

The mosquito, in a peculiar and dark twist, influenced history, contributing to the "scientific" and legal justifications for the African slavery in the southern United States. This factor played a role in the multiple causes of the American Civil War, a period when the mosquito metaphorically went on a feeding frenzy (Humphreys 2013). Thus, *Anopheles* and *Aedes* mosquitoes, unintentionally introduced in the Columbian Exchange, are deemed "Accidental Conquerors" for their significant impact on shaping the course of events.

The deadly yellow fever virus reached the Americas with African slaves and an imported *Aedes* mosquito breed that thrived on slave ships. The mosquitoes, breeding in barrels and pools on these ships, facilitated a continuous cycle of viral infection among European slave traders and their human cargo. The *Aedes* mosquito found a favourable environment in America, becoming a carrier of suffering and death (Fig. 1.4).

Yellow fever first appeared in America when a Dutch slaving vessel brought it to Barbados in 1647, resulting in over 6000 deaths (Gessner 2016). Following that, outbreaks rapidly killed 35% of the populations in Cuba, St. Kitts, and Guadeloupe in 6 months and then spread through Spanish Florida causing widespread devastation (Murphy 2003). The "Black Vomit", "Yellow Jack", or "Saffron Scourge" went viral, affecting Caribbean regions and coastal areas as far north as Halifax and Quebec City, Canada (Carrigan 1994). The virus appeared via the Royal Navy in British North America, reaching Boston in 1693. Before the American Revolution, it caused deaths in cities as diverse as Philadelphia, Charleston, and New York. Yellow fever, and chronic malaria, had a profound effect on the development of America. Port cities crucial for slaving and trading vessels, such as New Orleans, Charleston, Philadelphia, Boston, New York, and Memphis, faced deadly yellow fever epidemics, shaping the history of the Americas (Crosby 2006). These epidemics, considered the most lethal in American history, played a pivotal role in the continent's density and configuration.

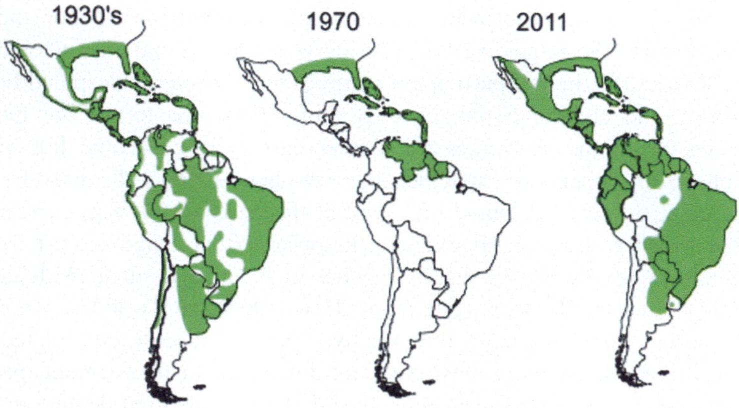

Fig. 1.4 Reinfestation of tropical America by Aedes aegypti, 1930–2011 (Gubler 2011). http://www.ncbi.nlm.nih.gov/pmc/articles/PMC3317603/figure/F3/

The link between slavery, economic imperialism, and territorial power projection has persisted throughout history. In antiquity, slavery was indiscriminate, involving people of various backgrounds. Unlike the localized and controlled form of slavery in earlier societies, the transatlantic slave trade resulted in a more widespread and brutal system. The impact of diseases, like yellow fever, in the Americas would have followed a different historical script without the African slave trade.

1.4 Buzzing Peril: Mosquito Threats Loom Over Our Military

Armed with the potential to transmit deadly diseases, mosquitoes have orchestrated a symphony of change that echoes through time. Mosquito-borne diseases have disrupted military campaigns, crippled economies, and influenced migration patterns (Petriello 2016). For instance, American Civil War (1861–1865) is glanced at by historians as the harmful effects of diseases on military performance (Humphreys 2008). The Civil War stands as America's deadliest and most polarizing conflict, with the Union Army facing off against the Confederate States of America. With more than a million cases of mosquito-borne diseases, both Union and Confederate soldiers had to face a common enemy in the form of disease-carrying mosquitoes, representing a significant challenge known as the Third Army in the Civil War (Bell 2010). During the 4-year war, soldiers from both the sides suffered from the dreaded black vomit. No medicine of that era could alleviate the severe symptoms of yellow fever.

Soldiers training or stationed in various climates face increased risk of mosquito-borne diseases (Gajda et al. 2023). Throughout history, epidemics during wars have greatly diminished the effectiveness of armies, leading to the suspension or abandonment of military operations and causing widespread harm to civilian populations. Before World War I, infectious diseases, especially those transmitted by vectors, like plague, yellow fever, malaria, typhoid, and relapsing fever, were the leading causes of illness and death among soldiers (Smallman-Raynor and Cliff 2004; Prinzing and Westergaard 1916). During World War I, malaria emerged as an unexpected foe, affecting all participating armies and causing widespread harm to both military and civilian populations (Brabin 2014). The term "war malaria" encapsulates the unique challenges of malaria control during wartime, highlighting distinctions from peacetime strategies (Christophers 1939). The prevalence of malaria caught military personnel off guard at the onset of the war, especially in western and Eastern Europe. Historians acknowledge that a significant portion of troop casualties resulted from diseases rather than direct combat (Mitchell and Smith 1931). Since the Russo-Japanese War (1904–1905) and World War I, advancements in weapons development, military hygiene, and disease control methods, such as vaccination, chemoprophylaxis, antibiotics, medical treatment, personal protection, and vector control measures, have led to a consistent decline in infectious diseases among soldiers and civilians (Hoeffler and Melton 1981). Despite these improvements, vector-borne diseases remained a threat during World War II (McCoy 1946). Even today, infectious diseases still pose a major concern in developing countries, affecting the health and well-being of populations (Gayer et al. 2007). Military activities in remote ecological areas, habitat destruction, population migration, and infrastructure disruption create conditions that favour the development and transmission of diseases in nature (Morse 1995).

About 2 million confirmed malaria cases and 1000 deaths due to malaria are reported annually. The World Health Organization (WHO) states that India contributes 77% of the total malaria cases in south-east Asia. We need to approach the challenges of mosquito-borne diseases with the same dedication and rigorous training as we do against enemy forces (Smith and Hooper 2016). Field Marshal Sir Archibald Wavell, the commander in chief of British forces in Burma during World War II, highlighted the stark reality that losses due to malaria and preventable diseases among allied forces in the China-Burma-India theatre surpassed casualties from enemy action. Today, as we face global challenges, like terrorism, neglecting environmental threats, including disease-carrying mosquitoes, could again weaken deployed forces. The significance of Field Marshal Wavell's warning became evident in August 2003 when a US Marine Corps team, engaged in stabilization operations in Liberia, faced an unexpected outbreak, with nearly 30 percent of deployed military personnel contacting malaria. This not only posed a health threat but also diverted military medical resources already committed to supporting operations in Iraq and Afghanistan.

The NATO (North Atlantic Treaty Organisation) armed forces, operating globally, regularly adapt strategies to manage infectious disease carriers when deploying military groups. Vector-borne diseases can impose a considerable burden on armed forces, prompting the development of vector control strategies by most western armies. Surveillance data on mosquitoes enables risk evaluation and planning of preventive measures. Information gathered from mosquito surveillance at military sites, encompassing entomological and epidemiological data, is crucial for predicting mosquito population changes and monitoring diseases transmitted by mosquitoes in these regions (Britch et al. 2008).

1.5 Battling Mosquitoes: A New Dawn in Mosquito Menace Mitigation

As the world moved into the modern era, scientific understanding of mosquito-borne diseases improved. The realization that mosquitoes were the vectors for the malaria parasite led to efforts to control the disease by targeting the insect. The use of quinine as a treatment and later the development of synthetic antimalarial drugs brought some relief, but malaria remained a global health challenge (Webb 2009).

The mosquito had surpassed the effectiveness and usefulness of her adversaries and no longer dreaded them. Faced with the threat of extermination, the mosquito and her realm of diseases retaliated, adapting and evolving during the silent periods of the 1960s (post publication of the book by Rachel Carson, *Silent Spring*, that outlined the indiscriminate use of pesticides and harm they cause to the environment. While mosquitoes built up a robust immunity during their exposure to DDT, malarial parasites developed resistance to chloroquine, consuming it between doses of other antimalarials (Jukes 1969). The unintended consequence of DDT's early success was a decline in research on malaria drugs and other pesticides during its potent reign. The widespread resistance to DDT left the world without effective tools to combat the resurgent enemy (McWilliams 2008). Between 1950 and 1972, significant resources were spent on DDT-based malaria control, but interest waned after the termination of the Malaria Eradication Programme in 1969 (Van den Berg 2008).

The decline in interest and studies during the late 1970s and the 1980s coincided with a global resurgence of mosquitoes and mosquito-borne diseases. With mosquitoes developing resistance against DDT, they are becoming stronger and more relentless. In 1968, When Sri Lanka ceased spraying DDT, it resulted in a malaria outbreak infecting 100,000 people. Mosquito-borne disease rates returned to pre-DDT levels in south and central America, the Middle East, and central Asia by the early 1970s. By 1975, India reported over 6.5 million cases of malaria. Africa continued to be heavily affected, and even Europe experienced a malaria outbreak in 1995 (McWilliams 2008). Currently, European clinics and hospitals treat eight times more malaria patients than in the 1970s, while malaria rates in central Asia and the Middle East have increased tenfold.

Relentless persistence of mosquitoes has driven innovations and public health initiatives. The quest to combat these teeny terrors has spurred advancement in science leading to breakthrough in understanding of infectious diseases. The race to develop effective mosquito control measures has given rise to innovations, such as insecticide-treated bed nets and genetically modified mosquitoes, showcasing humanity's resilience in the face of this persistent adversary.

The twentieth century saw concerted efforts to combat malaria through mosquito control programmes and public health initiatives. In recent decades, there has been progress in the fight against malaria, with increased investment in research, treatment, and prevention. Organizations, like the WHO and initiatives, like the Roll Back Malaria partnership have worked to coordinate global efforts against the disease. The development of new antimalarial drugs and the exploration of innovative tools, such as genetically modified mosquitoes, offer hope for further advancements.

Over time, our genetic defences against malaria have emerged through natural selection, and now, with CRISPR gene-editing technology, we are attempting to counter the mosquito's lethal impact (Gates 2018; Kozubek 2016; Ledford 2015). However, CRISPR, like other tools of extermination like DDT, may face challenges against the mosquito's evolving nature (Winegard 2019). For this, we need more behavioural study on mosquitoes to understand their evolutionary aspect.

1.6 Conclusion

Mosquitoes, inconspicuous yet influential, have played the role of silent architects in shaping the history. For instance, they aided Romans in building the glorious empire, mosquitoes carried diseases, contributing to health crisis, that ultimately weakened societies (Winegard 2019), thereby, playing a subtle yet impactful role in the eventual decline of once-mighty empires.

We are still fighting against mosquitoes. As survivors of extinction events and shapers of human history, mosquitoes continue to be our deadliest threat. Despite their small size, mosquitoes have been a significant force on Earth for 200 million years, influencing history by impacting empires, nations, and battles. Our genetic defences against malaria have evolved over time, and now, with CRISPR gene-editing technology, we are trying to fight back.

The current urgency surrounding the formidable destructive potential of mosquitoes affects us all. The mosquitoes, rather than bombs or bullets, have proven to be one of the most dangerous adversaries that the world may ever face, causing more casualties during the conflicts. Progress has been hard to achieve over the last 25 years, with a breakthrough in combating these disease vectors still distant, and billions of mosquito bites standing in the way. Globalization is facilitating the dissemination of a fresh wave of mosquito-borne diseases, previously restricted to tropical areas. Examples include ancient ailments, like dengue, existing for possibly a thousand years, and more recent discoveries, such as chikungunya and Zika, identified in humans only in 1952. Simultaneously, climate change is significantly broadening the territories where mosquitoes and the diseases they transmit can flourish. According to a recent study, in the next 50 years, approximately a billion more people could face exposure to mosquito-borne infections compared to the present scenario. Despite setbacks, research persists in developing new insecticides and repellents. Scientists worldwide are dedicated to advancing insecticides and repellents, delving into mosquito biology, genetics, and behaviour.

We, as contemporary individuals, are also at fault for assuming that our aspirations and technological advancements will shield us from the mechanisms of the natural world. Throughout the entire duration of human existence, mosquitoes have consistently served as evidence that we are not exempt from these natural dynamics.

Points to Remember

1. In the intricate tapestry of historical narratives, an often-overlooked player emerges, the mosquito.
2. From ancient civilizations to the modern world, the mosquito has been a potent force, shaping the course of history in ways both direct and indirect.
3. Mosquitoes have affected economies, preyed on armies, and outsmarted generals throughout history. While we have made some progress, global warming is expanding their reach, bringing mosquito-borne diseases to new areas.
4. A formidable and relentless legion of 110 trillion mosquitoes blankets virtually every corner of the globe, with the exception of Antarctica, Iceland, and a few minuscule micro-islands in French Polynesia.
5. Despite our efforts, mosquitoes remain a resilient and deadly predator.
6. Combination of genetic immunities and prior exposure have resulted in heightened resilience of African population against mosquito-borne diseases. Africans were thus preferred slaves in comparison to non-Africans.
7. Mosquitoes' tolerance against DDT lead to a world filled with DDT-resistant mosquitoes carrying malaria parasites resistant to the best drugs available.

References

Atkinson J, Truter E, Truter E (2009) Alexander's last days: malaria and mind games? Acta Classica 52:23–46

Bell AMI (2010) Mosquito soldiers: malaria, yellow fever, and the course of the American Civil War. Louisiana State University Press, Baton Rouge

Black J, Green A, Rickards T (1998) Gods, demons and symbols of ancient Mesopotamia: an illustrated dictionary. British Museum Press, London

Bose P (2003) Alexander the great's art of strategy: the timeless leadership lessons of history's greatest empire builder. Gotham Books, New York

Brabin BJ (2014) Malaria's contribution to World War One–the unexpected adversary. Malaria Journal, 13(1), 497

Brier B (2023) Tutankhamun and the Tomb that Changed the World. Oxford University Press

Britch SC, Linthicum KJ, Anyamba A et al (2008) Satellite vegetation index data as a tool to forecast population dynamics of medically important mosquitoes at military installations in the continental United States. Mil Med 173(7):677–683. https://doi.org/10.7205/milmed.173.7.677

Carrigan JA (1994) The Saffron scourge: a history of yellow fever in Louisiana, 1796–1905. University of Louisiana Press, Lafayette

Christophers SR (1939) Malaria and war. Trans R Soc Trop Med Hyg 33:277–304. https://doi.org/10.1016/S0035-9203(39)90190-2

Clark DP (2010) Germs, genes, and civilization: how epidemics shaped who we are today. FT Press, Upper Saddle River

Cline EH (2014) 1177 B.C.: the year civilization collapsed. Princeton University Press, Princeton

Cloudsley-Thompson JL (1976) Insects and history. St. Martin's Press, New York

Crosby MC (2006) The American plague: the untold story of yellow fever, the epidemic that shaped our history. Berkley, New York

Crosby MLA, Alfred W (1986) Ecological imperialism: the biological expansion of Europe, 900-1900. Cambridge University Press, Cambridge

D'Este C (2008) Bitter victory: the battle for Sicily, 1943. Harper Perennial, New York

Doherty P (2004) The death of Alexander the great: what—or who—really killed the young conqueror of the known world? Carroll & Graf, New York

Dowdeswell M (2022) Malaria: the ancient disease that likely killed Genghis Khan. https://www.thecollector.com/malaria-ancient-disease-killed-genghis-khan/

Drexler M (2003) Secret agents: the menace of emerging infections. Penguin Books, New York

Dubois L, Garrigus JD (eds) (2017) Slave revolution in the Caribbean, 1789–1804: a brief history with documents, 2nd edn. Bedford-St. Martin's, New York

DuPage County Health Department (2023) Fight the bite. https://www.dupagehealth.org/DocumentCenter/View/147/10-Things-You-Should-Know-About-Mosquitoes-PDF. Accessed 3 May 2023

Earle R (1996) 'A grave for Europeans'? Disease, death, and the Spanish-American revolutions. War History 3(4):371–383

Frankopan P (2017) The silk roads: a new history of the world. Vintage, New York

Freeman P (2011) Alexander the great. Simon & Schuster Paperbacks, New York

Gajda E, Krzowski Ł, Kowalczuk K, Pabin A, Maculewicz E (2023) Influence of mosquito-borne biological agents on health risks among soldiers and military personnel. Ann Agric Environ Med 30(1):2–8. https://doi.org/10.26444/aaem/155003

Gates B (2018) Gene editing for good: how CRISPR could transform global development. Foreign Aff 97(3):166–170

Gayer M, Legros D, Formenty P et al (2007) Conflict and emerging infectious diseases. Emerg Infect Dis 13(11):1625–1631. https://doi.org/10.3201/eid1311.061093

Geissler E, Guillemin J (2010) German flooding of the Pontine Marshes in World War II: biological warfare or total war tactic? Politics Life Sci 29(1):2–23

Gessner I (2016) Yellow fever years: an epidemiology of nineteenth century American literature and culture. Peter Lang, New York

Gillett JD (1971) The mosquito: its life, activities, and impact on human affairs. Doubleday, New York

Global health, CDC (2023) Fighting the World's Deadliest Animal. Global Health Center, U.S.A. https://www.cdc.gov/globalhealth/impact/fighting-the-worlds-deadliestanimal.html#:~:text=The%20mosquito%20is%20the%20world's,other%20creature%20in%20the%20world

Global Malaria Programme, World Health Organization – processed by Our World in Data (2021). Deaths from malaria" [dataset]. Global Malaria Programme, World Health Organization [original data]. https://ourworldindata.org/malaria

Goldsworthy A (2001) The Punic wars. Cassell, London

Goldsworthy A (2016) Pax Romana: war, peace and conquest in the roman world. Yale University Press, New Haven

Green P (1991) Alexander of Macedon, 356–323 B.C.: a historical biography. University of California Press, Berkeley

Gubler DJ (2011) Dengue, urbanization and globalization: the unholy trinity of the 21st century. Tropical medicine and health, 39(4SUPPLEMENT), S3–S11

Hammond NGL (1997) The genius of alexander the great. University of North Carolina Press, Chapel Hill

Harper K (2017) The fate of Rome: climate, disease, and the end of an empire. Princeton University Press, Princeton

Hawass Z (2013) Discovering Tutankhamun: from Howard Carter to DNA. American University in Cairo Press, Cairo

Hawass Z, Saleem SN (2018) Scanning the pharaohs: CT imaging of the new kingdom royal mummies. American University in Cairo Press, Cairo

Hoeffler DF, Melton LJ (1981) Changes in the distribution of Navy and Marine Corps casualties from World War I through the Vietnam conflict. Mil Med 146(11):776–779

Hong SC (2011) Malaria and economic productivity: a longitudinal analysis of the American case. J Econ Hist 71(3):654–671

Horwitz T (2008) A voyage long and strange: on the trail of Vikings, conquistadors, lost colonists, and other adventurers in Early America. Picador, New York

Hoyos D (2008) Hannibal: Rome's greatest enemy. Bristol Phoenix Press, Exeter
Hughes JD (2014) Environmental problems of the Greeks and Romans: ecology in the ancient Mediterranean, 2nd edn. Johns Hopkins University Press, Baltimore
Humphreys M (2008) Intensely human: the health of the black soldier in the American civil war. Johns Hopkins University Press, Baltimore
Humphreys M (2013) Marrow of tragedy: the health crisis of the American Civil War. Johns Hopkins University Press, Baltimore
Jackson P (2005) The Mongols and the West, 1221–1410. Routledge, New York
Jones WHS (1907) Malaria: a neglected factor in the history of Greece and Rome. Macmillan & Bowes, Cambridge
Jukes TH (1969) DDT: the chemical of social change. Toxicology 2(4):359–370
Karlen A (1996) Man and microbes: disease and plagues in history and modern times. Simon & Schuster, New York
Kozubek J (2016) Modern prometheus: editing the human genome with CRISPR-CAS9. Cambridge University Press, Cambridge
Lancel S (1999) Hannibal. Blackwell Publishers, Oxford
Ledford H (2015) CRISPR, the disruptor. Nature 522:20–24
Liu W et al (2014) African origin of the malaria parasite Plasmodium vivax. Nat Commun 5:3346
Martin T, Blackwell CW (2012) Alexander the great: the story of an ancient life. Cambridge University Press, Cambridge
McCoy OR (1946) Incidence of insect-borne diseases in US Army during World War II. Mosq News 6(4):214
McLynn F (2016) Genghis Khan: his conquests, his empire, his legacy. Da Capo Press, Cambridge
McNeill WH (1998) Plagues and peoples. Anchor, New York
McWilliams JE (2008) American pests: the losing war on insects from colonial times to DDT. Columbia University Press, New York
Micallef MJ (2016) The Roman fever: observations on the understanding of malaria in the ancient Roman world. Med J Aust 205(11):501–503
Mitchell TJ, Smith GM (1931) History of the great war based on official documents. Medical services. Casualties and medical statistics of the great war. His Majesty's Stationary Office, London
Morse SS (1995) Factors in the emergence of infectious diseases. Emerg Infect Dis 1(1):7–15. https://doi.org/10.3201/eid0101.950102
Murphy J (2003) An American plague: the true and terrifying story of the yellow fever epidemic of 1793. Clarion Books, New York
Norrie P (2016) A history of disease in ancient times: more lethal than war. Palgrave Macmillan, New York
O'Brien JM (1992) Alexander the great: the invisible enemy. A biography. Routledge, New York
O'Connell RL (2011) The ghosts of Cannae: Hannibal and the darkest hour of the Roman Republic. Random House, New York
Petriello DR (2016) Bacteria and bayonets: the impact of disease in American military history. Casemate, Oxford
Prinzing F, Westergaard H (1916) Epidemics resulting from wars. Clarendon Press, Oxford
Retief F, Cilliers L (2004) Malaria in Graeco-Roman times. Acta Classica 47:127–137. http://www.jstor.org/stable/24595382
Romm J (2012) Ghost on the throne: the death of alexander the great and the bloody fight for his empire. Vintage, New York
Sallares R (2002) Malaria and Rome: a history of malaria in ancient Italy. Oxford University Press, Oxford
Sherman IW (2007) Twelve diseases that changed our world. ASM Press, Washington
Slater LB (2014) War and disease: biomedical research on malaria in the twentieth century. Rutgers University Press, New Brunswick
Smallman-Raynor MR, Cliff AD (2004) Impact of infectious diseases on war. Infect Dis Clin N Am 18(2):341–368. https://doi.org/10.1016/j.idc.2004.01.009
Smith AM, Hooper C (2016) The mosquito can be more dangerous than the mortar round-the obligations of command. J Military Veterans Health 24(3):60–65

Spencer D (2010) Roman landscape: culture and identity. Cambridge University Press, Cambridge
Sverdrup CF (2017) The Mongol conquests: the military operations of Genghis Khan and Sübe'etei. Helion, Warwick
Van den Berg H (2008) Global status of DDT and its alternatives for use in vector control to prevent disease. United Nations Environment Programme: Stockholm Convention on Persistent Organic Pollutants UNEP/POPS/DDTBP.1/2, pp 1–31
Webb JLA (2009) Humanity's burden: a global history of malaria. Cambridge University Press, Cambridge
Winegard TC (2019) The mosquito: a human history of our deadliest predator. Penguin Random House LLC, New York

Mosquito: Origin and Evolution

2

Tripti Yadav, Omkar, and Geetanjali Mishra

Abstract

The Culicidae, the largest insect family globally, was explored in this chapter, with the Neotropical region exhibiting the highest biodiversity. Mosquitoes, notorious disease vectors, annually afflict millions with diseases, like malaria and dengue, posing formidable challenges to global health. With 3721 species, Culicidae traces its origins back 120 million years, undergoing substantial diversification during the Jurassic period. The exploration of mitogenomes offers crucial insights into the structural conservation and evolutionary dynamics across *Anopheles* species. Maintaining a consistent diploid chromosome count of six, mosquitoes exhibit distinct karyotypic variations between anopheline and culicine species. This chapter encapsulates a comprehensive exploration of mosquito studies, emphasizing the interdisciplinary nature of research blending biology, genetics, and evolution to decipher the enigmatic world of these resilient and often perilous insects.

Keywords

Mosquito · Culicidae · Evolution · Phylogeny

T. Yadav (✉) · Omkar · G. Mishra
Ladybird Research Laboratory, Department of Zoology, University of Lucknow, Lucknow, Uttar Pradesh, India

© The Author(s), under exclusive license to Springer Nature Singapore Pte Ltd. 2024

Omkar (ed.), *Mosquitoes*, https://doi.org/10.1007/978-981-97-4163-2_2

Learning Objectives

1. To understand the distribution patterns and ecological roles of mosquitoes worldwide
2. To analyze the scale and significance of mosquito-borne diseases for global health
3. To investigate the historical roots and pivotal papers shaping the field of mosquito genetics
4. To recognize the taxonomy and evolutionary importance of Culicidae subfamilies and genera
5. To explore the structural features and advancements in sequencing techniques for mosquito mitogenomes
6. To examine chromosomal karyotypes and genetic studies, emphasizing experimental approaches

2.1 Introduction

Mosquitoes, belonging to the family Culicidae, are the largest group of insects globally. This family is widespread, inhabiting temperate, tropical, and Arctic regions. The distribution of mosquito species varies worldwide, with the Neotropical region having the highest diversity (31%), followed by Oriental (30%), Afrotropical (22%), and Australasian (22%) regions. The Nearctic area, including the United States and Canada, has the least species diversity (5%). It consists of 3721 extant species (https://mosquito-taxonomic-inventory.myspecies.info. Accessed 31 December 2023), divided into two subfamilies (Anophelinae and Culicinae) and 113 genera (Reinert et al. 2004; Foster and Walker 2019). The subfamily Anophelinae includes three genera, while Culicinae comprises of 110 genera organized into 11 tribes: Aedeomyiini, Aedini, Culicini, Culisetini, Ficalbiini, Hodgesiini, Mansoniini, Orthopodomyiini, Sabethini, Toxorhynchitini, and Uranotaeniini (Table 2.1).

Mosquitoes pose a significant threat to human health, infecting 250–500 million people annually with diseases, like malaria, chikungunya, dengue, Zika, and West Nile Virus. Malaria alone is responsible for over a million deaths annually on a global scale. Approximately 20% of the world's population resides in areas endemic to the *Plasmodium* protozoan parasite, primarily transmitted by various species of *Anopheles* mosquitoes (Manguin et al. 2010; Wesolowski et al. 2015; Sato 2021). Consequently, malaria stands out in terms of economic losses and human suffering, making it unparalleled in impact compared to other diseases. Due to its significance, considerable attention has been dedicated to the study of crucial mosquito species that serve as vectors for human diseases (Schaffner et al. 2013; Manikandan et al. 2023). This focus is particularly prominent in the examination of *Aedes*, *Culex*, and *Anopheles*, which represent the principal genera. Research efforts have been concentrated on understanding the biology, ecology, and evolution of these mosquito species, with a specific emphasis on various genetic aspects.

Table 2.1 Classification of culicidae

Subfamily	Tribe	Genera
Anophelinae		*Anopheles* (An.), *Bironella* (Bi.), *Chagasia* (Ch.)
Culicinae	Aedeomyiini	*Aedeomyia* (Ad.)
	Aedini	*Aedes* (Ae.), *Armigeres* (Ar.), *Eretmapodites* (Er.), *Haemagogus* (Hg.), *Heizmannia* (Hz.), *Opifex* (Op.), *Psorophora* (Ps.), *Udaya* (Ud.), *Verrallina* (Ve.), *Zeugnomyia* (Ze.)
	Culicini	*Culex* (Cx.), *Deinocerites* (De.), *Galindomyia* (Ga.), *Lutzia* (Lu.)
	Culisetini	*Culiseta* (Cs.)
	Ficalbiini	*Ficalbia* (Fi.), *Mimomyia* (Mi.)
	Hodgesiini	*Hodgesia* (Ho.)
	Mansoniini	*Coquillettidia* (Cq.), *Mansonia* (Ma.)
	Orthopodomyiini	*Orthopodomyia* (Or.)
	Sabethini	*Isostomyia* (Is.), *Johnbelkinia* (Jb.), *Kimia* (Km.), *Limatus* (Li.), *Malaya* (Ml.), *Maorigoeldia* (Mg.), *Onirion* (On.), *Runchomyia* (Ru.), *Sabethes* (Sa.), *Shannoniana* (Sh.), *Topomyia* (To.), *Trichoprosopon* (Tr.), *Tripteroides* (Tp.), Wyeomyia (Wy.)
	Toxorhynchitini	*Toxorhynchites* (Tx.)
	Uranotaeniini	*Uranotaenia* (Ur.)

The classification of mosquitoes, dividing them into 2 subfamilies, 11 tribes within Culicinae, and 41 genera, relies on the framework established by Knight and Stone (1977), with subsequent adjustments made in accordance with updates from the online Systematic Catalog of the Culicidae on the Walter Reed Biosystematics Unit website. The two-letter generic abbreviations, acknowledged by the American Mosquito Control Association and adopted in various journals and books, are enclosed in parentheses (Foster and Walker 2019)

2.2 Mosquito Evolution: Culicidae Origins and Evolutionary Significance

2.2.1 Ancestral Origins

The family Culicidae, encompassing mosquitoes, has ancient origins dating back to the Jurassic or Cretaceous periods within the order Diptera (Bertone et al. 2008). Derived from primitive fly-like ancestors, early mosquito-like species exhibited distinctive features, including elongated proboscis and specialized mouthparts adapted for blood-feeding. Over millions of years, mosquitoes underwent significant evolutionary changes, diversifying into various genera and species, shaped by environmental shifts, ecological interactions, and selective pressures. The emergence of the family Culicidae marked a pivotal point in insect evolution, with mosquitoes developing unique attributes, such as specialized feeding habits and complex life cycles, contributing to their role as disease vectors.

The fossil records and zoogeographic evidences suggest that mosquitoes evolved around 210 million years ago (MYA) during the Jurassic, coinciding with the onset of continental drift (Edwards 1932; Darlington Jr 1964). The breakup of continents led to population fragmentation and geographical isolation, potentially promoting rapid speciation. A burst of Culicinae lineages is proposed to have occurred around 120 MYA (Ross 1951), with the generic composition of the family Culicidae well-established by the end of the Cretaceous, approximately 65 MYA.

New Zealand's isolation for the last 50 million years provides circumstantial evidence of the genus *Aedes* existing prior to the island's separation from Australia (Belkin 1968; Rick 1970). Fossils of the family Culicidae and its sister family Chaoboridae are well-documented from the Eocene and Oligocene periods, starting around 60 and 55 MYA, respectively (Borkent 1993; Munstermann and Conn 1997). The family is considered the youngest among blood-sucking Nematocerans. Fossil records reveal the oldest known mosquito in Late Cretaceous Canadian "amber," while Oligocene culicid fossils are more regularly found, indicating increased culicid abundance in post-Eocene times (Poinar et al. 2000; Rasnitsyn and Quicke 2007). The data, while suggesting potential vertebrate evolution, also highlighted larval adaptations in small temporary water bodies as a contributing factor. Notably, the Late Eocene Baltic amber fauna is distinctive for the rarity of Culicidae (Rasnitsyn and Quicke 2007).

Reidenbach et al. (2009) aimed to elucidate family Culicidae relationships using molecular and morphological data, successfully identifying certain clades but facing challenges in resolving deeper genera relationships due to ancient and rapid radiations. The chronogram (Fig. 2.1) and divergence time estimates (Table 2.2) indicated the earliest divergence within mosquitoes, between Culicinae and Anophelinae, around 226 MYA, aligning with Krzywinski et al. (2001) findings of a split between *Anopheles* and *Aedes* around 145–200 MYA based on mitochondrial DNA sequences. The preference for older divergence estimates is supported by evidence suggesting mosquito origins in the Jurassic, as proposed by Edwards (1923). Surprisingly, the split between *Anopheles* and *Bironella* appears relatively recent at

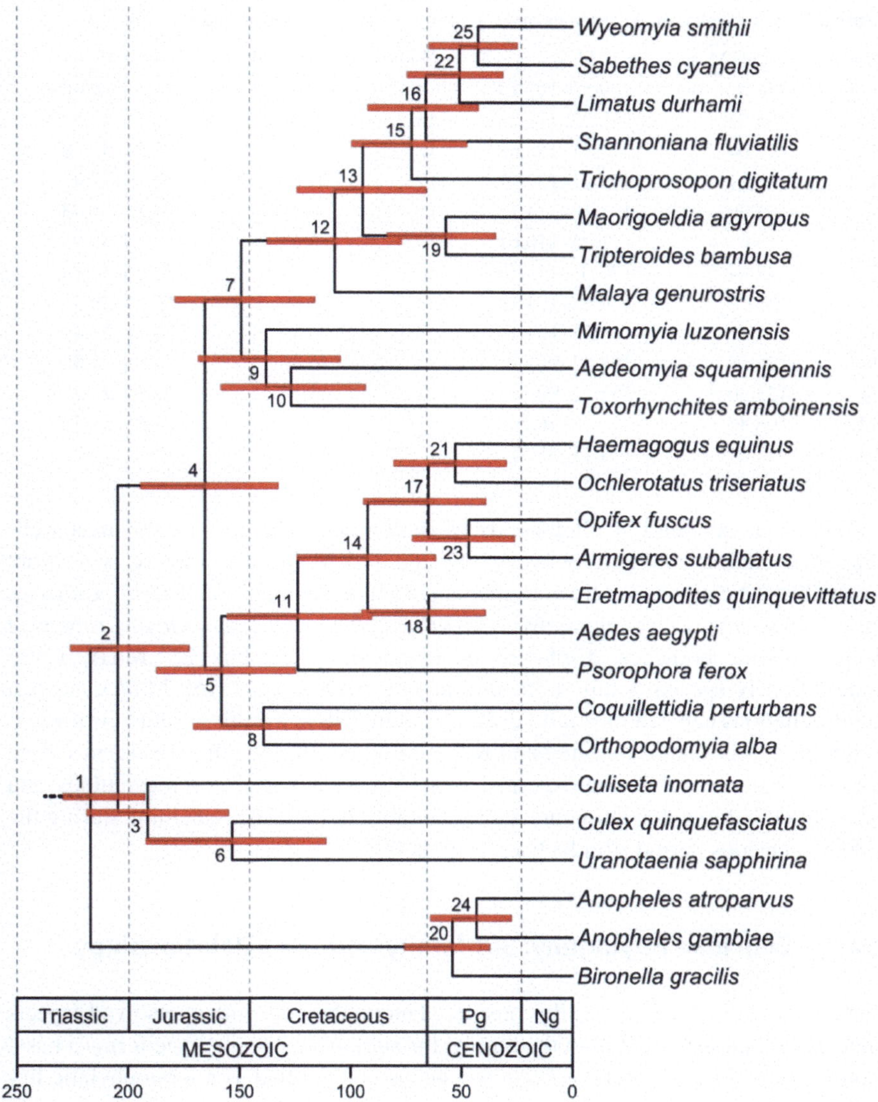

Fig. 2.1 Chronogram of mosquito age divergences with 95% confidence intervals (red bars) Numerical node ages and their 95% confidence intervals are presented in Table 2.1. Calibration points: Chaoboridae+Culicidae, 230–187 Ma (Borkent 1993; Bertone et al. 2008); *Toxorhynchites mexicanus* fossil, ≥16 Ma (Zavortink and Poinar 2008); *Culex winchesteri* fossil, ≥34 Ma (Cockerell 1919); *Anopheles dominicanus* fossil, ≥34 Ma (Zavortink and Poinar 2000)

approximately 51 MYA, contrasting Krzywinski et al. (2001) earlier estimate of a more extended divergence (90–106 MYA) between the *Anopheles* subgenera, *Anopheles* and *Cellia*.

Table 2.2 Divergence time estimates and credibility intervals (Ma) for nodes in Fig. 2.1

Node	Time (Ma)	CI (Ma)	Node	Time (Ma)	CI (Ma)
1	216.86	229.50–192.19	14	92.05	123.87–61.07
2	204.53	226.22–172.28	15	72.31	99.71–47.39
3	191.04	218.81–154.68	16	66.00	92.37–42.28
4	165.31	194.41–132.01	17	64.60	94.29–38.72
5	157.79	187.17–124.13	18	64.44	94.85–38.93
6	153.07	191.95–110.61	19	56.69	83.49–34.25
7	148.94	179.01–115.70	20	53.94	75.76–37.14
8	138.75	170.62–104.19	21	52.59	80.56–29.45
9	137.56	168.53–104.14	22	50.89	74.61–31.19
10	126.38	158.42–92.81	23	46.36	72.08–25.80
11	123.39	155.71–90.18	24	43.11	63.95–27.20
12	106.86	137.35–76.70	25	42.51	64.85–24.73
13	94.20	124.15–65.57			

Within the subfamily Culicinae, tribes Aedini and Sabethini are the most well-supported clades, emerging between 112 and 115 MYA, respectively, with more recent diversifications over time. Eight genera (*Aedeomyia*, *Coquillettidia*, *Culiseta*, *Culex*, *Mimomyia*, *Orthopodomyia*, *Toxorhynchites*, and *Uranotaenia*) represent deeper internal branches, displaying ancient nodes exceeding 127 MYA. These nodes, closely spaced within a 30-million-year interval (127–158 MYA), suggest rapid radiations early in mosquito diversification. While awaiting future corroboration with denser taxon sampling and fossil-based constraints, these estimates align with the emergence of angiosperms, which provide nectar for mosquitoes and water-filled habitats for immature stages, notably in the tribe Sabethini, during the early Cretaceous period (Reidenbach et al. 2009).

2.2.2 Culicidae Phylogeny: Subfamily and Tribe Relationships

The family Culicidae is generally considered monophyletic, with unresolved aspects in deeper relationships. Within the family, the subfamily Anophelinae forms a basal monophyletic lineage, and the *Chagasia* genus is identified as a monophyletic lineage at the base of Anophelinae. However, the Culicinae subfamily's monophyly concerning the genus *Toxorhynchites* lacks conclusive evidence. Certain tribes, including Aedini, Culicini, and Sabethini, are recognized as monophyletic, but others lack systematic testing, leading to uncertainty in their phylogenetic relationships. Despite a limited species analysis, De Oliveira Aragão et al. (2019) using mitochondrial DNA gene sequences (COX1, COX2, NADH4, NADH5, and CYOB genes) supports the monophyly of Anophelinae, Aedini, Culicini, and Sabethini. Further insights are available in works by Harbach and Kitching (1998) and Harbach (2007).

2.3 Molecular Evolution

The formal discipline of mosquito genetics has its historical roots in the influential paper titled "Mosquito genetics and cytogenetics," authored by Kitzmiller (1953). Remarkably, this emergence occurred more than half a century after the rediscovery of Mendelism and the establishment of genetics in 1900. Prior to the mid-1950s, when mosquito genetics began to be explored, entomology primarily engaged in descriptive studies, and formal genetics had limited presence within the entomological domain. This is noteworthy considering the well-established infrastructure of genetics in the early twentieth century, largely established through research on the fruit fly, *Drosophila melanogaster*. Since the mid-twentieth century, various researches have played pivotal roles in advancing interdisciplinary studies on mosquitoes, focusing on the structure, organization, and evolution of mosquito genomes.

2.3.1 Comparative Karyology and Evolution

Extensive research on the chromosomal karyotypes of mosquitoes within the family Culicidae has been conducted, covering over 20 genera and more than 300 species (Rai 1963, 1999; White 1980; Rao and Rai 1987; Rai and Black 1999). Notably, despite the family's ancient origin and substantial chromosomal repatterning, all examined species consistently exhibit a diploid chromosome number of six, with the exception of *Chagasia bathana* ($2n = 8$), which has three autosome pairs and a pair of heteromorphic sex chromosomes (Kreutzer 1978).

Among anopheline species, the majority possess two pairs of metacentric chromosomes and a pair of heteromorphic sex chromosomes (White 1980). In contrast, species within the Toxorhynchitinae and Culicinae subfamilies typically exhibit three pairs of homomorphic, metacentric, or slightly submetacentric chromosomes (Rai et al. 1982; Rai and Black 1999). In culicine mosquitoes, sex determination is generally governed by a single locus gene, with the shortest chromosome holding the sex locus (Gilchrist and Haldane 1947).

Species-specific studies, such as those on *Aedes*, reveal considerable variation in total chromosome length, with almost threefold differences within the *Aedes* genus and fivefold differences across the entire Culicidae family (Rai and Black 1999).

Studies utilizing Giemsa C-banding and other banding techniques have yielded significant insights into the linear differentiation and evolution of chromosomes. These analyses have revealed that karyotypic differentiation in mosquitoes involves alterations in the amounts, expression, and location of heterochromatin. Notably, the distribution of heterochromatin differs markedly between anopheline and culicine mosquitoes, with particular emphasis on the sex chromosome pair (Gatti et al. 1977; Motara and Rai 1978; Motara 1982; Rao and Rai 1987; Baimai 1988).

Motara and Rai (1978) identified two distinct types of heterochromatin in *Aedes*: constitutive and facultative. The latter demonstrated a unique capacity for differential expression under specific genetic backgrounds. In anopheline species, the Y chromosome may be entirely heterochromatic, with the X chromosome ranging from less than half to over three-fourths heterochromatic (Baimai 1988). Unlike the proposed pathway for sex chromosome evolution, where heteromorphic sex chromosomes develop from virtually identical homologues, Culicidae evolution suggests a different pattern. The subfamily Anophelinae, considered primitive, has heteromorphic sex chromosomes, while the more advanced Toxorhynchitinae and Culicinae subfamilies have homomorphic sex chromosomes. Rai and Black (1999) explored two alternative scenarios for sex chromosome evolution in the Culicoidea superfamily.

2.3.2 Genetic Studies on Mosquito Evolution

Extensive research has been conducted on the evolutionary genetics of the *Aedes scutellaris* group within the subgenus *Stegomyia*. This group comprises of two subgroups: the *scutellaris* subgroup, encompassing approximately 34 species primarily found in allopatric distributions on South Pacific islands, and the *albopictus* subgroup, which includes 11 described species predominantly found in sympatric distributions in South-East Asia (Rai et al. 1982).

The unique zoogeography of the *Aedes scutellaris* group makes these species valuable resources for evolutionary studies aimed at addressing fundamental questions related to reproductive isolation onset, as well as the patterns and rates of genetic differentiation based on their distinct distribution patterns (Rai et al. 1982). There has been considerable progress in both subgroups, especially in (a) experimental hybridization (Hilburn and Rai 1981; Dev and Rai 1982, 1985), (b) mate-choice tests indicating isolation intensity indices (McLain et al. 1985; McLain and Rai 1986), (c) chromosomal differentiation (Dev and Rai 1984; Sherron and Rai 1984), (d) population genetics (Pashley and Rai 1983; Pashley et al. 1985), and (e) molecular organization and evolution of genomes (McLain and Rai 1986; McLain et al. 1987; Black and Rai 1988).

Furthermore, by calibrating known geological events, such as the separation of islands in the South Pacific, these studies, along with related research, have provided estimates of the rates of genetic differentiation among various species within the group over evolutionary time (Pashley et al. 1985; Kambhampati and Rai 1991).

2.3.2.1 Genome Size: Inter- and Intraspecific Variations

Several studies have been conducted over the past three decades to determine the amount of haploid nuclear DNA in Culicoidea. A variety of methodologies have been used to accomplish this, such as quantitative cytophotometry and renaturation kinetics analysis (Jost and Mameli 1972; Kumar and Rai 1990; Black and Rai 1988; Warren and Crampton 1991; Besansky and Powell 1992).

The findings revealed a nearly eightfold variation in DNA amount per haploid genome across different mosquito taxa, ranging from 0.23 pg to 1.9 pg. Anophelinae exhibited a relatively small genome size (0.23 ± 0.28 pg per haploid genome), Toxorhynchitinae display an intermediate size (0.62 pg), and Culicinae demonstrated higher variation (0.54 ± 1.9 pg), suggesting a general increase in genome size during Culicidae evolution (Jost and Mameli 1972). At the generic level, the genus *Aedes* exhibits more than a threefold variation in nuclear DNA amounts (Rai and Black 1999).

Based on Black and Rai (1988), repetitive DNA sequences increased linearly with genomic size. The linear regression analysis, which included 28 species from 11 Culicoidea genera, established a noteworthy positive correlation between total chromosomal length and haploid genome size. Nevertheless, there was an eightfold variation in haploid genome size along with an approximately fivefold variation in the total chromosomal length. This suggests that DNA amounts increased nearly twice as much as chromosomal size.

DNA analysis of 47 *Aedes albopictus* geographic populations showed a 2.5-fold variation in DNA amounts (Ferrari and Rai 1989; Kumar and Rai 1990). Notably, the independence of genome size from geographic origin was highlighted in the study by Kumar and Rai (1990). Black and Rai (1988) illustrated that the intraspecific variation in the DNA content in *Aedes albopictus* stemmed primarily from highly repetitive DNA sequences. Additionally, McLain et al. (1987) demonstrated that populations exhibiting differences in DNA content also displayed variations in the frequency of highly repetitive DNA classes. This suggests that the diversity in genome size among *Aedes albopictus* populations is mainly attributed to the rapid changes in repetitive DNA sequences.

Moreover, a correlation between genome size and the duration of the generation time in *Aedes albopictus* populations was observed, implying a functional role of repetitive DNA (Ferrari and Rai 1989).

2.3.2.2 Genome Organization

Studies employing reassociation kinetics have identified distinct genome organizations in Culicinae and Anophelinae. *Culicine* species, such as *Culex pipiens*, *Aedes aegypti*, *Ae. albopictus*, and *Ae. triseriatus*, exhibit a "short-period interspersion" pattern characterized by alternating single-copy sequences (1000 ± 2000 bp) with short (200 ± 600 bp) and moderately long (1000 ± 4000 bp) repetitive DNA sequences. In contrast, *Anopheles quadrimaculatus*, *An. gambiae*, and the primitive species, *Chironomus tentans*, display a "long-period interspersion" pattern with long (>5600 bp) repeats alternating with very long (>13,000 bp) uninterrupted stretches of unique sequences (Warren and Crampton 1991; Besansky and Powell 1992).

Two hypotheses on Culicidae genome organization evolution are proposed. The first posits that the long-period interspersion was ancestral, retained in Anophelinae, while larger Culicinae genomes evolved with short-period-interspersed elements. The second hypothesis suggests that Culicidae evolved from a short-period-interspersed ancestor, maintained in Culicinae, while Anophelinae lost repetitive elements, organizing them into a long-period interspersion pattern (Rai and Black 1999).

Chromosomal organization in terms of the distribution of repetitive elements differs between anopheline and culicine species. Recombination frequencies, as deduced from linkage map lengths, do not correlate with known physical genome sizes. For instance, *An. gambiae*, with a smaller genome size, exhibits a linkage map size, like *Culicine* species with larger genome sizes. The frequency of recombination remains high in *An. gambiae* despite its smaller genome size, as DNA reassociation kinetics indicate lower repetitive sequences in *Anopheles* compared to *Culicine* species (Black and Rai 1988; Warren and Crampton 1991; Besansky and Powell 1992). This suggests that, although recombination is restricted in chromosomal regions rich in repeated DNA sequences, *Anopheles* species might have higher recombination rates overall. Despite a relatively small genome size, Anophelinae has a linkage map size comparable to that of the tribe Aedini, indicating relatively constant recombination frequencies over the course of culicid evolution (Rai and Black 1999).

2.3.3 Mitogenomes and Molecular Evolution

The advancement in nucleic acid sequencing technology over the past decade has facilitated a surge in genomic data across various species. Mitogenomes, encompassing the entire mitochondrial genome, serve as valuable molecular markers for studying metazoan species evolution, including population dynamics and phylogenetic relationships. Complete mitogenomes offer reliable insights into ecological and evolutionary processes, featuring genes with varying evolutionary rates. Notably, the widely used COI gene aids in precise species identification, enabling the accurate establishment of speciation events (Battaglia et al. 2016; Ding et al. 2018; Chu et al. 2018; Chen et al. 2023).

Mitogenomes exhibit structural conservation among mosquito species, comprising 37 genes with a defined order. Recent studies, primarily focused on *Anopheles* species, have contributed to understanding mosquito mitogenomes. Their sequencing, initially challenging with Sanger-based platforms, has become more accessible with the advent of next-generation sequencing (NGS). Various approaches, such as PCR/long-range PCR coupled with NGS, shotgun whole-genome sequencing, and RNA-Seq data, have emerged, providing alternative methods for mitogenome recovery (Da Silva et al. 2020). Bioinformatics tools tailored for mitogenome assembly and annotation further facilitate comprehensive analysis.

While *Anopheles* species dominate the available mitogenome data, there is a notable scarcity of genomes from other genera, like *Culex*, *Aedes*, *Haemagogus*, *Bironella*, *Sabethes*, and *Lutzia*. The expanded repertoire of sequencing techniques and bioinformatics tools has significantly enhanced our ability to explore mosquito mitogenomes, contributing to a deeper understanding of their evolutionary dynamics. The mitochondrion, vital in eukaryotic cells, harbors a distinct genome known as the mitogenome. With a small size, low recombination, and maternal inheritance, mitogenomes are pivotal for species identification and phylogenetic studies. In Culicidae, 125 species' mitogenomes, including 74 from *Anopheles*, have been sequenced. Dipteran mitogenomes, around 14–20 kb, feature 37 genes in a circular structure, with a unique "trnR-trnA" rearrangement in mosquitoes (Guo et al.

(2021). This molecular marker has proven effective in phylogenetic studies across various insect orders.

Guo et al. (2021) examined the complete mitogenomes of *Anopheles peditaeniatus* and *Anopheles nitidus*, contributing to the analysis of phylogenetic relationships among 76 *Anopheles* species. Despite typical insect mitogenome characteristics, a distinctive "trnR-trnA" rearrangement was observed. Genome variations focused on the control region, unveiling six new repeat regions. Monophyletic patterns were identified in subgenera and series, revealing novel insights into the phylogenetic relationships among *Anopheles* species. However, certain series and sections displayed polyphyletic or paraphyletic traits, emphasizing the need for further studies to enhance understanding.

2.4 Conclusion

To conclude, Culicidae, the world's largest insect family, exhibits varied species distribution, with the Neotropical region hosting the highest diversity. The ancient origins of these disease vectors can be explained by taxonomic insights within the Culicidae family and evolutionary narratives. With advances in sequencing techniques, mitogenomes and chromosomal karyotypes are being explored for understanding the molecular evolution of mosquitoes. The chapter highlights the interdisciplinary nature of mosquito studies, combining biology, genetics, and evolution to unravel the mysteries of these resilient and often perilous insects.

Points to Remember

1. Culicidae, the world's largest insect family, exhibits varied species distribution, with the Neotropical region hosting the highest diversity.
2. Mosquitoes, as prominent disease vectors, annually afflict millions with diseases, like malaria and dengue, posing significant challenges to global health.
3. With 3719 species, Culicidae traces its roots back 120 million years, experiencing substantial diversification during the Jurassic period.
4. Mitogenomes serve as vital markers, unraveling evolutionary dynamics and structural conservation across *Anopheles* species.
5. Maintaining a stable diploid chromosome, mosquitoes exhibit distinctive karyotypic differences between Anopheline and Culicine species.

References

Baimai V (1988) Constitutive heterochromatin differentiation and evolutionary divergence of karyotype in Oriental *Anopheles* (Cellia). Pac Sci 42:13–27

Battaglia V, Gabrieli P, Brandini S, Capodiferro MR, Javier PA, Chen XG, Olivieri A (2016) The worldwide spread of the tiger mosquito as revealed by mitogenome haplogroup diversity. Front Genet 7:208

Belkin JN (1968) Mosquito studies (Diptera: Culicidae). VII. The culicidae of New Zealand. Contrib Am Entomol Inst 3:1–182

Bertone MA, Courtney GW, Wiegmann BM (2008) Phylogenetics and temporal diversification of the earliest true flies (Insecta: Diptera) based on multiple nuclear genes. Syst Entomol 33(4):668–687

Besansky NJ, Powell JR (1992) Reassociation kinetics of *Anopheles gambiae* (Diptera: Culicidae) DNA. J Med Entomol 29(1):125–128

Black WC, Rai KS (1988) Genome evolution in mosquitoes: intraspecific and interspecific variation in repetitive DNA amounts and organization. Genet Res 51(3):185–196

Borkent A (1993) A world catalogue of fossil and extant Corethrellidae and Chaoboridae (Diptera), with a listing of references to keys, bionomic information and descriptions of each known life stage. Insect Syst Evol 24(1):1–24

Chen DH, He SL, Fu WB, Yan ZT, Hu YJ, Yuan H, Chen B (2023) Mitogenome-based phylogeny of mosquitoes (Diptera: Culicidae). Insect Sci 31(2):599–612

Chu H, Li C, Guo X, Zhang H, Luo P, Wu Z, Zhao T (2018) The phylogenetic relationships of known mosquito (Diptera: Culicidae) mitogenomes. Mitochond DNA Part A 29(1):31–35

Cockerell TDA (1919) The oldest mosquitoes. Nature 103(2577):44–44

Da Silva AF, Machado LC, de Paula MB, da Silva Pessoa Vieira CJ, de Morais Bronzoni RV, de Melo Santos MAV, Wallau GL (2020) Culicidae evolutionary history focusing on the Culicinae subfamily based on mitochondrial phylogenomics. Sci Rep 10(1):18823

Darlington PJ Jr (1964) Drifting continents and late Paleozoic geography. Proc Natl Acad Sci 52(4):1084–1091

De Oliveira Aragão A, Neto JPN, Cruz ACR, Casseb SMM, Cardoso JF, da Silva SP, Ishikawa EAY (2019) Description and phylogeny of the mitochondrial genome of *Sabetheschloropterus*, *Sabethesglaucodaemon* and *Sabethesbelisarioi* (Diptera: Culicidae). Genomics 111(4):607–611

Dev V, Rai KS (1982) Genetics of speciation in the *Aedes* (*Stegomyia*) *scutellaris* group (Diptera: Culicidae). I. Crossing relationships among six species. In: Evolutionary significance of insect polymorphism. Entomological Society of America, San Diego

Dev V, Rai KS (1984) Genetics of speciation in the *Aedes* (*Stegomyia*) *scutellaris* group (Diptera: Culicidae). V. Chromosomal relationships among five species. Genetica 64:83–92

Dev V, Rai KS (1985) Genetic relationships among certain species of the *Aedes* (*Stegomyia*) scuitellaris group (Diptera: Culicidae). Ann Trop Med Parasitol 79(3):325–331

Ding YR, Li B, Zhang YJ, Mao QM, Chen B (2018) Complete mitogenome of *Anophelessinensis* and mitochondrial insertion segments in the nuclear genomes of 19 mosquito species. PLoS One 13(9):e0204667

Edwards FW (1923) Oligocene mosquitoes in the British Museum; with a summary of our present knowledge concerning fossil Culicidae. Q J Geol Soc 79(1-4):139–155

Edwards FW (1932) Genera insectorum, Diptera. Fam. Culicidae. Fascicle 194:258

Ferrari JA, Rai KS (1989) Phenotypic correlates of genome size variation in *Aedesalbopictus*. Evolution 43(4):895–899

Foster WA, Walker ED (2019) Mosquitoes (Culicidae). In: Medical and veterinary entomology. Elsevier, Amsterdam, pp 261–325

Gatti M, Santini G, Pimpinelli S, Coluzzi M (1977) Fluorescence banding techniques in the identification of sibling species of the *Anopheles gambiae* complex. Heredity 38:105–108

Gilchrist BM, Haldane JBS (1947) Sex linkage and sex determination in a mosquito, *Culex molestus*. Hereditas 33(1-2):175–190

Guo J, Yan ZT, Fu WB, Yuan H, Li XD, Chen B (2021) Complete mitogenomes of *anopheles peditaeniatus* and *Anopheles nitidus* and phylogenetic relationships within the genus *anopheles* inferred from mitogenomes. Parasit Vectors 14:1–14

Harbach RE (2007) The Culicidae (Diptera): a review of taxonomy, classification and phylogeny. Zootaxa 1668(1):591–638

Harbach RE, Kitching IJ (1998) Phylogeny and classification of the Culicidae (Diptera). Syst Entomol 23(4):327–370

Hilburn LR, Rai KS (1981) Electrophoretic similarities and mating compatibility among four species of the *Aedes (Stegomyia) scutellaris* complex (Diptera: Culicidae). J Med Entomol 18(5):401–408

Jost E, Mameli M (1972) DNA content of nine species of Nematocera with special reference to the sibling species of the *Anopheles maculipennis* group and the *Culex pipiens* group. Chromosoma 37:201–208

Kambhampati S, Rai KS (1991) Variation in mitochondrial DNA of *Aedes* species (Diptera: Culicidae). Evolution 45(1):120–129

Kitzmiller JB (1953) Mosquito genetics and cytogenetics. Rev Bras Malariol Doencas Trop 5(4):285–359

Knight KL, Stone A (1977) A catalog of the mosquitoes of the world (Diptera: Culicidae). In: Thomas say found, vol VI, 2nd edn. Entomological Society of America, College Park, p 611. http://www.mosquitocatalog.org/main.asp

Kreutzer RD (1978) A mosquito with eight chromosomes: *Chagasia bathana* Dyar. Mosq News 38:554–558

Krzywinski J, Wilkerson RC, Besansky NJ (2001) Toward understanding Anophelinae (Diptera, Culicidae) phylogeny: insights from nuclear single-copy genes and the weight of evidence. Syst Biol 50(4):540–556

Kumar A, Rai KS (1990) Intraspecific variation in nuclear DNA content among world populations of a mosquito, *Aedes albopictus* (Skuse). Theor Appl Genet 79:748–752

Manguin S, Bangs MJ, Pothikasikorn J, Chareonviriyaphap T (2010) Review on global co-transmission of human *Plasmodium* species and *Wuchereriabancrofti* by *Anopheles* mosquitoes. Infect Genet Evol 10(2):159–177

Manikandan S, Mathivanan A, Bora B, Hemaladkshmi P, Abhisubesh V, Poopathi S (2023) A review on vector borne disease transmission: Current strategies of mosquito vector control. Indian J Entomol 85:503–513

McLain DK, Rai KS (1986) Reinforcement for ethological isolation in the southeast Asian *Aedes albopictus* subgroup (Diptera: Culicidae). Evolution 40:1346–1350

McLain DK, Rai KS, Rao PN (1985) Ethological divergence in allopatry and asymmetrical isolation in the south pacific *Aedes scutellaris* subgroup. Evolution 39:998–1108

McLain DK, Rai KS, Fraser MJ (1987) Intraspecific and interspecific variation in the sequence and abundance of highly repeated DNA among mosquitoes of the *Aedes albopictus* subgroup. Heredity 58(3):373–381

Motara MA (1982) Giemsa C-banding in four species of mosquitoes. Chromosoma 86:319–323

Motara MA, Rai KS (1978) Giemsa C-banding patterns in *Aedes* (*Stegomyia*) mosquitoes. Chromosoma 70(1):51–58

Munstermann LE, Conn JE (1997) Systematics of mosquito disease vectors (Diptera, Culicidae): impact of molecular biology and cladistic analysis. Annu Rev Entomol 42(1):351–369

Pashley DP, Rai KS (1983) Comparison of allozyme and morphological relationships in some *Aedes* (*Stegomyia*) mosquitoes (Diptera: Culicidae). Ann Entomol Soc Am 76(3):388–394

Pashley DP, Rai KS, Pashley DN (1985) Patterns of allozyme relationships compared with morphology, hybridization, and geologic history in allopatric island-dwelling mosquitoes. Evolution 39:985–997

Poinar GO, Zavortinik TJ, Pike T, Johnston PA (2000) *Paleoculicisminututs* (Diptera: Culicidae) n. Gen., n. Sp., from Cretaceous Canadian amber, with a summary of described fossil mosquitoes. Acta Geol Hisp 35:119–130

Rai KS (1963) A comparative study of mosquito karyotypes. Ann Entomol Soc Am 56(2):160–170

Rai KS (1999) Genetics of mosquitoes. J Genet 78:163–169

Rai KS, Black WC (1999) Mosquito genomes: structure, organization, and evolution. Adv Genet 41:1–33

Rai KS, Pashley DP, Munstermann LE (1982) Genetics of speciation in Aedine mosquitoes. In: Recent developments in the genetics of insect disease vectors. Springer, Cham, pp 84–128

Rao PN, Rai KS (1987) Comparative karyotypes and chromosomal evolution in some genera of nematocerous (Diptera: Nematocera) families. Ann Entomol Soc Am 80(3):321–332

Rasnitsyn AP, Quicke DL (eds) (2007) History of insects. Springer, Cham

Reidenbach KR, Cook S, Bertone MA, Harbach RE, Wiegmann BM, Besansky NJ (2009) Phylogenetic analysis and temporal diversification of mosquitoes (Diptera: Culicidae) based on nuclear genes and morphology. BMC Evol Biol 9:1–14

Reinert JF, Harbach RE, Kitching IJ (2004) Phylogeny and classification of Aedini (Diptera: Culicidae), based on morphological characters of all life stages. Zool J Linnean Soc 142:289–368

Rick EF (1970) Fossil history. In: Insects of Australia. Melbourne University Press, Melbourne

Ross HH (1951) Conflict with *Culex*. Mosq News 11:128–132

Sato S (2021) Plasmodium—a brief introduction to the parasites causing human malaria and their basic biology. J Physiol Anthropol 40(1):1–13

Schaffner F, Medlock JM, Van Bortel AW (2013) Public health significance of invasive mosquitoes in Europe. Clin Microbiol Infect 19(8):685–692

Sherron DA, Rai KS (1984) Genetics of speciation in the *Aedes* (*Stegomyia*) *scutellaris* subgroup (Diptera: Culicidae). 4. Chromosomal relationships of Aedes cooki with four sibling species. Can J Genet Cytol 26(3):237–248

Warren AM, Crampton JM (1991) The *Aedes aegypti* genome: complexity and organization. Genet Res 58(3):225–232

Wesolowski R, Wozniak A, Mila-Kierzenkowska C, Szewczyk-Golec K (2015) Plasmodium knowlesi as a threat to global public health. Korean J Parasitol 53(5):575

White GB (1980) Academic and applied aspects of mosquito cytogenetics. In: Blackman RL, Hewitt GM, Ashburner M (eds) Insect cytogenetics. Blackwell Scientific Publications, London

Zavortin TJ, Poinar GO (2000) *Anopheles (Nyssorhynchus) dominicanus* sp. n.(Diptera: Culicidae) from Dominican amber. Ann Entomol 93(6):1230–1235

Zavortink TJ, Poinar GO (2008) *Toxorhynchites (Toxorhynchites) mexicanus*, n. sp.(Diptera: Culicidae) from Mexican amber: a New World species with Old World affinities. Proc Entomol Soc Wash 110(1):116–125

Mosquito Morphology: Anatomy to Adaptation in the Shadow of Evolution

3

Susmita Das, Ida Anandi Mukhiya, Tameka Hazra, Sampurna Roy, and Amlan Das

Abstract

Mosquitoes are critical to public health, because they often transmit various deadly diseases. Among a well-diverse mosquito population, some of them cause severe health risks, which may sometime turn into fatality. As a result, numerous strategies are used to manage and limit their populations worldwide. However, it is noteworthy that they can exhibit highly distinctive adaptive morphological modifications to cope with new environmental conditions and crises. In this context, understanding the morphological diversity and adaptability of morpho-organs becomes essential for developing successful management techniques of these tiny terrors. The present chapter addresses mosquitoes' structural variations and modifications, including the intricate sensory organs, such as antennae and compound eyes, and the changes in morphological features in the head, thorax, and abdomen. It discusses the differences between species in head structure, thoracic shape, abdominal changes, and appendicular articulation, such as the number, color, and arrangement of sensory hairs, bristles, setae, and papillae. The structural complexity of the compound eyes to day-night vision, the antennary Johnston's organ, and its chemosensitive and mechanosensitive functions, as well as the variations and modifications of the respiratory siphon, comb scales, feeding apparatus, and proboscis, are all compared among several mosquito species. The scales over their body, particularly on the legs, exhibit unique differences in appearance and hues among the species. In addition, the color, shape, and structure of their wings and the number and arrangement of wing veins are also highly unique among the species. All these characteristics play critical roles in distinguishing species for taxonomic purpose and cladistic analysis. Mosquito morphologies and their evolution in medically important species are therefore essential for developing and implementing effective vector control strategies.

S. Das · I. A. Mukhiya · T. Hazra · S. Roy · A. Das (✉)
Entomology Laboratory, Department of Zoology, University of Calcutta, Kolkata, West Bengal, India

© The Author(s), under exclusive license to Springer Nature Singapore Pte Ltd. 2024
Omkar (ed.), *Mosquitoes*, https://doi.org/10.1007/978-981-97-4163-2_3

Keywords

Mosquito · Morphology · Evolution · Head · Thorax · Abdomen · Compound eyes

Learning Objectives

1. Disseminate information regarding mosquito morphology.
2. Understand species-specific morphologies for enhancing mosquito identification.
3. Comprehend evolutionary forces shaping the morphological diversity of mosquitoes.

3.1 Introduction

They are called *little fly*; it aptly describes these tiny flying insects, well-known for their buzzing sound, and for biting humans and nonhumans to feed on blood. The term "mosquito" derives from the Spanish word *mosca*, which means *fly*, and the diminutive suffix *-ito* is frequently used to signify something small or diminutive; therefore, *mosquito* literally translates to *little fly* (Kaulfers 1940). They are found everywhere, in different ecosystems, in different habitats, and in different countries around the world. Given their ubiquitous appearance, understanding their morphology, structure, and architecture is crucial for comprehending their biological makeup, behavioral conduct, and activity as a potent vector for transmitting fatal diseases (Lehrer 2010; Benelli et al. 2016). However, they can adapt to various climates and ecosystems, except in the polar regions, and so, their morphological features vary across the geographical scales. Their larvae are typically found everywhere in different aquatic habitats, from shallow water bodies to large lakes, from big ponds to cattle puddles, from snow-melt pools to rock pools, and from tree holes to bamboo stumps. They can be found in confined spaces, ranging from household storage containers to earthen teacups, municipal water tanks to cesspits, and discarded tires to jardinières.

Adult mosquitoes typically measure around 3–6 mm in length and can be distinguished from other Dipterans by three primary features: (1) a forwardly projected proboscis; (2) the presence of numerous closely packed scales on the thorax, legs, and abdomen; and (3) an arrangement of scales along the rear edges of the wings (Service 2012). Mosquitoes are classified into three subfamilies: Anophelinae (anophelines), Culicinae (culicines), and Toxorhynchitinae, comprising 41 genera and over 3600 species and subspecies worldwide (Kachhwaha et al. 2017) (Table 3.1).

3.1.1 Mosquito Subfamilies

3.1.1.1 Anophelinae

Anophelinae mosquitoes are distinguished by their long, slender palps, almost as long as their proboscis, and black and white speckled scales on their wings. The

Table 3.1 Worldwide mosquito diversity

Subfamily	Tribe	Genera
Anophelinae		*Anopheles (An.), Bironella (Bi.), Chagasia (Ch.)*
Culicinae	Aedeomyiini	*Aedeomyia (Ad.)*
	Aedini	*Aedes (Ae.), Ochlerotatus (Oc.), Verrallina (Ve.), Ayurakitia (Ay.), Armigeres (Ar.), Eretmapodites (Er.), Haemagogus (Hg.), Heizmannia (Hz.), Opifex (Op.), Psorophora (Ps.), Udaya (Ud.), Zeugnomyia (Ze.)*
	Culicini	*Culex (Cx.), Deinocerites (De.), Galindomyia (Ga.)*
	Culisetini	*Culiseta (Cs.)*
	Ficalbiini	*Ficalbia (Fi.), Mimomyia (Mi.)*
	Hodgesiini	*Hodgesia (Ho.)*
	Mansoniini	*Coquillettidia (Cq.), Mansonia (Ma.)*
	Orthopodomyiini	*Orthopodomyia (Or.)*
	Sabethini	*Sabethes (Sa.), Wyeomyia (Wy.), Phoniomyia (Ph.), Limatus (Li.), Trichoprosopon (Tr.), Shannoniana (Sh.), Runchomyia (Ru.), Johnbelkinia (Jb.), Isostomyia (Is.), Tripteroides (Tp.), Malaya (Ml.), Topomyia (To.), Maorigoeldia (Mg.)*
	Uranotaeniini	*Uranotaenia (Ur.)*
Toxorhynchitinae		*Toxorhynchites (Tx.)*

wings of the insect display pronounced costa, subcosta, radius, media, cubitus, and anal wing veins. Certain species may have dark spots or patches on their wings, although the number and arrangement of these markings differ among species. Species-specific lines and colors can be seen on their abdomens. They deposit eggs individually on still-water surfaces. They are typically more active during dawn, dusk, and night, engaging in nocturnal feeding activities.

Anophelinae mosquitoes are distinguished by their wings, which are decorated with scales that are speckled in black and white, as well as their palpi, which are of equal length to their proboscis. The *Orthopodomyia* genus is distinguished by its spherical abdomens, wide wings adorned with clear and dark scales, and a unique arrangement of white lines on the mesonotum. In addition, the front pair of legs bear a brief fourth segment (Kachhwaha et al. 2017; Rueda 2020; World Health Organization 2020; Wilkerson et al. 2021).

3.1.1.2 Culicinae

The Culicinae mosquitoes have remarkable wing venation, including notable costa, subcosta, radius, medium, cubitus, and anal veins. They possess elongated sensory appendages, or palps, located near the proboscis. Their slender yet elongated thorax and abdomen feature various bands and colorations unique to each species. Typically, they deposit their eggs in groups in still water or areas with high moisture levels. The larvae of Culicinae, also known as "wrigglers," have segmented bodies and are adapted for aquatic environments. They possess specialized siphons on the terminal part of abdomen that allow them to breathe on the water surface. On the other hand, the pupae, commonly called as "tumblers," have a comma-shaped appearance and are equipped with respiratory trumpets on the cephalothorax that enable them to breathe when submerged in water. Culicinae mosquitoes typically inhabit numerous freshwater settings, while certain species have adapted to urban contexts. They are predominantly active at dawn and dusk, although they may also exhibit activity during the night time.

The subfamily Culicinae is highly diverse, with a wide range of species identified all over the world. The diversity of Culicinae mosquitoes emphasizes their importance as disease vectors in ecology and epidemiology, and their adaptability to a wide range of environments and ecological niches. Members of Culicinae differ morphologically in many morpho-features. The genus *Uranotaenia* includes tiny mosquitoes with bluish-iridescent scales organized in rows or patches. The second marginal wing cell of this group is shorter than the stem, or petiole, distinguishing it from other genera. *Wyeomyia* has a prominent longitudinal stripe on its abdomen, where black dorsal scales meet white ventral scales. The postnotum is furnished with setae. *Aedes* and *Psorophora* can be distinguished by the shape of their abdomens, the placement of pale scale bands or lateral patches on their abdominal segments, and the presence or absence of prespiracular bristles. In *Aedes*, pale scales appear at the anterior end of the abdominal segment without prespiracular bristles, while in *Psorophora*, they appear posteriorly (apical) with prespiracular bristles. *Coquillettidia* is distinguished by its convex abdomen, large wing scales, and a white stripe on the hind tibia with no post-spiracular hairs. *Deinocerites* have a rounded abdomen and short wings, but their antenna is longer than their proboscis and has a visible second segment. *Mansonia* mosquitoes are distinguished by their large winged and blunt abdomens, the absence of a subapical ring with pale scales on the hind tibia, and post-spiracular bristles (Nugroho et al. 2021). Both *Culex* and *Culiseta* have slender wings and rounded abdomens. However, *Culex* lacks prespiracular bristles, whereas *Culiseta* does. *Culiseta* also has small wing scales and bristles on the subcostal wing vein, whereas *Culex* lacks them (Kachhwaha et al. 2017; Rueda 2020; World Health Organization 2020; Wilkerson et al. 2021).

3.1.1.3 Toxorhynchitinae

Toxorhynchitinae, also called "elephant mosquitoes," exhibit the lowest level of diversity among all subfamilies. They are relatively larger (with a maximum length of 2 cm) than the *Anophelinae* and *Culicinae* species. Instead of consuming blood meal, they obtain nourishment from nectars, plant saps, and other sugary substances. The larvae of Toxorhynchitinae are carnivorous and consume the larvae of other mosquitoes. Their mouthparts are adapted explicitly for collecting and digesting their prey. Adult Toxorhynchitinae has an extended proboscis, specifically designed for nectar consumption but not for piercing the mammalian skin. Unlike others, they do not feed on blood but are predators in both larval and adult stages. Typically, these mosquitoes have fewer or no mouthpart sensilla than blood-feeding mosquitoes, reflecting their non-blood-feeding activity. They are excellent fliers and often fly near flowers or foliage for food. They are capable of thriving in both artificial and natural aquatic environments. The adult Toxorhynchites mosquitoes are characterized by their long legs, vibrant coloration, and downward-curled proboscis employed for nectar feeding.

3.2 General Morphological Features of Mosquitoes

Mosquitoes undergo holometabolous development, which involves four distinct life stages: egg, larva, pupa, and adult. The first three stages primarily occur in water or soil near water sources, whereas the mature stage is determined by the insect's capacity to fly. They exhibit distinctive morphological attributes intricately tailored to their ecological niche. They have unique anatomical characteristics, including bristled antennae, forwardly projected curved proboscis, and compound intricate eyes. Furthermore, their unique proboscis traits can be distinguished from other species relative to thorax length. Mosquitoes can be distinguished from other true flies by the scales on their veins and wing margins (Becker et al. 2020). These modifications facilitate essential biological functions, such as prey recognition, food acquisition, successful reproduction, and navigation (Montell and Zwiebel 2016). Mosquitoes possess intricate mouthparts that are specifically designed for piercing and sucking. The proboscis, consisting of stylets and maxillae, enables efficient blood feeding while remaining concealed during rest periods. When a mosquito is prepared to feed, it inserts its proboscis into the host's skin, typically utilizing sensory organs on the tip of the labium to locate an appropriate blood vessel. Subsequently, the mosquito injects its saliva into the host, which serves the purpose of inhibiting blood coagulation and facilitating the absorption of blood. After consuming a blood meal, the mosquito retracts its proboscis and departs by flying. The proboscis of a mosquito is a specialized organ designed for extracting blood, a behavior exhibited by female mosquitoes to obtain the necessary nutrients for egg formation. Male mosquitoes possess proboscises primarily specialized for nectar feeding, as they do not consume blood. However, the mouthpart architecture and proboscis articulation vary depending on the species. These alterations and variances are crucial for their hematophagous behavior and transmission of diseases (Kong and Wu 2010).

Mosquitoes often have somewhat smaller and more fragile wings than other airborne insects. Flapping their wings at higher frequencies and lower amplitudes generates a distinctive trailing-edge vortex and rotational drag, enhancing their effectiveness during flight (Zhang and Huang 2019). This distinctive aerodynamic characteristic enhances their effectiveness during flight (Zhang and Huang 2019). Understanding these adaptations is crucial for comprehending population dynamics and distribution patterns. Reproduction involves the use of specialized structures, such as the ovipositor and genitalia. Sexual dimorphism is conspicuous among mosquitoes, as males with bushier antennae, generally exhibit specialized genital structures designed for reproduction (Steward and Atwood 1963). The female's ovipositor is adapted explicitly for precise egg deposition in different aquatic habitats (Montell and Zwiebel 2016). They generally use their antennal receptors to detect chemical cues that help them find hosts and suitable places to oviposit. They are capable of visual orientation with the help of their compound eyes. Moreover, the sensory hairs, or sensilla, dispersed throughout their bodies are essential for perceiving environmental cues (Steward and Atwood 1963). The general morphological features of mosquito genera are tabulated in Table 3.2.

Table 3.2 Morphological characteristics of mosquito genera

Genera	Head and thorax	Scutellum	Proboscis	Palpi	Antenna	Wing	Appendages/legs and claw
Aedeomyia	Broad scales are on pedicel Lower mesepimeral setae present Prespiracular and postspiracular setae lacking		Body covered with broad scales, superficially resemble tribe Aedini except antennal flagellomeres in females Two apical stout flagellomeres in males		Stout antennae with thick flagellomeres; flagellomere base with scale tufts Antenna with two thick flagellomeres in males	Wings broad with white/yellow/brown scales	Apical tuft scales on mid and hind femora Sub-erect scales present on femora apex
Aedes	Globular head convex laterally Postspiracular setae present Both sexes have flat-scaled, silvery white vertex Adults are thin, black; distinct light-dark scales on thorax and abdomen, light-dark color bands on legs Abdomen with sharp tip in females Maxillary palps are shorter than proboscis	Scutum with white spots Each lobe of the scutellum with white scales Mesepimeral and lower proepisternal scales present Postpronotum covers scales, but no scale on postspiracular area White scales on paratergite White scales on subspiracular region	Proboscis is dark, long, and straight Proboscis with white scale patch Male proboscis length 0.76 ± 0.04 mm, longer than females Length of female palps <0.4 mm of proboscis	Mouth with two maxillary palps Males with five white-scale bands; length 0.77 ± 0.06 mm Females with short, white apical scales	Reduced scape, antennae emerge from spherical pedicel Thirteen flagellar segmented antenna Male antenna longer (0.57 ± 0.03 mm) than females (0.52 ± 0.07 mm) Male antennal hairs bushy, plumose, female hairs are thinner Basal antennal lobes with white scale covering	Wings are membranous, thin, and flat in both sex Wings without white scales with unique venation Wing base narrow, middle is broad Posterior margin bears upright fringe scales, while anterior margin is flat and linear Wing base slightly convex, alula with tiny fringe scales Male wings: L: 2.93 ± 0.52 mm/W: 0.66 ± 0.08 mm; female wings: L: 3.09 ± 0.36 mm/W: 0.71 ± 0.16 mm	Legs with distinct femur, tibia, trochanter, and coxa Elongated, cylindrical first segment joins thorax with trochanter Each segment with white flat scale patches Hind coxa bears one scale patch, mid coxa bears two Femur joined by a small cylindrical trochanter Last tarsal segment bears claws Dark apical and subapical silvery specks on mid-hind femora Typically, claws bears subbasal teeth

Anopheles	Mesonotum extended to thorax Scutal integument grey/brown/black Scales distributed on thoracic pleura	Scutellum uni-lobed/trilobed/rounded Scutellum consistently setose and circular	Female palps short, equal; proboscis length not exceeding 0.7 mm	Maxillary palps with five palpomeres for both sex Silvery white/cream/yellow/dark brown/black scales Proboscis is correlated with palp length in females	Antenna with 13 extended flagellomeres Apical longer setae; shorter setae scattered over flagellomere Male antennae bear longer, spirally arranged setae	Wings black membranour Scale color matches with wing color Pale or dark scales Longitudinal veins present	Legs with dark/pale spirally arranged scales with rings and bands
Armigeres	Palp length approximately equal to proboscis in males Female proboscis downturned, laterally compressed Long, slender, elongated with three segments Two lengthy scale rows ventrally divide the eyes Postspiracular setae absent, scales present in postspiracular region Small mesomeron base equal Head scales are mainly broad and flat Dorsocentral and acrostichal setae are missing		Proboscis curved downward and compressed laterally. Proboscis length 0.4–0.7 mm	Palps extend beyond the clypeus		Alula and calypter margins resembling hair and somewhat scaled	

(continued)

Table 3.2 (continued)

Genera	Head and thorax	Scutellum	Proboscis	Palpi	Antenna	Wing	Appendages/legs and claw
Carrollia	Prothoracic lobes are very small A pair of coarse vertical bristles projecting between eyes	Scutellum trilobite	Proboscis moderate, slightly enlarged at the apex	Palpi short in female, long in male	In females, antennae with scanty hairs, filiform, and subequally jointed In male, short antennal joints with a high, curving rim that holds the hair whorl (long, dense, and plumose)	Compared to females, males have smaller wings Basal cross-vein and anterior cross-veins are distinct	Leg length is typical While female claws are equal and plain, male claws can occasionally have teeth
Coquillettidia	Bare clypeus, longer than broad					Wing scales are often symmetrical, noticeable, and a mixture of light and dark colors Wing scales are typically wide	PLegs are striking, post-tarsus with ungues No pulvillus is present Abdomen terga has metallic purple scales, whereas the wing has yellow scales Large and noticeable ungues of the rear legs Gonostylar claw present
Culex	Postspiracular or prespiracular setae are missing Prothoracic lobes are widely spaced apart	Trilobite scutellum	Proboscis length varies; it can be robust or thin, uniform or expanded at the apex, straight in males, and somewhat curved in females	Palpi are shorter than proboscis	Female antennal shaft has subequal filiform joints with basal whorls of relatively sparse hairs Male with last two shortest joints, with a curved, long, dense, plumose	Male wings are usually thinner than female wings The scales on the wings narrow Alula with a fringed thin scale	Moderate leg length A scraper with no more than five teeth on the hind tibiae Females have equal and simple claws, whereas males have uneven and sometimes toothed claws Small ungues of the rear legs are not very noticeable Pulvilli developed
Culiseta	There are prespiracular setae Postspiracular region typically devoid of scales Lower mesepimeral setae	The paratergite and postspiracular areas have pale scales				Anterior veins with brown and white scales At subcosta base, ventral setae present (excluding subgenus Theomyia)	Pulvillus not present

Deinocerites	Prothoracic lobes remote dorsally			Three-jointed palpi, short in both sexes, with the longest joint at the end	Long, slender antennae in both sexes, shaft with scattered hairs, and small, sparse hair whorls Fourteen times as long as wide is the second joint Third joint is two-third in length of the second, joints are progressively shorter In females, third joint is roughly one-third the length of the second joint		Long and thin legs Simple claws in females, males with toothed claws for some species
Dinanamesus	Dorsally reaching prothoracic lobes			Both sexes have short palpi	Long, slender antennae in females; second joint roughly eight times longer than width; third and fourth joints combined being longer than second; last joint narrow In male second joint is roughly nine times longer than it is wide Subequal joints with third almost as long as the second Terminal joint forms a little knob; whorls of hair were sparse, very short in sexes		Legs are thin and somewhat lengthy Female's claws are simple

(continued)

Table 3.2 (continued)

Genera	Head and thorax	Scutellum	Proboscis	Palpi	Antenna	Wing	Appendages/legs and claw
Dinomimetes	Dorsally reaching prothoracic lobes			Both sexes have small, three-jointed palpi; last joint longer than others	Short, slightly cryptic antennae, longer in males Male successive joints progressively get less in length		Long and thin legs are present In females, all claws are simple Males have one claw on front and middle legs
Ficalbia	Occiput with erect head scales Both sexes have slightly enlarged tips of their proboscis Scutellum with scaled in thorax Dorsocentral and acrostichal setae are present Absence of lower mesepimeral setae, prealar scales, and prespiracular setae Postspiracular area bare		Male proboscis swollen to distal end		Flagellomere on female antenna is around three times longer than males	Alula and upper calypter with narrow scale fringes Vein 1A ends above the mediocubital crossvein base	
Goeldia	Flat scales covering the head Mesothorax has larger, narrower-curved scales and flat, spindle-shaped ones Scutellum lanceolate in shape	Scutellum has scales that are flat	Compared to the body, the proboscis is shorter and thicker	Palpi are quite tiny and approximately one-third the length of the proboscis		Wing scales that finish asymmetrically, thick, and elongated like runchomyla	

Heizmannia	Erect scales present on the occiput Bright metallic scutum with wide, flat scales, meso-postnotum with a cluster of tiny setae, dorsocentral and acrostichal regions devoid of setae Head vertex has areas of broad white or silver scales between the eyes and laterally Postspiracular, upper mesokatepisternal, or prespiracular setae absent Mesepimeral setae on the bottom Scales present on the laterotergite in addition to the subspiracular, postspiracular, and paratergal scales					Beyond the medial cubital crossvein, vein 1A ends Broadly scaled alula with dorsal edge Upper calypter has fringed setae	Mostly dark
Hodgesia	Broad, flat head scales with silvery scales on vertex Scutum is typically bald; large, silvery scales on antepronotum Pleuron with few setae and a patch of silvery scales Prominent dorsocentral setae; acrostichal setae lacking					Emarginated tip scales on outer half of wings Vein 1A extended to wing edge at base of the mid-cubital cross-vein Distal wing membrane with scales; notched at tip; prominent microtrichia	Numerous small, robust, submedian setae on caudal border of tergum

(continued)

Table 3.2 (continued)

Genera	Head and thorax	Scutellum	Proboscis	Palpi	Antenna	Wing	Appendages/legs and claw
Isostomyia	Reflective green to blue-colored dorsal head scales Scutum flat with wide scales No posterior postpronotal setae Dorsal border of Hind-coxa is roughly level with mesepimeron base Both male and female heads are black Yellow thorax, somewhat darker mesonotum, covered by brown squamulae		Black proboscis extended to abdomen Palps are brown		Brown antennae Male plumosity is lengthy, thick, and dark. Terminal joint with short hairs. Females with moderate length verticils, shorter terminal joints	Wing veins squamulate; consistently dark-brown; about 4–5 mm length	Brown/blackish legs; purplish/greyish reflection in light; yellow femora Yellowish tibia beneath the tomentum Proboscis shorter
Limatus	Moderate-sized head with eyes Large, well-separated prothoracic lobes with coarse setae at frontal edges Violet/blue/golden scales cover scutum Presepractal scales present but no setae		Proboscis in females strong and short Males proboscis straight/swelling/curved; with expanded scales	Fused joints and short palpi in both sexes	Antennae subequal; more plumed in males than females.		Small claws for both sex legs; only one claw on hind pair Among Sabethines, unguis single
Lutzia	Antepronotum scales thin Mesepimeral setae present; 4–6 mm length Scutellum with slender scales	Scutellum with three lobes	Proboscis moderate in size	Proboscis is short in length; short palpi in females Palpi in males are long, acuminate	Female antennae slender; subequal joints and sporadic basal whorls; fragile hairs. Males with thick plumose have two lengthy terminal joints	Cross veins nearly at same level as vein base Prothoracic lobes well segregated	Legs are quite lengthy and slender Large, well-developed tarsi empodia Female claws are minimal Males with uneven, toothed claws on front legs Large pulvillus Femur and Tarsi with light scale patches

Malaya	Scutum with a median longitudinal white/silver stripe		Numerous upright, upturned, and apically enlarged setae on proboscis Proboscis hairy, tilted, and enlarged apically			Alula fringed, calypter naked on top wings Vein 1A intersects mediocubital cross-vein at tip	
Manosonia	Clypeus is naked Each thoracic lobe clearly divided		Proboscis is uniform; moderate length	Longer palpi in males, shorter in females	Prominent filiform antennae; subequal joints, basal whorls in females Males have plumose, last two joints are longer.	Wide dorsal scales, clearly asymmetrical wing veins Scales blended dark and white Wings are modest	Eight spinal rows in tibial scraper on hind leg Male with uneven, toothed front and middle claws; females with simple ones
Mimomyia	Scales on vertex and occiput; occiput erect Both dorsocentral and acrostichal setae on thorax					Alula bare/with broad decumbent dorsal scales Upper calypter with slender setae and hair-like scales Vein 1A terminates beyond mediocubital cross-vein	Leg pulvillus absent/ indistinct
Orthopodomyia	Thorax with white stripes; silver/gold scales; distinct mottling in some Postspiracular setae absent					Wing spotted	Tarsomeres are noticeably shorter than fore- and mid-tarsomeres Tarsomere I of fore legs longer than tarsomeres II to V Reduced tarsomere IV on fore legs

(continued)

Table 3.2 (continued)

Genera	Head and thorax	Scutellum	Proboscis	Palpi	Antenna	Wing	Appendages/legs and claw
Runchomyia (Lesticocampa)	Head scales bright/blue/ silver; dorsally arranged Prothoracic lobes moderate, well separated dorsally Scutal scales narrow and curved Mesepimeron lower edge situated above lower mesokatepisternal setae Postpronotal setae present Clypeus without setae and scales Eyes contiguous		Long, thin proboscis without any bulge at tip Proboscis length 1.2–1.4 times longer than forefemur	Females with short palpi, long in males	Antennae with subequal joints; slightly more plumose in males than females		Long, thin legs; hind legs often ciliated Short, equal, simple claws in females; Slender single claw in hind pair Males with uneven front and middle legs
Sabethoides	Head moderate, eyes large, contiguous at vertex Vertex with violet/bluish scales; proboscis mostly corresponds femur length Prothoracic lobes large, dorsally arranged Thorax antepronotum large, nearly contiguous dorsally Antepronotal and scutal scales metallic green/bluish, appearing golden in some lights Upper proepisternal and lower mesokatepisterna setae present Pre-alar setae absent Upper mesepimeral setae shortMesopostnotum large		Proboscis long and slender	Palpi very short and slender	Antennae with subequal joints; long hair whorls		Long and thin legs Simple, equal, and tiny claws Tarsomere length is greater than total length of femur and tibia Tarsa with ventral white-scales No leg paddles

Theobaldia	Prothoracic lobes well separated		Proboscis slender	Females with short palpi, long in males Thick, club-shaped; terminal ends non-acuminate	Filiform antennae; subequal joints; sparse basal whorls with short hairs in females Last two joints long in males; others short, thickened	Wings with roughly aligned cross-veins Cross-veins non-linearly arranged; narrower in males	Very long legs Simple claws in females; eight spines in hind tibial scraper
Topomyia	Proboscis not altered A central thin strip of broad white/silvery/brown scales on scutum		Slender, hairy upright proboscis			Bare upper calypter 1A vein terminates at mediocubital cross-vein base	
Toxorhynchites	Very large mosquito, often covered with intensely colored iridescent scales Head capsule long, tapered, and distinctly downward-bent proboscis Abdominal tip with scale tufts Mosquitoes with vivid colors	Circular scutellum, non-lobed No setae on acrostichal, dorsocentral, and prescutellar regions	Apical half of proboscis attenuated; sharply curved downward			Upper calypter and alula naked Distinct sclerotization in front of vein Cu in cell M4 Clearly emarginate wing margin close to vein CuA's termination Wing with emarginated posterior border little beyond anterior point of vein cubitus	

(continued)

Table 3.2 (continued)

Genera	Head and thorax	Scutellum	Proboscis	Palpi	Antenna	Wing	Appendages/legs and claw
Trichoprosopon (Goeldia)	Flat scales covering the head Mesothorax has scales that are lanceolate in form and have flat, spindle-shaped Proboscis not long as forefemur (Fe-I) Clypeus usually with lateral setae Antepronotum well-separated, small Long row of setae on the lower mesokatepisternum Lower margin of mesepimeron above level of lower mesokatepisternal setae Mesopostnotal setae present, postprocoxal scales absent, prespiracular setae present	Flat scales on scutellum Metanotum with chaetae and squamae	Proboscis short and thick, not as long as the body	Palpi in the nearly one-third the length of the proboscis, very small		Dense, elongated wing scales in Runchomyla; arranged asymmetrically	

Trichoprosopon (Joblotia)	Prothoracic lobes moderate, well separated dorsally Mesonotum without setae Eyes dorsally contiguous A pair of coarse bristles projecting forward at vertex Proboscis about equal length to forefemur Clypeus with lateral setae Antepronotum well-separated, small; setae on lower mesokatepisternum Lower margin of mesepimeron above level of lower mesokatepisternal setae Mesopostnotal setae present Postprocoxal scales absent, prespiracular setae present Clypeus with coarse or fine hairs		Proboscis long, slender; hardly swollen at the tip	Female with short palpi; male with long proboscis	Female with subequal jointed antennae Last two joints are long in males. Each joint with minute secondary subapical whorl Median joints shortened and thickened in males		Legs long and slender; cillate in some species Claws small, equal and simple in females Unequal claws on front and mid leg; claws are large, toothed
Tripteroides	Erect head scales on occiput Thorax with prespiracular setae, acrostichal setae missing					Above level of mediocubital crossvein; anal vein terminates Marginal setae on upper calypter Wing membrane with distinct microtrichia; anal vein beyond of mediocubital cross-vein	

(continued)

Table 3.2 (continued)

Genera	Head and thorax	Scutellum	Proboscis	Palpi	Antenna	Wing	Appendages/legs and claw
Udaya	Distinct interocular patch of silver scales extending to antennal pedicels Wide silver antepronotal scales divide the two groups of antepronotal setae Paratergite with a distinct region of large silver scales; postspiracular and postpronotal scales absent Acrostichal region devoid of setae					Broad marginal scales on alula	Upper calypter bare/few short setae; hind tarsomeres with conspicuous white basal bands Mosquitoes color yellow-brown with conspicuous silver bands/ patches
Uranotaenia	Tiny body for all species Enlarged proboscis for both sexes Wing cell small, short; palp length shorter than proboscis Iridescent blue scales on wings, thorax, and head					No spot on wings; faint microtrichia Anal veins marginal Anal vein sharply bent apically; ending slightly before cubitus	Pale scales present/absent on hind tarsomeres Black/whitish scales cover hind legs including tarsomeres Fore and mid tarasomeres longer than the rest
Verrallina	Broad vertex with decumbent scales Scutum with dorsocentral setae Sparse acrostichal setae Scutal and scutellar scales thin, bented Mesopostnotum usually nude No setae/scales on paratergite Postpronotum naked, all scales narrow, bented Postspiracular setae present; prespiracular region bare					Alula with fairly broad marginal scales	

Wyeomyia (Prosopolepis)	Head devoid of vertical scales Broad, dorsally well-separated thoracic lobes Eyes together at vertex Wedge-shaped, parallel bands separate eyes above antennae Clypeus without setae Scutum scales brilliantly golden/bronzish/metallic prespiracular, prealar setae present		Proboscis shorter than abdomen; apically swelled Male palpi short; female palpi undetermined; clypeus with scales	Palpi short, fused in both sexes	Antenna with subequal joints for both sexes Males with small, indistinct, subapical whorls on joints		Long, slender legs; equal-sized claws in females but uneven/single claw in males Tarsa dark

Genera	Abdomen	Male genitalia	Turgum	General surface covering	Diagram	References
Aedeomyia		Gonostylus with/without comb-like apical process			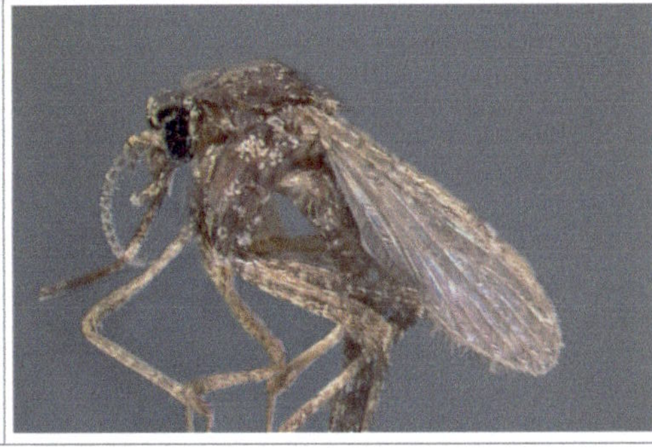	Rattanarithikul (1982), Burkett-Cadena and Blosser (2017), Boehmler (2022)

Genera	Abdomen	Male genitalia	Turgum	General surface covering	Diagram	References
Aedes	White dark scales on abdomen VIII abdominal segment reduced in females First abdominal segment with pale scales and patches; tergites dark brown Dorsal abdominal segments with white stripes Lateral abdominal segments with white scales Abdominal tip round in females; narrow in males Dorsolateral white scale patches on male terminal segments Larger abdomen in males (L: 2.94 ± 0.20 mm/W: 0.41 ± 0.06 mm) than females (L: 3.03 ± 0.18 mm/W: 0.51 ± 0.07 mm) Abdominal terga with silvery spots Postspiracular region with white scales Long, clearly noticeable cerci	Gonocoxite membranous at tergomesal surface; no gonostylar claw Phallus, prosophallus, and opisthophallus Aedeagus Setae absent in tergum Lack of claspette or inadequate development One/two apical teeth on paraproct Gonostylus apical spine is longer than width Gonostylus articulates subapically if spine is shorter than width Gonostylus splits at base if spine lacking			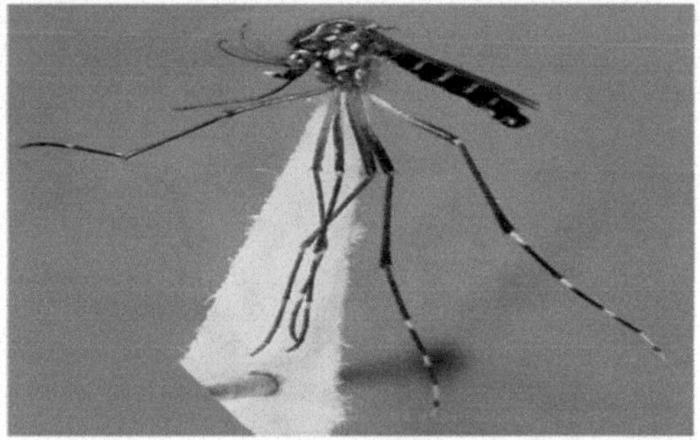	Rattanarithikul (1982), Bar and Andrew (2013), Becker et al. (2020)

Genera	Abdomen	Male genitalia	Turgum	General surface covering	Diagram	References
Anopheles	Scales abscent on abdominal segments Setae always cover abdomen Abdominal segments with posterolateral scale tufts Small abdominal sternum alikes metathorax	Aedeagus with apical tubular, finger-like leaflets Simple, nonfilaments or spines without leaflets Lobes do not exist in gonocoxite Gonostylus longer than gonocoxite Sclerotized paraproct absent Proctiger membranous, cone-shaped, rarely noticeable in males			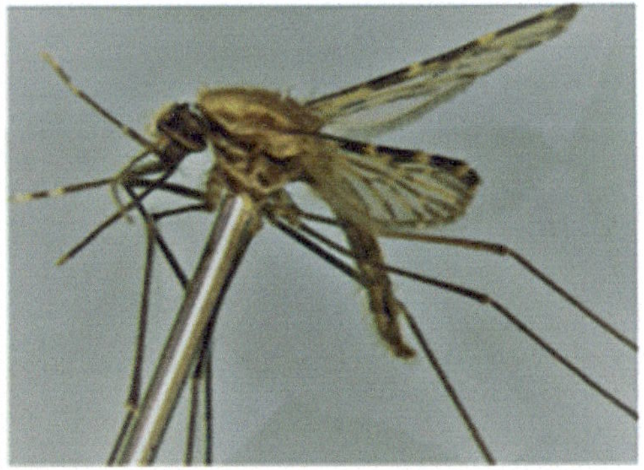	Rattanarithikul (1982), Sedaghat et al. (2003), Sallum et al. (2020)
Armigeres	Black dorsal scales and white ventral scales are present in postspiracular area	Distal comb-like structure may or may not exist in gonostylus				Rattanarithikul (1982), Anoopkumar et al. (2019)

Genera	Abdomen	Male genitalia	Turgum	General surface covering	Diagram	References
Carrollia	Abdomen of females is apically truncate, laterally compressed, and subcylindrical		Mesonotum lacks longitudinal rows of granular setae Postnotum is not covered Cerci are short and discrete In males, there is no lateral ciliation, the sides are compressed, and the tip is slightly expanded		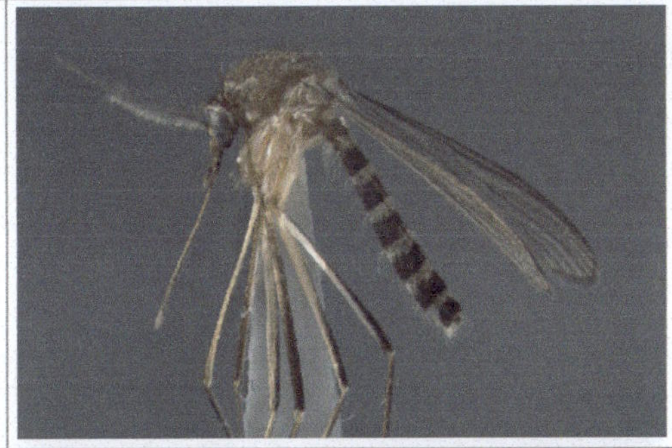	Liria and Navarro (2003), Sabino et al. (2023)
Coquillettidia	Blunt apex in females VIII-Te is unaltered, only the basal half has sclerotization	Cerci spatulate, IX-Te articulated IX-Te bilobed, and has a few lengthy setae Female has three spermathecae that are all the same size				Becker et al. (2020), Laojun et al. (2023a, b)

Genera	Abdomen	Male genitalia	Turgum	General surface covering	Diagram	References
Culex	Female's abdomen is subcylindrical, truncated at tip, with short, barely noticeable cerci In males, slightly or not enlarged near the tip, and sometimes depressed	Proctiger either poorly developed without spicules or with one to three powerful apical spicules, or powerfully developed with numerous fine apical spicules Claspette absent or present Gonocoxite has a single, subapically positioned lobe that is coated with setae and spines Paraproct crown, or apex, has many rows of denticles or a lot of spines	Rows of coarse setae on a disc form the mesonotum		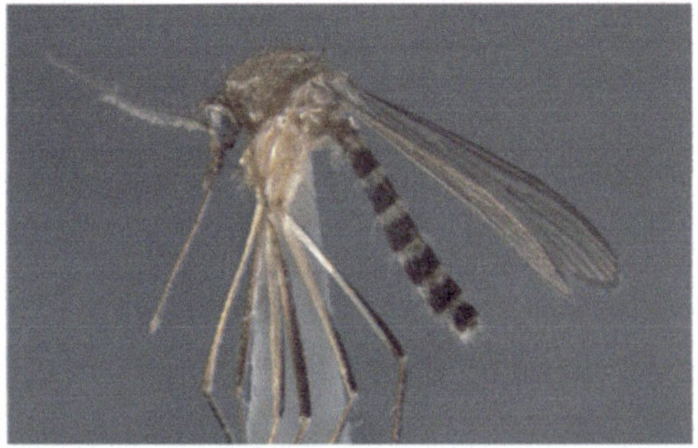	Dehghan et al. (2011), Laurito et al. (2018), Becker et al. (2020)
Culiseta	Abdominal terga with separate basal bands Narrow subapical band of pale scales on femora; large pale bands on hind tarsomeres	Gonostylus apices do not have fingers Spine noticeably shorter than male gonostylus width			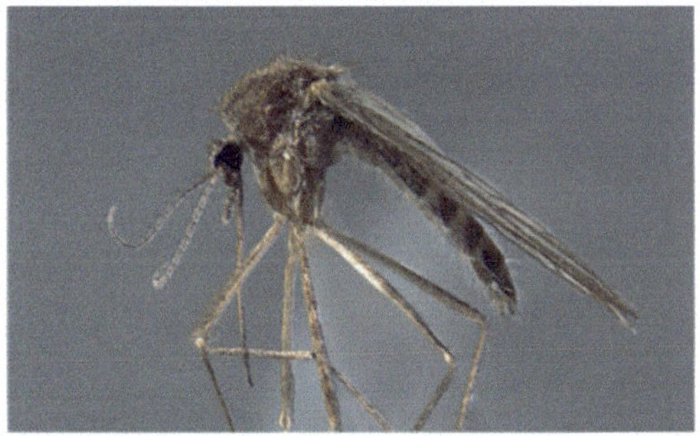	Owen (1985), Becker et al. (2020)

Genera	Abdomen	Male genitalia	Turgum	General surface covering	Diagram	References
Deinocerites	Compressed and shortened in female Cerci are robust, elliptic Male subcylindrical abdomen with strong claspers expanding at the tip		Longitudinal rows of coarse setae on mesonotum Postnotum has two-minute bristles inserted medially Bristles present		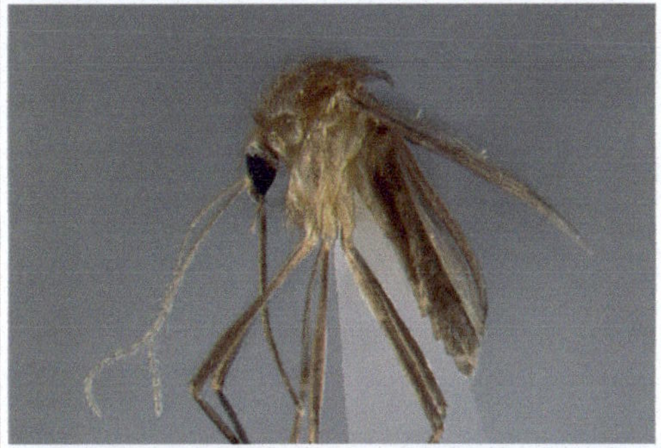	Harbach et al. (2012)
Dinanamesus	Female abdomen is truncated and squeezed In males, the tip is slightly enlarged strong claspers		Longitudinal rows of coarse setae on the mesonotum Postnotum in its natural state		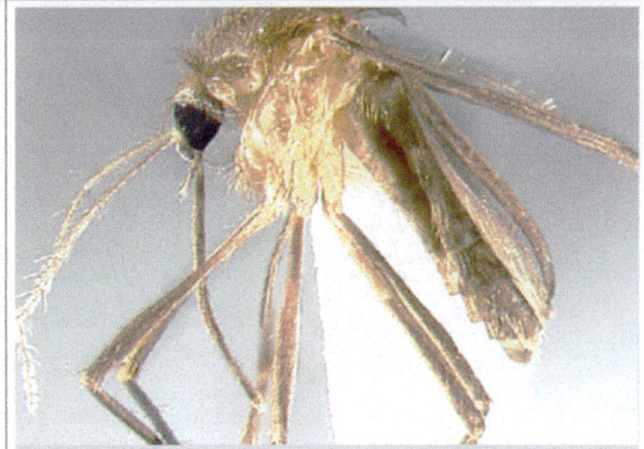	Belkin and Hogue (1959), Toro-Cantillo and Hoyos-López (2018)

Genera	Abdomen	Male genitalia	Turgum	General surface covering	Diagram	References
Dinomimetes	Female's abdomen is compressed and truncates at tip Cerci are robust, elliptic Male subcylindrical abdomen with strong claspers, somewhat enlarged at the tip		Longitudinal rows of coarse setae on mesonotum Postnotum placed medially with tiny bristles Constricted abdomen in female, terminated at tip Robust elliptical cerci		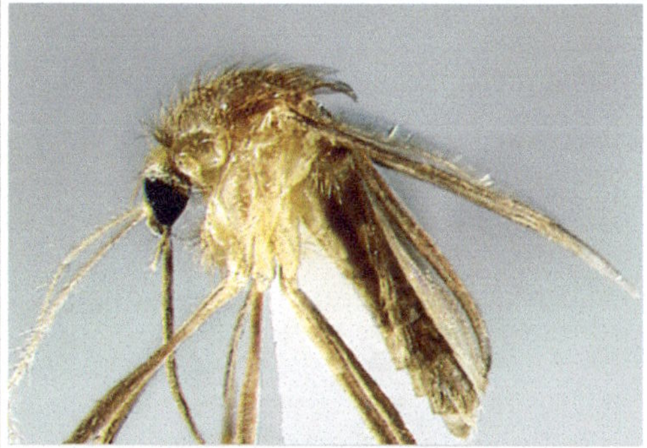	
Ficalbia		Presence of one median caudal lobe in tergum Aedeagus narrow; if broad, not emarginate apically			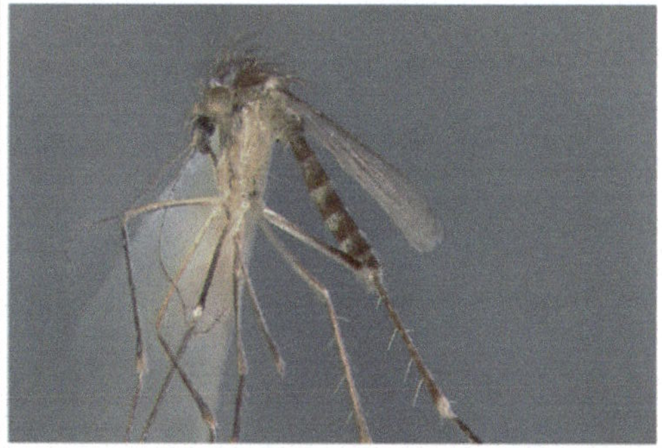	Rattanarithikul (1982), Rattanarithikul et al. (2006)

Genera	Abdomen	Male genitalia	Turgum	General surface covering	Diagram	References
Goeldia			Squamae and Chaetae are with the metanotum		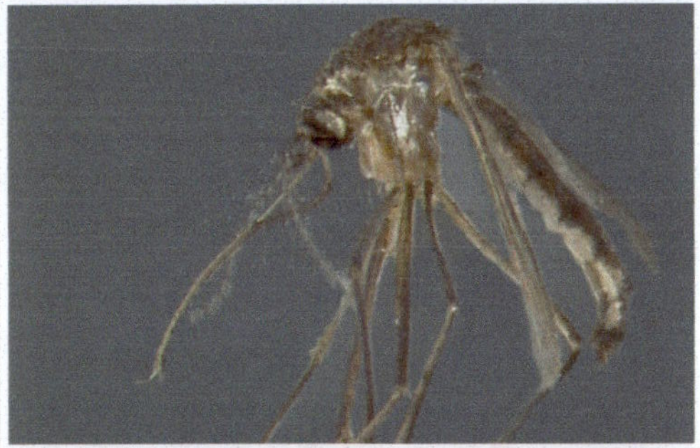	Almeida-Silva et al. (2009)
Heizmannia	Prominent triangular basolateral patterns in white or silver on the abdomen terga	Male has both apical setae and subapical leaflets Gonocoxite can have lobes with one or two strong setae			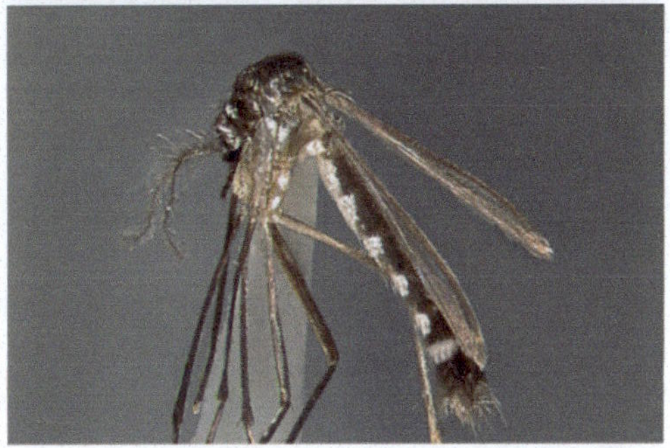	Rattanarithikul (1982), Rajavel et al. (2005), Yang et al. (2019)

Genera	Abdomen	Male genitalia	Turgum	General surface covering	Diagram	References
Hodgesia					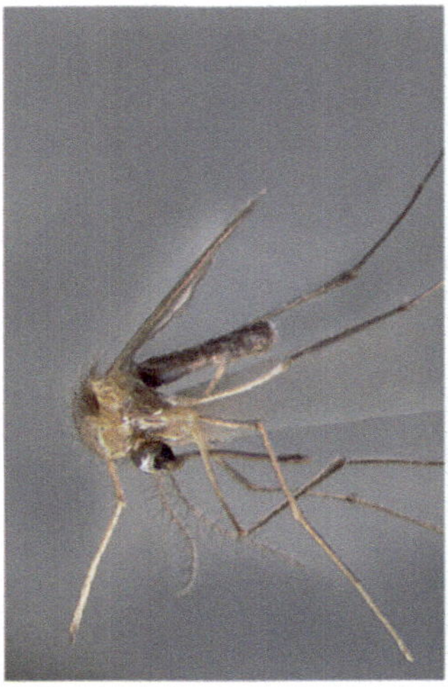	Rattanarithikul (1982), Rattanarithikul et al. (2006)
Isostomyia	Abdomen brown/ yellow/ brownish-yellow Hypopygium and terminal segments are brownish-black. Brown squamulae covered the upper surface				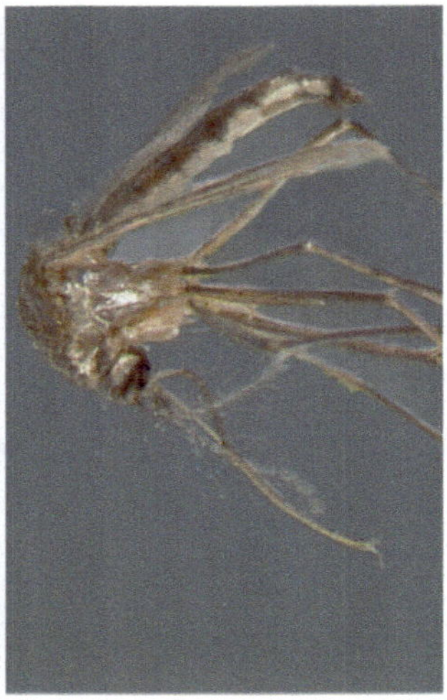	Marti et al. (2007), Albicócco et al. (2011)

Genera	Abdomen	Male genitalia	Turgum	General surface covering	Diagram	References
Limatus	Female abdomen blunt, subcylindrical			Flat appressed scales cover the vestiture, the mesonotum lacks sets on disc, and the occiput lacks erect forked scales	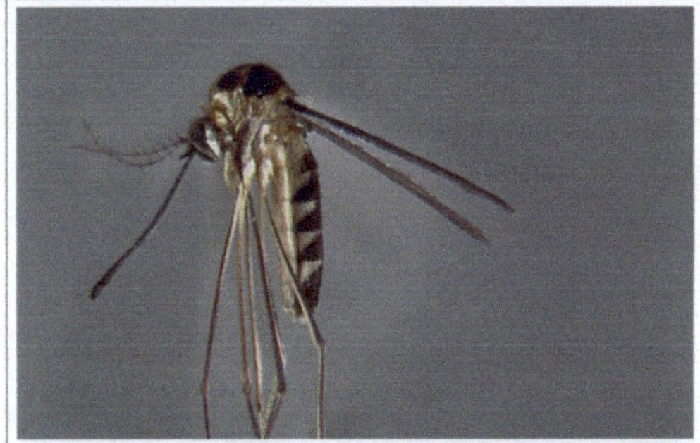	Lira-Vieira et al. (2013), Talaga et al. (2018)
Lutzia	Subcylindrical abdomen in females, truncated tip Long and depressed in male		Postnotum nude			Moirangthem et al. (2018), Laojun et al. (2023a, b)

Genera	Abdomen	Male genitalia	Turgum	General surface covering	Diagram	References
Malaya		Two submedian caudal lobes in tergum In male, aedeagus is large, extended, and emarginate apically				Rattanarithikul (1982), Shafie et al. (2016)
Manosonia	Depressed, and subcylindrical abdomen in female Elongated, lateral ciliation in males Female with three distinct receptaculum seminis		Postnotum nude			Sharma and Venkteash (2015)

Genera	Abdomen	Male genitalia	Turgum	General surface covering	Diagram	References
Mimomyia					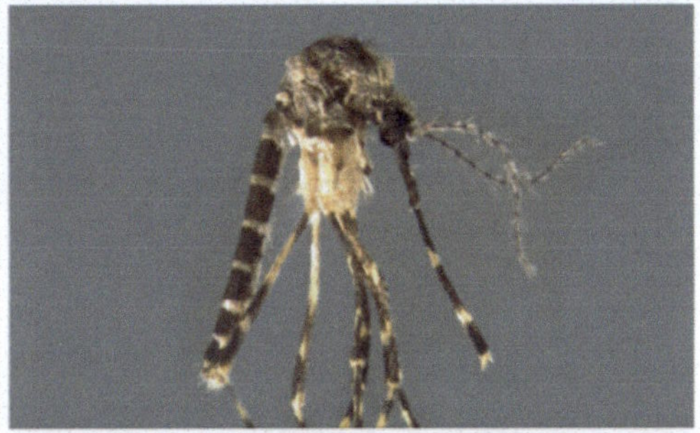	Rattanarithikul (1982), Rattanarithikul et al. (2006)
Orthopodomyia		Tergum caudal border with/without a prominent median lobe Gonostylus apical spine fingered at tip Long spine in male gonostylus			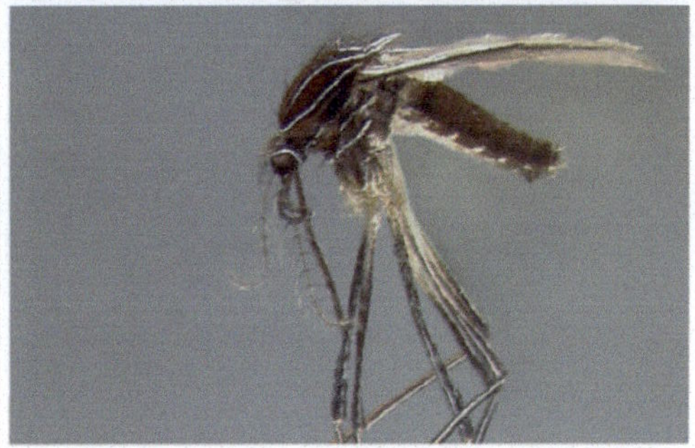	Rattanarithikul (1982), Huang and Rueda (2013), Becker et al. (2020)

Genera	Abdomen	Male genitalia	Turgum	General surface covering	Diagram	References
Runchomyia (Lesticocampa)	Subcylindrical abdominal tip is blunt		Setae absent in Mesonotum	Collar-like row of upright forked scales on head; appressed scales for vestiture structure	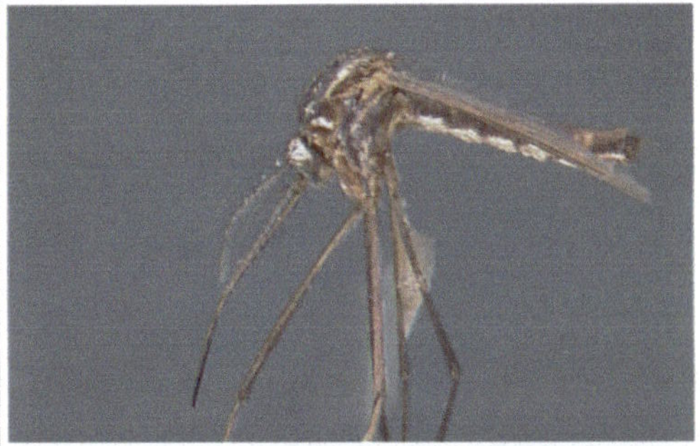	Dyar (1919)
Sabethoides	Abdomen with metallic scales; longer than femur Some tergite scales are metallic/ greenish-blue/violet/ pale yellow/ silvery-white Subcylindrical abdomen with blunt tip			Vestiture with broad appressed scales Occiput lacking in vertically erect scales Mesonotum devoid of setae	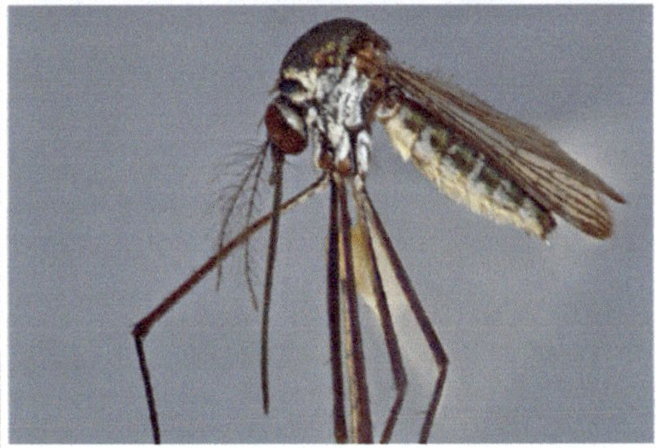	Morales et al. (2010)

Genera	Abdomen	Male genitalia	Turgum	General surface covering	Diagram	References
Theobaldia	Female abdomital tip blunt, subcylindrical Elongated, depressed, lengthy lateral ciliation in males		Postnotum nude		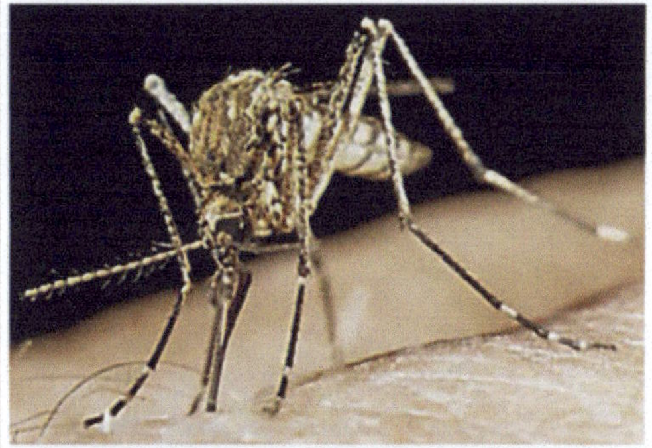	
Topomyia		Tergum with/without stout or fine setae Claspette absent/well developed in male			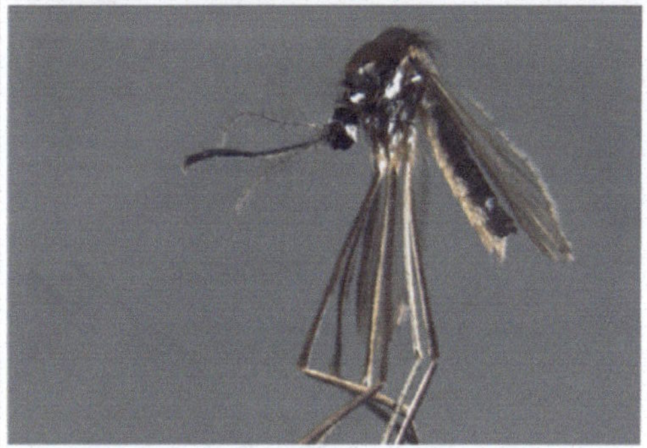	Rattanarithikul (1982)

Genera	Abdomen	Male genitalia	Turgum	General surface covering	Diagram	References
Toxorhynchites		Large, long, conical gonocoxite Segments VII–VII are substantially developed, with lateral marginal setae producing a prominent tuft, or segment IX is large and emarginate medially, with several long setae on each side			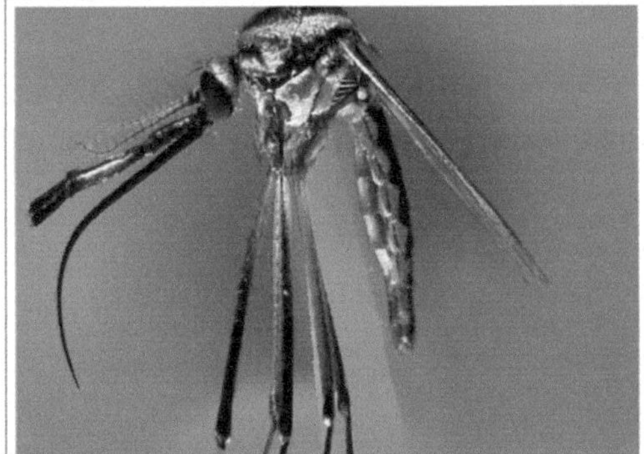	Rattanarithikul (1982), Schreiber (2007), Faithpraise et al. (2014)
Trichoprosopon (Goeldia)	Abdomen with distally sparsed scales		Metanotum with chaetae and squamae			Campos and Zavortink (2010)

Genera	Abdomen	Male genitalia	Turgum	General surface covering	Diagram	References
Trichoprosopon (Joblotia)	Subcylindrical abdomen, blunt tip Tergite distant with sparse scales			Flat vestiture, appressed scales; ertect collar-like rowed; forked scales at nape	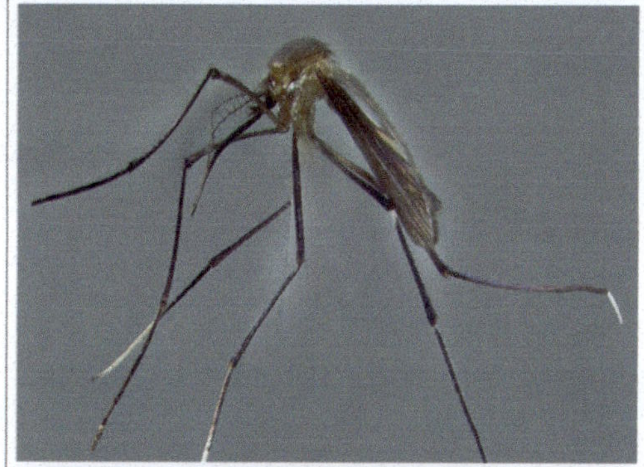	Judd (1996)
Tripteroides	Terga markings light metallic	Prominent lobe pair on tergum; lobe length equal to tergum Apical robust seate on each lobe in males				Rattanarithikul (1982), Chaves et al. (2019)

Genera	Abdomen	Male genitalia	Turgum	General surface covering	Diagram	References
Udaya		Paraproct with apical teeth			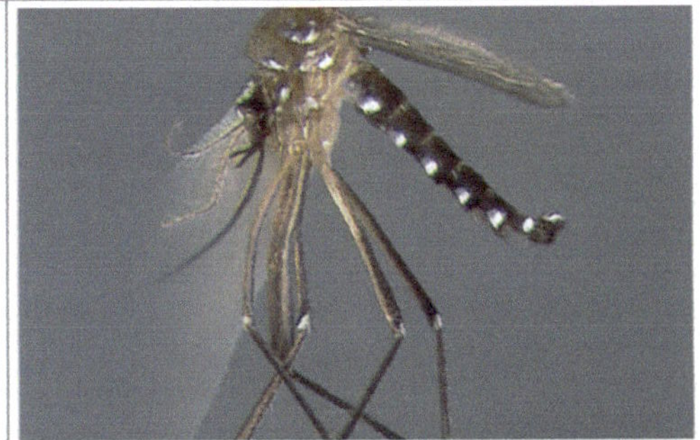	Rattanarithikul (1982), Kaur (2003), Saba et al. (2022)
Uranotaenia		Dorsoventral gonostylus widened and flattened Gonocoxite absent from membrane Gonocoxite short; plump with small flattened basal lobe Two lateral sclerotized plates on aedeagus with apical teeth Protoctige lobe-shaped, membranous in males			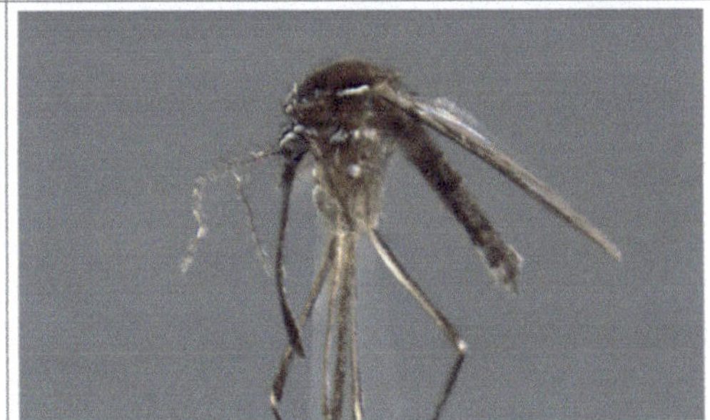	Rattanarithikul (1982), Junglen et al. (2009), Reeves et al. (2018), Becker et al. (2020)

Genera	Abdomen	Male genitalia	Turgum	General surface covering	Diagram	References
Verrallina					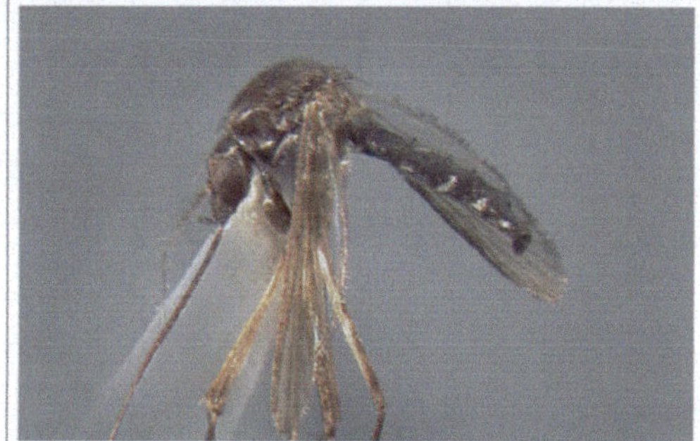	Rattanarithikul (1982), Webb et al. (2008), Mizuta (2011)
Wyeomyia (Prosopolepis)	Subcylindrical abdomen Male abdominal tip enlarged, blunt in females		Postnotum with setae Mesonotum devoid of setae on disk	Vestiture flat, appressed scales Occiput lacking in vertically erect scales Mesonotum on disc devoid of setae	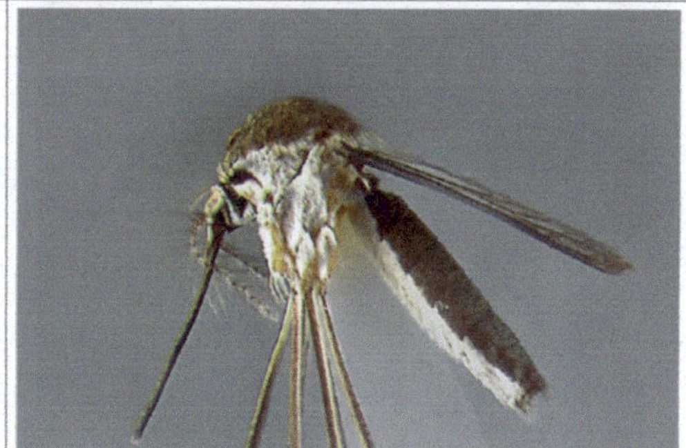	Bergland et al. (2005), Hoekman et al. (2007)

Image courtesy: WRBU (https://vectormap.si.edu)

3.2.1 Egg

Mosquitoes like to lay their eggs in water bodies, such as freshwater, floodplains, and snow-melt pools. The eggs are typically around, 1 mm long, and have a brown to blackish appearance (Service 2012). In just 2–4 days following a blood meal, a female mosquito can lay as many as 500 eggs (Becker et al. 2020). Depending on the species, mosquito eggs can have a variety of morphologies, such as rhomboid, spherical, fusiform, elliptical, sausage-shaped, columnar, oval, or spindle shaped. Some species from *Anopheles*, *Toxorhynchites*, *Wyeomyia*, *Aedes*, *Ochlerotatus*, *Psorophora*, and *Haemagogus* lay their eggs independently, whereas species from *Culex*, *Culiseta*, *Coquillettidia*, and *Mansonia* lay their eggs adherently in clusters or clumps, forming a floating egg raft (Foster and Walker 2019).

It is acknowledged that their eggs' dimensions, shape, and color may serve as taxonomic keys for distinguishing the species. This attribute was initially uncovered by Theobald in 1901, and later, (Mitchell and Dupree 1907) employed the sculpturing pattern of the outermost layer of the egg, known as chorion, for identification purposes (Kalpage and Brust 1968). For example, Anopheline eggs are characterized by a boat-shaped structure with air-filled sacs called "floats" on either side; in contrast, Culicinae eggs do not have any "floats" (Service 2012). According to deposition trends, their eggs can be categorized into two groups: rapid-hatch eggs, which are laid immediately in the waterbody, and delayed-hatch eggs, which are laid on moist ground above the waterline. After emerging from their eggs, the larvae display voracious appetites for the nutrients in the aquatic environment, where the eggs were laid (Hawkes and Hopkins 2022). Like many other insects, egg hatching is primarily influenced by temperature, and the embryo undergoes development for a period ranging from 2 to over 7 days, depending on the species. For example, the eggs of *Cx. pipiens* hatch within 24 h at a temperature of 30 °C, within 72 h at a temperature of 20 °C, and within 240 h at a temperature of 10 °C. However, the larvae of *Ae. vexans* can hatch within 96 h at a temperature of 25 °C and within 192 h at a temperature of 20 °C (Becker et al. 2020).

3.2.2 Larva

The setae of mosquito larvae are typically unique, consisting of approximately 222 pairs (Forattini 1996). They can be simple, aciculate, branched, stellate, pinnate, plumose, palmate, laterally branched, fan-like, or dendriform (Hawkes and Hopkins 2022), and thus, being of different forms and structures, they critically play an essential role in identifying species. For instance, the *Anopheles* and *Chagasia* genera can be distinguished by the palmate setae; therefore, species identification relies on the unique combination of setae found in each species. Typically, the larval forms of most mosquito species are characterized by having lateral palatal bristles, except for the predatory larvae from Toxorhynchitinae individuals. Additionally, the larval forms of mosquitos are also characterized by a tubular siphon for respiration and a fused or expanded thorax.

Mosquito larvae undergo four instar stages before pupation (Hawkes and Hopkins 2022). In tropical climates, larval development typically takes 5–14 days; however, it can last longer in temperate areas, ranging from several weeks to months. In certain situations, larval development can even continue over the winter season (Service 2012). Since mosquito larvae are detritivores and a desirable food source for many carnivorous insects, they are crucial components of the food chain (Hawkes and Hopkins 2022). The carnivore Toxorhynchitinae larvae are employed as vector control agents to combat larvae of other mosquito species due to their adult phytophagous nature (Becker et al. 2020).

3.2.2.1 Larval Characteristics

3.2.2.1.1 Larval Head

Typically, the widths of mosquito larval heads are significantly more comprehensive than their lengths, and this ratio generally varies across different species (Becker et al. 2020). The larvae typically possess 18 sets of symmetrically paired setae on their heads, commonly labeled from 0 °C to 17 °C; among them, setae numbered 2–7 °C are critical indicators for identifying the species. The antennae of mosquito larvae are comparatively less complex than adults. The larvae primarily utilize their antennae for sensory perception to detect changes in their surroundings, such as shifts of chemical properties in water, sources of nourishment, the presence or absence of predators and possible partners. They are also involved in detecting temperature changes in the water, which can impact the behavioral attitudes, dynamicity, and development of larvae. The larval antennae are diminutive, filamentous appendages located in close proximity to the mouthparts. Although not as intricate as the antennae of fully grown mosquitoes, they are crucial for their survival and capacity to navigate aquatic habitats. In the genus *Aedeomyia*, the larval antennae are noticeably broad and flattened; however, in the genus *Mimomyia*, the antennae are segmented with a freely articulated distal section (movable end part). Six setae are often seen on each antenna, ranging from 1-A to 6-A (Becker et al. 2020).

3.2.2.1.2 Larval Thorax

The mosquito larval thorax is the narrow region between the head and abdomen. Despite its small shape compared to the abdomen, it contains several unique morphological features. The larvae breathe through thoracic spiracles, tiny apertures on the thoracic wall. Since mosquito larvae are aquatic and need access to the air surface for respiration, they use these spiracles to draw in oxygen. The larvae feature three pairs of bristle-covered, short, fringed thoracic appendages for movement. The larvae's thoracic muscles control their locomotion, enabling them to swim wagglingly. The thoracic muscles propel the larva through the water with its appendages. The larvae's thoracic nerve ganglions govern and coordinate several sensory and motor functions, allowing them to respond efficiently to their living environments.

As a larva develops in its aquatic habitat, the thorax is crucial in enabling respiration, locomotion, and sensory processing. The thorax may have three different pairs of setae and may be merged without pro-, meso-, or metathoracic segments.

However, when the segments are divided into three sections, the characteristics can be transmitted to the adults. The setae 1-P to 3-P are used as identifying keys for taxonomic identifications among the setae arrangements in prothorax (0-P to 14-P), mesothorax (1-M to 14-M), and metathorax (1-T to 13-T) (Becker et al. 2020).

3.2.2.1.3 Larval Abdomen

The abdomen of the mosquito larva is the posterior part of its body, following the thorax, which usually encloses multiple significant structures. It primarily consists of the digestive system of the larva, hence consisting of an eliminating waste aperture with several Malpighian tubules. The respiratory siphon is positioned on the upper side of the abdomen. The abdomen of larvae can be divided into ten distinct segments, with a distinguishing number of setae distributed in each segment. For example, the first abdominal segment contains 13 pairs of setae, whereas segments II to VII contain 15. The cylindrical respiratory organ siphon is at the tip of the abdomen, which often bears a few to many pairs of simple setae. At the base of the siphon are two projections on either side, known as "acus." The siphon and spiracular plate contain approximately 13 pairs of setae, designated 1-S through 13-S. A row of scales, collectively called "comb," is found in segment VIII and is positioned posteriorly. However, segment IX is rudimental. The spiracular plate, a valve with five lobes, encloses the spiracles and covers the aperture, while the larvae are immersed in water. Two pairs of sausage-shaped anal papillae surround the anus in segment X to help osmoregulation (Service 2012; Becker et al. 2020).

3.2.3 Pupa

The head and thorax of the pupa often merge to form a cephalothorax, resulting in a large spherical frontal lobe. Despite being narrower than the thorax, the pupal abdomen consists of nine segments or folds that mold the pupa into a comma-shaped structure, often called tumblers. Generally, the abdomen is embellished with paired setae on its dorsal and ventral surfaces. In the last abdominal segment, unique oval-shaped paddles help them move; however, these swimmerets may vary according to species. For example, the pupae of the Culicinae subfamily have a visible midrib on their paddles, whereas in Anophelinae, the mid- ridge may be absent. Pupae breathe through specialized cylindrical tubes called "respiratory trumpets" on the dorsal mesothorax; however, the morphology of these trumpets may also differ among species. For instance, in the genera *Mansonia* and *Coquillettidia*, the trumpets are structurally tapering at the apex, but in *Anopheles*, the trumpets are short and broad-headed. Typically, the pupal cuticle layer is translucent, and the antennae are visible, protruding out of the upper region of the compound eye. The tarsi are curled beneath the wing apex, and the leg-covering sheaths are observed between the thoracic and head coverings (Becker et al. 2020).

Pupae are nonfeeding and are commonly observed floating and breathing, using their respiratory trumpets in most species. However, the pupae of *Mansonia* and *Coquillettidia* exhibit exceptional behavior by remaining submerged in water. This is possible, because their modified trumpets enable them to obtain oxygen by piercing aquatic plants. Pupae in tropical climates generally have a short lifespan of 2–3 days, but pupal development lasts for 9–12 days in temperate regions. Following a split across the dorsal region of the cephalothorax, the adult mosquito emerges (Service 2012).

3.2.4 Adult

Mature mosquitoes comprise three primary anatomical divisions: the head, thorax, and abdomen. The head serves as the sensory hub, housing a pair of compound eyes, a pair of brushy antennae, and a projected proboscis for liquid feeding. Though tiny, the head is the home to numerous vital organs, and their diverse structures are often very species-specific. Their antennae are elongated but slender, typically exhibiting greater length in males than females, and serve to detect various atmospheric stimuli. Their prominent compound eyes are highly adapted for perceiving motion, light, and other visual stimuli. Female mosquito mouthparts are specialized for piercing and sucking blood, with a slender proboscis and elongated styles; however, male mouthparts are not specialized for liquid drawing. The thorax is the middle segment of the mosquito's body, functioning as the site of attachment for the wings and legs. The thorax consists of three body segments: the prothorax, mesothorax, and metathorax, each with a pair of hairy, articulated appendages. In addition, the mesothorax is equipped with a pair of membranous, translucent wings with definite venations, whereas the metathorax possesses a pair of balancing organs known as halteres. Their legs are though long than many Dipterans, they are slender, and often modified with various adaptations for walking, jumping, and perching. Mosquito abdomen is typically longer and slimmer than the thorax; it comprises the rear part of the mosquito's body. Structurally, the abdomen is cylindrical, oval, or elliptical, but after consuming a blood meal, female mosquitoes experience abdominal swelling and elongation as their stomach fills with liquid. The abdomen consists of ten body segments and houses the digestive, respiratory, and reproductive systems of the insect.

The types and distribution of different sensory units are immensely diverse in mosquitoes. Sensilla, also known as sensory hairs, together with other specialized organs, such as tagmo- and chemoreceptors, are distributed throughout the body of these organisms. They are crucial in detecting various environmental cues, including moisture, heat, and chemical signals. Adult mosquitoes often have unique coloration and markings on their bodies and wings, which can vary significantly between species. These markings serve multiple functions, including camouflage and mate detection. The physical traits of mosquitoes facilitate their adaptation to various habitats and lifestyles, encompassing their dietary habits and reproductive strategies.

3.2.4.1 Adult Characteristics

3.2.4.1.1 Head

The adult mosquito head is characterized by an elliptical, oval, or oblong shape and integrates two prominent compound eyes, two antennae, two maxillary palpi, and a proboscis. The annulate antennae of mosquitoes assist in host-seeking, finding oviposition sites, and sensing the temperature and humidity of the surroundings through its specialized receptors (Hawkes and Hopkins 2022). The antennal chordotonal organ, or *Johnston's organ*, is a sensory structure found in the pedicel of mosquito antennae and was first recognized as a "sound-receptor" organ by Christopher Johnston in 1855. Positioned near the second segment of the antenna, this doughnut-shaped mechano-sensitive organ was subsequently found to be capable of differentiating sounds with minute variations in frequency. A key component of insect sense perception, *Johnston's organ*, is responsible for receiving mechanical stimuli, like vibrations and sound waves. The composition of this structure includes sensory neurons known as chordotonal neurons, supporting cells called "scolopale," and many accessory structures. Stretched between the cuticle of the pedicel and a flexible internal structure, the "scolopale," the chordotonal neurons are sensory cells that detect mechanical stimuli and sound vibrations. When the antennae are oscillated, the "scolopale" undergoes deformation, and subsequently activate chordotonal neurons. Being elastic, the "scolopale" connects sensory neurons to the cuticular pedicel and thus performs as a mechanical amplifier and thereby enhances the organ's sensitivity to minute vibration. The tiny sensory hairs, or sensilla, surround the antennas convert mechanical external inputs into electrical impulses (McLver 1982). On analyzing sound-vibrational-response in different species of mosquitoes, it has been evidenced that acoustic emissions from mosquito wings help identify their species and sex (Lapshin and Vorontsov 2017).

According to Service (2012), their kidney-shaped compound eyes are also highly responsive to many light sources and levels of light intensities day and night. For instance, nocturnal blood-feeding mosquitoes, like *Anopheles gambiae*, have the eyes that are differently adapted to low light than diurnal nectar-feeding mosquitoes, like *Toxorhynchites brevipalpis* (Lang et al. 2018). It is widely hypothesized that night-active mosquitoes have large, hollow cone cells that are densely packed with photopigments within their rhabdoms, which are equipped with ommatidial lenses that focus on distal ends and capture light at an angle of up to 60°. This results in a ninefold improvement in light-gathering power over the more common cylinder-shaped rhabdom at an acceptance angle of less than 25° (Land et al. 1997, 1999). Therefore, the compound eyes of mosquitoes own to have a high degree of light sensitivity, allowing them to adjust to the illumination conditions in their surroundings during the day or at night.

A combination of multiple mouthparts, including the labrum, labium, mandible and maxilla, and hypopharynx, make up their feeding apparatus, or proboscis. These structures have evolved into tightly fitting styles, constituting a *fascicle* (a group or cluster of sensory structures or appendages) penetrating the host's skin. Typically, the *fascicle* is surrounded by the labium and has a pair of gustatory lobes at its tip. The mandibles and hypopharynx have narrow endings on the stylet's *fascicle*, whereas the maxillae culminate in serrated blades. During penetration, the relatively sharp mandibles puncture the host tissue, while the serrated and curved teeth of the maxillae assist in attaching to the tissue. A salivary duct that runs the length of the hypopharynx transports saliva, which includes anticoagulants and anti-hemostatic enzymes, causing hematomas in the skin to aid in the absorption of blood feeds. Blood or nectar is extracted via the labrum, a food channel that assumes a laterally curved configuration during ingestion (Foster and Walker 2019).

3.2.4.1.2 Thorax

The thorax is covered with scales on its dorsal and lateral surfaces, displaying various colors ranging from pale to glossy white, brown, or black. The thorax comprises three segments: the prothorax, mesothorax, and metathorax. The prothorax is furnished with forelimbs, whereas the mesothorax has a pair of wings and a pair of mid-limbs. The metathorax articulates the hindlegs and two knobbed structures, called halteres, which aid in flight control, commonly referred to as balancers. The legs are elongated and slender, studded with tiny scales, and arranged in different configurations, often appearing as circular bands. The coloration of leg scales can vary between brown, black, and white, with or without dotted marks. Wing veins are elongated, thick, and constructed in a particular shape tailored for the forewings, allowing them to provide stiffness. The wings are covered with little scales in many colors, including black and light, and may or may not have spots and bands. The prominent veins on the wing consist of Costa, Subcosta, Radial, Medial, Cubital, and Anal, which are placed and divided in specific patterns and orientations unique to each species. The wings also feature a series of elongated, slender setae on the apical and posterior ends, collectively called the wing fringe. Although wings usually facilitate flight, their unique morpho-features, including size, shade, and pattern, help identify species. For instance, species from genera *Mansonia*, *Orthopodomyia*, and *Coquillettidia* have featured broad wing scales with various shades, while *Deinocerites*, *Culex*, and *Culiseta* have thin and small wing scales. On the other hand, the wings of the *Toxorhynchites* genus are known for being larger in size and having vibrant colors (Burkett-Cadena 2013). There are five separate parts to a mosquito's legs: the coxa, trochanter, femur, tibia, and a five-segmented tarsus. Species identification is occasionally determined by the measurements of the tarsal segments and the existence of scales on the hind tibiae. For example, the hind tibia of *Coquillettidia* is characterized by a conspicuous white band, whereas the foreleg of *Orthopodomyia* is distinguished by a short fourth tarsal segment, and the hind tibia of *Mansonia* lacks the subapical pale band (Burkett-Cadena 2013).

3.2.4.1.3 Abdomen

The ten-segmented abdomen consists of two dorsoventral plates: the dorsal plate called tergites and the ventral plate called sternites. Both the tergite and sternite are connected by a flexible exoskeleton called plurite. The abdominal segments, five through eight, diminish toward the posterior end; therefore, the tip of the abdomen can have either a pointed shape, as seen in *Aedes* and *Psorophora*, or a rounded and blunt shape, as observed in *Orthopodomyia*, *Mansonia*, *Deinocerites*, *Culex*, *Culiseta*, and *Coquillettidia* (Burkett-Cadena 2013). The female post-genital lobe, the cerci, and the male claspers are located in segment nine. The seminal vesicles (house mature sperm) and testes (house developing sperm) are located at the posterior end of the abdomen. The two seminal vesicles nestle together and combine to form the ejaculatory duct and link to the gonopore. The female reproductive system consists of a pair of ovaries (with several dozen to over 200 polytrophic ovarioles) and other associated components. Each ovary has a posteriorly extending duct, which combines to form a common oviduct and opens through the gonopore via a vaginal chamber. One to three spermathecae, small accessory glands, and the seminal bursa, which in most *Anopheles*, lacks open into the genital chamber by tiny ducts (Foster and Walker 2019). While feeding, the abdomen appears red and dilated, while it appears white when the female develops eggs (Service 2012).

3.3 Morphological Adaptation

Mosquitoes exhibit several morphological adaptations that promote their flight aptitude, food acquisition, and breeding success. Thus, physical modifications have evolved to improve their chances of survival by increasing their reproductive capacity and physiological fitness (Burkett-Cadena and Blosser 2017). The morphological adaptations boost their capacity to live, obtain nutrition, and achieve reproductive success. Mosquitoes possess key morphological modifications that enhance their ability to survive and reproduce more effectively.

Their specialized mouthparts, constituted with proboscis and stylets, have evolved to pierce the host epidermis and operate as a feeding tube for blood drainage. Thus, the mouthparts are adapted to penetrate the vertebrate skin to extract blood efficiently. The long and slender legs are also specifically adapted to support their bodies on water surfaces, enabling them to breed in aquatic habitats. Moreover, the fine hairs and pads in the legs provide an even distribution of their body weight by generating surface tension, thereby preventing them from sinking. Their wings are tailored for flight and are distinctively slim, lengthy, and lightweight; such unique wing characteristics allow them to fly rapidly for precise navigation with high accuracy. The ability to locate potential partners, identify suitable hosts, and evade predators confers survival, reproduction, and widespread distribution advantages to mosquitoes (Wilkerson et al. 2021). Antennae and mouthpart sensilla are among the highly developed sensory organs that allow them to detect host smells, warmth, carbon dioxide, and other environmental cues. It enables them to effectively locate, recognize, and consequently choose and feed on precise hosts. Some mosquitoes have evolved cryptic coloration, letting them blend in with their surroundings and become inconspicuous to both prey and predators. Such camouflage makes it easier for them to deceive victims and stay hidden while at rest (Lorenz et al. 2017; Burkett-Cadena and Blosser 2017).

Female mosquitoes feature distinctive physical characteristics, such as an ovipositor for egg-laying. Species variation may, however, result in mosquitoes depositing eggs in rafts or clusters on the water surface by adhering to genital frothy materials or ovipositing singly on the surface by attaching adjacent aquatic vegetation. Therefore, morphological adaptations in mosquitoes improve their disease-carrying ability and allow them to thrive in many environmental settings worldwide.

3.4 Morphological Evolution

Mosquitoes have a significant impact on human health and general welfare around the world. Although they are modest in size, they act as carriers for numerous lethal diseases, rendering this a highly relevant issue in clinical studies. A list of therapeutically useful species, along with a description of their unique physical traits, is given in Table 3.3. There are more than 476 species of *Anopheles*, but only approximately 70 of them are capable of transmitting diseases to humans, and only around 40 of them are considered noteworthy (Service 2012). *Anopheles gambiae*, *Anopheles coluzzii*, and *Aedes aegypti* are the primary vectors of dengue, chikungunya, and yellow fever, whereas *Culex* mosquitoes transmit Nile fever and Japanese encephalitis. When it comes to the transmission of lethal diseases, mosquitoes exhibit distinctive adaptations and ways of feeding on human hosts by dodging several chemical pesticides. Thus, from the perspective of vector management, it is imperative to identify the factors influencing their evolution.

Mosquitoes's olfactory receptors are so precisely attuned to human odor that their correlations are critical in determining mosquitoes' host preferences (McBride 2016). Three main gene families make up the olfactory receptors on the antennal sensilla of mosquitoes: odorant receptors (ORs), ionotropic receptors (IRs), and gustatory receptors (GRs) (Steinbrecht 1996; Suh et al. 2014). ORs recognize esters, alcohols, and ketones, while antennal IRs detect amines and organic acids, and GRs detect carbon dioxide. *An. gambiae* and *Ae. aegypti* are highly receptive to carbon dioxide, lactic acid, and ammonia emitted from human skin as a result of sweating (Healy and Copland 1995; Dekker et al. 2005). An adaptive mutation, known as Aaeg-OR4, was found in the OR, making it more receptive to sulcatone, an element presents in human odor. This suggests that mosquitoes' increased sensitivity to sulcatone may be influencing their preference for humans as hosts (McBride et al. 2014).

The wing shape and architecture of mosquitoes exhibit significant diversity and variability, making them an important focal point in microevolutionary investigations. As a result, understanding microevolution in mosquitoes is essential for improving vector control efforts (Suesdek 2019). *Aedes albopictus* and *Aedes aegypti* have been shown to change the shape of their wings in about 4 years (Vidal et al. 2012; Louise et al. 2015). Adaptive microevolution is also supportive of the changing of wing structures in different microgeographic populations of *Culex nigripalpus* (Carvalho et al. 2017) and *Anopheles cruzii* (Lorenz et al. 2014). Microevolution is when the number of alleles changes within a population of the same species over a short period (Provorov et al. 2022). As a result, speciation frequently occurs due to environmental selective pressures. Due to their tremendous adaptability and short generation times, mosquitoes are especially prone to microevolutionary processes. However, many variables may drive microevolution in mosquito populations; among them, the following events are relevant.

Microevolution in mosquito populations can be driven by a number of reasons; the following are most important: (1) pesticide resistance (repeated exposure to insecticides can cause resistance), (2) host preference (changes in host availability or behavior can alter selection pressure on mosquito populations, causing shifts in

Table 3.3 Morphological characteristics and their variations across medically important mosquito species

Species	Characteristic features	Responsible disease
Aedes aegypti	Head palps covered with while/pale scales Scutum and legs covered with scales and patches Dark-scaled proboscis Clypeus and pedicel with pale scales/patches on ventro/lateral surface Scutum with lyre-shaped white markings Scutellum with broad white scales Lower pro-episternal and mesepimeral scales present Postpronotal scales present Postspiracular area without scales Paratergite and Subspiracular area with broad white scales Legs (Ta III, 1–5) with pale basal bands Legs (Ta-III, 5) mostly white Gonostyles narrow, tapering with apical spines	Chikungunya—a viral disease transmitted to humans via infected *Aedes aegypti*
Aedes albopictus	Head palpus with white scales at apex Proboscis entirely dark-scaled Pedicel with scales on lateral surfaces Scutum with median longitudinal stripe, antealar area with patch of broad pale scales Mesepimeron with lower scales Paratergite and postpronotal areas with scales Postspiracular scales absent and Subspiracular area with broad white scales Proepisternal scales present and scutal angles without pale scales Legs (Ta-I–III, 1–5) with only basal bands and Silvery or white scale patches Tergal scales basal, often not connected with lateral pale scales Setae below subapical spine of gonostyl usually forming a row Claspette lobes covered with long spine-like setae and curved at the apex Median part of tergum IX pointed	Chikungunya—a viral disease also transmitted to humans via infected *Aedes albopictus*

Table 3.3 (continued)

Species	Characteristic features	Responsible disease
Aedes caspius	Scutum with golden scales and narrow dorsocentral lines of white scales Terga with median pale stripe, or all pale-scaled Wings speckled with dark and pale scales and costa predominantly dark-scaled Legs (Ta-III, 1–4) with basal and apical bands Basal lobe gradually arising from gonocoxite, spines situated close together, larger spine strongly curved apically (usually tip extending backwards to almost the middle of the spine)	Concerning vector competence for Zika virus (ZIKV).
Aedes geniculatus	Proboscis and palpus black-scaled Proboscis distinctly longer than Fe-I; Ti-I–III and Ta-I–III all dark; Fe-I–III with apical pale spots Scutum with distinctive white and blackish patterns dominated by two dorsocentral blackish stripes Scutellar scales are narrow and yellowish Pleuron with patches of broad, whiteish scales. Abdomen Abdomen with distinct lateral triangular white patches dorsally Female cercus short, blunt. Setae of gonocoxite are short and less dense	These are notorious for arboviral diseases chikungunya
Aedes vexans	Proboscis brown with numerous white scales ventrally vertex with median narrow scales Scutellum with pale, narrow scales Wing scales are sparse dark and pale. No pale fringe scales along posterior margin Legs (Fe-II, III) anterior surfaces of speckled with pale scales. Ta-II, III with basal white bands Frontal setae are arranged in a triangular pattern. Median setae of labral brush serrated apically	Aedes vexans can transmitted various diseases such as eastern and western equine encephalitis, Japanese encephalitis and California encephalitis
Anopheles atroparvus	Thorax elongated and represented mostly by the mesonotum Scutal integument varies from blackish to brownish to grayish Scales can be absent or present	Causative agent of malaria

(continued)

Table 3.3 (continued)

Species	Characteristic features	Responsible disease
Anopheles claviger	Vertex with distinct tuft of broad white scales, lateral and posterior scales dark brown Labellum and proboscis dark, concolorous Palpomere MPlp5 short, not longer than half length of MPlp4 Scutum with narrow pale scales on broad median area Lower proepisternal setae present Wing scales uniformly distributed, without concentrations of dark scale Legs (Ta-I1) long, about equal in length (Ta-I, 2–5) Gonocoxite with 3 parabasal setae, inner one simple, the outer two apically branched	It is suspected of having contributed to malaria transmission
Anopheles dirus	Proboscis slightly longer than forefemur, apex not pale; palpus with ≥4 pale bands Antepronotal scales present; presector dark spot of vein R with 1-4 pale interruptions, most frequently 2 Bright shiny white scales on presector pale and sector pale spots, white scales on the halter, distinctly contrasting with yellowish scales of the subcostal pale and preapical pale spots Apex of Ti-III and base of Ta-III 1 with large white band Ti-III apical pale band without dark interruption; Ta-III4 always with conspicuous basal pale patch on dorsal surface Sternum VI (VI-S) without posteromedian patch of dark scales	*Anopheles dirus* is a vector of malaria in Asian forested zones
Anopheles gambiae	Palps usually with 3 pale bands. Legs Legs (Ta-III, 1–5) not entirely pale, legs sparsely speckled Costa and veins R–R1 with ≥4 dark spots, preapical dark spot on vein R1 with or without pale interruption; wing with pale fringe spots up to CuA or 1A Abdomen without posterolateral scale tufts; sparse or no dorsal abdominal scales	Malaria is transmitted to humans by female mosquitoes of the genus Anopheles

Table 3.3 (continued)

Species	Characteristic features	Responsible disease
Anopheles maculipennis	Scutum laterally darkened, with a distinct anterior tuft of pale scales Predominantly dark scales at crossveins and bifurcations form dark spots Claspette lobes not well defined, bearing only spine-like setae of variable length and shape	The Anopheles maculipennis complex was responsible for most of the malaria transmission in EU countries
Anopheles pulcherrimus	Palps with 4 pale bands Leading edge of wings with 4 distinct dark areas Legs (Ti-I–III and Fe-I–III) with sparse pale scales; Fe-II, III with longitudinal white lines terminating in an oval spot; Ta-III2 with at least apical 0.5 pale; Ta-III3–5 all pale Terga densely covered with broad pale scales; II–VII-Te with prominent posterolateral dark scale tufts	Act as a host of malarial parasite
Anopheles sergentii	Palpus with 3 pale bands; MPlp5 entirely pale Vertex with broad pale scales. Thorax: Scutal fossa without scales Scutum with broad silvery pruinose stripe with pale scales Upper proepisternal setae present; vertex with broad pale scales, lateral and posterior scales dark brown Remigium pale-scaled Presector pale spot present Prehumeral pale and humeral pale spots absent; fringe-spots well marked at all veins except anal vein (1A); vein R without basal dark spot distal to humeral crossvein; vein R4+5 dark-scaled except for small pale spots at base and apex Legs (Ta-III 5) dark. Longest spine-like seta of claspette of similar length as spatula-like seta. The longest leaflet of aedeagus is distinctly longer than the second longest Leaflets strongly serrated	*Anopheles sergentii* is often referred to as the "desert malaria vector" in northern Africa, the Mediterranean basin and the Middle East

(continued)

Table 3.3 (continued)

Species	Characteristic features	Responsible disease
Anopheles superpictus	Palps with 3 distinct pale bands, MPlp5 all pale Vertex with broad white scales, posterior and lateral scales dark brown Scutal fossa without scales; scutum with many narrow pale scales; upper proepisternal setae absent One or two pale spots at base of costa, humeral pale and presector pale; presector pale spot present; accessory sector dark spot absent due to fusing of sector pale and accessory sector pale spots; pale fringe spot absent between cubital vein and vein 1A; vein 1A with two dark spots, distal spot much longer; most of vein R4+5 pale-scaled, with small basal and apical dark spots Femura and tibia without speckling; Ta-III dark, some tarsomeres may be slightly pale apically; Fe-II,III without longitudinal white stripe Terga not pale scaled, without distinct dark posterolateral scale tufts. Leaflets of aedeagus short, distinctly shorter than half the length of the aedeagus	Primary and secondary vectors of malaria in Turkey and are both widely distributed throughout the Anatolian plateau
Anophesles darlingi	Palpomere MPlp4 with at least some white or cream-colored scales on mediolateral surface; MPlp5 all pale Anterior mesepimeron with small patch of apressed scales; upper mesepimeron without pale scales Terga with posterolateral scale tufts, covered with scales; I-S bare, rarely with few, sparse scales Accessory sector pale spot about 0.4 from wing base, conspicuously closer to sector pale than to subcostal pale spot; prehumeral dark spot well-developed, 3.0–4.0× length of the humeral pale spot; vein 1A with fewer than five pale and five dark alternating spots Leg (Fe-III and Ti-III) dark with sparse pale scales, not speckled; Ta-III4,5 entirely white-scaled	Malaria cases are caused by *Plasmodium* and are usually transmitted by *Anopheles*

Table 3.3 (continued)

Species	Characteristic features	Responsible disease
Anophesles funestus	Palps with three pale bands, including one apically No pale spots on the costa basal to sector pale spot; accessory sector pale spot continuous to costa; preapical dark spot without pale interruption, longer than preapical pale spot; vein R all pale basally; no pale fringe at end of vein 1A; vein M3+4 with a pale spot Legs not speckled; Ta-III1–5 with very narrow apical pale bands	The malaria vector *Anopheles funestus* is increasingly recognized as a dominant vector of residual transmission in many African countries
Anophesles plumbeus	Aedeagus without leaflets Parabasal setae of approximately the same length, tubercles weakly sclerotised	*Anopheles plumbeus* is an aggressive biter and a potential vector of malaria parasites and West Nile virus
Anophesles stephensi	Maxillary palps with pale bands; MPlp5 entirely pale; erect head scales broad, white on vertex and dark brown laterally and posteriorly Scutum with obvious pale scales in addition to setae; scutal fossa with scattered pale scales Vein 1A with 3 dark spots; wing with pale spots present on nearly all veins. Abdomen: V–VIII-S usually with pale scales; II–VII-Te without dark scale tufts	*Anopheles stephensi* is a primary mosquito vector of malaria in urban India
Armigeres subalbatus	Proboscis curved downward and laterally compressed Eyes ventrally with 2 rows of scales Scutum without pair of submedian golden stripes, lateral margins with complete line of pale scales Prescutellar, postspiracular and lower anterior mesepimeral setae present Terga with lateral pale patches not extending dorsally; III–VI-S with apical dark bands Legs (Fe-III) posteroventral surface with broad white band extending to or nearly to apex	It is a natural vector for filarial worms such as zoonotic *Brugia pahangi*, and *Wuchereria bancrofti*, which cause filariasis to humans

(continued)

Table 3.3 (continued)

Species	Characteristic features	Responsible disease
Culex gelidus	Proboscis with broad median pale band; vertex with erect white scales Distinct double row of acrostichal setae present Scutum white-scaled anterior to prescutellar area Pleuron with distinct scale patches; lower mesepimeral setae absent Legs (Ta-I–III) with pale bands at articulations; Fe-II,III without pale speckling on anterior surfaces Terga with broad basal pale bands	*Culex* has emerged as a major vector of Japanese encephalitis virus (JEV) in India, Southeast Asian countries and Australia
Culex pipiens	Ventral arm of aedeagus slender, sickle shaped	The virus caused numerous cases of disease in birds, particularly North American crows (*Corvus brachyrhynchos*), domestic horses and humans
Culex quinquefasciatus	Ventral arm of aedeagus broad, leaflike	Transmission of avian malaria and avian poxvirus led to the extinction of several endemic bird species, with many more at risk
Culex theileri	Proboscis without pale bands Postspiracular and prealar scales present Lower mesepimeron with at least one seta Scales of upper and lower mesokatepisternum and posterior mesanepisternum (prealar area) merged Legs (Ta-I–III) dark Tergal pale bands widened medially	Causative agent of dirofilariasis
Culex tritaeniorhynchus	Proboscis with broad median pale band, vertex with erect brownish scales Scutum with unicolorous dark scales; acrostichal setae present Pleuron with distinct scale patches; postspiracular and mesepimeral setae absent; lower mesokatepisternal scales present Legs (Fe-I,II) dark anteriorly; Ta-I–III with pale bands Wing dark-scaled; vein R2+3 shorter than R2 Terga only with basal pale bands	Japanese encephalitis (JE) virus, a flavivirus, is closely related to West Nile and St. Louis encephalitis viruses JE virus is transmitted to humans through the bite of infected *Culex* species mosquitoes, particularly *Culex tritaeniorhynchus*

Table 3.3 (continued)

Species	Characteristic features	Responsible disease
Culex univittatus	Proboscis entirely dark-scaled Postspiracular and prealar scales present; upper and lower mesokatepisternal scale patches separate; lower mesepimeron with one or two setae Base of costa with line of pale scales; vein 2A present Legs (Fe-II) with complete anterior pale stripe; Ta-I–III not banded Terga with basal bands narrow, of even width; abdominal basal pale bands slightly expanded mesally	It has been associated with a human disease of rash, fever, and muscle and joint pain in Uganda, South Africa, and Australia
Culex vishnui	Proboscis with distinct median band not extending basally onto ventral surface; vertex pale brown-scaled Scutum pale brown-scaled; acrostichal setae present; pleuron with distinct scale patches; postspiracular scales absent; lower mesepimeron setae absent Terga with only basal bands Vein R2+3 shorter than vein R2 Legs (Ta-I–III) pale-banded; Fe-I,II without anterior speckling; Fe-III with apical dark band	*Culex vishnui* is one of the main vectors of Japanese encephalitis virus (JEV) This virus is of major medical importance in 24 countries of the Asia-Pacific region and can lead to severe encephalitis with sequels and deaths, especially in children
Mansonia indiana	Palps (MPlp) with apical white scales Scutum without well-defined round spots or greenish stripes Pleuron not contrasting with scutum or coxae (C-I–III); postpronotum with appressed white scales Wing scales mixed dark and pale, broad and asymmetrical. Leg (Fe-III) with c. 5 bands of pale scales; pale bands incomplete on all tarsomeres Abdomen III–VII-S mostly pale-scaled; VIII-Te with strong chitinized hooks	*Mansonia* species as vectors of lymphatic filariasis (LF) in Ghana and in West Africa since that of 1958 in Guinea
Mansonia uniformis	Palps with apical yellowish scales Postpronotum with narrow curved scales; coxae C-I–III not contrasting with pleuron or scutum; scutum with pair of longitudinal greenish stripes, without well-defined round spots Abdomen III–VII-S mostly pale-scaled; VIII-Te with strong chitinized hooks Legs (Fe-III) with approximately five bands of pale scales; Ta-III1–5 pale bands complete posteriorly Wing scales broad and asymmetrical, mixed dark and pale	*Mansonia uniformis* can be a vector of human diseases, such as Ross River virus, Kunjin virus, Murray Valley encephalitis, and lymphatic filariasis

(continued)

Table 3.3 (continued)

Species	Representative individual	References
Aedes aegypti	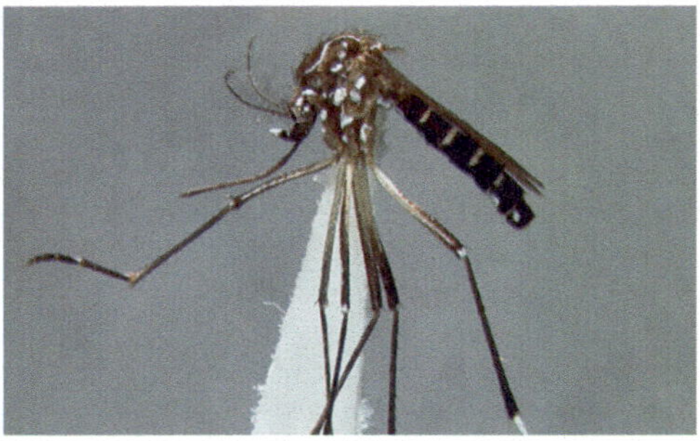	Jansen and Beebe (2010), Powell (2018), Leta et al. (2018), Becker et al. (2020)
Aedes albopictus		Leta et al. (2018), Becker et al. (2020)
Aedes caspius	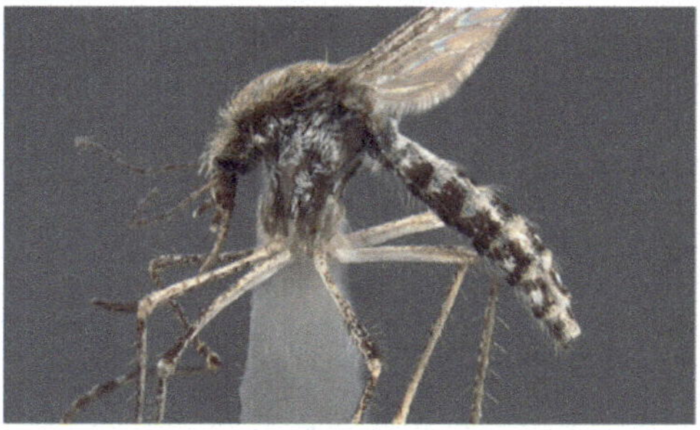	Núñez et al. (2019), Becker et al. (2020)
Aedes geniculatus	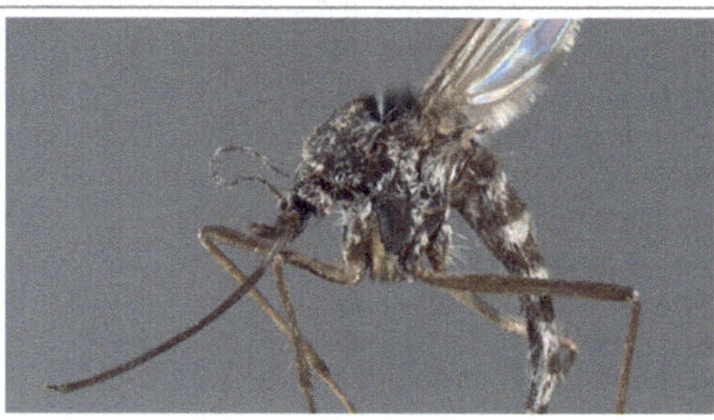	Prudhomme et al. (2019), Becker et al. (2020), Nikookar et al. (2023)

Table 3.3 (continued)

Species	Representative individual	References
Aedes vexans	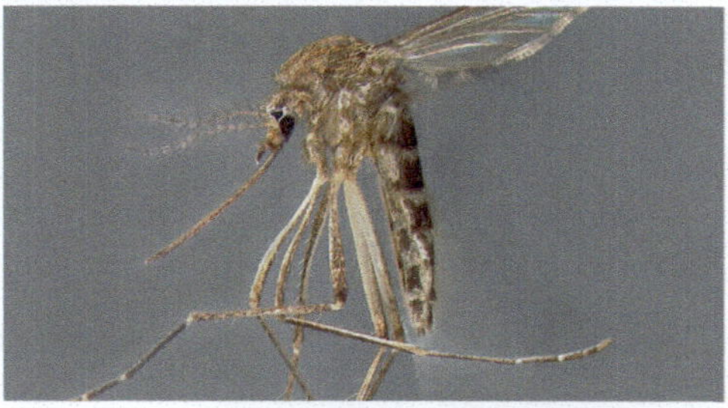	Moradi-Asl et al. (2019), Becker et al. (2020)
Anopheles atroparvus		Constantlnescu and Negulici (1967), Capinha et al. (2009)
Anopheles claviger	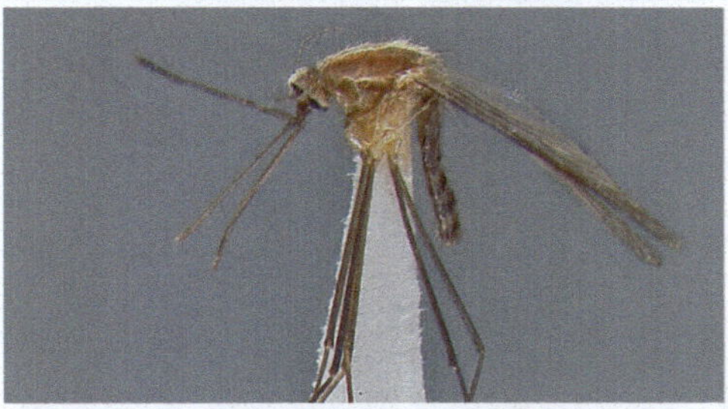	Glick (1992), Tagliapietra et al. (2019), Becker et al. (2020)
Anopheles dirus	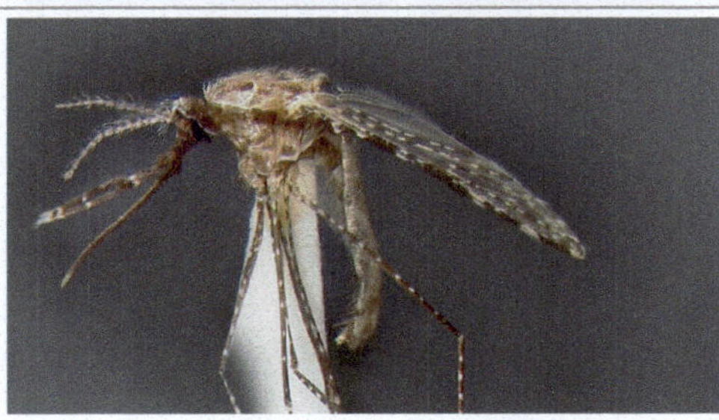	Tagliapietra et al. (2019)

(continued)

Table 3.3 (continued)

Species	Representative individual	References
Anopheles gambiae	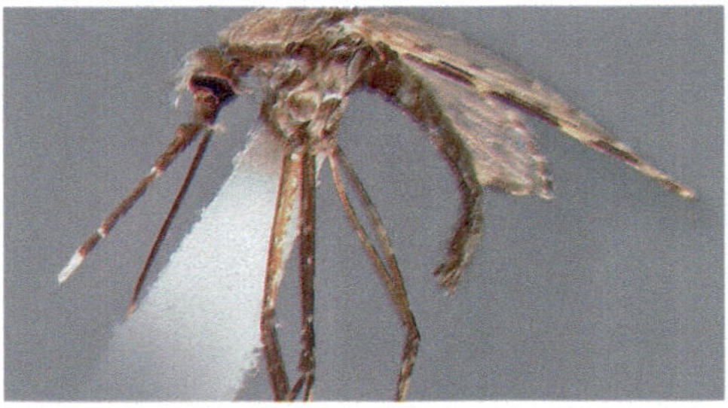	Gouagna et al. (2004)
Anopheles maculipennis	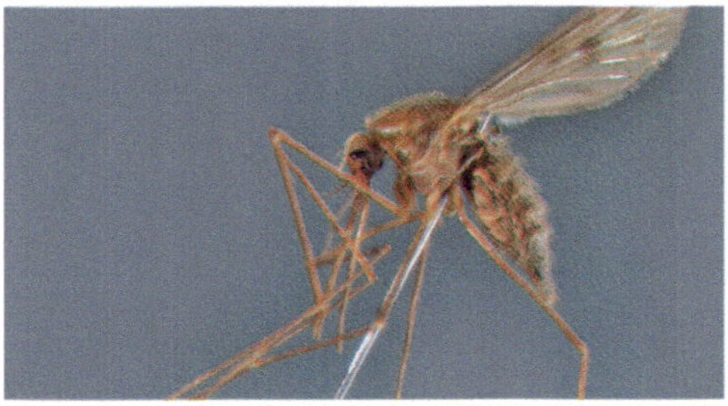	Kampen et al. (2016)
Anopheles pulcherrimus	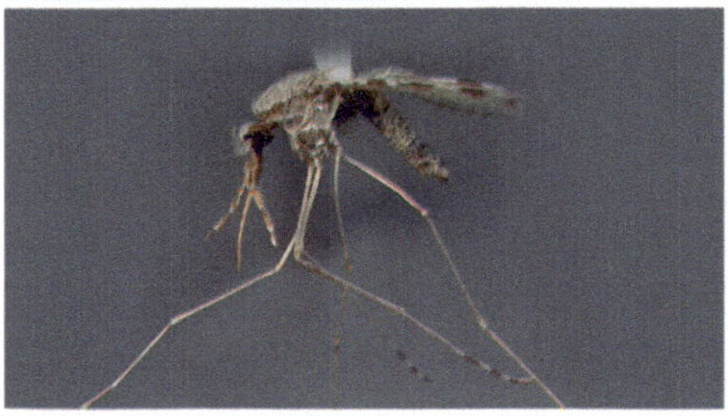	Habirov et al. (2012)
Anopheles sergentii	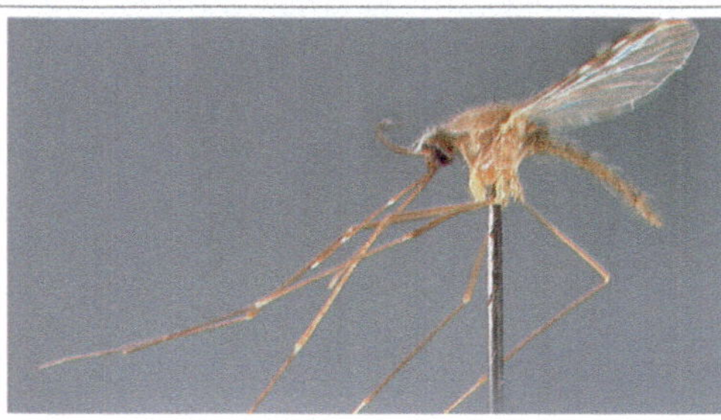	Becker et al. (2020), Adugna et al. (2021)

Table 3.3 (continued)

Species	Representative individual	References
Anopheles superpictus	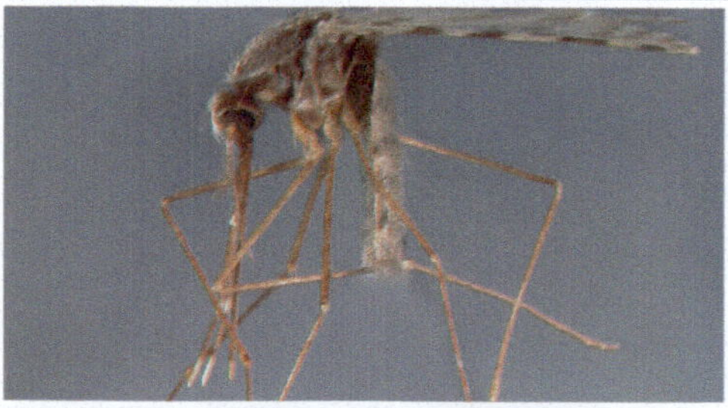	Ramsdale et al. (2001), Hanafi-Bojd et al. (2018), Becker et al. (2020)
Anophesles darlingi	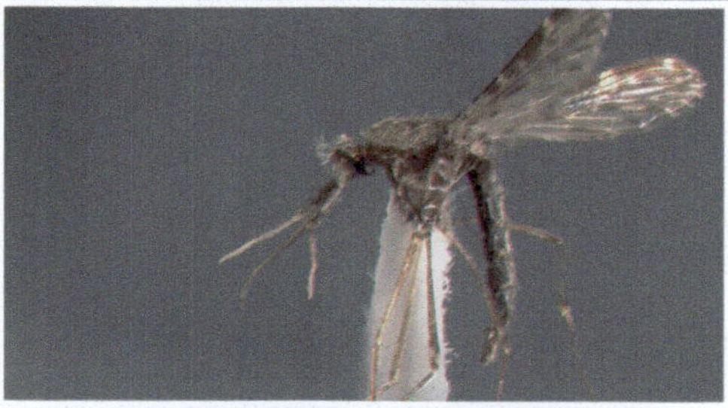	Póvoa et al. (2003)
Anophesles funestus	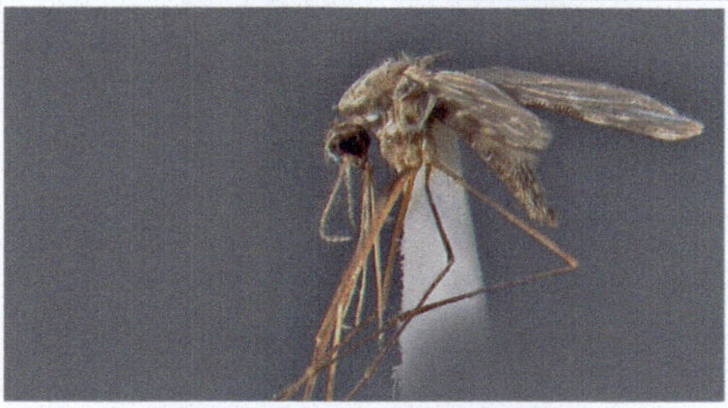	Kweka et al. (2013)
Anophesles plumbeus	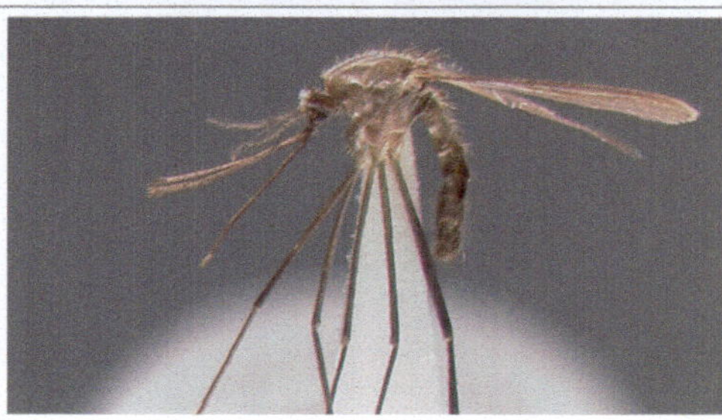	Bueno-Marí and Jiménez-Peydró (2011), Becker et al. (2020)

(continued)

Table 3.3 (continued)

Species	Representative individual	References
Anophesles stephensi	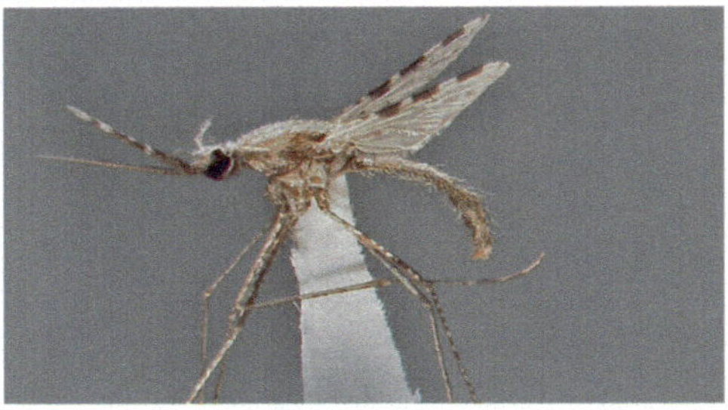	Sor-suwan (2015)
Armigeres subalbatus		Aliota et al. (2010)
Culex gelidus	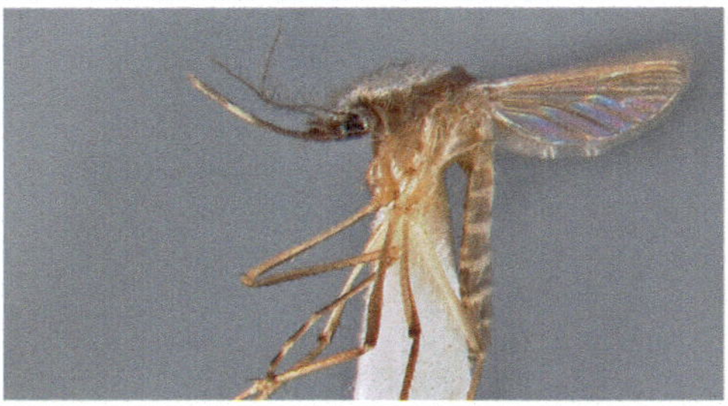	Williams et al. (2005), Sudeep (2014)
Culex pipiens	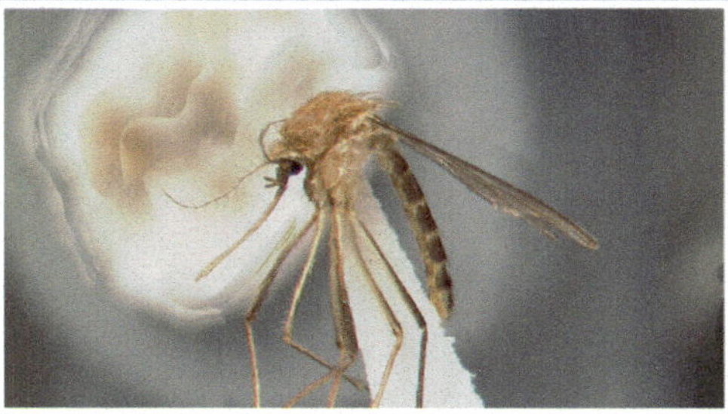	Farajollahi et al. (2011), Becker et al. (2020)

Table 3.3 (continued)

Species	Representative individual	References
Culex quinquefasciatus		Whiteman et al. (2005), Becker et al. (2020), Harvey-Samuel et al. (2021)
Culex theileri		de la Puente et al. (2014)
Culex tritaeniorhynchus		Longbottom et al. (2017)
Culex univittatus		Miller et al. (2000)

(continued)

Table 3.3 (continued)

Species	Representative individual	References
Culex vishnui		Mixão et al. (2016)
Mansonia indiana	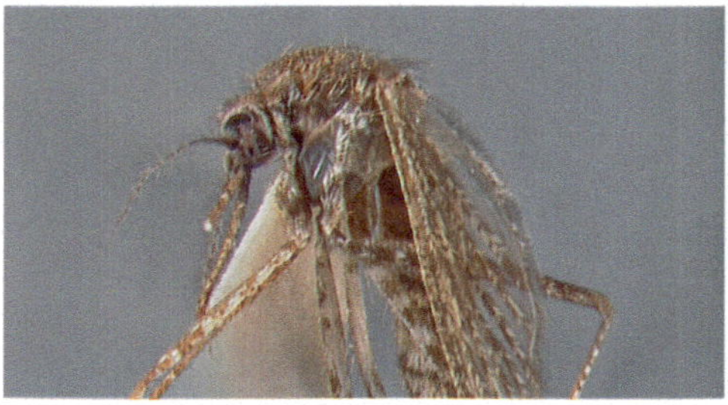	Gopalakrishnan et al. (2014)
Mansonia uniformis	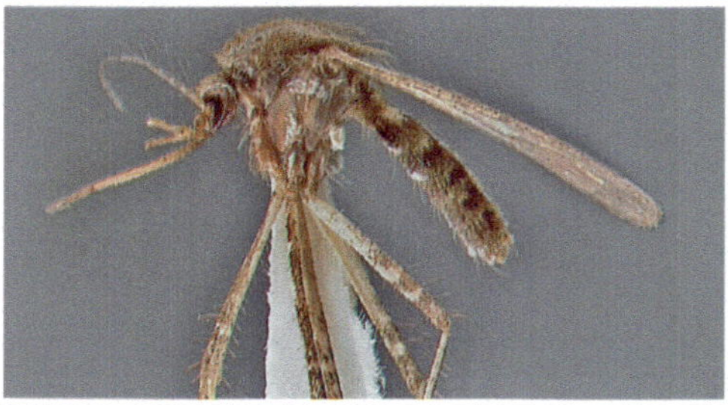	Onapa et al. (2007)

Image courtesy: WRBU (https://vectormap.si.edu)

host preference over time), (3) adaptation to new environments (mosquito populations can undergo microevolutionary changes in response to changes in land use, climate, and other environmental factors), (4) seasonal variation (periodic fluctuations in temperature, precipitation, and other environmental conditions can influence mosquito survival, reproduction, and behavior, resulting in microevolutionary changes within populations), and (5) viral evolution (mosquito-borne viruses, such as dengue and Zika), can exert selective pressure on mosquito populations (Barrio-Nuevo et al. 2019). However, fluctuations in viral strains and transmission dynamics can influence alterations in mosquito vulnerability, vector proficiency, and virus transmission, resulting in coevolutionary interactions between mosquitoes and the viruses. The microevolutionary changes in mosquitos can have serious consequences for public health, agriculture, and ecosystems. The development of successful solutions for mosquito control and disease prevention depends on our understanding of these mechanisms.

Ectoparasites often modify their body form to enhance their mobility on the host, but they can also hide well when they are not on it. Arthropods that feed on blood and come into close contact with their hosts for a short period, such as skin-piercing mosquitoes, typically do not have any dorsoventral or lateral compression in their bodies. Solenophages are hematophagous arthropods with highly specialized mouthparts capable of penetrating blood capillaries and imbibing directly on host blood. One of the best examples of solenophages, extremely specialized in blood feeding, is mosquito. They have a unique mouth apparatus called styletiform, specifically adapted for drawing blood without leaving any noticeable puncture marks on the skin at the biting site.

Mosquitoes also exhibit specialized sensory structures for locating and recognizing their hosts; however, such sensory organs may change over time. These organs can sense movements, temperature, moisture, carbon dioxide, and numerous chemical compounds the host releases. Therefore, the signals associated with the hosts are distinct for them to detect and feed upon. It is true that insects primarily rely on their eyes and antennae as their primary sensory organs, but hematophagous Dipterans, mainly the blood-feeding insects, like mosquitoes, have specialized antennae equipped with receptors to sense chemicals and aromatic compounds emitted by the host dermis and exhaled air. Compounds that attract mosquitoes include fatty acids, amino acids, octanol, carbon dioxide, and lactic acid. The sensory receptors are usually located on the basal region of their antennae, which are extensively developed and diverse. One of them constitutes the Johnston's organ, which is responsible for detecting vibrations in the air. Therefore, mosquitoes can perceive a low amplitude of vibrations from their host movements. Since these receptors are grouped close to the tip of the rostrum and proboscis, they can determine the exact location of blood capillaries just below the host's dermis.

Mosquitoes have a specialized feeding apparatus called the proboscis, which comprises five inflated organs: the labrum, labium, maxillae, mandibles, and hypopharynx. These are the most common types of mouthparts in solenophage species. The labium of the mosquito serves as a protective covering and route for the fascicle, composed of delicate and slender oral organs that penetrate the host's epidermis, deliver saliva, and transport blood. The labrum, maxillae, mandibles, and hypopharynx are all parts of the fascicle, and fine stylets connect them all. The labrum, which resembles the "upper lip" of many insects, encloses the blood-feeding tube of mosquitoes. In mosquitoes, the mandibles and maxillae, generally used for chewing and masticating foods, evolved to puncture the host's skin. The hypopharynx, an elongated and needle-shaped organ seen in insects with chewing mouthparts, is used to penetrate the host's epidermis (Burkett-Cadena and Blosser 2017).

3.5 Conclusion

Studies on the morphology of mosquitoes are crucial not only for species identification but also for the progress of vector control and the prevention of epidemics. The area of eco-morphology enables the study of the evolutionary changes in an organism's physical properties and the correlation between its form and the surrounding environment. The presence of morphological changes in mosquitoes, such as cuticular thickening and enhanced sensory proteins, poses a significant challenge to the effectiveness of conventional vector control strategies (Samal and Kumar 2021). In general, the morphological characteristics of mosquitoes are of significant importance in a multitude of disciplines, such as ecology, taxonomy, vector control, disease prevention, and evolutionary biology.

Points to Remember

The morphology of mosquitoes is significant for several reasons, including the following:

1. Various species of mosquitoes possess unique physical characteristics, including the arrangement of scales on their wings, the morphology of their mouthparts, and the distribution of hairs on their bodies. These characteristics are essential for taxonomists and researchers to precisely identify and categorize mosquito species.
2. Effective vector control programs require a thorough understanding of mosquito morphology. An understanding of the morphology of mosquito larvae aids in the identification of breeding areas and the implementation of specific control methods, such as application of larvicides or habitat modification.
3. The structure of mosquito mouthparts affects their capacity to spread diseases. Mosquitoes with extended mouthparts, such as the *Anopheles* species, are more effective carriers of malaria, because they can efficiently penetrate human skin to extract blood.
4. Analyzing the physical characteristics of mosquitoes offers valuable information on their evolutionary past, ability to adjust to diverse habitats, and interactions with other living beings. An instance of this is the examination of the morphological disparities of closely related species, which can provide insights into their gradual divergence over time.
5. The ecological role of mosquitoes is influenced by their form, which is crucial for understanding their function within ecosystems. Specific morphological adaptations, such as mouthparts specialized for consuming nectar or piercing

skin to extract blood, have a role in their interactions with other creatures and their overall ecological influence.

Acknowledgments The authors are thankful to the Head, Department of Zoology, University of Calcutta, DST-FIST, and UGC DRS-SAP for providing necessary laboratory facilities and support.

References

Research Papers and Review Articles

Adugna F, Wale M, Nibret E (2021) Review of Anopheles mosquito species, abundance, and distribution in Ethiopia. J Trop Med 2021:1–7

Albicócco AP, Carbajo AE, Vezzani D (2011) Mosquito community structure in phytotelmata from a South American temperate wetland. J Vector Ecol 36(2):437–446

Aliota MT, Fuchs JF, Rocheleau TA, Clark AK, Hillyer JF, Chen CC, Christensen BM (2010) Mosquito transcriptome profiles and filarial worm susceptibility in Armigeres subalbatus. PLoS Negl Trop Dis 4(4):e666

Almeida-Silva LM, Brescovit AD, Dias SC (2009) A new species of Goeldia (Araneae: Titanoecidae) with notes on its natural history. Zoologia 26:363–368

Anoopkumar AN, Puthur S, Rebello S, Aneesh EM (2019) Molecular characterization of Aedes, Culex, Anopheles, and Armigeres vector mosquitoes inferred by mitochondrial cytochrome oxidase I gene sequence analysis. Biologia 74:1125–1138

Bar A, Andrew J (2013) Morphology and morphometry of Aedes aegypti larvae. Ann Res Rev Biol 3(1):1–21

Barrio-Nuevo KM, Medeiros-Sousa AR, Ceretti W, Fernandes A, Rocco IM, Cunha MS, Luchs A, Mucci LF, de Souza RP, Marrelli MT (2019) Zika and dengue viruses infecting wild-caught mosquitoes in an environmental protection area in Brazil. bioRxiv, p 2019-12

Belkin JN, Hogue CL (1959) A review of the crabhole mosquitoes of the genus Deinocerites (Diptera, Culicidae). University of California Press, Berkeley

Benelli G, Lo Iacono A, Canale A (2016) Mosquito vectors and the spread of cancer: an overlooked connection? Parasitol Res 115:2131–2137. https://doi.org/10.1007/s00436-016-5037-y

Bergland AO, Agotsch M, Mathias D, Bradshaw WE, Holzapfel CM (2005) Factors influencing the seasonal life history of the pitcher-plant mosquito, Wyeomyia smithii. Ecol Entomol 30(2):129–137

Bueno-Marí R, Jiménez-Peydró R (2011) Anopheles plumbeus Stephens, 1828: a neglected malaria vector in Europe. Malaria Rep 1(1):e2

Boehmler M (2022) Aedeomyia (Aedeomyia) Squamipennis—new genus and species county record for Monroe County, Florida, USA. J Am Mosq Control Assoc 38(4):280–282

Burkett-Cadena ND, Blosser EM (2017) Aedeomyia squamipennis (Diptera: Culicidae) in Florida, USA, a new state and country record. J Med Entomol 54(3):788–792

Chaves LF, Friberg MD, Jian JY, Moji K (2019) Landscape and environmental factors influencing stage persistence and abundance of the bamboo mosquito, Tripteroides bambusa (Diptera: Culicidae), across an altitudinal gradient. Insects 10(2):41

Campos RE, Zavortink TJ (2010) Description of the larva and pupa and redescription of the adults of Isostomyia paranensis (Brèthes) (Diptera: Culicidae). Zootaxa 2689(1):27–36

Constantlnescu P, Negulici E (1967) The experimental transmission of Plasmodium malariae to Anopheles labranchiae atroparvus. Trans R Soc Trop Med Hyg 61(2):182–188

Capinha C, Gomes E, Reis E, Rocha J, Sousa CA, Do Rosário VE, Almeida AP (2009) Present habitat suitability for Anopheles atroparvus (Diptera, Culicidae) and its coincidence with former malaria areas in mainland Portugal. Geospat Health 3(2):177–187

Carvalho GC, Vendrami DP, Marrelli MT, Wilke ABB (2017) Wing variation in Culex nigripalpus (Diptera: Culicidae) in urban parks. Parasit Vectors 10:423

Dehghan H, Sadraei J, Moosa-Kazemi S (2011) The morphological variations of Culex pipiens (Diptera: Culicidae) in central Iran. Asian Pac J Trop Med 4(3):215–219

Dekker T, Geier M, Carde RT (2005) Carbon dioxide instantly sensitizes female yellow fever mosquitoes to human skin odours. J Exp Biol 208:2963–2972

Dyar HG (1919) A note on Lesticocampa and a new species. Insecutor Inscitiae Menstruus 7:9–11

Faithpraise FO, Idung J, Usibe B, Chatwin CR, Young R, Birch P (2014) Natural control of the mosquito population via Odonata and Toxorhynchites. Int J Innov Res Sci Eng Technol 3(5):12898–12911

Farajollahi A, Fonseca DM, Kramer LD, Kilpatrick AM (2011) "Bird biting" mosquitoes and human disease: a review of the role of Culex pipiens complex mosquitoes in epidemiology. Infect Genet Evol 11(7):1577–1585

Foster WA, Walker ED (2019) Mosquitoes (culicidae). In: Medical and veterinary entomology. Elsevier, London, pp 261–325. https://doi.org/10.1016/b978-0-12-814043-7.00015-7

Forattini OP (1996) Culicidologia medica. In: Principios gerais morfologia glossario taxonomico, vol 1. Editora da Universidade de Sao Paulo-Edusp, São Paulo

Glick JI (1992) Illustrated key to the female Anopheles of southwestern Asia and Egypt (Diptera: Culicidae). Mosquito Syst 24(2):125–153

Gopalakrishnan R, Baruah I, Veer V (2014) Monitoring of malaria, Japanese encephalitis and filariasis vectors. Med J Armed Forces India 70(2):129–133

Gouagna LC, Ferguson HM, Okech BA, Killeen GF, Kabiru EW, Beier JC, Githure JI, Yan G (2004) Plasmodium falciparum malaria disease manifestations in humans and transmission to Anopheles gambiae: a field study in Western Kenya. Parasitology 128(3):235–243

Habirov Z, Kadamov D, Iskandarov F, Komilova S, Cook S, McAlister E, Harbach RE (2012) Malaria and the Anopheles mosquitoes of Tajikistan. J Vector Ecol 37(2):419–427

Hanafi-Bojd AA, Sedaghat MM, Vatandoost H, Azari-Hamidian S, Pakdad K (2018) Predicting environmentally suitable areas for Anopheles superpictus Grassi (sl), Anopheles maculipennis Meigen (sl.) and Anopheles sacharovi Favre (Diptera: Culicidae) in Iran. Parasit Vectors 11:1–12

Harbach RE, Kitching IJ, Culverwell CL, Dubois J, Linton YM (2012) Phylogeny of mosquitoes of tribe Culicini (Diptera: Culicidae) based on morphological diversity. Zool Scr 41(5):499–514

Harvey-Samuel T, Ant T, Sutton J, Niebuhr CN, Asigau S, Parker P, Sinkins S, Alphey L (2021) Culex quinquefasciatus: status as a threat to island avifauna and options for genetic control. CABI Agric Biosci 2:1–21

Healy TP, Copland MJ (1995) Activation of Anopheles gambiae mosquitoes by carbon dioxide and human breath. Med Vet Entomol 9:331–336

Hoekman D, Terhorst C, Bauer A, Braun S, Gignac P, Hopkins R, Joshi S, Laskis K, Sanscrainte N, Travis J, Miller TE (2007) Oviposition decreased in response to enriched water: a field study of the pitcher-plant mosquito, Wyeomyia smithii. Ecol Entomol 32(1):92–96

Huang YM, Rueda LM (2013) A new species of the genus Orthopodomyia from Ototomo, Cameroon, with notes on other Orthopodomyia group-species. J Am Mosq Control Assoc 29(3):270–274

Jansen CC, Beebe NW (2010) The dengue vector Aedes aegypti: what comes next. Microbes Infect 12(4):272–279

Judd D (1996) Review of the systematics and phylogenetic relationships of the Sabethini (Diptera: Culicidae). Syst Entomol 21(2):129–150

Junglen S, Kopp A, Kurth A, Pauli G, Ellerbrok H, Leendertz FH (2009) A new flavivirus and a new vector: characterization of a novel flavivirus isolated from uranotaenia mosquitoes from a tropical rain forest. J Virol 83(9):4462–4468

Kalpage KS, Brust RA (1968) Mosquitoes of manitoba. I. Descriptions and a key to Aedes eggs (Diptera: Culicidae). Can J Zool 46(4):699–718. https://doi.org/10.1139/z68-098

Kampen H, Schäfer M, Zielke DE, Walther D (2016) The Anopheles maculipennis complex (Diptera: Culicidae) in Germany: an update following recent monitoring activities. Parasitol Res 115(9):3281–3294
Kaulfers WV (1940) Spanish expressions found in English. Hispania 23(2):175–184
Kaur R (2003) An update on the distribution of mosquitoes of the tribe Aedini in India (Diptera: Culicidae). Orient Insects 37(1):439–455
Kong XQ, Wu CW (2010) Mosquito proboscis: an elegant biomicroelectromechanical system. Phys Rev E 82(1):011910
Kweka EJ, Kamau L, Munga S, Lee MC, Githeko AK, Yan G (2013) A first report of Anopheles funestus sibling species in western Kenya highlands. Acta Trop 128(1):158–161
Laojun S, Changbunjong T, Sumruayphol S, Chaiphongpachara T (2023a) Molecular and morphometric differentiation of secondary filariasis vector Coquillettidia mosquitoes (Diptera: Culicidae) in Thailand. Infect Genet Evol 112:105452
Land MF, Gibson G, Horwood J (1997) Mosquito eye design: conical rhabdoms are matched to wide aperture lenses. Proc R Soc Lond B 264:1183–1187. https://doi.org/10.1098/rspb.1997.0163
Land MF, Gibson G, Horwood J, Zeil J (1999) Fundamental differences in the optical structure of the eyes of nocturnal and diurnal mosquitoes. J Comp Physiol A 185:91–103
Lang BJ, Idugboe S, McManus K, Drury F, Qureshi A, Cator LJ (2018) The effect of larval diet on adult survival, swarming activity and copulation success in male Aedes aegypti (Diptera: Culicidae). Journal of medical entomology 55(1), pp.29–35
Laojun S, Changbunjong T, Chaiphongpachara T (2023b) Evaluation of modern techniques for species identification of Lutzia mosquitoes (Diptera: Culicidae) in Thailand: geometric morphometrics and DNA barcoding. Insects 14(1):78
Lapshin DN, Vorontsov DD (2017) Frequency organization of the Johnston's organ in male mosquitoes (Diptera, Culicidae). J Exp Biol 220(21):3927–3938
Laurito M, Briscoe AG, Almiron WR, Harbach RE (2018) Systematics of the Culex coronator complex (Diptera: Culicidae): morphological and molecular assessment. Zool J Linnean Soc 182(4):735–757
Lehrer S (2010) Association between malaria incidence and all cancer mortality in fifty U.S. states and the District of Columbia. Anticancer Res 30:1371–1373
Liria J, Navarro JC (2003) Morphology of the larval mandible and maxilla in the subgenus Carrollia Lutz of Culex L. (Diptera: Culicidae). Zootaxa 186(1):1–8
Leta S, Beyene TJ, De Clercq EM, Amenu K, Kraemer MU, Revie CW (2018) Global risk mapping for major diseases transmitted by Aedes aegypti and Aedes albopictus. Int J Infect Dis 67:25–35
Lira-Vieira AR, Gurgel-Goncalves R, Moreira IM, Yoshizawa MAC, Coutinho ML, Prado PS, Souza JLD, Chaib AJDM, Moreira JS, Castro CND (2013) Ecological aspects of mosquitoes (Diptera: Culicidae) in the gallery forest of Brasilia National Park, Brazil, with an emphasis on potential vectors of yellow fever. Rev Soc Bras Med Trop 46:566–574
Longbottom J, Browne AJ, Pigott DM, Sinka ME, Golding N, Hay SI, Moyes CL, Shearer FM (2017) Mapping the spatial distribution of the Japanese encephalitis vector, Culex tritaeniorhynchus Giles, 1901 (Diptera: Culicidae) within areas of Japanese encephalitis risk. Parasit Vectors 10:1–12
Louise C, Vidal PO, Suesdek L (2015) Microevolution of Aedes aegypti. PLoS One 10:e0137851
Lorenz C, Marques TC, Mureb Sallum MA, Suesdek L (2014) Altitudinal population structure and microevolution of the malaria vector Anopheles cruzii (Diptera: Culicidae). Parasit Vectors 7:581
Lorenz C, Almeida F, Almeida-Lopes F, Louise C, Pereira SN, Petersen V, Suesdek L (2017) Geometric morphometrics in mosquitoes: what has been measured? Infect Genet Evol 54:205–215. https://doi.org/10.1016/j.meegid.2017.06.029
Marti GA, Micieli MV, Macia A, Lounibos LP, Garcia JJ (2007) Seasonality and abundance of the mosquito Isostomyia paranensis from phytotelmata in temperate Argentina. J Am Mosq Control Assoc 23(3):252–258

de la Puente J, Moreno-Indias I, Enrique Hernández-Castellano L, Argüello A, Ruiz S, Soriguer R, Figuerola J (2014) Host-feeding pattern of Culex theileri (Diptera: Culicidae), potential vector of Dirofilaria immitis in the Canary Islands. J Med Entomol 49(6):1419–1423

McLver SB (1982) Sensilla of mosquitoes (Diptera: Culicidae). J Med Entomol 19(5):489–535. https://doi.org/10.1093/jmedent/19.5.489

McBride CS, Baier F, Omondi AB, Spitzer SA, Lutomiah J, Sang R, Ignell R, Vosshall LB (2014) Evolution of mosquito preference for humans linked to an odorant receptor. Nature 515:222–227

McBride SC (2016) Genes and odors underlying the recent evolution of mosquito preference for humans. Curr Biol 26:R41–R46. https://doi.org/10.1016/j.cub.2015.11.032

Miller BR, Nasci RS, Godsey MS, Savage HM, Lutwama JJ, Lanciotti RS, Peters CJ (2000) First field evidence for natural vertical transmission of West Nile virus in Culex univittatus complex mosquitoes from Rift Valley province, Kenya. Am J Trop Med Hyg 62(2):240–246

Mitchell EG, Dupree JW (1907). Mosquito life: the habits and life cycles of the known mosquitoes of the United States; methods for their control; and keys for easy identification of the species in their various stages (Vol. 23). GP Putnam

Mixão V, Bravo Barriga D, Parreira R, Novo MT, Sousa CA, Frontera E, Venter M, Braack L, Almeida APG (2016) Comparative morphological and molecular analysis confirms the presence of the West Nile virus mosquito vector, Culex univittatus, in the Iberian Peninsula. Parasit Vectors 9:1–13

Mizuta H (2011) Occurrence of two rare mosquito species, Aedes bekkui and Verrallina nobukonis (Diptera: Culicidae) in southern Osaka Prefecture, Japan. Med Entomol Zool 62(3):195–198

Moirangthem BD, Singh SN, Singh DC (2018) Lutzia tigripes (Diptera: Culicidae, Metalutzia) for the mosquito larval control: a new prospect of mosquito control. IJMR 5(6):1–4

Montell C, Zwiebel LJ (2016) Mosquito sensory systems. In: Progress in mosquito research. Elsevier, Amsterdam, pp 293–328. https://doi.org/10.1016/bs.aiip.2016.04.007

Moradi-Asl E, Vatandoost H, Adham D, Emdadi D, Moosa-Kazemi H (2019) Investigation on the occurrence of Aedes species in borderline of Iran and Azerbaijan for control of arboviral diseases. J Arthropod Borne Dis 13(2):191

Morales AIO, Mis Avila P, Elizondo-Quiroga A, Harbach RE, Siller-Rodríguez QK, Fernández-Salas I (2010) The mosquitoes of Quintana Roo State, Mexico (Diptera: Culicidae). Acta Zool Mex 26(1):33–46

Nikookar SH, Charkame A, Nezammahalleh A, Moradi-Asl E, Enayati A, Fazeli-Dinan M, Sedaghat MM, Zaim M (2023) Entomological surveillance of invasive Aedes mosquitoes in Mazandaran Province, northern Iran from 2014 to 2020. Sci Rep 13(1):8683

Núñez AI, Talavera S, Aranda C, Birnberg L, Rivas R, Pujol N, Verdún M, Failloux AB, Busquets N (2019) European Aedes caspius mosquitoes are experimentally unable to transmit Zika virus. Parasit Vectors 12:1–7

Nugroho SS, Mujiyono M, Ayuningrum FD (2021) Species distribution update of Mansonia blanchard, 1901 mosquitoes (Diptera: Culicidae) in Indonesia with the illustrated key for female mosquito. Treubia 48(2):117–128

Onapa AW, Pedersen EM, Reimert CM, Simonsen PE (2007) A role for Mansonia uniformis mosquitoes in the transmission of lymphatic filariasis in Uganda? Acta Trop 101(2):159–168

Owen WB (1985) Morphology of the head skeleton and muscles of the mosquito, Culiseta inornata (Williston) (Diptera: Culicidae). J Morphol 183(1):51–85

Póvoa MM, Conn JE, Schlichting CD, Amaral JC, Segura MNO, Silva AND, Santos CCD, Lacerda RN, De Souza RT, Galiza D, Santa Rosa EP (2003) Malaria vectors, epidemiology, and the re-emergence of Anopheles darlingi in Belém, Pará, Brazil. J Med Entomol 40(4):379–386

Powell JR (2018) Mosquito-borne human viral diseases: why Aedes aegypti? Am J Trop Med Hyg 98(6):1563

Provorov NA, Andronov EE, Kimeklis AK, Onishchuk OP, Igolkina AA, Karasev ES (2022) Microevolution, speciation and macroevolution in rhizobia: genomic mechanisms and selective patterns. Front Plant Sci 13:1026943

Prudhomme J, Fontaine A, Lacour G, Gantier JC, Diancourt L, Velo E, Bino S, Reiter P, Mercier A (2019) The native European Aedes geniculatus mosquito species can transmit chikungunya virus. Emerg Microbes Infect 8(1):962–972

Ramsdale CD (2001) Internal taxonomy of the Hyrcanus Group of Anopheles (Diptera: Culicidae) and its bearing on the incrimination of vectors of malaria in the west of the Palaearctic Region. Eur Mosq Bull 10:1–8

Rajavel AR, Natarajan R, Vaidyanathan K (2005) Mosquito collections in the Jeypore hill tracts of Orissa, India, with notes on three new country records, Culex (Lophoceraomyia) pilifemoralis, Culex (Lophoceraomyia) wilfredi, and Heizmannia (Heizmannia) chengi. J Am Mosq Control Assoc 21(2):121–127

Rattanarithikul R (1982) A guide to the genera of mosquitoes (Diptera: Culicidae) of Thailand with illustrated keys, biological notes and preservation and mounting techniques. Mosquito Syst 14(3):139–208

Rattanarithikul R, Harrison BA, Panthusiri P, Peyton EL, Coleman RE (2006) Illustrated keys to the mosquitoes of Thailand III. Genera aedeomyia, ficalbia, mimomyia, hodgesia, coquillettidia, mansonia, and uranotaenia. Southeast Asian J Trop Med Public Health 37(S1):1

Reeves LE, Holderman CJ, Blosser EM, Gillett-Kaufman JL, Kawahara AY, Kaufman PE, Burkett-Cadena ND (2018) Identification of Uranotaenia sapphirina as a specialist of annelids broadens known mosquito host use patterns. Commun Biol 1(1):92

Rueda LM (2020) Key to the species of adult female mosquitoes of Texas. In: Mosquitoes, communities, and public health in Texas. Elsevier, Amsterdam, pp 177–192. https://doi.org/10.1016/b978-0-12-814545-6.00004-3

Saba S, Naeem-Ullah U, Saeed S, Rajwana IA, Khan AA (2022) First record of one Genus and four Species of mosquitoes (Diptera: Culicidae) from Pakistan. Int J Trop Insect Sci 42(6):3649–3657

Sabino EM, Marquez-Ocaña OP, Otiniano-Moreno GA, Condezo GKD, Chuquiyauri-Talenas MA, Melo T, Seixas G, Parreira R, de Almeida APG (2023) First report of Culex (Culex) camposi Dyar, 1925 and Culex (Carrollia) bonnei Dyar, 1921 via male genitalia and CoxI barcoding sequence analyses in Huanuco, Peru

Samal RR, Kumar S (2021) Cuticular thickening associated with insecticide resistance in dengue vector, Aedes aegypti L. Int J Trop Insect Sci 41(1):809–820

Sallum S, Eisner JA, Stone JM, Dietrich J, Hinz P, Spalding E (2020) ELT Imaging of MWC 297 from the 23 m LBTI: Complex Disk Structure and a Companion Candidate. The Astronomical Journal 161(1), p.28.

Schreiber ET (2007) Toxorhynchites. J Am Mosq Control Assoc 23(2):129–132

Sedaghat MM, Linton YM, Nicolescu G, Smith L, Koliopoulos G, Zounos AK, Oshaghi MA, Vatandoost H, Harbach RE (2003) Morphological and molecular characterization of Anopheles (Anopheles) sacharovi Favre, a primary vector of malaria in the Middle East. Syst Entomol 28(2):241–256

Shafie A, Roslan MA, Ngui R, Lim YAL, Sulaiman WYW (2016) Mosquito biology and mosquito-borne disease awareness among island communities in Malaysia. J Am Mosq Control Assoc 32(4):273–281

Sharma RS, Venkteash S (2015) Mosquito vectors of human diseases: public health challenge in India. In: Entomology for Sustainable agriculture. Ludhiana Printed at Foil Printers, Ludhiana

Sor-suwan S (2015). Proteomic and transcriptomic analyses of salivary glands of anophesles campestris-like mosquito, a potential vector of Plasmodium vivax (Doctoral dissertation, เชียงใหม่: บัณฑิต วิทยาลัย มหาวิทยาลัย เชียงใหม่)

Steward CC, Atwood CE (1963) The sensory organs of the mosquito antenna. Can J Zool 41(4):577–594. https://doi.org/10.1139/z63-030

Steinbrecht RA (1996) Structure and function of insect olfactory sensilla. Ciba Found Symp 200:158–174; discussion 174-7. https://doi.org/10.1002/9780470514948.ch13

Sudeep AB (2014) Culex gelidus: an emerging mosquito vector with potential to transmit multiple virus infections. J Vector Borne Dis 51(4):251–258

Suh E, Bohbot J, Zwiebel LJ (2014) Peripheral olfactory signaling in insects. Curr Opin Insect Sci 6:86–92

Suesdek L (2019) Microevolution of medically important mosquitoes–a review. Acta Trop 191:162–171

Tagliapietra V, Arnoldi D, Di Luca M, Toma L, Rizzoli A (2019) Investigation on potential malaria vectors (Anopheles spp.) in the Province of Trento, Italy. Malar J 18:1–5

Talaga S, Dejean A, Mouza C, Dumont Y, Leroy C (2018) Larval interference competition between the native Neotropical mosquito Limatus durhamii and the invasive Aedes aegypti improves the fitness of both species. Insect Sci 25(6):1102–1107

Toro-Cantillo A, Hoyos-López R (2018) Molecular identification of Deinocerites atlanticus (Adames, 1971) (Diptera: Culicidae) using cytochrome oxidase I. J Vector Borne Dis 55(1):63–65

Vidal PO, Carvalho E, Suesdek L (2012) Temporal variation of wing geometry in Aedes albopictus. Mem Inst Oswaldo Cruz 107:1030–1034

Webb CE, Doggett SL, Ritchie SA, Russell RC (2008) Vector competence of three Australian mosquitoes, Verrallina carmenti, Verrallina lineata, and Mansonia septempunctata (Diptera: Culicidae), for Ross River virus. J Med Entomol 45(4):737–740

Whiteman NK, Goodman SJ, Sinclair BJ, Walsh TIM, Cunningham AA, Kramer LD, Parker PG (2005) Establishment of the avian disease vector Culex quinquefasciatus Say, 1823 (Diptera: Culicidae) on the Galápagos Islands, Ecuador. Ibis 147(4):844–847

Williams CR, Ritchie SA, Whelan PI (2005) Potential distribution of the Asian disease vector Culex gelidus Theobald (Diptera: Culicidae) in Australia and New Zealand: a prediction based on climate suitability. Aust J Entomol 44(4):425–430

Yang X, Zhuoma Y, Duoji Z, Xu H, Li H, Ciren W, Wang J, Yang G, Zhou H, Gong Z (2019) Taxonomy and fauna of mosquitoes in Tibet III. Discovery of Heizmannia and its three species in Tibet for the first record. Chin J Vect Biol Contr 30(1):72–74

Zhang JD, Huang WX (2019) On the role of vortical structures in aerodynamic performance of a hovering mosquito. Phys Fluids 31(5):051906

Book Chapter

Hawkes FM, Hopkins RJ (2022) The mosquito: an introduction. In: Hall M, Tamïr D (eds) Mosquitopia: the place of pests in a healthy world. Routledge, New York. https://doi.org/10.4324/9781003056034-3

Kachhwaha N, Rajpurohit A, Meena GD (2017) Morphological patterns of different mosquito. Apex Publishing House, Udaipur

Books

Burkett-Cadena ND (2013) Mosquitoes of the southeastern United States. University of Alabama Press, Tuscaloosa

Becker N, Petrić D, Zgomba M, Boase C, Madon MB, Dahl C, Kaiser A, Becker N, Petrić D, Zgomba M, Boase C (2020) Morphology of mosquitoes. In: Mosquitoes: identification, ecology and control. Springer, Cham

Service (2012) Medical entomology for students, 5th edn. Cambridge University Press, New York

Wilkerson RC, Linton YM, Strickman D (2021) Mosquitoes of the world, vol 1. Johns Hopkins University Press, Baltimore

World Health Organization (2020) Pictorial identification key of important disease vectors in the WHO South-East Asia Region

Mosquito Diversity and Mosquito-Borne Diseases

4

Susobhan Datta Mudi, Dipanwita Das, Sampa Banerjee, Milita Roy, Shreya Brahma, Soujita Pramanik, Goutam K. Saha, and Gautam Aditya

Abstract

Mosquitoes (Diptera: Culicidae) are known for their irritating biting habits and disease transmitting vectors. An approximately 3600 species of mosquitoes are recorded worldwide under the 2 major subfamilies Anophelinae and Culicini. The subfamily Culicini comprises 11 tribes with varied number of genera under them, while the subfamily Anophelinae includes 3 genera and 500 species roughly. In almost all regions of the world, mosquitoes are regarded as the vectors of diseases with the ability to exploit varieties of habitat as breeding places. Among the various habitats, phytotelmata of diverse types are exploited by the mosquitoes for breeding. The species diversity of the mosquitoes varies with the zoogeographical realm with the Neotropical realm carrying more species than that of the Oriental realm. The Nearctic realm appears to have the least number of species. A correspondence of the mosquito diversity and the physical space is observed, facilitating the estimate of the species richness of mosquitoes with precision. In contrast to the species richness of the mosquitoes (~3600), on a world wide scale, selected species (10%) bear the significance as disease vectors. Most of the vector mosquitoes are generalists in nature with the ability to host multiple pathogens and parasites and vice versa. Selected species like *Aedes aegypti*, *Ae. albopictus*, and *Culex quinquefasciatus* are more prominent as dis-

S. Datta Mudi · S. Banerjee · S. Brahma · S. Pramanik · G. K. Saha · G. Aditya (✉)
Department of Zoology, University of Calcutta, Kolkata, West Bengal, India
e-mail: gazoo@caluniv.ac.in

D. Das
Department of Zoology, Bagnan College, Bagnan, West Bengal, India

M. Roy
Department of Zoology, Bejoy Narayan Mahavidyalaya, Itachuna, West Bengal, India

© The Author(s), under exclusive license to Springer Nature Singapore Pte Ltd. 2024

Omkar (ed.), *Mosquitoes*, https://doi.org/10.1007/978-981-97-4163-2_4

ease- transmitting agents in different parts of the world, in addition to the malaria vectors, *Anopheles* spp.

Keywords

Culicidae · *Anopheles* · *Culex* · *Aedes* · Phytotelmata

Learning Objectives

1. The features and origin of mosquitoes
2. Taxonomic features of mosquitoes
3. Diversity of the mosquito larval habitats
4. Species diversity of mosquitoes worldwide
5. Diversity of the vectors and diseases transmitted
6. Conclusion
7. Points to remember

4.1 Introduction

Mosquitoes (Diptera: Culicidae) are one of the most studied blood-sucking insects responsible for the transmission of several parasitic and pathogenic diseases of human and various vertebrates. As a carrier and host, the mosquitoes provide a feasible biotope for the pathogens and the parasites to breed and propagate and subsequently transmitted to the secondary hosts. Many species of mosquitoes utilize birds, reptiles, and amphibians as hosts for the blood meal and, in course, transmit the causative agents of diseases. Owing to the painful biting habits, the mosquitoes qualify as a nuisance pest apart from being the vectors for several diseases (Silver 2008; Becker et al. 2020). However, the mosquitoes of the genus *Toxorhynchites* are nonbiting in the adult stages and are predatory in the larval stages consuming mostly other mosquito larvae (Steffan and Evenhuis 1981; Collins and Blackwell 2000). Mosquitoes being holometabolous insects complete their life cycle through the aquatic immature (egg, larva, and pupa) and aerial adult stages. The growth and the development in the larval stages determines the various life history traits of mosquito, including the ability as a disease-transmitting agent. The resources acquired during the larval stages are significant in shaping the adult size, longevity, and the fecundity, which in turn determine the fitness of the individual mosquito. Larval development time for various species of mosquitoes may be completed within 5 days in hot weather condition, while in low temperature, diapause may be observed. The adult longevity varies with the temperature and the relative humidity, in situations with availability of the carbohydrate source and the blood meal. While the adult stage of the mosquito bears significance in the disease transmission and public health, the larval stages bear importance in the ecological regulation of the detritus and algae. The size of the mosquitoes varies considerably, depending on the species concerned, with the wing length of *Toxorhynchites* being 44mm and the standard length being 18mm at one end to the tiny *Aedes* mosquitoes that are 4.4mm in wing span. The size and the biomass of the mosquitoes depend on the resource acquisition pattern by the larval stages and the carbohydrate consumption by the adult stage. The blood meal is essential for the mosquitoes for egg development and thus fecundity. Although the females mate only once post emergence as adult, mosquito oviposition may occur in multiple times with varying number of eggs being deposited on the moist soil or water surface. One of the invariant features in mosquitoes is the chromosome number, $2n = 6$, with the chromosomes being metacentric or submetacentric. In *Anopheles*, the sex determination is carried out through the heteromorphic chromosomes (X or Y) differing in morphology, but in Culicinae mosquito, the homomorphic sex chromosome is observed (Matthews and Munstermann 1994). In general, the mosquitoes may be considered as the insects with short generation time; still, there are species that can extend the life history for a considerable time period and exhibit overwintering. Variations in the life history features of the mosquitoes correspond to several physiological, genetic, and environmental conditions, but the unique adaptation, redundant in all, is the host-seeking behavior. A complex mechanism inclusive of the carbon dioxide trail, odor of the host and olfaction mechanism of the mosquito, enables successful landing and blood feeding from the host (Takken and Verhulst 2013).

On a world wide scale, the mosquitoes are more abundant in the hot and humid climatic conditions, but the distribution ranges from the cold region of Alaska to the hot humid equatorial region, as well as the high altitudes of Darjeeling Himalayas (Aditya et al. 2006) and the arid regions of Sahara (Becker et al. 2020; Wilkerson et al. 2021). An estimated origin of the order Diptera is considered to be around the late Permian (260–251 Mya) with the major lineages established around the late Triassic (235–202 Mya) (Becker et al. 2020) including the superfamily Culicoidea (Knudson et al. 2002). A recent estimate based on the phylogenomic assessment comprising 709 single-copy orthologue groups of 256 species of mosquitoes indicated that the mosquitoes originated about 217 Mya (Soghigian et al. 2023). On the basis of the paleontological evidence, the fossils are mostly from the Oligocene period, and the family Culicidae is assumed to be established by 58–36 Mya, coinciding with the Eocene epoch (Knudson et al. 2002). The studies on the distribution and the evolutionary history of the mosquitoes are reflected in the biogeographical features with higher number of species observed in the hot humid zones and least species in the temperate zones. As small-sized insects with variant life cycle forms and short generation time (Honěk 1993; Crans 2004; Danks 2006, 2007), mosquitoes provide ample adaptation to complete their life history with minimal requirement of the water and the resources therein. Studies on the mosquito remained significant due to their role in the disease transmission bearing implications in medical and veterinary sciences. Several diseases linked with the mosquito as a carrier spread as epidemics in different parts of the world. Owing to the range expansion of various mosquito species, the possibilities and risks of the mosquito-borne diseases have increased correspondingly. As an effort to promote the successful management of the mosquito vectors, deductions about the species diversity and the phylogeny are prerequisite. Adequate knowledge of the concerned species is required to highlight the prospective chance of its abundance in a geographical region. On a worldwide scale, contributions of the various researchers on mosquiotoes are noteworthy. Of particular relevance are the works on the phylogeny and the diversity of the Culicidae (Harbach and Kitching 1998), Culicinae (Harbach et al. 2012), and Anophelinae (Harbach 2013; Harbach and Kitching 2016) which are noteworthy. The diversity and biogeography of the mosquitoes (Foley et al. 2007) including the diversity of the mosquito larval habitats (Rueda 2008) have provided an overview of the mosquito systematics and ecology on a comparative worldwide scale. The species diversity, general biology, and ecology of the mosquitoes have been dealt in classical literatures by Clements (1992, 1999, 2012), Silver (2008), Becker et al. (2020), and Wilkerson et al. (2021), apart from the region-specific works of Edwards (1912, 1921), Barraud (1923, 1934), Christophers (1924, 1933), and Belkin (1962), and the researchers of the present day like Sallum et al. (2000), Harbach (2013), Rattanarithikul and Panthusiri (1994), and Rattanarithikul et al. (2005, 2006a, b, 2007). With the increasing need of the management of the mosquito-borne diseases, the studies on the mosquito species diversity have been on increasing demand among the different researchers. A narrative on the mosquito species diversity and the disease vectors is highlighted to serve as a prelude to the biology and management of the mosquitoes.

4.2 Mosquito Systematics

Mosquitoes belong to the order Diptera comprising 150,000 diverse species commonly termed as flies (Cranston and Gullan 2009). The presence of two wings and the intricate number and arrangement of veins in the wings, the features of the mouthparts, and the antennae are bases for the taxonomic classification of the Diptera. Further, the order Diptera includes two suborders: Nematocera and Brachycera. At the family level, the Nematocera includes, Tipulidae (craneflies), Bibionidae (march flies), Mycetophilidae (fungus gnats), Sciaridae (dark-winged fungus gnats), Psychodidae (moth flies, sand flies), Chaoboridae (phantom midges), Ceratopogonidae (biting midges), Simuliidae (black flies), Chironomidae (chironomid midges), and Culicidae (mosquitoes). Insects under these families are featured by the long multisegmented antenna and bear fine, long-legged flies, as well as stout-bodied flies. In contrast, the insects under Brachycera bear short antennae like the houseflies (Muscidae). The family Culicidae is further divided into two subfamilies, Anophelinae and Culicinae. The subfamily Culicinae consists of 11 tribes and a total of 3566 extant species (Wilkerson et al. 2021; Soghigian et al. 2023). The tribes under Culicinae are Aedeomyiini, Aedini, Culicini, Culisetini, Ficalbiini, Hodgesiini, Mansoniini, Orthopodomyiini, Sabethini, Toxorhynchitini, and Uranotaeniini. There are 41 genera under Anophelinae and Culicinae, few of which are shown in Table 4.1 (Knight and Stone 1977; Harbach and Kitching 1998, 2016; Harbach 2013; Mullen and Durden 2019). In India, there are about 393 species known with 31 species linked with the transmission of the diseases (Bhattacharyya et al. 2014). Along with the general biology (Silver 2008; Mullen and Durden 2019;

Table 4.1 The genera of mosquitoes worldwide (based on the information in Harbach and Kitching 1998 and Mullen and Durden 2019; as per WRBU, USA)

Species number	Subfamily	Tribe	Genus
1	Anophelinae		*Anopheles* Meigen, 1818
2			*Bironella* Theobald, 1905
3			*Chagasia* Cruz, 1906
4	Culicinae	Aedeomyiini	*Aedeomyia* Theobald, 1901
5		Aedini	*Aedes* Meigen, 1818
6			*Armigeres* Theobald, 1901
7			*Eretmapodites* Theobald, 1901
8			*Haemagogus* Williston, 1896
9			*Heizmannia* Ludlow, 1905
10			*Opifex* Hutton, 1902
11			*Psorophora* Robineau-Desoidy, 1827
12			*Udaya* Thurman, 1954
13			*Verrallina* Theobald, 1903
14			*Zeugnomyia* Leicester, 1908
15		Culicini	*Culex* Linnaeus, 1758
16			*Galindomyia* Stone & Barreto, 1969
17			*Deinocerites* Theobald, 1901
18			*Lutzia* Theobald, 1903
19		Culisetini	*Culiseta* Felt, 1904
20		Ficalbiini	*Ficalbia* Theobald, 1903
21			*Mimomyia* Theobald, 1903
22		Hodgesini	*Hodgesia* Theobald, 1904
23		Mansoniini	*Coquillettidia* Dyar, 1905
24			*Mansonia* Blanchard, 1901
25		Orthopodomyiini	*Orthopodomyia* Theobald, 1904
26		Sabethini	*Isostomyia* Coquillett, 1906
27			*Johnbelkinia* Zavortink, 1979
28			*Limatus* Theobald, 1901
29			*Malaya* Leicester, 1908
30			*Maorigoeldia* Edwards, 1930
31			*Phoniomyia* Theobald, 1903
32			*Runchomyia* Theobald, 1903
33			*Sabethes* Robineau-Desvoidy, 1827
34			*Shannoniana* Lane & Cerqueira, 1942
35			*Topomyia* Leicester, 1908
36			*Trichoprosopon* Theobald, 1901
37			*Tripteroides* Giles, 1904
38			*Kimia* Vu Duc Huong & Harbach, 2007
39			*Wyeomyia* Theobald, 1901
40		Uranotaeniini	*Uranotaenia* Lynch Arribálzaga, 1891
41		Toxorhynchitini	*Toxorhynchites* Theobald, 1901

Becker et al. 2020), the taxonomy and the phylogeny of mosquitoes (Harbach 2007, 2013) supplement the identification (Wilkerson et al. 2021) and the management of vector (WHO 2023) and pest mosquitoes (Fig. 4.1).

Mosquito species richness is considerably high in a few countries like Brazil, Indonesia, Malaysia, and Thailand (Foley et al. 2007). In terms of endemic species of mosquito, Australia, the Philippines, and Indonesia host vast number of endemic species compared to other nations (Foley et al. 2007). On a comparative scale, the taxonomic resolutions of the mosquitoes were quite low when the countries from the Middle East Asia were considered. However, on a biogeographical scale, the Oriental realm and the Neotropical realm comprise 25 and 24 genera of mosquitoes compared to the Australian with 22 genera, the Palearctic realm with 19 genera, Afrotropical with 15 genera, and Nearctic with 24 genera (Rueda 2008). The diversity of the mosquitoes is considered to be highest in the Neotropical realm (~31% of

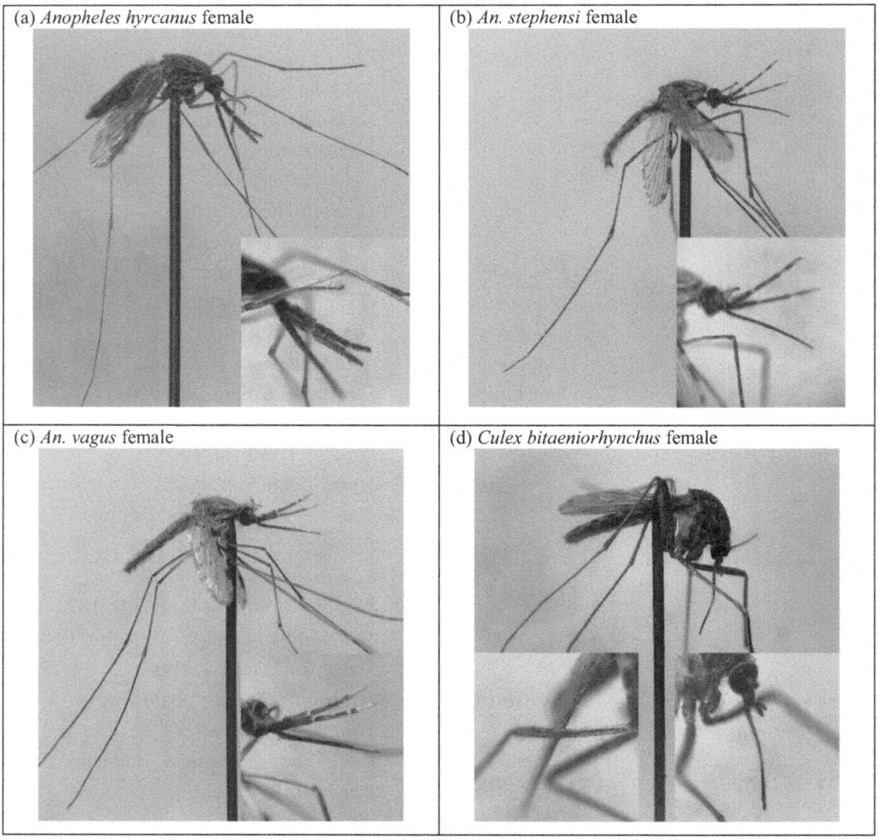

Fig. 4.1 Representing mosquitoes with their identify characters in the inset. (**a**) *Anopheles hyrcanus* female. (**b**) *An. stephensi* female. (**c**) *An. vagus* female. (**d**) *Culex bitaeniorhynchus* female. (**e**) *Cx. quinquefasciatus* female. (**f**) *Cx. tritaeniorhynchus* female. (**g**) *Aedes aegypti* female. (**h**) *Ae. albopictus* female. (**i**) *Lutzia fuscana* female. (**j**) *Toxorhynchites splendens* male

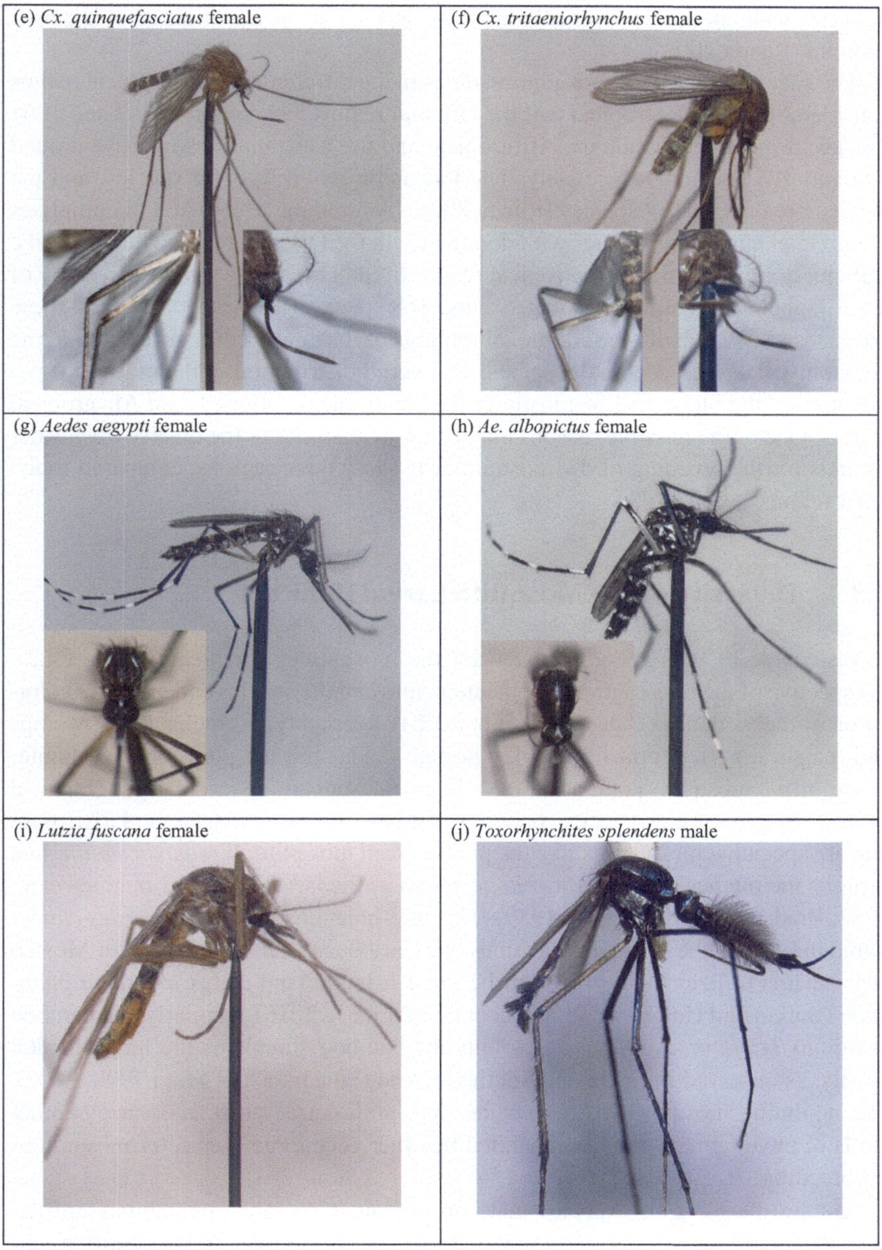

Fig. 4.1 (continued)

known species) followed by the Oriental realm (~30% of all known species) and Afrotropical and Australian realms with 22% of all the known species (Rueda 2008). In the Oriental realm, the tribe Aedini exhibits the highest number of species richness, while in the Neotropical realm, the species richness is highest for the tribe Culicini (Rueda 2008).

The species richness of mosquitoes varies in the different biogeographical realms considerably. The Neotropical and the Oriental realms have around 1069 and 1061 species, respectively, while the Afrotropical and the Australian realms have around 795 and 794 species, respectively. The Palearctic realm has 492 species, and the Nearctic realm has 178 species (Rueda 2008). Among the tribes, Aedini comprises of maximal number of species with dominance in the Oriental realm, while the tribe Culicini dominates in the Neotropical realm. In the Oriental realm, the diversity of the Anophelinae is maximum with around 168 species (Rueda 2008). While these estimates may vary with the additional findings of the new species or revision of the taxonomy of certain genera or species, the overall distribution of the mosquito species around the globe will be mostly in the Neotropical, Oriental, and Afrotropical realms. One of the possible reasons may be the availability of the diverse freshwater habitats for the breeding of the mosquitoes in these three regions, compared to the other realms.

4.3 Diversity of the Mosquito Larval Habitats

A vast range of habitats are utilized by the mosquitoes for breeding (Fig. 4.2). Irrespective of the biogeography of the mosquito and the species endemicity, exploitation of the wetlands (Dale and Knight 2008) of varied types, inclusive of swamps and marshes, agricultural fields, like the rice fields, and the pools enable maintenance of the mosquito population in different seasons in a particular geographical area. Although species-specific breeding habitats are observed particularly, those that are specialist mosquitoes like the pitcher plant mosquito, *Wyeomyia smithii* that exploits the pitcher plant *Sarracenia purpurea* as the larval habitat (Lounibos et al. 1982; Bradshaw and Holzapfel 1989). The crab-hole mosquito *Deinocerites cancer* found in Florida, the USA, the Bahamas, the Caribbean islands, and parts of Mexico and Panama utilizes the burrows of the great Atlantic land crab *Cardisoma guanhumi* (Belkin and Hogue 1959; Betancourt and Loaiza 2016). Similarly, the bamboo mosquito *Tripteroides bambusa* exploits the bamboo stumps as the larval habitat mostly, as observed in Japan and Southeast Asia (Sunahara and Mogi 1998, 2002). The nonbiting mosquito *Toxorhynchites* spp. prefers tree holes and various other kinds of phytotelmata that are inhabited by other vector and nonvector mosquitoes for breeding.

Among the mosquito species that can exploit wide range of habitats are *Ae. aegypti*, *Ae. albopictus*, *Cx. quinquefasciatus*, and *An. stephensi*. The ability to utilize wide range of resources by the larval stages enables these mosquitoes to breed in varied type of water bodies. As shown in Table 4.2, a wide range of habitats are exploited by the mosquitoes for breeding, which makes it difficult for vector management. Among the various factors that influence the selection of the breeding habitats are the physical environmental factors, including temperature, water quality and resource abundance, the presence of the predators, and the habitat

Fig. 4.2 Representing mosquito larval habitats (**a**) pool water with the macrophytes in the surrounding, (**b**) puddles formed by human footsteps, (**c**) leaf axil of banana plant, and (**d**) extended pond bank

Table 4.2 Diversity of the mosquito larval habitats exploited for breeding by the mosquitoes (adapted from Liehne (1991), Rueda (2008), and Wilkerson et al. (2021))

Habitat	Mosquitoes species	Remarks
(A) Stagnant/flowing natural water bodies		
1. Streams	*Culex fuscocephala*, *Cx. gelidus*, *Cx. annulus*, *Cx. bitaeniorhynchus*, *Anopheles kochi*, *Lutzia fuscana*	Rural, countryside
2. Lakes	*An. farauti*, *An. maculipennis*, *An. quadrimaculatus*, *An. pseudopunctipennis*, *Cx. annulirostris*, *Cx. squamosus*	Rural, urban, suburban, and even cold regions and high altitude

(continued)

Table 4.2 (continued)

Habitat	Mosquitoes species	Remarks
3. Swamps and marshes	*An. farauti*, *An. gambiae*, *An. kochi*, *An. punctulatus*, *An. sinensis*, *Cx. annulus*, *Cx. bitaeniorhynchus*, *Cx. gelidus*, *Cx. sitiens*, *Cx. tritaeniorhynchus*; *Lz. fuscana*	Rural, countryside, forests
4. Pools	*Aedes longirostris*, *An. kochi*, *An. sinensis*, *Cx. gelidus*, *Cx. tritaeniorhynchus*, *Mansonia uniformis*, *Mimomyia chamberlaini*	Cosmopolitan, both temporary and permanent, including rock pools in high-altitude
(B) Artificial water bodies: anthropogenic		
1. Rice fields	*An. vagus*, *An. subpictus*, *An. kochi*, *An. annularis*, *Cx. fuscocephala*, *Cx. gelidus*, *Cx. tritaeniorhynchus*, *Cx. bitaeniorhynchus*, *Cx. quinquefasciatus*, *Lz. fuscana*	Rural agricultural landscape
2. Puddles	*Cx. tritaeniorhynchus*, *Cx. gelidus*, *Cx. quinquefasciatus*, *An. subpictus*, *An. baimaii*	Rural agricultural landscape, urban gardens
3. Excavations: roadside ditches	*Cx. restuans*, *An. subpictus*	Urban
(C) Phytotelmata		
1. Bamboo stumps	*Ae. albopictus*, *Tripteroides bambusa*, *Ae. japonicus*, *Cx. kyotoensis*, *Uranotaenia novobscura*, *Topomyia yanbarensis*, *Armigeres subalbatus*, *Ochlerotatus japonicus*, *Cx. quinquefasciatus*	Cosmopolitan, including ornamental bamboo plants in urban areas
2. Tree hole	*Cx. bicornutus*, *Ae. japonicus*, *Ae. aureostriatus*, *Ae. riversi*, *Orthopodomyia anopheloides*, *Tr. bambusa*, *Ur. novobscura*, *Tx. manicatus*, *Wyeomyia smithii*	Cosmopolitan
3. Ornamental bromeliads	*Ae. aegypti*, *Wy. mitchellii*, *Wy. vanduzeei*, *Cx. quinquefasciatus*, *Cx. biscaynensis*	Cosmopolitan including the pineapple cultivated lands
4. Leaf axils, fruit shells, fallen spathe, fallen leaves	*Ae. albopictus*, *Cx. quinquefasciatus*, *Oc. notoscriptus*, *Ar. subalbatus*	Both urban and rural
5. Pitcher plant *Sarracenia purpurea*	*Wy. smithii*	North America
6. Pitcher plant *Nepenthes alata*	*Cx. navalis*, *Cx. curtipalpis*, *Tripteroides* spp.	Japan
(D) Built habitats		
1. Water storage	*An. stephensi*, *Aedes* sp.	Mostly urban
2. Septic tank	*Cx. quinquefasciatus*, *Cx. decens*, *Cx. cinereus*, *Cx. tigripes*, *Ae. aegypti*, *Ae. vittatus*, *Ar. subalbatus*	Urban human settlements in Mexico, India
3. Discarded tyres	*Cx. quinquefasciatus*, *Ar. subalbatus*, *Ae. aegypti*, *Ae. albopictus*, *An. stephensi*, *Tx. splendens*	Mostly urban

Table 4.2 (continued)

Habitat	Mosquitoes species	Remarks
4. Porcelain and glass containers	*Cx. quinquefasciatus*, *Ar. subalbatus*, *Ae. aegypti*, *Ae. albopictus*, *An. stephensi*, *Tx. splendens*	Mostly urban
5. Household waste containers	*Cx. quinquefasciatus*, *Ar. subalbatus*, *Ae. aegypti*, *Ae. albopictus*, *An. stephensi*, *Tx. splendens*	Mostly urban
(E) Water with high organic load		
1. Sewage water	*Cx. quinquefasciatus*, *Ar. subalbatus*	Mostly urban
2. Drains	*Cx. quinquefasciatus*, *Ar. subalbatus*	Both urban and rural
(F) Brackish water	*An. sundaicus*, *Cx. sitiens*, *An. culicifacies*	Coastal areas of Florida
(G) Crab holes	*Oc. baisasi*, *Deinocerites cancer*	Coastal region of central America

permanence. Irrespective of the size of the habitats, the time period for the larval development is a pivotal factor that should correspond with the habitat permanence. A shortage in the time of the habitat permanence will bear detrimental consequences for the mosquitoes. Apart from the habitat permanence, the resource quality and quantity determine the successful larval development of the mosquitoes. In addition, the presence of the predators may deter the mosquitoes from selection of a particular larval habitat.

4.4 Diversity of Mosquitoes: A Synoptic View

The studies on the diversity of the mosquitoes have been carried out from different geographical regions. A correspondence of prevalence of the mosquito-borne diseases and the species richness of the mosquitoes can obviously be deduced from the presentation shown in Table 4.3. In Indian context, around 393 species of mosquitoes are recorded (Bhattacharyya et al. 2014), of which selected species (fewer than 20) are responsible for the spread of the diseases. The vector mosquito species bear severe economic burden and public health concerns (Reddya Naik et al. 2023). Studies on the species richness of the mosquitoes enable highlighting the possibility of the mosquito-borne diseases, particularly in areas endemic for filariasis, malaria, dengue, and Japanese encephalitis. As presented in Table 4.3, the mosquito species diversity is higher from the tropical and subtropical regions indicating the correspondence between the mosquito species richness and disease prevalence. Recent researches however suggest more of the social, economic, and ecological factors are linked with the prevalence of the mosquitoes and mosquito-borne diseases (Rodríguez-González et al. 2024) in the urban areas (Rodríguez-González et al. 2023).

Table 4.3 Summary of the selected works representing the mosquito distribution and the diversity world wide with emphasis on the geographical range, habitat, and species composition

Place and Country	Habitat	Mosquito genera	Species	Citation
Singapore	Freshwater ponds, ponded streams	7 genera, 22 species	*Anopheles*-3; *Uranotaenia*-2; *Ficalbia*-1; *Taeniorhynchus*-1; *Aedes*-3; *Armigeres*-1; *Culex*-11	Laird (1959)
Southeastern United States	Mix habitat	11 genera, 77 species	*Aedes*-24; *Anopheles*-11; *Culex*- 15; *Culiseta*-2; *Deinocerites*-1; *Mansonia*-3; *Orthopodomyia*-2; *Psorophora*-12; *Toxorhynchites*-2; *Uranotaenia*-2; *Wyeomyia*-3	King et al. (1960)
Colorado, USA	Different altitudinal region	3 genera, 21 species	*Aedes*-15; *Culex*-3; *Culiseta*-3	Baker (1961)
Uganda	Forest	4 genera, 12 species	*Toxorhynchites*-2; *Eretmapodites*-1; *Aedes*-6; *Culex*-3	Corbet (1964)
South Africa	Sewage effluent disposal area	2 genera, 7 species	*Culex*-3; *Aedes*-4	Jupp (1967)
Samoa and Tonga Island	Island	4 genera, 13 species	*Aedes*-9; *Culex*-4; *Coquillettidia*-1; *Toxorhynchites*-1	Ramalingam (1976)
St. Joseph County, Indiana, USA	Mix habitat	8 genera, 35 species	*Aedes*-13; *Anopheles*-5; *Coquillettidia*-1; *Culex*-6; *Culiseta*-4; *Orthopodomyia*-2; *Psorophora*-3; *Uranotaenia*-1	Shroyer et al. (1976)
Louisiana, USA	Rice field	5 genera, 9 species	*Psorophora*-3; *Anopheles*-2; *Uranotaenia*-1; *Culex*-2; *Aedes*-1	Chambers et al. (1979)
Peradeniya, Sri Lanka	Bamboo traps	3 genera, 7 species	*Aedes*-5; *Armigeres*-1; *Culex*-1	Amerasinghe and Alagoda (1984)
Central Nigeria	Septic tank	2 genera, 6 species	*Culex*-5; *Aedes*-1	Irving-Bell et al. (1987)
Nagasaki, Japan	Forest foothill	5 genera, 7 species	*Aedes*-2; *Tripteroides*-1; *Armigeres*-1; *Orthopodomyia*-1; *Culex*-2	Iriarte et al. (1991)
Nepal	Artificial containers, ditches, and different phytotelmata	3 genera, 14 species	*Aedeomyia*-1; *Aedes*-12; *Tripteroides*-1	Darsie Jr. et al. (1993)
Murray Valley, southeast Australia	Residential areas	4 genera, 20 species	*Culex*-8; *Aedes*-10; *Anopheles*-1; *Coquillettidia*-1	Dhileepan (1996)

Table 4.3 (continued)

Place and Country	Habitat	Mosquito genera	Species	Citation
Southeastern Tanzania	Rural house	1 genus, 12 species	*Anopheles*-12	Charlwood 1997
Darjeeling Himalaya, India	Cattle sheds	1 genus, 9 species	*Anopheles*-9	Saha et al. (2000)
Mwea, Kenya	Riceland agroecosystem	5 genera, 25 species	*Anopheles*-8; *Aedes*-9; *Culex*-5; *Coquillettidia*-2; *Mansonia*-1	Muturi et al. (2006)
Darjeeling Himalayas, India	Aquatic bodies of Darjeeling town	4 genera, 6 species	*Aedes*-1; *Armigeres*-2; *Culex*-2; *Toxorhynchites*-1	Aditya et al. (2006)
West Bengal, India	Rural and urban	1 genus, 5 species	*Anopheles*-5	Pramanik et al. (2006)
Mwea, Kenya	Riceland agroecosystem	6 genera, 25 species	*Anopheles*-9; *Aedes*-7; *Coquillettidia*-1; *Culex*-5; *Ficalbia*-1; *Mansonia*-2	Muturi et al. (2008)
Happy Valley Tea Garden, Darjeeling, India	Bamboo stumps	2 genera, 2 species	*Aedes*-1; *Culex*-1	Aditya et al. (2008)
Northern Adelaide region of south Australia	Coastal suburbs	3 genera, 8 species	*Aedes*-3; *Anopheles*-1; *Culex*-4	Williams et al. (2009)
Kolkata, India	Urban aquatic habitat	5 genera, 9 species	*Aedes*-1; *Anopheles*-3; *Culex*-3; *Toxorhynchites*-1; *Lutzia*-1	Banerjee et al. (2010)
Southwest Chicago, USA	Mix habitat	8 genera, 19 species	*Aedes*-5; *Anopheles*-4; *Coquillettidia*-1; *Culiseta*-1; *Culex*-4; *Orthopodomyia*-1; *Psorophora*-2; *Uranotaenia*-1	Chaves et al. (2011)
Malindi, Kenya	Urban, periurban, and rural	5 genera, 16 species	*Anopheles*-4; *Culex*-5; *Aedes*-4; *Mansonia*-2; *Coquillettidia*-1	Mwangangi et al. (2012)
Kerala, India	Rubber plantation	6 genera, 12 species	*Aedes*-6; *Armigeres*-1; *Culex*-2; *Heizmannia*-1; *Toxorhynchites*-1; *Verrallina*-1	Sumodan (2012)
Kenya	Human settlement	11 genera, 101 species	*Aedes*-35; *Culex*-27; *Anopheles*-15; *Coquilettidia*-7; *Eretmapodites*-4; *Aedomyia*-3; *Ficalbia*-3; *Mansonia*-2; *Harpagomyia*-1; *Theobaldia*-1; *Uranotaenia*-1	Lutomiah et al. (2013)

(continued)

Table 4.3 (continued)

Place and Country	Habitat	Mosquito genera	Species	Citation
Central Thailand	Mix habitat	5 genera, 10 species	*Aedes*-2; *Culex*-5; *Armigeres*-1; *Anopheles*-1; *Mansonia*-1	Thongsripong et al. (2013)
Mexico	Temporary and permanent aquatic sites of the Pacific Coast	15 genera, 74 species	*Aedes*-9; *Aedeomyia*-1; *Coquillettidia*-2; *Culex*-35; *Culiseta*-1; *Deinocerites*-4; *Haemagogus*-2; *Limatus*-1; *Mansonia*-5; *Ochlerotatus*-16; *Psorophora*-13; *Sabethes*-1; *Uranotaenia*-7; *Wyeomyia*-5; *Anopheles*-19	Bond et al. (2014)
Hanoi, North Vietnam	Rice field	4 genera, 13 species	*Culex*-6; *Anopheles*-5; *Mansonia*-1; *Mimomyia*-1	Ohba et al. (2015)
Mizoram, Northeastern Himalayan region, India	Rural	5 genera, 20 species	*Culex*-6; *Anopheles*-9; *Aedes*-2; *Culiseta*-2; *Toxorhynchites*-1	Vanlalruia et al. (2014)
Veracruz, Mexico	Forest, mainly rural	8 genera, 10 species	*Aedes*-1; *Anopheles*-1; *Culex*-1; *Wyeomyia*-2; *Uranotaenia*-1; *Coquillettidia*-1; *Mansonia*-1; *Sabethes*-1	Abella-Medrano et al. (2015)
Nagpur, Maharashtra, India	Urban and rural	5 genera, 19 species	*Aedes*-3; *Anopheles*-8; *Culex*-5; *Armigeres*-2; *Mansonia*-2	Karlekar and Andrew (2015)
Republic of Moldova	Urban, rural, natural habitat	5 genera, 20 species	*Anopheles*-4; *Aedes*-5; *Uranotaenia*-1; *Culiseta*-2; *Culex*-6	Sulesco et al. (2015)
Puerto Iguazú, Misiones, Argentina	Urban, periurban and wild	1 genus, 6 species	*Anopheles*-6	Ramirez et al. (2016)
North Carolina, USA	Forest and grassland	8 genera, 23 species	*Aedes*-2; *Anopheles*-3; *Coquillettidia*-1; *Culex*-3; *Ochlerotatus*-7; *Psorophora*-5; *Stegomyia*-1; *Uranotaenia*-1	Reiskind et al. (2017)
West Bengal, India	Rice field	2 genera, 5 species	*Culex*-2; *Anopheles*-3	Roy et al. (2016)
Sweden, Italy, and Netherlands, Europe	Farm, periurban and wetland habitats	6 genera, 40 species	*Aedes*-19; *Anopheles*-5; *Coquillettidia*-1; *Culex*-7; *Culiseta*-7; *Uranotaenia*-1	Möhlmann et al. (2017)
Sri Lanka	Ornithophilic mosquitoes	8 genera, 25 species	*Culex*-13; *Coquillettidia*-1; *Aedes*-4; *Anopheles*-1; *Mansonia*-2; *Armigeres*-1; *Orthopodomyia*-2; *Aedeomyia*-1	Chathuranga et al. (2018)

Table 4.3 (continued)

Place and Country	Habitat	Mosquito genera	Species	Citation
Costa Rica	Tropical rainforest	7 genera, 9 species	*Aedes*-1; *Culex*-2; *Limatus*-2; *Sabethes*-1; *Toxorhynchites*-1; *Trichoprosopon*-1; *Wyeomyia*-1	Chaverri et al. (2018)
Cruzeiro do Sul, Brazil	Urban	1 genus, 9 species	*Anopheles*-9	Martins et al. (2018)
Ondo State, Nigeria	Urban	4 genera, 13 species	*Aedes*-5; *Anopheles*-1; *Culex*-6; *Toxorhynchites*-1	Afolabi et al. (2019)
United Arab Emirates	Artificial wetlands, periurban and protected natural habitat	4 genera, 11 species	*Aedes*-1; *Anopheles*-2; *Culex*-7; *Culiseta*-1	Camp et al. (2019)
Republic of Kosovo, southeastern Europe	Urban	6 genera, 13 species	*Anopheles*-4; *Aedes*-3; *Culex*-1; *Culiseta*-2; *Coquillettidia*-2; *Uranotaenia*-1	Muja-Bajraktari et al. (2019)
Baringo County, Kenya	Rural aquatic habitat	4 genera, 17 species	*Anopheles*-5; *Culex*-8; *Aedes*-3; *Mansonia*-1	Ondiba et al. (2019)
Bure district, northwestern Ethiopia	Rural	1 genus, 9 species	*Anopheles*-9	Adugna et al. (2020)
State of Amazonas, Brazil	Rio Pardo rural settlement	10 genera, 20 species	*Aedes*-1; *Culex*-4; *Haemagogus*-1; *Limatus*-2; *Ochlerotatus*-1; *Orthopodomyia*-1; *Sabethes*-7; *Toxorhynchites*-1; *Trichoprosopon*-1; *Wyeomyia*-1	Almeida et al. (2020)
West Azerbaijan Province, northwestern Iran	Forest	4 genera, 7 species	*Culex*-4; *Culiseta*-1; *Anopheles*-1; *Aedes*-1	Amini et al. (2020)
Kandal province, Cambodia	Rural and periurban	10 genera, 30 species	*Aedes*-3; *Anopheles*-5; *Armigeres*-3; *Coquillettidia*-1; *Culex*-8; *Ficalbia*-2; *Lutzia*-3; *Mansonia*-3; *Mimomyia*-1; *Uranotaenia*-1	Boyer et al. (2020)

(continued)

Table 4.3 (continued)

Place and Country	Habitat	Mosquito genera	Species	Citation
Rio de Janeiro, Brazil	Urban, periurban and rural	10 genera, 39 species	*Aedes*-7; *Anopheles*-3; *Uranotaenia*-4; *Coquillettidia*-2; *Culex*-6; *Limatus*-3; *Wyeomyia*-5; *Mansonia*-3; *Haemagogus*-1; *Psorophora*-4	Câmara et al. (2020)
Western Burkina Faso	Rural	4 genera, 10 species	*Anopheles*-7; *Culex*-1; *Aedes*-1; *Mansonia*-1	Epopa et al. (2020)
Lagos State, Nigeria	Urban	3 genera, 5 species	*Anopheles*-2; *Culex*-1; *Aedes*-2	Fagbohun et al. (2020)
Bajo Cauca region, Colombia	Rural human localities	1 genus, 10 species	*Anopheles*-10	Hernández-Valencia et al. (2020)
South Africa	Nature reserves, horse farms	10 genera, 94 species	*Aedeomyia*-1; *Aedes*-35; *Anopheles*-15; *Culex*-25; *Coquillettidia*-4; *Mansonia*-3; *Ficalbia*-3; *Culiseta*-1; *Mimomyia*-6; *Uranotaenia*-1	Johnson et al. (2020)
Hainan Island, China	Mix habitat	5 genera, 23 species	*Aedes*-4; *Armigeres*-1; *Culex*-11; *Mansonia*-1; *Anopheles*-6	Li et al. (2020a)
Hainan Island, China	Urban, suburban and rural	4 genera, 9 species	*Anopheles*-4; *Aedes*-1; *Culex*-3; *Armigeres*-1	Li et al. (2020b)
West region of Cameroon	Urban, periurban and rural	8 genera, 37 species	*Aedes*-14; *Anopheles*-1; *Culex*-12; *Coquillettidia*-3; *Eretmapodites*-3; *Lutzia*-1; *Mimomyia*-1; *Uranotaenia*-2	Mayi et al. (2020)
Taita-Taveta County, Kenya	Rural	4 genera, 18 species	*Aedes*-7; *Culex*-8; *Coquillettidia*-1; *Mansonia*-2	Munyao et al. (2020)
Amazon rainforest, Brazil	Rural settlements	17 genera, 137 species	*Aedeomyia*-1; *Aedes*-11; *Psorophora*-8; *Anopheles*-23; *Chagasia*-2; *Kerteszia*-1; *Nyssorhynchus*-25; *Stethomyia*-1; *Culex*-37; *Lutzia*-1; *Coquillettidia*-8; *Mansonia*-7; *Johnbelkinia*-1; *Limatus*-1; *Trichoprosopon*-2; *Wyeomyia*-1; *Uranotaenia*-9	Chaves et al. (2021)
Cambodia	Rural	6 genera, 25 species	*Anopheles*-9; *Culex*-12; *Ficalbia*-1; *Mansonia*-4; *Mimomyia*-2; *Uranotaenia*-1	Maquart et al. (2021)

Table 4.3 (continued)

Place and Country	Habitat	Mosquito genera	Species	Citation
Southeastern Mexico	Deciduous forest	6 genera, 22 species	*Anopheles*-1; *Aedes*-8; *Culex*-7; *Culiseta*-1; *Psorophora*-3; *Uranotaenia*-2	Barrientos-Roldán et al. (2022)
Baltimore City, Maryland, USA	Mix habitat	4 genera, 11 species	*Aedes*-5; *Culex*-4; *Orthopodomyia*-1; *Toxorhynchites*-1	Zettle et al. (2022)
Cambodia	Forest	13 genera, 85 species	*Aedeomyia*-1; *Aedes*-17; *Anopheles*-11; *Armigeres*-10; *Coquillettidia*-1; *Culex*-16; *Heizmannia*-5; *Lutzia*-1; *Mansonia*-3; *Mimomyia*-3; *Tripteroides*-3; *Toxorhynchites*-1; *Uranotaenia*-13	Rakotonirina et al. (2023)
Southwestern Ethiopia	Rural	1 genus, 14 species	*Anopheles*-14	Assa et al. (2023)
Western regions, Burkina Faso	Urban	5 genera, 33 species	*Aedes*-12; *Anopheles*-9; *Culex*-9; *Mansonia*-2; *Ficalbia*-1	Kaboré et al. (2023)
Northwestern Spain	Aquatic habitat	5 genera, 10 species	*Aedes*-1; *Anopheles*-1; *Coquillettidia*-1; *Culex*-4; *Culiseta*-3	Martínez-Barciela et al. (2024)

4.5 Diversity of Mosquitoes and Mosquito- Borne Diseases

Mosquitoes transmit a wide variety of pathogens that include viruses, bacteria, and parasites which establish relationships that, in certain cases, may be specific but not exclusive between the vector and the pathogen (WHO 2024). However, with more than 3500 species described worldwide (Wang and Zhu 2023; Wilkerson et al. 2021), only 150 species bear significance as disease vectors (Becker et al. 2020). Considering the transmission of the disease in the host, the abundance of the vector mosquitoes, the infected hosts, and the environmental conditions are crucial factors. In comparison to the vector mosquitoes, the nonvector mosquitoes may require a different set of conditions for completion of their life history. As an extension, it may be possible that the environmental and the social factors that influence the population of the vector and nonvector mosquitoes are different (Rodríguez-González et al. 2023, 2024). Addressing this issue may help to frame strategies for the regulation of the mosquitoes with higher precision retaining the ecological roles of the nonvector mosquitoes (Rodríguez-González et al. 2023, 2024). Apparently, the nonvector mosquitoes are considerably larger in terms of species richness, and only about 10% of the described species are responsible for the transmission of disease pathogens and parasites (Kain et al. 2022; Yee et al. 2022; Soghigian et al. 2023). On a world wide scale, there are about 500 *Anopheles* species of which around 30–40 species are linked with the disease transmission (Nicoletti 2020); others, even if depending on human blood meal, are not capable of hosting the malarial parasites. In North Africa, 83 mosquito species are recorded, 59 from Culicinae and 24 under Anophelinae, of which only 26 mosquitoes are incriminated as vectors of various diseases (Nebbak et al. 2022). In India, around 31 species are considered as the vectors out of the 393 species of mosquitoes recorded (Bhattacharyya et al. 2014). Despite the low number of the vector species, the impact on the disease transmission remains quite high, particularly due to the generalist nature of both the parasites/pathogens and the mosquito vectors. In order to highlight the significance of the vector mosquitoes in disease transmission, ecological network models are used with the vector mosquitoes, parasites/pathogens as the key elements (Kain et al. 2022; Yee et al. 2022). An overview of the pathogen-mosquito network enables identification of the crucial links that can be targeted to reduce the menace of mosquito-borne diseases with higher effectiveness (Kain et al. 2022; Yee et al. 2022). The ecological network analysis (Yee et al. 2022) reflected that around 331 out of 3578 mosquito species can be considered relevant to the human diseases. Among various disease pathogens transmitted by mosquitoes, vast majority are arboviruses (belonging to the family Bunyaviridae, Flaviviridae, Togaviridae, Reoviridae, Picornaviridae), while the rest are helminth parasites (Onchocercidae, Filariidae) and the protozoan *Plasmodium* spp. (Yee et al. 2022). A correspondence of these disease pathogens and the vector mosquitoes was elaborated through the ecological network, which indicated that the three species, *Culex quinquefasciatus*, *Aedes aegypti*, and *Ae. albopictus*, are generalist in nature, with worldwide distribution and linkage to multiple disease-causing pathogens and parasites (Yee et al. 2022). A similar observation was made through the ecological network analysis involving viruses and the mosquito vectors in Australia (Kain et al. 2022). An appraisal of the different disease pathogens transmitted by the mosquitoes is shown in Table 4.4, emphasizing the major genera of mosquito as the vectors. The generalist nature of both the pathogen/parasite and the vector mosquitoes is observed, which would mean that the parasites/pathogens can opt for more than one mosquito for thriving. Similarly, for a mosquito hosting, more than one parasite/pathogen species would augment its vector potential. As a consequence, the population regulation of the vectors as well as the parasites/pathogens becomes more critical in reducing the risks of epidemics with multiple disease infestations. In addition, a change in the sylvatic or feral cycle of the parasite can induce a newer intensity of affecting the hosts. Nonetheless, at the local scale as well as at the regional scale, the mosquito disease transmission pattern may differ based on the environmental factors and the mosquito species and the pathogen strains.

Table 4.4 An appraisal of the different disease pathogens transmitted by the mosquitoes

Genus	Species	Type of pathogen	References
Aedes	*albopictus*	Chikungunya virus, Dengue virus, Yellow fever virus, ZIKA virus	Reiter et al. (2006), Chouin-Carneiro et al. (2016), Ferreira-de-Lima and Lima-Camara (2018), Kamgang et al. (2019)
	aegypti	Chikungunya virus, Dengue virus, Yellow fever virus, ZIKA virus	Black IV et al. (2002), Morrison et al. (2008); Chouin-Carneiro et al. (2016), Ferreira-de-Lima and Lima-Camara (2018), Kamgang et al. (2019)
	serratus	Yellow fever virus	Cardoso et al. (2010)
	cantans	Rhabdovirus, Ohlsdorf virus	Shahhosseini et al. (2017)
	caspius	ZIKA virus, West Nile virus	Ben Ayed et al. (2019)
	tahoensis	Jamestown Canyon virus	Hardy et al. (1993)
	hexodontus	Jamestown Canyon virus	Hardy et al. (1993)
	cataphylla	Jamestown Canyon virus	Hardy et al. (1993)
	cinereus	West Nile virus	Medlock et al. (2005)
	communis	Jamestown Canyon virus	Boromisa and Grayson (1990)
	detritus	ZIKA virus, West Nile virus	Ben Ayed et al. (2019)
	geniculatus	Chikungunya virus, *Dirofilaria* filarial nematode	Silaghi et al. (2017), Prudhomme et al. (2019)
	scapularis	Arbovirus, *Dirofilaria* filarial nematode	Petersen et al. (2022)
	desmotes	*Wuchereria bancrofti* filarial nematode	Pothikasikorn et al. (2008)
	fulvidorsalis	Tahyna virus	Cao et al. (2011)
	japonicus	Chikungunya virus, *Dirofilaria* filarial nematode	Silaghi et al. (2017), Jansen et al. (2018)
	provocans	Jamestown Canyon virus	Boromisa and Grayson (1990)
	intrudens	Jamestown Canyon virus	Boromisa and Grayson (1990)
	punctor	Jamestown Canyon virus	Boromisa and Grayson (1990)
	taeniorhynchus	Rift Valley fever virus, Venezuelan equine encephalitis virus	Turell et al. (1992)
	sollicitans	Venezuelan equine encephalitis virus	Turell et al. (1992)

(continued)

Table 4.4 (continued)

Genus	Species	Type of pathogen	References
Culex	*annulus*	Japanese encephalitis virus	Mitchell and Chen (1973)
	bitaeniorhynchus	Japanese encephalitis virus	Kanojia (2007)
	fuscocephala	Japanese encephalitis virus, Getah IV virus	Mitchell and Chen (1973), Sam et al. (2022)
	gelidus	Japanese encephalitis virus, Getah IV virus, West Nile virus	Simpson et al. (1975), Gajanana et al. (1997), Sudeep (2014)
	modestus	West Nile virus	Golding et al. (2012)
	pallidothorax	*Wuchereria bancrofti* filarial nematode	Hu (1940)
	vishnui	Japanese encephalitis virus	Reuben et al. (1988), Sunish et al. (2006)
	pseudovishnui	Japanese encephalitis virus, Kunjin virus	Bowen et al. (1970), Reuben et al. (1988)
	quinquefasciatus	Japanese encephalitis virus, West Nile virus, ZIKA virus, *Wuchereria bancrofti* filarial nematode	Ramaiah and Das (1992), Triteeraprapab et al. (2000), Kanojia (2007), Kent et al. (2010), Guedes et al. (2017)
	sitiens	Japanese encephalitis virus	Vythilingam et al. (2002)
	theileri	*Dirofilaria* filarial nematode	de la Puente et al. (2014)
	tritaeniorhynchus	Japanese encephalitis virus, Rift Valley fever virus	Mitchell and Chen (1973), Eifan et al. (2021)
	whitmorei	West Nile virus	Khan et al. (2017)
	pipiens	*Wuchereria bancrofti* filarial nematode, *Dirofilaria* nematode, Flavivirus, West Nile virus	Rajagopalan et al. (1977), Saiyasombat et al. (2011), Hamer et al. (2014), Bravo-Barriga et al. (2016)
	univittatus	West Nile virus	Miller et al. (2000)
	poicilipes	Rift Valley fever virus	Biteye et al. (2018)
	annulus	Japanese encephalitis virus	Cates and Detels (1969)
	nigripalpus	West Nile virus	Mores et al. (2007)
	erraticus	Eastern equine encephalitis virus	Estep et al. (2010)
	coronator	West Nile virus	Alto et al. (2014)
	perexiguus	Bagaza virus, Barkedji virus	Camp et al. (2019)
	tarsalis	West Nile virus	Reisen et al. (2006)

Table 4.4 (continued)

Genus	Species	Type of pathogen	References
Anopheles	*coluzzii*	*Plasmodium* spp. Protozoa	WHO (2023)
	hispaniola	*Plasmodium* spp. Protozoa	WHO (2023)
	labranchiae	*Plasmodium* spp. Protozoa	WHO (2023)
	melas	*Plasmodium* spp. Protozoa	WHO (2023)
	moucheti	*Plasmodium* spp. Protozoa	WHO (2023)
	multicolor	*Plasmodium* spp. Protozoa	WHO (2023)
	nili	*Plasmodium* spp. Protozoa	WHO (2023)
	pharoensis	*Plasmodium* spp. Protozoa, Bangui virus	Gordon et al. (1992), WHO (2023)
	arabiensis	*Plasmodium* spp. Protozoa, *Wuchereria bancrofti* filarial nematode	Bryan et al. (1990), WHO (2023)
	culicifacies	*Plasmodium* spp. Protozoa	WHO (2023)
	d'thali	*Plasmodium* spp. Protozoa	WHO (2023)
	fluviatilis	*Plasmodium* spp. Protozoa	WHO (2023)
	funestus	*Plasmodium* spp. Protozoa, Bwamba virus, *Wuchereria bancrofti* filarial nematode	Bryan et al. (1990), Lutwama et al. (2002), WHO (2023)
	gambiae	*Plasmodium* spp. Protozoa, Anopheles C virus, Anopheles cypovirus, Anopheles gambiae densovirus, Anopheles gambiae flavivirus, Anopheles totivirus, Ilesha virus, *Wuchereria bancrofti* filarial nematode	Digoutte et al. (1980), Bryan et al. (1990), Ren et al. (2008), Carissimo et al. (2016), Fauver et al. (2016), Villinger et al. (2017), WHO (2023)
	hyrcanus	*Plasmodium* spp. Protozoa	WHO (2023)
	maculipennis	*Plasmodium* spp. Protozoa, West Nile virus, Anopheles-associated C virus, Batai virus	Cook et al. (2013), Huhtamo et al. (2013), Mancini et al. (2017), WHO (2023)
	pulcherrimus	*Plasmodium* spp. Protozoa	WHO (2023)
	rhodesiensis	*Plasmodium* spp. Protozoa	WHO (2023)
	sacharovi	*Plasmodium* spp. Protozoa	WHO (2023)
	sergentii	*Plasmodium* spp. Protozoa	WHO (2023)
	stephensi	*Plasmodium* spp. Protozoa, Cypovirus	Bird et al. (1972), WHO (2023)
	superpictus	*Plasmodium* spp. Protozoa	WHO (2023)
	leesoni	*Plasmodium* spp. Protozoa	WHO (2023)
	rivulorum	*Plasmodium* spp. Protozoa	WHO (2023)
	vaneedeni	*Plasmodium* spp. Protozoa	WHO (2023)
	albimanus	*Plasmodium* spp. Protozoa	WHO (2023)
	albitarsis	*Plasmodium* spp. Protozoa, Western equine encephalitis virus	Mitchell et al. (1987), WHO (2023)
	dirus	*Plasmodium* spp. Protozoa, *Wuchereria bancrofti* filarial nematode	Pothikasikorn et al. (2008), WHO (2023)

(continued)

Table 4.4 (continued)

Genus	Species	Type of pathogen	References
Anopheles	*aquasalis*	*Plasmodium* spp. Protozoa	WHO (2023)
	argyritarsis	*Plasmodium* spp. Protozoa	WHO (2023)
	braziliensis	*Plasmodium* spp. Protozoa	WHO (2023)
	cruzii	*Plasmodium* spp. Protozoa	WHO (2023)
	darlingi	*Plasmodium* spp. Protozoa	WHO (2023)
	neivai	*Plasmodium* spp. Protozoa	WHO (2023)
	nuneztovari	*Plasmodium* spp. Protozoa	WHO (2023)
	pseudopunctipennis	*Plasmodium* spp. Protozoa	WHO (2023)
	punctimacula	*Plasmodium* spp. Protozoa	WHO (2023)
	annularis	*Plasmodium* spp. Protozoa	WHO (2023)
	balabacensis	*Plasmodium* spp. Protozoa	WHO (2023)
	barbirostris	*Plasmodium* spp. Protozoa	WHO (2023)
	hinesorum	Anopheles hinesorum orbivirus, Australian Anopheles totivirus	Colmant et al. (2017a)
	farauti	*Plasmodium* spp. Protozoa	WHO (2023)
	lesteri	*Plasmodium* spp. Protozoa	WHO (2023)
	maculatus	*Plasmodium* spp. Protozoa, *Wuchereria bancrofti* filarial nematode	Pothikasikorn et al. (2008), WHO (2023)
	minimus	*Plasmodium* spp. Protozoa, *Wuchereria bancrofti* filarial nematode, Anopheles minimus virus	Trung et al. (2004), Pothikasikorn et al. (2008), WHO (2023)
	peditaeniatus	*Plasmodium* spp. Protozoa, Japanese encephalitis virus	Mourya et al. (1989), WHO (2023)
	philippinensis	*Plasmodium* spp. Protozoa	WHO (2023)
	pseudowillmori	*Plasmodium* spp. Protozoa	WHO (2023)
	punctulatus	*Plasmodium* spp. Protozoa	WHO (2023)
	sinensis	*Plasmodium* spp. Protozoa, Japanese encephalitis virus	Kuwata et al. (2015), WHO (2023)
	sundaicus	*Plasmodium* spp. Protozoa	WHO (2023)
	tessellatus	*Plasmodium* spp. Protozoa	WHO (2023)
	vagus	*Plasmodium* spp. Protozoa	WHO (2023)
	varuna	*Plasmodium* spp. Protozoa	WHO (2023)
	yatsushiroensis	*Plasmodium* spp. Protozoa	WHO (2023)
	anthropophagus	*Plasmodium* spp. Protozoa	WHO (2023)
	donaldi	*Plasmodium* spp. Protozoa	WHO (2023)
	epiroticus	*Plasmodium* spp. Protozoa	WHO (2023)
	flavirostris	*Plasmodium* spp. Protozoa	WHO (2023)
	jeyporiensis	*Plasmodium* spp. Protozoa	WHO (2023)
	koliensis	*Plasmodium* spp. Protozoa	WHO (2023)
	litoralis	*Plasmodium* spp. Protozoa	WHO (2023)
	mangyanus	*Plasmodium* spp. Protozoa	WHO (2023)
	quadrimaculatus	Cache Valley virus, Coot Bay virus	Blackmore et al. (1998), Contreras et al. (2017)

Table 4.4 (continued)

Genus	Species	Type of pathogen	References
Anopheles	*squamosus*	Rift Valley fever virus, Anopheles squamosus flavivirus	Ratovonjato et al. (2011), Villinger et al. (2017)
	coustani	Rift Valley fever virus, Eilat virus, Wesselsbron virus	Samina et al. (1986), Ratovonjato et al. (2011), Villinger et al. (2017)
	hackeri	*Plasmodium* spp. Protozoa	Antinori et al. (2013)
	boliviensis	Anopheles A virus, Anopheles B virus	Reagan et al. (1953)
	annulipes	Anopheles annulipes orbivirus, Australian Anopheles totivirus, Haslams Creek virus, Tilligerry virus	Belaganahalli et al. (2012), Colmant et al. (2017a, b)
	bancrofti	Dairy swamp virus	Colmant et al. (2017b)
	tesselatus	Kampung Karu virus	Guzman et al. (2018)
	meraukensis	Karumba virus, Leanyer virus	Stuckly and Wright (1983), Colmant et al. (2017b)
	crucians	Long Pine key virus	Guzman et al. (2018)
	pauliani	West Nile virus	Maquart et al. (2016)

Mosquito name	Types of pathogen	References
Mansonia sp.	Lymphatic filarial nematode; Venezuelan equine encephalitis virus; West Nile virus; St. Louis encephalitis virus	Sudia et al. (1971), Unlu et al. (2010), Ughasi et al. (2012), Beranek et al. (2018)
Aedeomyia sp.	Gamboa virus, Venezuelan equine encephalitis virus	Mitchell et al. (1985), Dutary et al. (1989)
Armigeres sp.	Brugian filariasis nematode, Japanese encephalitis virus, ZIKA virus, *Dirofilaria* nematode, Bancroftian filariasis nematode	Iyangar (1938), Edeson et al. (1960), Cheong et al. (1981), Chen et al. (2000), Lee et al. (2007), Yang et al. (2022)
Uranotaenia sp.	Eastern equine encephalitis virus, West Nile virus	Cupp et al. (2003), Andreadis et al. (2004), Armstrong and Andreadis (2010)
Eretmapodites sp.	Rift Valley fever virus	Cêtre-Sossah et al. (2023)
Haemagogus sp.	Yellow fever virus	Cardoso et al. (2010)
Psorophora sp.	Rift valley fever virus	Turell et al. (2015)
Culiseta sp.	West Nile virus	Andreadis et al. (2001)
Coquillettidia sp.	West Nile virus, Cache Valley virus	Blackmore et al. (1998), Sardelis et al. (2001)
Ochlerotatus sp.	Marisma virus	Vázquez et al. (2012)

4.6 Conclusion

Mosquitoes (Diptera: Culicidae) are common in all the geographical regions of the world with more than 3500 species recorded globally. Mosquitoes are mostly recognized as disease vectors, with about 100 species from 11 different genera are responsible for the disease transmission in vertebrates. Mosquitoes are classified under the family Culicidae, comprising Anophelinae and Culicinae as the 2 subfamilies with 11 tribes under Culicinae. The tribe Aedini is most speciose followed by the tribe Culicini and Sabethini. Around 500 species under three genera represent the subfamily Anophelinae. In almost all regions of the world, the dominance of the mosquitoes varies with the available breeding habitats. Mosquitoes exploit diverse types of water bodies for breeding and larval development that may include temporary pools, flowing streams, and rice fields, as the bigger larval habitats and the containers as the smaller larval habitats. Diverse types of phytotelmata are also exploited by the mosquitoes for breeding. Around the globe, considerable number of research work documents the species diversity of the mosquitoes, with the Neotropical, Oriental, and the Afrotropical realms hosting significant number of species per genera. Species under the tribe Aedini are observed in diverse number in the Oriental realm, while the species under the tribe Culicini are diverse in the Neotropical realm. A comparative assessment reflects huge amount of work on the diversity of the mosquitoes in the background of different habitat conditions. However, the diversity of mosquitoes is quite high in the tropics and the subtropics compared to other regions of the world. Among the disease vectors, the genera *Anopheles*, *Aedes*, *Culex*, and *Mansonia* are relevant in transmitting more pathogens and parasites than the rest of species. The factors like the environmental factors and the strains of the pathogens and parasites and the mosquito species availability determine the risks of the mosquito-borne diseases in human at the local as well as regional scales.

Points to Remember

1. The mosquitoes (Diptera: Culicidae) originated around 217 million years ago, during the Triassic period. As a holometabolous insect, mosquitoes complete the life cycle through the freshwater larval and pupal stages and the aerial adult stages. Female mosquitoes of almost all the species require blood meal for the completion of the life cycle. In course of the blood meal, the exchange of the pathogens and parasites between the hosts and the mosquitoes is the significant concern for the dynamics of mosquito-borne diseases.
2. The suborder Culicidae comprises of two subfamilies. The family Culicidae is further divided into two subfamily, Anophelinae and Culicinae. The subfamily Culicinae consists of 11 tribes and a total of 3566 extant species. There are 11 tribes under Culicinae, namely, Aedeomyiini, Aedini, Culicini, Culisetini, Ficalbiini, Hodgesiini, Mansoniini, Orthopodomyiini, Sabethini, Toxorhynchitini, and Uranotaeniini. There are 42 genera under Anophelinae and Culicinae, with the dominance in Oriental and Neotropical realms. The diversity of the mosquitoes is considerably high in the tropics and subtropics, but there are common species, like *Aedes aegypti*, *Ae. albopictus*, and *Culex quinquefascia-*

tus, that are distributed worldwide. Species diversity of Anophelinae is highest in the Oriental realm with 168 species, followed by the Afrotropical realm with 148 species and Neotropical realm with 125 species. The Neotropical realms host 944 species under Culicinae, followed by the Oriental realm with 893 species.

3. Almost all types of freshwater bodies are exploited by the mosquitoes for breeding including the slowly flowing streams to the puddles to the specialized habitats, like the phytotelmata. Many specialist species are observed that breed in specific habitats, like the bamboo stump specialist mosquito, *Tripteroides bambusa*, breeding in the stagnant water held by the bamboo stumps. The crab-hole mosquito, *Deinocerites cancer*, breeds in the burrows of the land crab. The pitcher plant mosquito, *Wyeomyia smithii*, uses the pitcher plant *Sarracenia purpura* as the larval habitat. Majority of the mosquito species are generalists and utilize all possible types of habitats, but the selection depends on the habitat permanence and the resources available therein.
4. In comparison to the species diversity of the mosquito (more than 3500 species described worldwide), only selected species qualify as the disease vectors. Among the ~500 *Anopheles* species observed, around 35–40 species are incriminated as the disease vectors. In India, there are about 393 species that are known with 31 species linked with the transmission of the diseases. In North African countries, 26 out of 83 mosquito species are incriminated as the vectors of various diseases. An estimated 10% of mosquitoes species out of the known 3578 described species are vectors of 78 different pathogens and parasites on a world wide scale. A number of factors including the disease vectors, the strains of the pathogens and parasites, and the environmental factors are significant for the possible mosquito-borne disease epidemics.

Acknowledgments The authors acknowledge the Head, Department of Zoology, University of Calcutta, for providing the facilities in compiling this chapter, including the facilities of UGC DRS-SAP and DST-FIST, Government of India. SDM acknowledges the financial support of UGC, Government of India [Ref. no.: 685/(CSIR-UGC NET JUNE 2018), Roll-341095, dated 15/04/2019]. DD acknowledges the Principal, Bagnan College, Bagnan, Howrah, and Department of Science and Technology and Biotechnology, Government of West Bengal, for the facilities provided including the project sanction no. 208(Sanc.)/ST/P/S&T/1G-73/2017 dated 21.03.2018. MR acknowledges the Principal, Bejoy Narayan Mahavidyalaya, Itachuna, Hooghly, for the facilities provided.

References

Abella-Medrano CA, Ibáñez-Bernal S, MacGregor-Fors I, Santiago-Alarcon D (2015) Spatiotemporal variation of mosquito diversity (Diptera: Culicidae) at places with different land-use types within a neotropical montane cloud forest matrix. Parasit Vectors 8:487

Aditya G, Pramanik MK, Saha GK (2006) Larval habitats and species composition of mosquitoes in Darjeeling Himalayas, India. J Vector Borne Dis 43(1):7–15

Aditya G, Tamang R, Sharma D, Subba F, Saha GK (2008) Bamboo stumps as mosquito larval habitats in Darjeeling Himalayas, India: a spatial scale analysis. Insect Sci 15(3):245–249

Adugna T, Getu E, Yewhalaw D (2020) Species diversity and distribution of *Anopheles* mosquitoes in Bure district, Northwestern Ethiopia. Heliyon 6:e05063

Afolabi OJ, Akinneye JO, Igiekhume AMA (2019) Identification, abundance, and diversity of mosquitoes in akure south local government area, Ondo state, Nigeria. J Basic Appl Zool 80:39

Almeida JF, Belchior HCM, Ríos-Velásquez CM, Pessoa FAC (2020) Diversity of mosquitoes (Diptera: Culicidae) collected in different types of larvitraps in an Amazon rural settlement. PLoS One 15(10):e0235726

Alto BW, Connelly CR, O'Meara GF, Hickman D, Karr N (2014) Reproductive biology and susceptibility of Florida *Culex coronator* to infection with West Nile virus. Vector Borne Zoonotic Dis 14(8):606–614

Amerasinghe FP, Alagoda TSB (1984) Mosquito oviposition in bamboo traps, with special reference to *Aedes albopictus*, *Aedes albopictus* and *Armigeres subalbatus*. Int J Trop Insect Sci 5:493–500

Amini M, Hanafi-Bojd AA, Aghapour AA, Chavshin AR (2020) Larval habitats and species diversity of mosquitoes (Diptera: Culicidae) in West Azerbaijan Province, Northwestern Iran. BMC Ecol 20:60

Andreadis TG, Anderson JF, Vossbrinck CR (2001) Mosquito surveillance for West Nile virus in Connecticut, 2000: isolation from *Culex pipiens*, *Cx. restuans*, *Cx. salinarius*, and *Culiseta melanura*. Emerg Infect Dis 7(4):670–674

Andreadis TG, Anderson JF, Vossbrinck CR, Main AJ (2004) Epidemiology of West Nile virus in Connecticut: a five-year analysis of mosquito data 1999–2003. Vector Borne Zoonotic Dis 4(4):360–378

Antinori S, Galimberti L, Milazzo L, Corbellino M (2013) *Plasmodium knowlesi*: the emerging zoonotic malaria parasite. Acta Trop 125(2):191–201

Armstrong PM, Andreadis TG (2010) Eastern equine encephalitis virus in mosquitoes and their role as bridge vectors. Emerg Infect Dis 16(12):1869–1874

Assa A, Eligo N, Massebo F (2023) *Anopheles* mosquito diversity, entomological indicators of malaria transmission and challenges of morphological identification in southwestern Ethiopia. Trop Med Health 51:38

Baker M (1961) The altitudinal distribution of mosquito larvae in the Colorado Front Range. Trans Am Entomol Soc 87(4):231–246

Banerjee S, Aditya G, Saha N, Saha GK (2010) An assessment of macroinvertebrate assemblages in mosquito larval habitats-space and diversity relationship. Environ Monit Assess 168:597–611

Barraud PJ (1923) A revision of the culicine mosquitoes of India. Part I. Indian J Med Res 10(3):772–788

Barraud PJ (1934) The Fauna of British India, including Ceylon and Burma. Diptera. In: Family Culieldae. Tribes Megarhinini and Culicini, vol 5. Taylor and Francis, London

Barrientos-Roldán MJ, Abella-Medrano CA, Ibáñez-Bernal S, Sandoval-Ruiz CA (2022) Landscape anthropization affects mosquito diversity in a deciduous forest in southeastern Mexico. J Med Entomol 59(1):248–256

Becker N, Petrić D, Zgomba M, Boase C, Madon MB, Dahl C, Kaiser A (2020) Mosquitoes: identification, ecology and control. Springer, Cham, p 570

Belaganahalli MN, Maan S, Maan NS, Nomikou K, Pritchard I, Lunt R, Kirkland PD, Attoui H, Brownlie J, Mertens PPC (2012) Full genome sequencing and genetic characterization of Eubenangee viruses identify Pata virus as a distinct species within the genus *Orbivirus*. PLoS One 7(3):e31911

Belkin JN (1962) The mosquitoes of the South Pacific (Diptera, Culicidae) [sic]. University of California Press, Berkeley

Belkin JN, Hogue CL (1959) A review of the crabhole mosquitoes of the genus *Deinocerites* (Diptera, Culicidae). Univ Calif Publ Entomol 14(6):411–458

Ben Ayed W, Amraoui F, M'ghirbi Y, Schaffner F, Rhaim A, Failloux AB, Bouattour A (2019) A survey of *Aedes* (Diptera: Culicidae) mosquitoes in Tunisia and the potential role of *Aedes detritus* and *Aedes caspius* in the transmission of Zika Virus. J Med Entomol 56(5):1377–1383

Beranek MD, Gallardo R, Almirón WR, Contigiani MS (2018) First detection of *Mansonia titillans* (Diptera: Culicidae) infected with St. Louis encephalitis virus (Flaviviridae: Flavivirus)

and Bunyamwera serogroup (Peribunyaviridae: Orthobunyavirus) in Argentina. J Vector Ecol 43(2):340–343

Betancourt A, Loaiza JR (2016) An effective sampling tool for adult crabhole inhabiting *Deinocerites* mosquitoes. J Vector Ecol 41(1):200–203

Bhattacharyya DR, Rajavel AR, Natarajan R, Mohapatra PK, Jambulingam P, Mahanta J, Prakash A (2014) Faunal richness and the checklist of Indian mosquitoes (Diptera: Culicidae). Check List 10(6):1342–1358

Bird RG, Draper CC, Ellis DS (1972) A cytoplasmic polyhedrosis virus in midgut cells of *Anopheles stephensi* and in the sporogonic stages of *Plasmodium berghei yoelii*. Bull World Health Organ 46(3):337–343

Biteye B, Fall AG, Ciss M, Seck MT, Apolloni A, Fall M, Tran A, Gimonneau G (2018) Ecological distribution and population dynamics of Rift Valley fever virus mosquito vectors (Diptera, Culicidae) in Senegal. Parasit Vectors 11:27

Black WC IV, Bennett KE, Gorrochótegui-Escalante N, Barillas-Mury CV, Fernández-Salas I, de Lourdes Muñoz M, Farfán-Alé JA, Olson KE, Beaty BJ (2002) Flavivirus susceptibility in *Aedes aegypti*. Arch Med Res 33(4):379–388

Blackmore CGM, Blackmore MS, Grimstad PR (1998) Role of *Anopheles quadrimaculatus* and *Coquillettidia perturbans* (Diptera: Culicidae) in the transmission cycle of Cache Valley virus (Bunyaviridae: Bunyavirus) in the midwest, USA. J Med Entomol 35(5):660–664

Bond JG, Casas-Martínez M, Quiroz-Martínez H, Novelo-Gutiérrez R, Marina CF, Ulloa A, Orozco-Bonilla A, Muñoz M, Williams T (2014) Diversity of mosquitoes and the aquatic insects associated with their oviposition sites along the Pacific coast of Mexico. Parasit Vectors 7:41

Boromisa RD, Grayson MA (1990) Incrimination of *Aedes provocans* as a vector of Jamestown Canyon virus in an enzootic focus of northeastern New York. J Am Mosq Control Assoc 6(3):504–509

Bowen ETW, Simpson DIH, Platt GS, Way HJ, Gordon Smith CE, Ching CY, Casals J (1970) Arbovirus infections in Sarawak: the isolation of Kunjin virus from mosquitoes of the *Culex pseudovishnui* group. Ann Trop Med Parasitol 64(3):263–268

Boyer S, Peng B, Pang S, Chevalier V, Duong V, Gorman C, Dussart P, Fontenille D, Cappelle J (2020) Dynamics and diversity of mosquito vectors of Japanese encephalitis virus in Kandal province, Cambodia. J Asia Pac Entomol 23(4):1048–1054

Bradshaw WE, Holzapfel CM (1989) Life-historical consequences of density-dependent selection in the pitcher-plant mosquito, *Wyeomyia smithii*. Am Nat 133(6):869–887

Bravo-Barriga D, Parreira R, Almeida APG, Calado M, Blanco-Ciudad J, Serrano-Aguilera FJ, Pérez-Martín JE, Sánchez-Peinado J, Pinto J, Reina D, Frontera E (2016) *Culex pipiens* as a potential vector for transmission of *Dirofilaria immitis* and other unclassified Filarioidea in Southwest Spain. Vet Parasitol 223:173–180

Bryan JH, McMahon P, Barnes A (1990) Factors affecting transmission of *Wuchereria bancrofti* by anopheline mosquitoes. 3. Uptake and damage to ingested microfilariae by *Anopheles gambiae*, *An. arabiensis*, *An. merus* and *An. funestus* in East Africa. Trans R Soc Trop Med Hyg 84(2):265–268

Câmara DCP, da Silva Pinel C, Rocha GP, Codeço CT, Honório NA (2020) Diversity of mosquito (Diptera: Culicidae) vectors in a heterogeneous landscape endemic for arboviruses. Acta Trop 212:105715

Camp JV, Karuvantevida N, Chouhna H, Safi E, Shah JN, Nowotny N (2019) Mosquito biodiversity and mosquito-borne viruses in the United Arab Emirates. Parasit Vectors 12:153

Cao Y, Fu S, Tian Z, Lu Z, He Y, Wang H, Wang J, Guo W, Tao B, Liang G (2011) Distribution of mosquitoes and mosquito-borne arboviruses in Inner Mongolia, China. Vector Borne Zoonotic Dis 11(12):1577–1581

Cardoso JC, de Almeida MAB, dos Santos E, da Fonseca DF, Sallum MAM, Noll CA, Monteiro HAO, Cruz ACR, Carvalho VL, Pinto EV, Castro FC, Nunes Neto JP, Segura MNO, Vasconcelos PFC (2010) Yellow fever virus in *Haemagogus leucocelaenus* and *Aedes serratus* mosquitoes, southern Brazil, 2008. Emerg Infect Dis 16(12):1918–1924

Carissimo G, Eiglmeier K, Reveillaud J, Holm I, Diallo M, Diallo D, Vantaux A, Kim S, Ménard D, Siv S, Belda E, Bischoff E, Antoniewski C, Vernick KD (2016) Identification and characterization of two novel RNA viruses from *Anopheles gambiae* species complex mosquitoes. PLoS One 11(5):e0153881

Cates MD, Detels R (1969) Japanese encephalitis virus in Taiwan: preliminary evidence for *Culex annulus* Theob. as a vector. J Med Entomol 6(3):327–328

Cêtre-Sossah C, Lebon C, Rabarison P, Cardinale E, Mavingui P, Atyame C (2023) Evidence of *Eretmapodites subsimplicipes* and *Aedes albopictus* as competent vectors for Rift Valley fever virus transmission in Mayotte. Acta Trop 239:106835

Chambers DM, Steelman CD, Schilling PE (1979) Mosquito species and densities in Louisiana ricelands. Mosq News 39(3):658–668

Charlwood JD (1997) Vectorial capacity, species diversity and population cycles of anopheline mosquitoes (Diptera: Culicidae) from indoor light-trap collections in a house in southeastern Tanzania. Afr Entomol 5(1):93–101

Chathuranga WGD, Karunaratne SHPP, Fernando BR, De Silva WAPP (2018) Diversity, distribution, abundance, and feeding pattern of tropical ornithophilic mosquitoes. J Vector Ecol 43(1):158–167

Chaverri LG, Dillenbeck C, Lewis D, Rivera C, Romero LM, Chaves LF (2018) Mosquito species (Diptera: Culicidae) diversity from ovitraps in a Mesoamerican tropical rainforest. J Med Entomol 55(3):646–653

Chaves LF, Hamer GL, Walker ED, Brown WM, Ruiz MO, Kitron UD (2011) Climatic variability and landscape heterogeneity impact urban mosquito diversity and vector abundance and infection. Ecosphere 2(6):70. https://doi.org/10.1890/ES11-00088.1

Chaves LSM, Bergo ES, Conn JE, Laporta GZ, Prist PR, Sallum MAM (2021) Anthropogenic landscape decreases mosquito biodiversity and drives malaria vector proliferation in the Amazon rainforest. PLoS One 16(1):e0245087

Chen WJ, Dong CF, Chiou LY, Chuang WL (2000) Potential role of *Armigeres subalbatus* (Diptera: Culicidae) in the transmission of Japanese encephalitis virus in the absence of rice culture on Liu-chiu islet, Taiwan. J Med Entomol 37(1):108–113

Cheong WH, Mak JW, Naidu S, Mahadevan S (1981) *Armigeres subalbatus* incriminated as an important vector of dog heartworm *Dirofilaria immitis* and bird cardiofilaria in urban Kualalumpur. Southeast Asian J Trop Med Public Health 12(4):611–612

Chouin-Carneiro T, Vega-Rua A, Vazeille M, Yebakima A, Girod R, Goindin D, Dupont-Rouzeyrol M, Lourenço-de-Oliveira R, Failloux AB (2016) Differential susceptibilities of *Aedes aegypti* and *Aedes albopictus* from the Americas to Zika virus. PLoS Negl Trop Dis 10(3):e0004543

Christophers SR (1924) Provisional list and reference catalogue of the Anophelini. Indian Med Res Memoirs 3:1–105

Christophers SR (1933) The fauna of British India, including Ceylon and Burma. Diptera. In: Family culicidae. Tribe anophelini, vol. IV. Taylor and Francis, London

Clements AN (1992) The biology of mosquitoes: volume 1. Development, nutrition and reproduction. Chapman and Hall, London, p 509

Clements AN (1999) The biology of mosquitoes: volume 2. Sensory reception and behavior. CABI, Wallingford, p 740

Clements AN (2012) The biology of mosquitoes: volume 3. Transmission of viruses and interactions with bacteria. CABI, Wallingford, p 571

Collins LE, Blackwell A (2000) The biology of *Toxorhynchites* mosquitoes and their potential as biocontrol agents. Biocontrol News Inf 21(4):105N–116N

Colmant AMG, Etebari K, Webb CE, Ritchie SA, Jansen CC, van den Hurk AF, Bielefeldt-Ohmann H, Hobson-Peters J, Asgari S, Hall RA (2017a) Discovery of new orbiviruses and totivirus from *Anopheles* mosquitoes in Eastern Australia. Arch Virol 162:3529–3534

Colmant AMG, Hobson-Peters J, Bielefeldt-Ohmann H, van den Hurk AF, Hall-Mendelin S, Chow WK, Johansen CA, Fros J, Simmonds P, Watterson D, Cazier C, Etebari K, Asgari S, Schulz BL, Beebe N, Vet LJ, Piyasena TBH, Nguyen HD, Barnard RT, Hall RA (2017b) A new clade

of insect-specific flaviviruses from Australian *Anopheles* mosquitoes displays species-specific host restriction. mSphere 2(4):e00262
Contreras MA, Eastwood G, Guzman H, Popov V, Savit C, Uribe S, Kramer LD, Wood TG, Widen SG, Fish D, Tesh RB, Vasilakis N, Walker PJ (2017) *Almendravirus*: a proposed new genus of rhabdoviruses isolated from mosquitoes in tropical regions of the Americas. Am J Trop Med Hyg 96(1):100–109
Cook S, Chung BY-W, Bass D, Moureau G, Tang S, McAlister E, Culverwell CL, Glücksman E, Wang H, Brown TDK, Gould EA, Harbach RE, de Lamballerie X, Firth AE (2013) Novel virus discovery and genome reconstruction from field RNA samples reveals highly divergent viruses in dipteran hosts. PLoS One 8(11):e80720
Corbet PS (1964) Observations on mosquitoes ovipositing in small containers in Zika Forest, Uganda. J Anim Ecol 33(1):141–164
Crans WJ (2004) A classification system for mosquito life cycles: life cycle types for mosquitoes of the northeastern United States. J Vector Ecol 29(1):1–10
Cranston PS, Gullan PJ (2009) Phylogeny of insects. In: Resh VH, Cardé RT (eds) Encyclopedia of insects, 2nd edn. Academic, London, pp 780–793
Cupp EW, Klingler K, Hassan HK, Viguers LM, Unnasch TR (2003) Transmission of eastern equine encephalomyelitis virus in central Alabama. Am J Trop Med Hyg 68(4):495–500
Dale PER, Knight JM (2008) Wetlands and mosquitoes: a review. Wetl Ecol Manag 16:255–276
Danks HV (2006) Key themes in the study of seasonal adaptations in insects II. Life-cycle patterns. Appl Entomol Zool 41(1):1–13
Danks HV (2007) How aquatic insects live in cold climates. Can Entomol 139(4):443–471
Darsie RF Jr, Courtney GW, Pradhan SP (1993) Notes on the mosquitoes of Nepal: III. Additional new records in 1992 (Diptera: Culicidae). Mosquito Syst 25(3):186–191
de la Puente J, Moreno-Indias I, Hernández-Castellano LE, Argüello A, Ruiz S, Soriguer R, Figuerola J (2014) Host-feeding pattern of *Culex theileri* (Diptera: Culicidae), potential vector of *Dirofilaria immitis* in the Canary Islands, Spain. J Med Entomol 49(6):1419–1423
Dhileepan K (1996) Mosquito seasonality and arboviral disease incidence in Murray Valley, southeast Australia. Med Vet Entomol 10(4):375–384
Digoutte JP, Salaun JJ, Robin Y, Bres P, Cagnard VJ (1980) Minor arboviral diseases in Central and West Africa. Med Trop 40(5):524–533
Dutary BE, Petersen JL, Peralta PH, Tesh RB (1989) Transovarial transmission of Gamboa virus in a tropical mosquito, *Aedeomyia squamipennis*. Am J Trop Med Hyg 40(1):108–113
Edeson JFB, Wharton RH, Laing ABG (1960) A preliminary account of the transmission, maintenance and laboratory vectors of *Brugia phangi*. Trans R Soc Trop Med Hyg 54(5):439–449
Edwards FW (1912) A synopsis of the species of African Culicidae, other than Anopheles. Bull Entomol Res 3(1):1–53
Edwards FW (1921) A revision of the mosquitos [sic] of the Palaearctic Region. Bull Entomol Res 12:263–351
Eifan S, Hanif A, Nour I, Alqahtani S, Eisa ZM, Dafalla O, Kohl A (2021) Distribution and molecular identification of *Culex pipiens* and *Culex tritaeniorhynchus* as potential vectors of Rift Valley fever virus in Jazan, Saudi Arabia. Pathogens 10(10):1334
Epopa PS, Millogo AA, Collins CM, North AR, Benedict MQ, Tripet F, O'Loughlin S, Dabiré RK, Ouédraogo GA, Diabaté A (2020) *Anopheles gambiae* (s.l.) is found where few are looking: assessing mosquito diversity and density outside inhabited areas using diverse sampling methods. Parasit Vectors 13:516
Estep LK, Burkett-Cadena ND, Hill GE, Unnasch RS, Unnasch TR (2010) Estimation of dispersal distances of *Culex erraticus* in a focus of eastern equine encephalitis virus in the southeastern United States. J Med Entomol 47(6):977–986
Fagbohun IK, Idowu ET, Awolola TS, Otubanjo OA (2020) Seasonal abundance and larval habitats characterization of mosquito species in Lagos State, Nigeria. Sci Afr 10:e00656
Fauver JR, Grubaugh ND, Krajacich BJ, Weger-Lucarelli J, Lakin SM, Fakoli LS III, Bolay FK, Diclaro JW II, Dabiré KR, Foy BD, Brackney DE, Ebel GD, Stenglein MD (2016) West

African *Anopheles gambiae* mosquitoes harbor a taxonomically diverse virome including new insect-specific flaviviruses, mononegaviruses, and totiviruses. Virology 498:288–299

Ferreira-de-Lima VH, Lima-Camara TN (2018) Natural vertical transmission of dengue virus in *Aedes aegypti* and *Aedes albopictus*: a systematic review. Parasit Vectors 11:77

Foley DH, Rueda LM, Wilkerson RC (2007) Insight into global mosquito biogeography from country species records. J Med Entomol 44(4):554–567

Gajanana A, Rajendran R, Philip Samuel P, Thenmozhi V, Tsai TF, Kimura-Kuroda J, Reuben R (1997) Japanese encephalitis in south Arcot district, Tamil Nadu, India: a three-year longitudinal study of vector abundance and infection frequency. J Med Entomol 34(6):651–659

Golding N, Nunn MA, Medlock JM, Purse BV, Vaux AGC, Schäfer SM (2012) West Nile virus vector *Culex modestus* established in southern England. Parasit Vectors 5:32

Gordon SW, Tammariello RF, Linthicum KJ, Dohm DJ, Digoutte JP, Calvo-Wilson MA (1992) Arbovirus isolations from mosquitoes collected during 1988 in the Senegal River basin. Am J Trop Med Hyg 47(6):742–748

Guedes DRD, Paiva MHS, Donato MMA, Barbosa PP, Krokovsky L, Rocha SWS, Saraiva KLA, Crespo MM, Rezende TMT, Wallau GL, Barbosa RMR, Oliveira CMF, Melo-Santos MAV, Pena L, Cordeiro MT, Franca RFO, de Oliveira ALS, Peixoto CA, Leal WS, Ayres CFJ (2017) Zika virus replication in the mosquito *Culex quinquefasciatus* in Brazil. Emerg Microbes Infect 6:e69

Guzman H, Contreras-Gutierrez MA, Travassos da Rosa APA, Nunes MRT, Cardoso JF, Popov VL, Young KI, Savit C, Wood TG, Widen SG, Watts DM, Hanley KA, Perera D, Fish D, Vasilakis N, Tesh RB (2018) Characterization of three new insect-specific flaviviruses: their relationship to the mosquito-borne flavivirus pathogens. Am J Trop Med Hyg 98(2):410–419

Hamer GL, Anderson TK, Donovan DJ, Brawn JD, Krebs BL, Gardner AM, Ruiz MO, Brown WM, Kitron UD, Newman CM, Goldberg TL, Walker ED (2014) Dispersal of adult *Culex* mosquitoes in an urban West Nile virus hotspot: a mark-capture study incorporating stable isotope enrichment of natural larval habitats. PLoS Negl Trop Dis 8(3):e2768

Harbach RE (2007) The Culicidae (Diptera): a review of taxonomy, classification and phylogeny. Zootaxa 1668(1):591–638

Harbach RE (2013) The phylogeny and classification of *Anopheles*. In: Manguin S (ed) *Anopheles* mosquitoes-new insights into malaria vectors. IntechOpen, Rijeka

Harbach RE, Kitching IJ (1998) Phylogeny and classification of the Culicidae (Diptera). Syst Entomol 23(4):327–370

Harbach RE, Kitching IJ (2016) The phylogeny of Anophelinae revisited: inferences about the origin and classification of *Anopheles* (Diptera: Culicidae). Zool Scr 45:34–47

Harbach RE, Kitching IJ, Culverwell CL, Dubois J, Linton YM (2012) Phylogeny of mosquitoes of tribe Culicini (Diptera: Culicidae) based on morphological diversity. Zool Scr 41:499–514

Hardy JL, Eldridge BF, Reeves WC, Schutz SJ, Presser SB (1993) Isolations of Jamestown canyon virus (Bunyaviridae: California serogroup) from mosquitoes (Diptera: Culicidae) in the western United States, 1990–1992. J Med Entomol 30(6):1053–1059

Hernández-Valencia JC, Rincón DS, Marín A, Naranjo-Díaz N, Correa MM (2020) Effect of land cover and landscape fragmentation on anopheline mosquito abundance and diversity in an important Colombian malaria endemic region. PLoS One 15(10):e0240207

Honěk A (1993) Intraspecific variation in body size and fecundity in insects: a general relationship. Oikos 66(3):483–492

Hu SMK (1940) *Culex pallidothorax* Theobald as a Carrier of *Wuchereria bancrofti* Cobbold. Lingnan Sci J 19(4):543–547

Huhtamo E, Lambert AJ, Costantino S, Servino L, Krizmancic L, Boldorini R, Allegrini S, Grasso I, Korhonen EM, Vapalahti O, Lanciotti RS, Ravanini P (2013) Isolation and full genomic characterization of Batai virus from mosquitoes, Italy 2009. J Gen Virol 94(6):1242–1248

Iriarte WLZ, Tsuda Y, Wada Y, Takagi M (1991) Distribution of mosquitoes on a hill of Nagasaki city, with emphasis to the distance from human dwellings. Trop Med 33(3):55–60

Irving-Bell RJ, Okoli EI, Diyelong DY, Lyimo EO, Onyia OC (1987) Septic tank mosquitoes: competition between species in central Nigeria. Med Vet Entomol 1(3):243–250

Iyangar MOT (1938) Studies on epidemiology of filariasis in Travancore. Indian Med Res Memoirs 30:65–69

Jansen S, Heitmann A, Lühken R, Jöst H, Helms M, Vapalahti O, Schmidt-Chanasit J, Tannich E (2018) Experimental transmission of Zika virus by *Aedes japonicus japonicus* from southwestern Germany. Emerg Microbes Infect 7(1):192

Johnson T, Braack L, Guarido M, Venter M, Gouveia Almeida AP (2020) Mosquito community composition and abundance at contrasting sites in northern South Africa, 2014–2017. J Vector Ecol 45(1):104–117

Jupp PG (1967) Larval habitats of culicine mosquitoes (Diptera: Culicidae) in a sewage effluent disposal area in the South African Highveld. J Entomol Soc South Afr 30(2):242–250

Kaboré DPA, Soma DD, Gil P, Kientega M, Sawadogo SP, Ouédraogo GA, Van de Perre P, Baldet T, Gutierrez S, Dabiré RK (2023) Mosquito (Diptera: Culicidae) populations in contrasting areas of the western regions of Burkina Faso: species diversity, abundance and their implications for pathogen transmission. Parasit Vectors 16(1):438

Kain MP, Skinner EB, Athni TS, Ramirez AL, Mordecai EA, van den Hurk AF (2022) Not all mosquitoes are created equal: a synthesis of vector competence experiments reinforces virus associations of Australian mosquitoes. PLoS Negl Trop Dis 16(10):e0010768. https://doi.org/10.1371/journal.pntd.0010768

Kamgang B, Vazeille M, Yougang AP, Tedjou AN, Wilson-Bahun TA, Mousson L, Wondji CS, Failloux AB (2019) Potential of *Aedes albopictus* and *Aedes aegypti* (Diptera: Culicidae) to transmit yellow fever virus in urban areas in Central Africa. Emerg Microbes Infect 8(1):1636–1641

Kanojia PC (2007) Ecological study on mosquito vectors of Japanese encephalitis virus in Bellary district, Karnataka. Indian J Med Res 126(2):152–157

Karlekar S, Andrew R (2015) Mosquito diversity and vector species status in and around Nagpur city of Maharashtra state, India. Int J Mosquito Res 2(2):19–23

Kent RJ, Crabtree MB, Miller BR (2010) Transmission of West Nile virus by *Culex quinquefasciatus* say infected with *Culex* Flavivirus Izabal. PLoS Negl Trop Dis 4(5):e671

Khan SA, Chowdhury P, Choudhury P, Dutta P (2017) Detection of West Nile virus in six mosquito species in synchrony with seroconversion among sentinel chickens in India. Parasit Vectors 10:13

King WV, Bradley GH, Smith CN, McDuffie WC (1960) A handbook of the mosquitoes of the Southeastern United States. United States Department of Agriculture, Washington

Knight KL, Stone A (1977) A catalog of the mosquitoes of the world (Diptera: Culicidae). Entomological Society of America, Annapolis, p 611

Knudson DL, Brown SE, Severson DW (2002) Culicine genomics. Insect Biochem Mol Biol 32(10):1193–1197

Kuwata R, Sugiyama H, Yonemitsu K, Van Dung N, Terada Y, Taniguchi M, Shimoda H, Takano A, Maeda K (2015) Isolation of Japanese encephalitis virus and a novel insect-specific flavivirus from mosquitoes collected in a cowshed in Japan. Arch Virol 160:2151–2159

Laird M (1959) Parasites of Singapore mosquitoes, with particular reference to the significance of larval epibionts as an index of habitat pollution. Ecology 40(2):206–221

Lee SE, Kim HC, Chong ST, Klein TA, Lee WJ (2007) Molecular survey of *Dirofilaria immitis* and *Dirofilaria repens* by direct PCR for wild caught mosquitoes in the Republic of Korea. Vet Parasitol 148(2):149–155

Li S, Jiang F, Lu H, Kang X, Wang Y, Zou Z, Wen D, Zheng A, Liu C, Liu Q, Kang L, Xia Q, Cui F (2020a) Mosquito diversity and population genetic structure of six mosquito species from Hainan Island. Front Genet 11:602863

Li Y, Zhou G, Zhong S, Wang X, Zhong D, Hemming-Schroeder E, Yi G, Fu F, Fu F, Cui L, Cui G, Yan G (2020b) Spatial heterogeneity and temporal dynamics of mosquito population density and community structure in Hainan Island, China. Parasit Vectors 13:444

Liehne PFS (1991) An atlas of the mosquitoes of Western Australia. Health Department of Western Australia, Perth

Lounibos LP, Van Dover C, O'meara, G.F. (1982) Fecundity, autogeny, and the larval environment of the pitcher-plant mosquito, *Wyeomyia smithii*. Oecologia 55:160–164

Lutomiah J, Bast J, Clark J, Richardson J, Yalwala S, Oullo D, Mutisya J, Mulwa F, Musila L, Khamadi S, Schnabel D, Wurapa E, Sang R (2013) Abundance, diversity, and distribution of mosquito vectors in selected ecological regions of Kenya: public health implications. J Vector Ecol 38(1):134–142

Lutwama JJ, Rwaguma EB, Nawanga PL, Mukuye A (2002) Isolations of Bwamba virus from south central Uganda and north eastern Tanzania. Afr Health Sci 2(1):24–28

Mancini G, Montarsi F, Calzolari M, Capelli G, Dottori M, Ravagnan S, Lelli D, Chiari M, Santilli A, Quaglia M, Federici V, Monaco F, Goffredo M, Savini G (2017) Mosquito species involved in the circulation of West Nile and Usutu viruses in Italy. Vet Ital 53(2):97–110

Maquart M, Boyer S, Rakotoharinome VM, Ravaomanana J, Tantely ML, Heraud JM, Cardinale E (2016) High prevalence of West Nile virus in domestic birds and detection in 2 new mosquito species in Madagascar. PLoS One 11(1):e0147589

Maquart PO, Sokha C, Boyer S (2021) Mosquito diversity (Diptera: Culicidae) and medical importance, in a bird sanctuary inside the flooded forest of Prek Toal, Cambodia. J Asia Pac Entomol 24(4):1221–1227

Martínez-Barciela Y, Polina A, Garrido J (2024) Ecology and diversity of mosquito larvae in ponds and lagoons of Northwestern Spain. Med Vet Entomol 38(1):38–47

Martins LMO, David MR, Maciel-de-Freitas R, Silva-do-Nascimento TF (2018) Diversity of *Anopheles* mosquitoes from four landscapes in the highest endemic region of malaria transmission in Brazil. J Vector Ecol 43(2):235–244

Matthews TC, Munstermann LE (1994) Chromosomal repatterning and linkage group conservation in mosquito karyotypic evolution. Evolution 48(1):146–154

Mayi MPA, Bamou R, Djiappi-Tchamen B, Fontaine A, Jeffries CL, Walker T, Antonio-Nkondjio C, Cornel AJ, Tchuinkam T (2020) Habitat and seasonality affect mosquito community composition in the west region of Cameroon. Insects 11(5):312

Medlock JM, Snow KR, Leach S (2005) Potential transmission of West Nile virus in the British Isles: an ecological review of candidate mosquito bridge vectors. Med Vet Entomol 19(1):2–21

Miller BR, Nasci RS, Godsey MS, Savage HM, Lutwama JJ, Lanciotti RS, Peters CJ (2000) First field evidence for natural vertical transmission of West Nile virus in *Culex univittatus* complex mosquitoes from Rift Valley province, Kenya. Am J Trop Med Hyg 62(2):240–246

Mitchell CJ, Chen PS (1973) Ecological studies on the mosquito vectors of Japanese encephalitis. Bull World Health Organ 49(3):287–292

Mitchell CJ, Monath TP, Sabattini MS, Cropp CB, Daffner JF, Calisher CH, Jakob WL, Christensen HA (1985) Arbovirus investigations in Argentina, 1977-1980. II. Arthropod collections and virus isolations from Argentine mosquitoes. Am J Trop Med Hyg 34(5):945–955

Mitchell CJ, Monath TP, Sabattini MS, Daffner JF, Cropp CB, Calisher CH, Darsie RF Jr, Jakob WL (1987) Arbovirus isolations from mosquitoes collected during and after the 1982-1983 epizootic of western equine encephalitis in Argentina. Am J Trop Med Hyg 36(1):107–113

Möhlmann TWR, Wennergren U, Tälle M, Favia G, Damiani C, Bracchetti L, Koenraadt CJM (2017) Community analysis of the abundance and diversity of mosquito species (Diptera: Culicidae) in three European countries at different latitudes. Parasit Vectors 10:510

Mores CN, Turell MJ, Dohm DJ, Blow JA, Carranza MT, Quintana M (2007) Experimental transmission of West Nile virus by *Culex nigripalpus* from Honduras. Vector Borne Zoonotic Dis 7(2):279–284

Morrison AC, Zielinski-Gutierrez E, Scott TW, Rosenberg R (2008) Defining challenges and proposing solutions for control of the virus vector *Aedes aegypti*. PLoS Med 5(3):e68

Mourya DT, Ilkal MA, Mishra AC, Jacob PG, Pant U, Ramanujam S, Mavale MS, Bhat HR, Dhanda V (1989) Isolation of Japanese encephalitis virus from mosquitoes collected in Karnataka state, India from 1985 to 1987. Trans R Soc Trop Med Hyg 83(4):550–552

Muja-Bajraktari N, Zhushi-Etemi F, Dikolli-Velo E, Kadriaj P, Gunay F (2019) The composition, diversity, and distribution of mosquito fauna (Diptera: Culicidae) in Kosovo. J Vector Ecol 44(1):94–104

Mullen GR, Durden LA (2019) Medical and veterinary entomology, 3rd edn. Academic Press, Starkville, p 769

Munyao V, Karisa J, Munyao CM, Ngari M, Menza N, Peshu N, Rono M, Mbogo C, Mwangangi J (2020) Surveillance of culicine mosquitoes in six villages of Taita-Taveta County, Kenya, with host determinations from blood-fed females. J Med Entomol 57(6):1972–1982

Muturi EJ, Shililu J, Jacob B, Gu W, Githure J, Novak R (2006) Mosquito species diversity and abundance in relation to land use in a riceland agroecosystem in Mwea, Kenya. J Vector Ecol 31(1):129–137

Muturi EJ, Shililu JI, Jacob BG, Mwangangi JM, Mbogo CM, Githure JI, Novak RJ (2008) Diversity of riceland mosquitoes and factors affecting their occurrence and distribution in Mwea, Kenya. J Am Mosq Control Assoc 24(3):349–358

Mwangangi JM, Midega J, Kahindi S, Njoroge L, Nzovu J, Githure J, Mbogo CM, Beier JC (2012) Mosquito species abundance and diversity in Malindi, Kenya and their potential implication in pathogen transmission. Parasitol Res 110:61–71

Nebbak A, Almeras L, Parola P, Bitam I (2022) Mosquito vectors (diptera: culicidae) and mosquito-borne diseases in North Africa. Insects 13(10):962. https://doi.org/10.3390/insects13100962

Nicoletti M (2020) Three scenarios in insect-borne diseases. In: Insect-borne diseases in the 21st century. Elsevier, Amsterdam, pp 99–251

Ohba SY, Van Soai N, Van Anh DT, Nguyen YT, Takagi M (2015) Study of mosquito fauna in rice ecosystems around Hanoi, Northern Vietnam. Acta Trop 142:89–95

Ondiba IM, Oyieke FA, Athinya DK, Nyamongo IK, Estambale BBA (2019) Larval species diversity, seasonal occurrence and larval habitat preference of mosquitoes transmitting Rift Valley fever and malaria in Baringo County, Kenya. Parasit Vectors 12:295

Petersen V, Santana M, Alves JMP, Suesdek L (2022) Genetic and morphological polymorphisms of *Aedes scapularis* (Diptera: Culicidae), vector of filariae and arboviruses. Infect Genet Evol 97:105193

Pothikasikorn J, Bangs MJ, Boonplueang R, Chareonviriyaphap T (2008) Susceptibility of various mosquitoes of Thailand to nocturnal subperiodic *Wuchereria bancrofti*. J Vector Ecol 33(2):313–320

Pramanik MK, Aditya G, Raut SK (2006) A survey of anopheline mosquitoes and malarial parasite in commuters in a rural and an urban area in West Bengal, India. J Vector Borne Dis 43(4):198–202

Prudhomme J, Fontaine A, Lacour G, Gantier JC, Diancourt L, Velo E, Bino S, Reiter P, Mercier A (2019) The native European *Aedes geniculatus* mosquito species can transmit chikungunya virus. Emerg Microbes Infect 8(1):962–972

Rajagopalan PK, Kazmi SJ, Mani TR (1977) Some aspects of transmission of *Wuchereria bancrofti* and ecology of the vector *Culex pipiens fatigans* in Pondicherry. Indian J Med Res 66(2):200–215

Rakotonirina A, Maquart PO, Flamand C, Sokha C, Boyer S (2023) Mosquito diversity (Diptera: Culicidae) and medical importance in four Cambodian forests. Parasit Vectors 16:110

Ramaiah KD, Das PK (1992) Seasonality of adult *Culex quinquefasciatus* and transmission of Bancroftian filariasis in Pondicherry, south India. Acta Trop 50(4):275–283

Ramalingam S (1976) An annotated checklist and keys to the mosquitoes of Samoa and Tonga. Mosquito Syst 8(3):298–318

Ramirez PG, Stein M, Etchepare EG, Almiron WR (2016) Diversity of anopheline mosquitoes (Diptera: Culicidae) and classification based on the characteristics of the habitats where they were collected in Puerto Iguazú, Misiones, Argentina. J Vector Ecol 41(2):215–223

Ratovonjato J, Olive MM, Tantely LM, Andrianaivolambo L, Tata E, Razainirina J, Jeanmaire E, Reynes JM, Elissa N (2011) Detection, isolation, and genetic characterization of Rift Valley fever virus from *Anopheles (Anopheles) coustani*, *Anopheles (Anopheles) squamosus*, and *Culex (Culex) antennatus* of the Haute Matsiatra region, Madagascar. Vector Borne Zoonotic Dis 11(6):753–759

Rattanarithikul R, Panthusiri P (1994) Illustrated keys to the medically important mosquitos of Thailand. Southeast Asian J Trop Med Public Health 25:1–66

Rattanarithikul R, Harrison BA, Panthusiri P, Coleman RE (2005) Illustrated keys to the mosquitoes of Thailand I. Background; geographic distribution; lists of genera, subgenera, and species; and a key to the genera. Southeast Asian J Trop Med Public Health 36:1–80
Rattanarithikul R, Harrison BA, Harbach RE, Panthusiri P, Coleman RE (2006a) Illustrated keys to the mosquitoes of Thailand. IV. *Anopheles*. Southeast Asian J Trop Med Public Health 37:1–128
Rattanarithikul R, Harrison BA, Panthusiri P, Peyton EL, Coleman RE (2006b) Illustrated keys to the mosquitoes of Thailand III. Genera *Aedeomyia, Ficalbia, Mimomyia, Hodgesia, Coquillettidia, Mansonia*, and *Uranotaenia*. Southeast Asian J Trop Med Public Health 37:1–85
Rattanarithikul R, Harbach RE, Harrison BA, Panthusiri P, Coleman RE (2007) Illustrated keys to the mosquitoes of Thailand V. Genera *Orthopodomyia, Kimia, Malaya, Topomyia, Tripteroides*, and *Toxorhynchites*. Southeast Asian J Trop Med Public Health 38:1–65
Reagan RL, Strand N, Brueckner AL (1953) Comparison by electron microscopy of anopheles "A" and Anopheles "B" viruses. Tex Rep Biol Med 11:508–511
Reddya Naik B, Tyagi BK, Xue R-D (2023) Mosquito-borne diseases in India over the past 50 years and their global public health implications: a systematic review. J Am Mosq Control Assoc 39(4):258–277
Reisen WK, Fang Y, Martinez VM (2006) Effects of temperature on the transmission of West Nile virus by *Culex tarsalis* (Diptera: Culicidae). J Med Entomol 43(2):309–317
Reiskind MH, Griffin RH, Janairo MS, Hopperstad KA (2017) Mosquitoes of field and forest: the scale of habitat segregation in a diverse mosquito assemblage. Med Vet Entomol 31(1):44–54
Reiter P, Fontenille D, Paupy C (2006) *Aedes albopictus* as an epidemic vector of chikungunya virus: another emerging problem? Lancet Infect Dis 6(8):463–464
Ren X, Hoiczyk E, Rasgon JL (2008) Viral paratransgenesis in the malaria vector *Anopheles gambiae*. PLoS Pathog 4(8):e1000135
Reuben R, Kaul HN, Soman RS (1988) Mosquitoes of arboviral importance in India. Mosquito-Borne Dis Bull 5:48–54
Rodríguez-González S, Portela Salomão R, Córdoba-Aguilar A (2023) Research trends in mosquito studies in urban areas. Acta Trop 241:106888. https://doi.org/10.1016/j.actatropica.2023.106888
Rodríguez-González S, Izquierdo-Suzán M, Rocha-Ortega M, Córdoba-Aguilar A (2024) Vector mosquito distribution and richness are predicted by socio-economic, and ecological variables. Acta Trop 254:107179. https://doi.org/10.1016/j.actatropica.2024.107179
Roy M, Pramanik S, Chatterjee S, Aditya G (2016) Observations on the pupal productivity of *Culex tritaeniorhynchus* in rice fields of West Bengal, India: implications for vector management (Diptera: Culicidae). Fragmenta Entomol 48(1):69–76
Rueda LM (2008) Global diversity of mosquitoes (Insecta: Diptera: Culicidae) in freshwater. Hydrobiologia 595:477–487
Saha GK, Malakar P, Das S (2000) Diversity of anopheline mosquitoes at different altitudes of Darjeeling Himalaya. J Parasit Dis 24(2):215–216
Saiyasombat R, Bolling BG, Brault AC, Bartholomay LC, Blitvich BJ (2011) Evidence of efficient transovarial transmission of *Culex* flavivirus by *Culex pipiens* (Diptera: Culicidae). J Med Entomol 48(5):1031–1038
Sallum MAM, Schultz TR, Wilkerson RC (2000) Phylogeny of Anophelinae (Diptera Culicidae) based on morphological characters. Ann Entomol Soc Am 93:745–775
Sam SS, Mohamed-Romai-Noor NA, Teoh BT, Hamim ZR, Ng HY, Abd-Jamil J, Khor CS, Hassan SS, Ahmad H, AbuBakar S (2022) Group IV Getah virus in *culex* mosquitoes, Malaysia. Emerg Infect Dis 28(2):475–477
Samina I, Margalit J, Peleg J (1986) Isolation of viruses from mosquitoes of the Negev, Israel. Trans R Soc Trop Med Hyg 80(3):471–472
Sardelis MR, Turell MJ, Dohm DJ, O'Guinn ML (2001) Vector competence of selected North American *Culex* and *Coquillettidia* mosquitoes for West Nile virus. Emerg Infect Dis 7(6):1018–1022

Shahhosseini N, Lühken R, Jöst H, Jansen S, Börstler J, Rieger T, Krüger A, Yadouleton A, de Mendonça Campos R, Cirne-Santos CC, Ferreira DF, Garms R, Becker N, Tannich E, Cadar D, Schmidt-Chanasit J (2017) Detection and characterization of a novel rhabdovirus in *Aedes cantans* mosquitoes and evidence for a mosquito-associated new genus in the family Rhabdoviridae. Infect Genet Evol 55:260–268

Shroyer DA, Beach RF, Munstermann L, Peloquin J, Petersen JL, Smith RP, Taylor DB (1976) Mosquito diversity in St. Joseph County, Indiana (Diptera: Culicidae). Proc Indiana Acad Sci 86:238–241

Silaghi C, Beck R, Capelli G, Montarsi F, Mathis A (2017) Development of *Dirofilaria immitis* and *Dirofilaria repens* in *Aedes japonicus* and *Aedes geniculatus*. Parasit Vectors 10:94

Silver JB (2008) Mosquito ecology. Field sampling methods. Springer, Dordrecht, p 1494

Simpson DIH, Way HJ, Platt GS, Bowen ETW, Hill MN, Kamath S, Bendell PJE, Heathcote OHU (1975) Arbovirus infections in Sarawak, October 1968–February 1970: GETAH virus isolations from mosquitoes. Trans R Soc Trop Med Hyg 69(1):35–38

Soghigian J, Sither C, Justi SA, Morinaga G, Cassel BK, Vitek CJ, Livdahl T, Xia S, Gloria-Soria A, Powell JR, Zavortink T, Hardy CM, Burkett-Cadena ND, Reeves LE, Wilkerson RC, Dunn RR, Yeates DK, Sallum MA, Byrd BD, Trautwein MD, Linton YM, Reiskind MH, Wiegmann BM (2023) Phylogenomics reveals the history of host use in mosquitoes. Nat Commun 14:6252

Steffan WA, Evenhuis NL (1981) Biology of *Toxorhynchites*. Annu Rev Entomol 26(1):159–181

Stuckly KG, Wright PJ (1983) Characterization of leanyer virus: resemblance to bunyavirus. Austr J Exp Biol Med Sci 61(2):193–200

Sudeep AB (2014) *Culex gelidus*: an emerging mosquito vector with potential to transmit multiple virus infections. J Vector Borne Dis 51(4):251–258

Sudia WD, Lord RD, Newhouse VF, Miller DL, Kissling RE (1971) Vector-host studies of an epizootic of Venezuelan equine encephalomyelitis in Guatemala, 1969. Am J Epidemiol 93(2):137–143

Sulesco TM, Toderas LG, Uspenskaia IG, Toderas IK (2015) Larval habitats diversity and distribution of the mosquito (Diptera: Culicidae) species in the Republic of Moldova. J Med Entomol 52(6):1299–1308

Sumodan PK (2012) Species diversity of mosquito breeding in rubber plantations of Kerala, India. J Am Mosq Control Assoc 28(2):114–115

Sunahara T, Mogi M (1998) Distribution and turnover of a mosquito (*Tripteroides bambusa*) metapopulation among bamboo groves. Ecol Res 13:291–299

Sunahara T, Mogi M (2002) Variability of intra-and interspecific competitions of bamboo stump mosquito larvae over small and large spatial scales. Oikos 97(1):87–96

Sunish IP, Reuben R, Rajendran R (2006) Natural survivorship of immature stages of *Culex vishnui* (Diptera: Culicidae) complex, vectors of Japanese Encephalitis Virus, in rice fields in Southern India. J Med Entomol 43(2):185–191

Takken W, Verhulst NO (2013) Host preferences of blood-feeding mosquitoes. Annu Rev Entomol 58:433–453

Thongsripong P, Green A, Kittayapong P, Kapan D, Wilcox B, Bennett S (2013) Mosquito vector diversity across habitats in central Thailand endemic for dengue and other arthropod-borne diseases. PLoS Negl Trop Dis 7(10):e2507

Triteeraprapab S, Kanjanopas K, Suwannadabba S, Sangprakarn S, Poovorawan Y, Scott AL (2000) Transmission of the nocturnal periodic strain of *Wuchereria bancrofti* by *Culex quinquefasciatus*: establishing the potential for urban filariasis in Thailand. Epidemiol Infect 125(1):207–212

Trung HD, Van Bortel W, Sochantha T, Keokenchanh K, Quang NT, Cong LD, Coosemans M (2004) Malaria transmission and major malaria vectors in different geographical areas of Southeast Asia. Trop Med Int Health 9(2):230–237

Turell MJ, Ludwig GV, Beaman JR (1992) Transmission of Venezuelan equine encephalomyelitis virus by *Aedes sollicitans* and *Aedes taeniorhynchus* (Diptera: Culicidae). J Med Entomol 29(1):62–65

Turell MJ, Britch SC, Aldridge RL, Xue RD, Smith ML, Cohnstaedt LW, Linthicum KJ (2015) Potential for *Psorophora columbiae* and *Psorophora ciliata* mosquitoes (Diptera: Culicidae) to transmit Rift Valley fever virus. J Med Entomol 52(5):1111–1116

Ughasi J, Bekard HE, Coulibaly M, Adabie-Gomez D, Gyapong J, Appawu M, Wilson MD, Boakye DA (2012) *Mansonia africana* and *Mansonia uniformis* are vectors in the transmission of *Wuchereria bancrofti* lymphatic filariasis in Ghana. Parasit Vectors 5:89

Unlu I, Kramer WL, Roy AF, Foil LD (2010) Detection of West Nile virus RNA in mosquitoes and identification of mosquito blood meals collected at alligator farms in Louisiana. J Med Entomol 47(4):625–633

Vanlalruia K, Senthilkumar N, Gurusubramanian G (2014) Diversity and abundance of mosquito species in relation to their larval habitats in Mizoram, North Eastern Himalayan region. Acta Trop 137:1–18

Vázquez A, Sánchez-Seco MP, Palacios G, Molero F, Reyes N, Ruiz S, Aranda C, Marqués E, Escosa R, Moreno J, Figuerola J, Tenorio A (2012) Novel flaviviruses detected in different species of mosquitoes in Spain. Vector Borne Zoonotic Dis 12(3):223–229

Villinger J, Mbaya MK, Ouso D, Kipanga PN, Lutomiah J, Masiga DK (2017) Arbovirus and insect-specific virus discovery in Kenya by novel six genera multiplex high-resolution melting analysis. Mol Ecol Resour 17(3):466–480

Vythilingam I, Tan SB, Krishnasamy M (2002) Susceptibility of *Culex sitiens* to Japanese encephalitis virus in peninsular Malaysia. Trop Med Int Health 7(6):539–540

Wang J, Zhu Z (2023) Novel paradigm of mosquito-borne disease control based on self-powered strategy. Front Public Health 11:1115000

WHO (2023) World malaria report 2023. WHO, Geneva, p 283

WHO (2024) Water and sanitation interventions to prevent and control mosquito-borne diseases: focus on emergencies. World Health Organization, Geneva, p 15

Wilkerson RC, Linton Y-M, Strickman D (2021) Mosquitoes of the world, vol 1. Johns Hopkins University Press, Baltimore, p 599

Williams CR, Williams SR, Nicholson J, Little SM, Riordan J, Fricker SR, Kokkinn MJ (2009) Diversity and seasonal succession of coastal mosquitoes (Diptera: Culicidae) in the northern Adelaide region of South Australia. Aust J Entomol 48(2):107–112

Yang W, Zhao S, Xie Y, Liu T, Kong L, Guo Y, Xie Z, Liu P, Chen XG (2022) *Armigeres subalbatus* is a potential vector for Zika virus but not dengue virus. Infect Dis Poverty 11:62

Yee DA, Dean Bermond C, Reyes-Torres LJ, Fijman NS, Scavo NA, Nelsen J, Yee SH (2022) Robust network stability of mosquitoes and human pathogens of medical importance. Parasit Vectors 15:216

Zettle M, Anderson E, LaDeau SL (2022) Changes in container-breeding mosquito diversity and abundance along an urbanization gradient are associated with dominance of arboviral vectors. J Med Entomol 59(3):843–854

Biology of Mosquitoes

5

Himmat Singh, Nasreen Akhtar, and Sanjeev Kumar Gupta

Abstract

Mosquitoes, long recognized as vectors of disease, have profoundly impacted human health, economies, and ecosystems throughout history. The multifaceted dynamics of human-mosquito interactions emphasize the enduring struggle against mosquito-borne diseases. Despite significant scientific advancements, these diseases persist, necessitating ongoing research, prevention and control efforts. With over 3500 species across 41 genera, mosquitoes present a complex challenge, compounded by their role in transmitting diseases to both humans and wildlife. Mosquitoes undergo a four-stage life cycle: egg, larva, pupa and adult. Eggs are laid in water or moist areas, hatching within days to a year, depending on species and environmental conditions. Larvae feed and grow in water, undergoing four instars before pupating. Pupae, active in the water, transform into adults. Only females bite for blood needed to develop eggs, potentially transmitting diseases like malaria, dengue, and Zika. Each species has specific modes of infection; for example, *Anopheles* transmits malaria, *Culex* spreads Japanese Encephalitis, and *Aedes* carries diseases like dengue and Zika. Understanding mosquito biology, including its life cycle, behaviours, and ecological roles, is paramount for developing effective control strategies while minimizing environmental impact. Moreover, recognizing the socioeconomic toll of mosquito-borne diseases underscores the urgency of comprehensive intervention strategies. By integrating knowledge of mosquito biology with interdisciplinary approaches, stakeholders can work collaboratively to address the root causes of mosquito-borne diseases and promote global health and well-being. Regrettably, there are currently no vaccines or drugs to treat most of these diseases, making vector control the primary method of prevention. So, there is a need of mosquito control through source reduction and many newer techniques as well.

H. Singh (✉) · N. Akhtar · S. K. Gupta
ICMR-National Institute of Malaria Research, New Delhi, India

© The Author(s), under exclusive license to Springer Nature Singapore Pte Ltd. 2024
Omkar (ed.), *Mosquitoes*, https://doi.org/10.1007/978-981-97-4163-2_5

Keywords

Mosquito · Mosquito-borne diseases · Breeding habitats · Stages of mosquito

Learning Objectives

1. To describe the anatomical features of mosquitoes, including the head, thorax, abdomen, wings and specialized mouthparts
2. To explain the four stages of the mosquito life cycle: egg, larva, pupa and adult with detail the environmental conditions required for successful mosquito reproduction
3. To explore the mechanisms involved in blood-feeding and nectar-feeding and to understand the role of mosquitoes as both pollinators and vectors of diseases
4. To investigate the mating behaviours of mosquitoes, including courtship rituals and reproductive strategies
5. To identify major mosquito-borne diseases and the pathogens responsible

5.1 Introduction

The crucial role that mosquitoes play in the lives of humans is universally acknowledged. For approximately a century, the recognition and incrimination of mosquitoes as a major nuisance to human health have shaped our understanding of their significance. Throughout human history, the persistent nuisance caused by mosquitoes and the diseases they transmit has inflicted immeasurable suffering and substantial economic losses. Mosquitoes have had an indelible influence on human health from the time immemorial. Since the dawn of society, humans have struggled with the complex difficulties offered by mosquito-borne diseases, with far-reaching effects on both individuals and civilizations as a whole. The spectrum of diseases transmitted by mosquitoes encompasses some of the most perilous and widespread afflictions, including malaria, filariasis, Japanese encephalitis, yellow fever, Zika, dengue, chikungunya, and an array of other infections. The toll of these diseases has been staggering, claiming the lives of millions and leaving a lasting imprint on the fabric of human health.

Mosquito-borne disease persistence extends beyond the obvious health concerns, rippling across economies and communities. The economic losses sustained as a result of medical expenses, reduced productivity, and the societal burden of caring for the sick highlight the profound effects of mosquitoes' involvement in human health. Furthermore, the complicated network of biological and environmental factors controlling mosquito populations complicates the continuous battle to reduce the effect of these tiny but powerful opponents. Reflecting on the history of human-mosquito interactions reveals that the fight against diseases transmitted by mosquitoes has been a long one. Despite substantial advances in scientific knowledge and medical treatments, the issues remain, necessitating continued investment in research, preventive and control strategies. The complex interaction between mosquitoes and the diseases they carry needs a multidisciplinary and interdisciplinary approach encompassing sciences, such as entomology, epidemiology, and environmental science. In today's world, when global travel helps the quick transmission of diseases, combating the mosquito scourge has taken on new significance. Combating diseases caused by mosquitoes requires not only the discovery of effective vaccinations and treatments but also the implementation of long-term mosquito control measures and the promotion of public awareness and education.

The family Culicidae, whose members commonly known as mosquitoes, comprised of 41 recognized genera, incorporating about 3500 distinct species. Because of their importance as vectors, many mosquito genera and species have been the subject of extensive cytological and genetic investigations over the last 40 years (Kitzmiller 1953, 1976; Rai et al. 1982; Rai 1991). Adding complexity to this scenario, many of the diseases affecting humans find common ground with those that impact various wild animal reservoirs. This interplay between human and wildlife pathogens underscores the complex dynamics of mosquito-borne diseases. Beyond their role as disease vectors, mosquitoes pose additional challenges to human and animal well-being. Their presence is not merely a matter of annoyance; their bites can elicit allergic reactions, and in cases of large mosquito populations, they can cause significant blood loss. These irksome insects, with their itchy bites, have long been a source of discomfort for communities worldwide.

Mosquito biology encompasses the taxonomy, life cycle and behaviours of these tiny yet impactful insects. Their metamorphic life cycle includes egg, larva, pupa and adult stages, showcasing adaptability to diverse environments. Female mosquitoes, notorious for blood-feeding, play a crucial role in disease transmission, while males primarily feed on nectar. Understanding mosquito biology is essential for effective disease control, ecological balance preservation and the development of innovative strategies to address the global health challenges posed by these ubiquitous insects.

Knowledge of mosquito biology is essential for unravelling the complex dynamics of disease transmission, allowing scientists and healthcare professionals to develop targeted strategies for disease prevention and control. Moreover, the intricate life cycle and feeding habits of mosquitoes contribute to their adaptability, making them resilient to various environmental conditions. Beyond their role as disease vectors, mosquitoes play vital ecological roles. As pollinators and a source of food for other organisms, their presence contributes to the delicate balance within ecosystems. Disrupting this balance through unguided control measures may have unintended consequences on biodiversity and ecosystem stability. Therefore, a comprehensive understanding of their biology is crucial for implementing sustainable mosquito control strategies that minimize environmental impact.

Furthermore, mosquitoes have a significant socioeconomic impact. Mosquito-borne diseases affect millions of people worldwide, leading to substantial healthcare costs, loss of productivity and economic burdens on affected communities. By delving into the biology of mosquitoes, researchers and policymakers can work collaboratively to develop innovative and targeted interventions, addressing the root causes of mosquito-borne diseases and contributing to global health and well-being. In essence, knowledge of mosquito biology is a linchpin for devising effective, sustainable and ethically responsible strategies to navigate the intricate interactions between mosquitoes, humans and the environment.

Scientific classification

Kingdom	:	Animalia
Phylum	:	Arthropoda
Class	:	Insecta
Order	:	Diptera
Sub-order	:	Nematocera
Infra-order	:	Culicomorpha
Family	:	Culicidae
Subfamily	:	Anophelinae Culicinae Toxorhynchitinae

5.2 Morphological Description

The life cycle of adult mosquito has four distinct stages: adult, egg, larvae and pupa. The adult stage is aerial, and the other three stages are aquatic. The terminology used here is taken from Harbach and Knight (1980) (Fig. 5.1).

5.2.1 Adult

The body of mosquito comprises the head, the thorax and the abdomen. The main part of the body and their appendages are briefly described below:

5.2.1.1 Head

Head is the smallest part of the mosquito body, is almost spherical and is connected to thorax through a narrow and membranous neck. The head consists of a pair of antennae, a proboscis, a pair of maxillary palps and a pair of compound eyes. The eyes occupy the bulk of the surface area of the head, which is characterized by the presence of white and black scales (Fig. 5.1).

(a) **Antennae**: Antennae are sense organs, which arise on the front of the head between the eyes and consist of a flagellum of approx. 13 flagellomeres in females and approx. 12 flagellomeres in males. In the male mosquito, the antennae are plumose owing to the presence of whorls of long hairs, while in females, the hairs are smaller, fewer and less conspicuous (pilose).

(b) **Proboscis**: The proboscis projects downward and forward from the lower front margin of the head and generally is continuously dark in colour. The length of

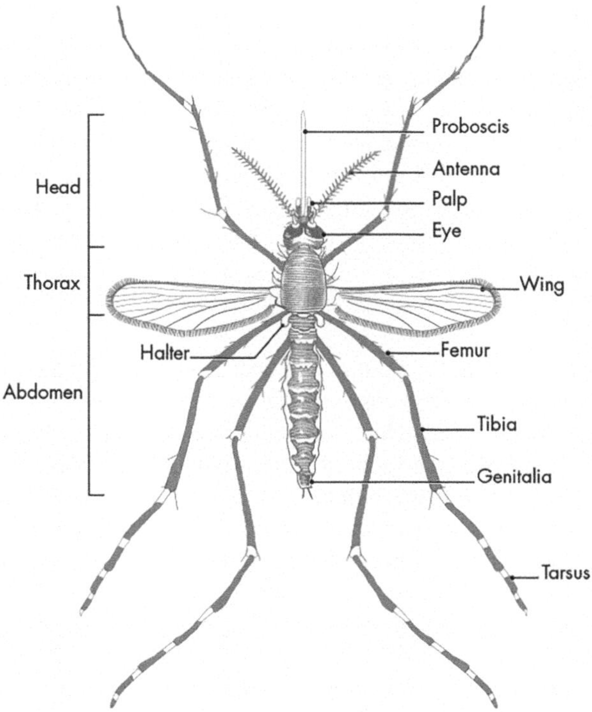

Fig. 5.1 Generalized diagram of a female adult mosquito (Source: Centers for Disease Control and Prevention 2024; https://www.cdc.gov/mosquitoes/about/what-is-a-mosquito.html)

the proboscis is measured by comparing it with that of the femur of the foreleg and is a very useful character to differentiate different species of mosquito. The proboscis also consists of a sheath- like structure, known as labium, enclosing a group of six styles. The labium serves as a protective sheath for the styles but does not enter into the wound when the mosquito is biting. The styles enter into the host skin and form a duct through which saliva is injected into the wound and liquid food is sucked in.

(c) **Maxillary palpi (palpi)**: A pair of palpi arise from the lower front margin of the head near the proboscis. The palpus has five palpomeres. The small basal segment is designated as the first palpomere, which is fused with the second and is vestigial. The distal segment, which is away from the head, is known as the fifth palpomere. The palpi in the female are slender, while they are club shaped toward the apical end in males. The palpi are also covered with scales except along their inner surface. The thickness of palpi depends on the type of scales, i.e. whether they are erect or flat. The pale scales on the palpi form the bands at the joints 2–3 and 3–4 and at the apex. The absence or presence of the pale bands, their number, their width and the distance between the two bands are important characters for species identification.

5.2.1.2 Thorax

The thorax is the middle part of the body. It bears two pairs of wings (one pair membranous wings on mesothorax, second pair of wings are reduced into halteres and work as balancing organ during flight) and three pairs of legs and has three distinct parts—pronotum, scutum and metanotum—each part consisting of one pair of legs. The pronotum is reduced to form a pair of antepronotal lobes lying dorsally on either side of the neck with a post pronotary area. The scutum is the largest part of the thorax. In Indian anophelines, the thorax has very little importance in the identification of species.

5.2.1.3 Wings

A pair of wings arise from the scutum and bear a number of important characters, which help to identify the subgenus and species. The wing membrane consists of a number of longitudinal veins, which are connected with each other through small cross veins. The thickened anterior border of the wing is known as costa and parallel to it is the subcosta, which fuses with the costa in the middle of the anterior border of the wing. Besides these two important veins, six longitudinal veins are also present. The orientation of the wings or scaling on the wings is of great importance in identification.

5.2.1.4 Legs

There are three pairs of legs originating from the lower side of thorax: the forelegs, midlegs and hindlegs. Each leg consists of conical coxa, a small hinge- like trochanter, a long femur, a slender tibia and a five segmented tarsus. The fifth tarsomere, which is toward the end of the leg, is smallest in size, while the first tarsomere, which is attached with the tibia, is longest in size in female mosquito. But in male mosquito, the fifth tarsomere is often longer than the fourth tarsomere of the foreleg. The fifth tarsomere in each leg consists of claws in female, but the foreleg of the male consists of very few spurred claws with some stout hairs (Fig. 5.2).

5.2.1.5 Abdomen

The abdomen of anopheline is composed of ten segments, but externally only eight segments can be seen. The last two segments are modified into genital organs. The dorsal part of abdomen is called dorsum, and the segments are known as tergites; the ventral side is known as venter, and segments are called sternites. The tergites and sternites join laterally through a membrane known as pleural membrane anteriorly and as intersegmental membranes posteriorly. In the case of male anopheline, the dorsal surface of the eighth segment is made up of sternite instead of tergite as after emergence this segment rotates with genitalia at an angle of 180°. The presence or absence of scaling, hair, coxites, or ceri on the abdomen is a useful feature in the identification of some Indian anophelines (Fig. 5.3).

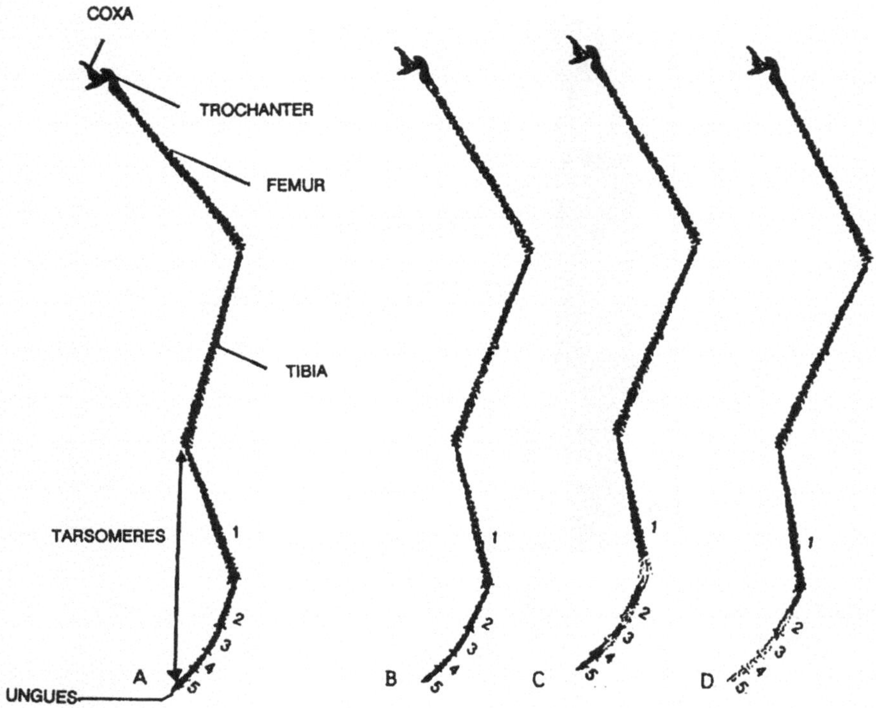

Fig. 5.2 Legs of an adult mosquito showing different parts (**A**); speckling on femur, tibia and tarsomeres (**B**); foreleg tarsomeres with broad pale bands (**C**); hind tarsomeres 5, 4 and 3 portion of second completely white (**D**); hind tarsomeres 2, 3, 4 and 5 are completely white (Source: Nagpal and Sharma 1995)

5.2.2 Eggs

There are three types of mosquito eggs: (1) those laid singly on the surface of the water, e.g. *Anopheles*; (2) those glued together to form rafts that floats on the water surface, e.g. *Culex* and *Mansonia*; and (3) those laid singly out of water on wet surface, e.g. *Aedes*. Generally, eggs are laid on water surface hatch within 2–3 days. Eggs of some species belonging to genera *Aedes* survive up to 3–4 years on wet grounds till the surface become flooded. Eggs of anopheline are white in colour when first deposited on the surface of water but darken after a few hours. The eggs are laid singly on the water surface and lie lengthwise on the water. These eggs are elongated, oval and pointed at one end (boat-shaped) and provided with a pair of lateral floats. The average length of an egg is about 1 mm. The inner and outer walls of the egg are known as endochorion and the exochorion, respectively. The general appearance of an anopheline egg is shown in Fig. 5.4.

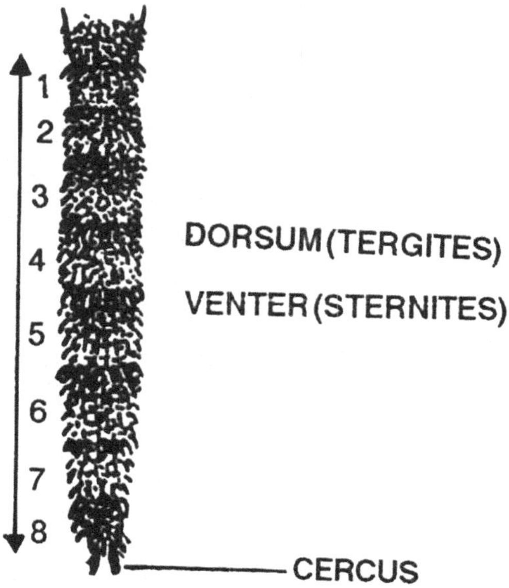

Fig. 5.3 Abdomen of an adult anopheline mosquito (source: Nagpal and Sharma 1995)

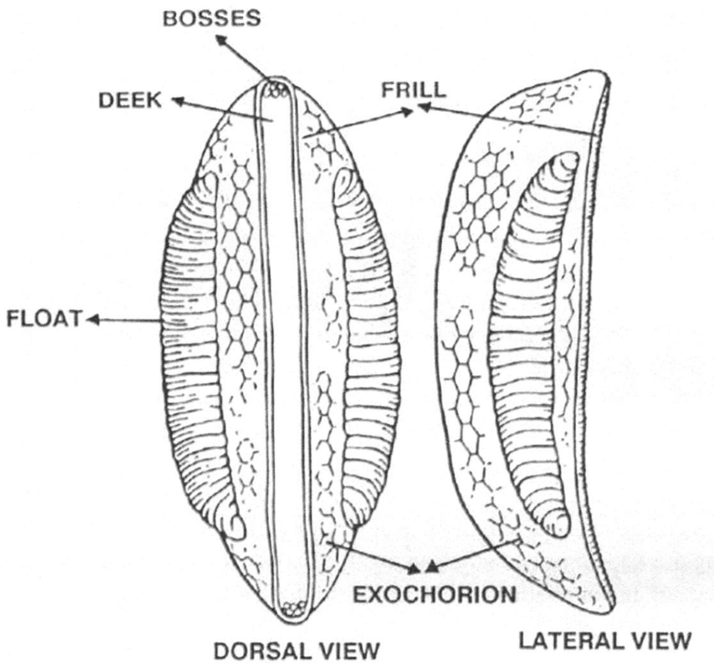

Fig. 5.4 Egg of an anopheline mosquito (Nagpal and Sharma 1995)

5.2.3 Larvae

Larvae hatch from the eggs and live in water. Mosquitoes have a strong preference for particular breeding sites. Some larvae develop in permanent waters, like canals, ponds, wells and riverbed pools, and some in temporary waters, like pools, rainwater collections, tree holes or leaf axils. For example, *An. sundaicus* larvae always develop in brackish waters, and *An. minimus* larvae in streams. The larvae get their food from the water and breathe by coming to the surface of water. All mosquito larvae undergo three moultings during their developments (7–10 days depending upon the environmental factors) when they shed the chitinous skin. The successive moults are known as I, II, III and IV instars or stages. The fourth instar is a mature stage, and with the fourth moult the pupa appears.

5.2.4 Pupae

The pupae of anophelines are aquatic and very active. The pupa is a non-feeding stage but swims to the surface of the water to breath. There is a large cephalothorax. Dorsally consists of a pair of trumpets and ventrally the developing mouth parts, eyes, legs and wings of the adults. The abdomen consists of eight freely movable segments with a pair of paddles at the tip. The pupae move with considerable speed because of the vigorous movement of the abdomen (Fig. 5.5).

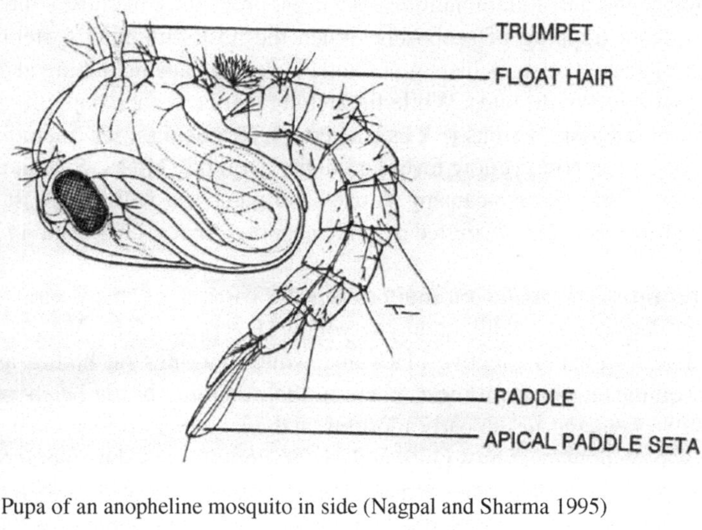

Fig. 5.5 Pupa of an anopheline mosquito in side (Nagpal and Sharma 1995)

5.3 Important Species of Mosquitoes

Mosquitoes, with their unique adaptive mechanisms, can thrive in diverse environments. Any aquatic habitat worldwide can serve as a breeding ground for mosquitoes. They inhabit both temporary and permanent water bodies, from highly polluted to clean, and colonize areas of varying sizes. Even the smallest accumulations, such as water-filled buckets, flower vases, tires, hoof prints and leaf axes, can be potential breeding sites for mosquitoes.

In areas with temporary flooding, like along rivers or lakes with water fluctuations, floodwater mosquitoes, such as *Aedes vexans* or *Ochlerotatus sticticus*, proliferate in large numbers. With a flight range of several miles, they can become a significant nuisance even in locations far from their breeding sites (Schäfer et al. 1997). In swampy woodlands, snow-melt mosquitoes, such as *Ochlerotatus cantans*, *Oc. communis*, *Oc. punctor*, *Oc. hexodontus*, and *Oc. cataphylla*, find optimal conditions for development in pools created after snow melts or heavy rainfall. Female mosquitoes primarily feed on the blood of vertebrates including humans, mammals, birds, reptiles and others, while males feed on plant juices, nectar, etc.

Mosquito species exhibit a preference for either humans or certain animals as the source of their blood meal. They are attracted by body odours, carbon dioxide and heat emitted by individuals or animals. The majority of mosquito's species tend to bite during specific times, such as dusk or dawn. Additionally, various species differ in their preferred locations for feeding or resting; some prefer natural vegetative habitats, while others are inclined towards urban settings, especially areas with trash or receptacles in yards. During the process of obtaining blood meal from a host, the female mosquito injects some of its saliva into the skin. Mosquitos saliva contains anticoagulants and anti-inflammation substances, preventing the host's blood from clotting to avoid trapping of proboscis. When the host's immune system detects foreign substances, it releases histamines and cytokines, causing itching and wheals associated with mosquito bites. While the itching can be bothersome, the mosquitoes pose a much more serious risk as carriers of various diseases, including Zika virus, dengue, West Nile, yellow fever and malaria, among others. Specific species of mosquitoes carry disease-causing viruses and parasites, with the most notable mosquito-borne diseases transmitted by three genera—*Aedes*, *Culex* and *Anopheles*.

***Aedes* Mosquito** The *Aedes* mosquitoes are the carriers of many viral diseases including Zika, dengue, chikungunya, yellow fever and Rift Valley disease. Recognizable by their distinctive black and white markings on bodies and legs, *Aedes* mosquitoes differ from most others in their daytime biting behaviour, with peak activity occurring in the early morning and evening. Two important species, *Aedes aegypti* (yellow fever mosquito) and *Aedes albopictus* (Asian tiger mosquito), play crucial roles in transmitting viruses. *Ae. aegypti*, prevalent in urban areas, is active both indoors and outdoors, displaying a preference for humans as the source of its blood meal. On the other hand, *Ae. albopictus*, often found in vegetated areas, primarily bites domestic and wild animals, in addition to humans. Due to their proximity to human habitats and preference for feeding on people, *Ae. aegypti* poses a higher risk of spreading viral diseases compared to *Ae. albopictus*. Originally confined to specific tropical and subtropical regions, *Ae. Aegypti* in Africa and *Ae. albopictus* in Southeast Asia, both species are now distributed globally in tropical and subtropical areas. In the United States, *Ae. aegypti* is prevalent in southern states, while *Ae. albopictus*, adaptable to colder temperatures, is found further north.

***Culex* Mosquito** *Culex* mosquitoes, commonly referred to as common house mosquitoes, are the principal vectors for spreading the viruses leading to West Nile fever, St. Louis encephalitis, and Japanese encephalitis. Additionally, they are responsible for transmitting viral diseases affecting birds and horses. Beyond viruses, *Culex* mosquitoes are capable of transmitting the parasitic disease lymphatic filariasis and the bacterial disease tularaemia. Widely distributed across tropical and temperate regions worldwide, excluding extreme northern latitudes, *Culex* mosquitoes exhibit a nocturnal feeding pattern on both humans and animals. Their presence extends to both indoor and outdoor environments, making them versatile vectors with significant public health implications.

***Anopheles* Mosquito** Mosquitoes are very efficient vectors of human diseases, like malaria, dengue, chikungunya and yellow fever. *Anopheles* mosquitoes are renowned for their role in transmitting malaria, despite their capability of spreading other diseases is notable. There are approximately 420 recognized species of *Anopheles* mosquitoes throughout the world, out of which only about 70 species are considered to be the main vectors of malaria. So far, 58 Anopheline species have been reported from India. Out of 58 Anopheline species, 6 were established as malaria vectors (Nagpal and Sharma 1995). Indian species of *Anopheles*, which are involved in malaria transmission, belongs to the subgenus *Cellia*. The malaria vectors of India are *Anopheles culicifacies*, *An. stephensi*, *An. fluviatilis*, *An. minimus*, *An. sundaicus* and *An. dirus* (formerly *An. balabacensis*) *sensulato* (Dash et al. 2007). They are active between sunrise and sunset and can be found both indoors and outdoors. The females feed on the blood of both humans and animals, but some species have a preference for one over the other; species that favour humans pose the greatest risk of transmitting malaria.

5.4 Malaria Vector Species

1. *Anopheles culicifacies* (*An. Culicifacies*) (Giles 1901): It prefers to rest in cattle sheds and houses during the day. The species was also collected from straw, mud cakes, near stables and from dense vegetation, under the bushes and in tree holes, an indication of outdoor resting. They generally bite throughout night, but peak biting occurs between 19.00 and 4.00 h with variations from place to place (Fig. 5.6).

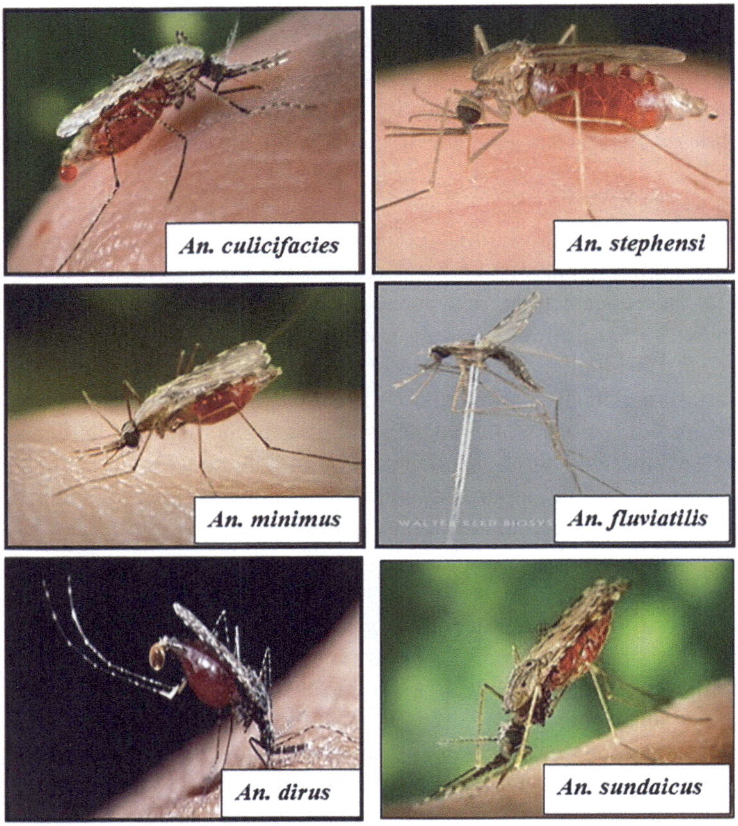

Fig. 5.6 Major malaria vectors (Sources: https://creativecommons.org/publicdomain/zero/1.0, http://www.scientistsagainstmalaria.net/sites/default/files/styles/full_post/public/anopheles_dirus.jpg?itok=7cXRLYeL)

2. *Anopheles stephensi* (*An. stephensi*) (Liston 1901): Adults rest in houses, cattle sheds and barracks, in rural and urban areas. The immature form predominantly breeds in wells, overhead or ground-level water tanks, cisterns, tanks, coolers, roof gutters and other artificial containers. Scantly breeding is also reported from rice fields and polluted and brackish water habitats. Larvae are shade lovers. Peak biting activity is generally between 22.00 and 24.00 h but varies from area to area and also from season to season (Fig. 5.6).
3. *Anopheles minimus* (*An. minimus*) (Theobald 1901): Rests in houses and cattle shed during the day. It rests mostly outdoors and collected from up to 1600 m area. Larvae are generally found in streams, ditches, channels in tea gardens and drains, which have a perceptible but slow flow of water. Occasionally, it also breeds in borrow pits, rice fields and seepages. The species prefers shady places. Peak biting time is between 18.00 and 19.00 h outdoors and between 24.00 and 2.00 h indoors. Biting time varies from locality and from season to season (Fig. 5.6).

4. *Anopheles fluviatilis* (old name *An. listoni* Liston) (*An. fluviatilis*) (James 1902): Adults of the species usually rest in human dwellings and to a lesser extent in cattle sheds during the day. The species also rests outdoors, collected from an altitude of 1060–3340 m in Gum Valley of Nepal. Peak biting activity occurs between 20.00 and 24.00 h, but timing may differ depending on the season and location (Fig. 5.6).
5. *Anopheles dirus/An. balabacensis* (*An. balabacensis*) (Peyton and Harrison 1979; Baisas 1936): A pool breeder in dense forest. They breed in pools by the side of rocky streams, disused wells, rainwater collections, borrow pits, drains in densely shaded foliage and stagnant water in forest. Exophagic, bites from 19.00 to 4.00 h but the peak biting activity is between 22.00 and 2.00 h (Fig. 5.6).
6. *Anopheles sundaicus* (old name *An. Ludlow* ivar *sundaicus*, Rodenwaldt) (*An. sundaicus*) (Rodenwaldt 1925): Mainly a brackish water breeder but also breeds in freshwater. Major breeding places are swamps and pits along bunds containing stagnant brackish water. Also breeds in saltwater lagoons, creeks, wells (brackish water or freshwater) overhead tanks and freshwater pools in coastal areas. It can tolerate salinity from 0.08 to 2.6% and pH from 7.7 to 8.5 (Fig. 5.6).

5.5 Dengue, Zika and Chikungunya Vector Species

1. *Aedes aegypti*: Although *Ae. aegypti* is currently distributed in urban areas throughout the tropical regions of Africa, Asia, Australia, South Pacific, Americas and some parts of the Middle East, the origin of the species is considered to be Africa (Mattingly 1957). Larvae breed from a variety of artificial containers, such as jars, discarded cans, flower vases, cement tanks, ant traps, used tires and plastic buckets around human dwellings (Christophers 1960; Chan et al. 1971; Tsuzuki et al. 2009). Many domestic animals are also available for blood meals, a situation often seen in tropical countries, but *Ae. aegypti* is highly anthropophilic and feeds in the daytime. Its preference for humans as a host is an important factor for transmission (Ponlawat and Harrington 2005; Suwonkerd et al. 2006). In addition, the species prefers dark places with moisture for blood feeding and resting (Christophers 1960). The indoors and urban areas apparently provide environments, which *Ae. aegypti* prefers over the outdoors and suburban/rural areas (Fig. 5.7).
2. *Aedes albopictus* (Skuse): *Ae. albopictus* originates in the Orient and is distributed throughout tropical and temperate areas (Huang 1972; Tanaka et al. 1979). Unlike *Ae. aegypti*, *Ae. albopictus* eggs have the ability to diapause during the winter season in temperate zone Asia (Mori and Oda 1981; Hawley et al. 1987). Larvae breed from a wide variety of natural and artificial habitats, such as bamboo stumps, tree holes, discarded tires and flower vases (Huang 1972; Hawley 1988; Chan et al. 1971; Toma and Miyagi 1986). Since hosts of *Ae. albopictus* blood feeding include not only humans but also a wide range of amphibians, reptiles, birds and mammals in forests and domestic environments, the species has been regarded as a bridge vector of zoonosis pathogens (Huang 1972;

Fig. 5.7 Vector of dengue, zika virus and chikungunya (Source: Centers for Disease Control and Prevention 2024; https://www.cdc.gov/dengue/transmission/index.html)

Hawley 1988; Ponlawat and Harrington 2005; Savage et al. 1993; Niebylski et al. 1994; Gratz 2004; Kyoko et al. 2010). In the domestic environment, *Ae. albopictus* prefers vegetation for feeding and rests outdoors (Niebylski and Craig 1994; Takagi et al. 1995a, b, c; Higa et al. 2001). Because of the preference for vegetation, the density of *Ae. albopictus* is usually high in rural and suburban areas (Chan et al. 1971; Tsuda et al. 2002, 2006; Rey et al. 2006; Maciel-de-Freitas et al. 2006; Braks et al. 2003; Rodhain and Rosen 1997). Biting usually takes place at dawn and dusk, although night-time biting is sometimes also observed (Hawley 1988; Higa et al. 2000) (Fig. 5.7).

5.6 Life Cycle of Mosquito

There are four distinct stages in the life cycle of a mosquito: egg, larva, pupa and adult, each characterized by a unique appearance. The entire cycle generally takes 1–2 weeks, but the time can vary depending on the species and environmental conditions, such as temperature.

The Egg The adult female can produce eggs at intervals throughout her life, but she requires proteins from a blood meal for the eggs to develop. All mosquito species share a common characteristic: their eggs are laid either in free water or on a moist surface. The *Aedes* mosquito, for example, generally lays 100 eggs at a time. Initially, the eggs are white, but they darken to a black or dark brown colour within 12–24 h. These eggs are approximately 1/50 inch (0.5 mm) in length and appear similar to the naked eye for most species (Fig. 5.8 (1)), except for the *Anopheles* spp., which have floats attached to each side of the egg (Fig. 5.8 (3)). Some species lay eggs individually, while others lay them together to form rafts (Fig. 5.8 (2)). The incubation period, which is the time between egg deposition and hatching, can vary significantly among species. For permanent-water mosquitoes, whose eggs are deposited on the water surface, hatching may occur within 1–3 days depending on the temperature. Floodwater species, on the other hand, deposit their eggs on moist soil or another wet substrate and have a wide range of incubation periods. These eggs hatch only when submerged by rising water, such as rainfall or melting snow in the spring, or during other floodwater events. Depending on the species and environmental conditions, these eggs may hatch during the next flooding event, as soon as 10 days, or even a year or more later. Adult mosquitoes lay their eggs on the surface of the water. The egg floats for a few days (2–4 days), before hatching, allowing larvae to hatch.

Larvae The larvae of mosquitoes, also known as wigglers or wrigglers, inhabit water and undergo four distinct developmental periods, known as instars. These instars are referred to as first, second, third and fourth, with each subsequent stage being larger than the previous one. At the end of each instar, the larva undergoes moulting, a process in which it sheds its skin. During the larval stage, the larvae actively feed on organic material present in the water. Most species of larvae possess a breathing tube (respiratory siphon) and periodically come to the water's surface to obtain oxygen. The duration of the larval stage varies depending on the species and water temperature, with some species completing their development in as little as 5 or 6 days. Once the larvae reach maturity, the fourth instar moults into the pupal stage. The larvae are aquatic and actively feed and grow, undergoing multiple moults until reaching maturity. At this point, a remarkable transformation occurs as the larva sheds its skin and transforms into a comma-shaped pupa (Fig. 5.9 (1–3)).

Pupae The mosquito pupa, unlike most other insects, is highly active and resides in water, just like the larva. However, it exhibits significant differences in both shape and appearance compared to the larva. The pupa possesses a comma-shaped body that can be divided into two distinct regions. The front region, which includes the

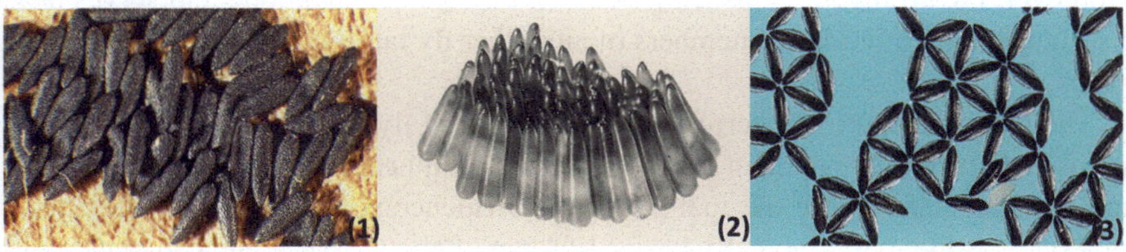

Fig. 5.8 (1–3) Eggs of different mosquito species 1) *Aedes*, 2) *Culex*, and 3) *Anopheles* (Source: Centers for Disease Control and Prevention)

Fig. 5.9 (1–3) Larvae of different mosquito 1) *Aedes*, 2) *Culex*, and 3) *Anopheles* (Source: Centers for Disease Control and Prevention)

Fig. 5.10 (1–3) Pupal stage of different mosquito species 1) *Aedes*, 2) *Culex*, and 3) *Anopheles* (Source: Centers for Disease Control and Prevention)

head and thorax (cephalothorax), is considerably enlarged and features a pair of respiratory trumpets on its upper surface. To obtain oxygen, the pupa must periodically rise to the water's surface. The second region is the abdomen, consisting of segments that can move freely and is equipped with paddle-like appendages at the tip. During the pupal stage, feeding does not occur. This stage is relatively short, lasting only a few days, and serves as the period in which all larval tissues transform into adult tissues. During the pupal stage, the insect undergoes significant anatomical changes, whereby the masticatory mouth parts of the larva are converted into the suctorial apparatus of the adult insect or imago. After a certain day, the pupal case ruptures, and the adult insect is enclosed, furnished with wings and legs adapted for a life in the air (Fig. 5.10 (1–3)).

Adult Mosquitoes can be readily differentiated from other flies based on distinct characteristics. Firstly, they are identifiable by their elongated piercing and sucking proboscis. Additionally, the wings of mosquitoes exhibit a distinctive feature, the presence of scales on the veins. This scaling sets them apart from other fly species. Moreover, the arrangement of wing veins in mosquitoes is unique and serves as a distinguishing feature from members of any other fly family.

The adult mosquito is a terrestrial insect that can fly long distances. Both male and female mosquitoes obtain energy by feeding on nectar. After emerging, males and females mate within the first 3–5 days. Evidence suggests that females mate only once, while males typically have a lifespan of about a week and can mate multiply (Cator et al. 2021). It is only the females that feed on blood, which occurs when they bite. Although blood meals provide little nourishment, females require them to develop eggs. Many mosquito species feed on warm-blooded birds or mammals, while some prefer cold-blooded animals. Certain species, like *Culex* spp., primarily feed on birds and rarely bite mammals. These mosquitoes are known to transmit the West Nile virus (WNV). Unfortunately, many species of mosquitoes feed on a wide range of warm-blooded mammals, including humans. Once a female mosquito has fully engorged herself with blood, she seeks a shaded environment to allow her eggs to develop, which usually takes 3–5 days. Once the eggs are ready, the female, now called a gravid female, searches for a suitable location to lay eggs. If she survives this process, she will soon begin searching for another blood meal and lay another batch of eggs. Unlike male mosquitoes, which will continue to mate

Fig. 5.11 (1–3) Adult stage of different mosquitoes 1) *Aedes*, 2) *Culex*, and 3) *Anopheles* (Source: Centers for Disease Control and Prevention)

until they die, most female mosquitoes will mate just once. They store sperm for the rest of their lives to fertilize tons of eggs, the amount of which depends on the ambient temperature and blood (a source of energy for egg development). Alongside the genetic material that allows sexual reproduction to take place, various proteins are also transferred from the male to the female during mating. This transmitted chemical concoction triggers changes in female behaviour, switching from mate-seeking to searching for blood to nourish her eggs (Hawkes and Hopkins 2022). Typically, a female mosquito lives long enough to lay one to three batches of eggs. Most mosquito species actively seek a blood meal during the evening hours, from just before dusk until 2–3 h after dark. During the daytime, females rest in cooler vegetated areas with higher humidity to avoid drying out. However, if humans or animals disturb these wooded areas, females may bite during the daytime. *Aedes albopictus*, on the other hand, is an aggressive mosquito species that prefers to feed during daylight hours and is often a nuisance in urban areas. When a female mosquito bites an individual infected with the Zika virus or another virus, it acquires the virus. Within the mosquito, the virus undergoes a reproductive phase, during which it moves from the mosquito's gut through the circulatory system to reach the salivary gland. Subsequently, the mosquito has the potential to transmit the virus to another individual during a subsequent biting episode (Fig. 5.11 (1–3)).

5.7 Mode of Infection

Mode of infection varies from species to species of mosquito. The details of *Anopheles* mosquito, *Aedes* mosquito and *Culex* mosquito are as follows:

5.7.1 Mode of Infection by *Anopheles* Mosquito

Malaria is caused by protozoan parasites of the genus *Plasmodium* and is a major cause of mortality and morbidity worldwide. These parasites have a complex life cycle in their mosquito vector and vertebrate hosts. Malaria parasite is transmitted through the bite of an infected female *Anopheles* mosquito. Of the approximately 400° species of *Anopheles* throughout the world, about 60 are malaria vectors under natural conditions, 30 of which are of major importance (Fig. 5.12).

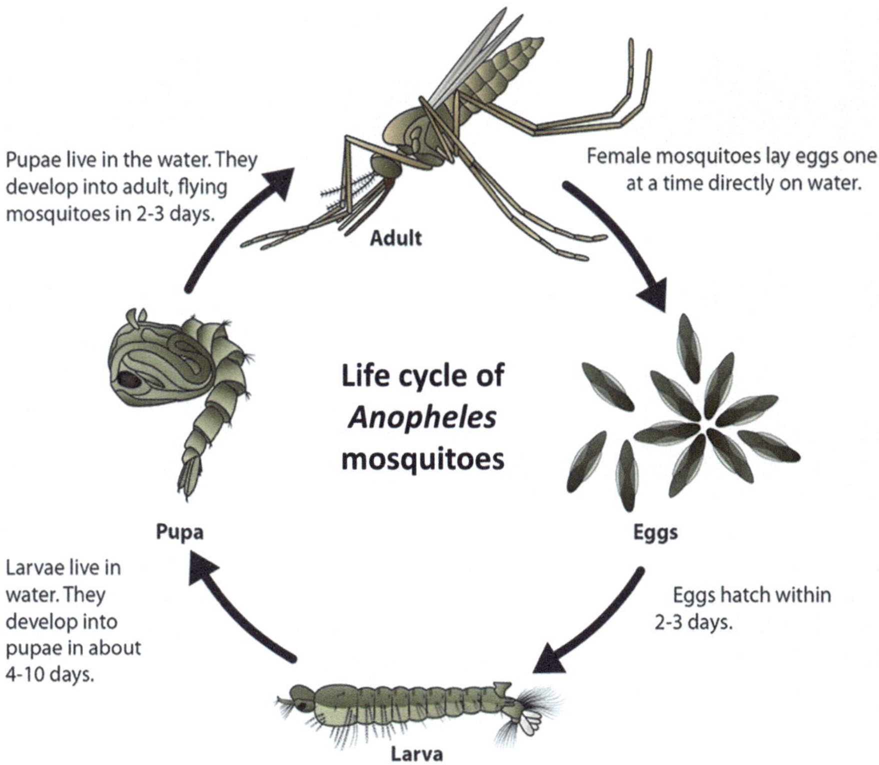

Fig. 5.12 Life cycle of *Anopheles* mosquitoes (Source: Centers for Disease Control and Prevention 2024; https://www.cdc.gov/mosquitoes/about/life-cycle-of-anopheles-mosquitoes.html)

5.7.2 Mode of Infection by *Culex* Mosquito

Arthropod-borne viruses, arboviruses, cause disease across the globe and throughout all climate zones. The transmission of arboviruses is dependent on an active arthropod vector that can transfer the virus to a vertebrate host, via so-called horizontal transmission. In some regions, vector populations will go through periods of low activity, where blood feeding ceases due to environmental circumstances, and horizontal transmission is decreased to a minimum or is completely halted. Mosquito-borne arboviruses affected by vector hibernation include both flaviviruses, e.g. West Nile virus (WNV) and Japanese encephalitis virus (JEV), and alphaviruses, e.g. Sindbis virus (SINV). These three viruses, WNV, JEV and SINV, have a tropical origin but have also become established in temperate regions. They are all transmitted by mosquitoes of the genus *Culex*, and it has been shown that they can persist throughout northern winters in hibernating adult *Culex* females (Rosen 1986; Rudolf et al. 2017; Bergman et al. 2020). These females emerge as adults in late summer, mate and feed on nectar before they enter a suitable

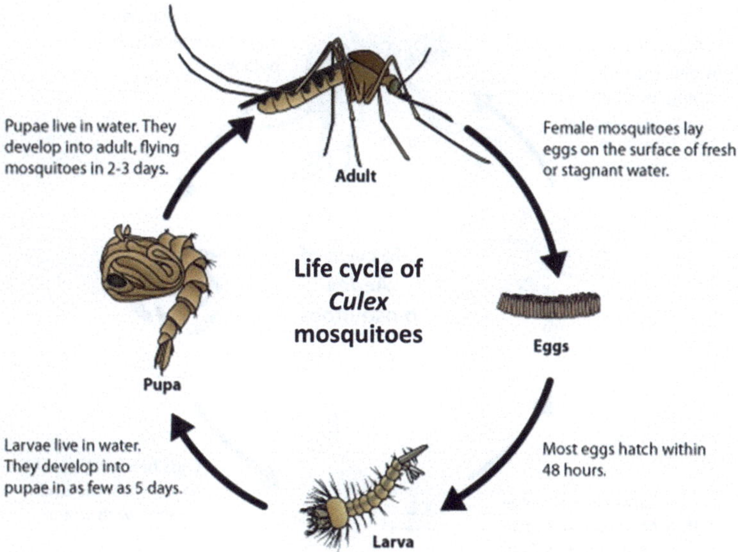

Fig. 5.13 Life cycle of *Culex* mosquitoes (Source: Centers for Disease Control and Prevention 2024; https://www.cdc.gov/mosquitoes/about/life-cycle-of-culex-mosquitoes.html)

hibernation site, which will provide shelter with temperatures close to freezing, such as cellars or caves (Service 1968). The most common hypotheses are that these females became infected either through the process of vertical transmission, i.e. when an infected female transmits the virus to her offspring, or through venereal transmission, i.e. during mating. Vertical transmission is a prerequisite for venereal transmission as males do not feed the blood and is the most commonly investigated type of alternative virus transmission. Vertical transmission has been shown experimentally in the laboratory for both alphaviruses and flaviviruses, transmitted by both *Culex* and *Aedes* vector species (Lequime et al. 2016) (Fig. 5.13).

5.7.3 Mode of Infection by *Aedes* Mosquito

The genus *Flavivirus* comprises a diverse group of viruses that phylogenetically cluster based on host or vector usage and includes insect-specific flaviviruses (ISFs), dual-host tick-borne flaviviruses, dual-host mosquito-borne flaviviruses (MBFVs) and viruses with no known vector (Gould et al. 2003). Flaviviruses within the MBFVs group, such as dengue virus (DENV 1–4), West Nile virus (WNV), yellow fever virus, Japanese encephalitis virus (JEV) and Zika virus (ZIKV) are the most recognized, since they are the causative agents of disease in humans and animals worldwide(Weaver and Reisen 2010; Weaver et al. 2018). Vertical transmission of Zika virus in *Ae. aegypti* has been demonstrated, but its role in maintaining the virus is unknown (Ciota et al. 2017; Thangamani et al. 2016) (Fig. 5.14).

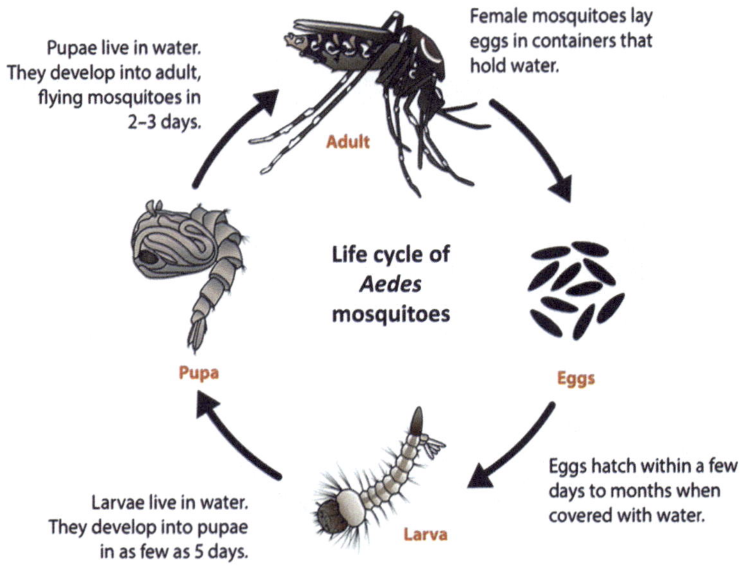

Fig. 5.14 Life cycle of *Aedes* mosquitoes (Source: Centers for Disease Control and Prevention; https://www.cdc.gov/mosquitoes/about/life-cycle-of-aedes-mosquitoes.html)

5.8 Conclusion

Mosquitoes are the carrier of pathogens that cause common human diseases, such as malaria, dengue, chikungunya, yellow fever, Zika and West Nile. They are holometabolous insects. They possess four stages in their life cycle: egg, larva, pupa and adult. Female adults feed on human blood, needed for the development and deposition of eggs. During feeding, she spreads the disease- causing organisms to humans.

Points to Remember

1. Understanding mosquito biology, including its life cycle, behaviours and feeding habits, is crucial for disease control and ecosystem preservation.
2. Mosquito life cycle: egg, larva, pupa and adult.
3. Eggs laid in water or moist areas, hatching time varies.
4. Larvae feed and grow in water and undergo four instars.
5. Pupae transform into adults, active in water.
6. Only female mosquitoes bite for blood to develop eggs.
7. Female mosquitoes transmit diseases like malaria, dengue and Zika.
8. *Anopheles* transmits malaria; *Culex* spreads Japanese encephalitis; and *Aedes* carries diseases, like dengue, chikungunya and Zika.
9. Different mosquito species have specific modes of infection.
10. Mosquitoes pose significant health risks worldwide.
11. Effective control measures are essential to mitigate mosquito-borne diseases.

References

Baisas FE (1936) Notes on Philippines Mosquitos, IV. The Pupal and certain Adult Characters of some rare Species of Anopheles, *Philippine Journal of Science* 59(1), Corpus ID: 85899946 (1980)

Bergman A, Dahl E, Lundkvist Å, Hesson JC (2020) Sindbis virus infection in non-blood-fed hibernating *Culex pipiens* Mosquitoes in Sweden. Viruses 12(12):1441. https://doi.org/10.3390/v12121441. PMID: 33327649; PMCID: PMC7765111

Braks MA, Honório NA, Lourençqo-De-Oliveira R, Juliano SA, Lounibos LP (2003) Convergent habitat segregation of Aedes aegypti and Aedes albopictus (Diptera: Culicidae) in southeastern Brazil and Florida. J Med Entomol 40(6):785–794. https://doi.org/10.1603/0022-2585-40.6.785. PMID: 14765654.

Cator LJ, Wyer CAS, Harrington LC (2021) Mosquito sexual selection and reproductive control programs. Trends Parasitol 37(4):330–339. https://doi.org/10.1016/j.pt.2020.11.009. Epub 2021 Jan 7. PMID: 33422425; PMCID: PMC8454880

Centers for Disease Control and Prevention (2024) National Center for Emerging and Zoonotic Infectious Diseases (NCEZID), Division of Vector-Borne Diseases (DVBD). https://www.cdc.gov/dengue/transmission/index.html. Accessed April 2024

Chan KL, Ho BC, Chan YC (1971) *Aedes aegypti* (L.) and *Aedes albopictus* (Skuse) in Singapore City. 2. Larval habitats. Bull World Health Organ 44:629–633

Christophers R (1960) *Aedesaegypti* (L.) the yellow fever mosquito. Cambridge University Press, London

Ciota AT, Bialosuknia SM, Ehrbar DJ, Kramer LD (2017) Vertical transmission of Zika virus by Aedes aegypti and Ae. albopictus mosquitoes. Emerg Infect Dis 23(5):880–882. https://doi.org/10.3201/eid2305.162041. Epub 2017 May 15. PMID: 28277199; PMCID: PMC5403030

Dash AP, Adak T, Raghavendra K, Singh OP (2007) The biology and control of malaria vectors in India. Curr Sci 92(11):1571–1578. http://www.jstor.org/stable/24097721

Giles GM (1901) Description of four new species of Anopheles from India. Entomol Monthly Magaz 37:196–198

Gould EA, de Lamballerie X, Zanotto PM, Holmes EC (2003) Origins, evolution, and vector/host coadaptations within the genus Flavivirus. Adv Virus Res 59:277–314. https://doi.org/10.1016/s0065-3527(03)59008-x. PMID: 14696332

Gratz NG (2004) Critical review of the vector status of Aedes albopictus. Med Vet Entomol 18(3):215–227. https://doi.org/10.1111/j.0269-283X.2004.00513.x. PMID: 15347388

Harbach RE, Knight KL (1980) Taxonomists glossary of mosquito anatomy. Plexus Marlton, Medford

Hawkes FM, Hopkins RJ (2022) The mosquito: an introduction. In: Hall M, Tamïr D (eds) Mosquitopia: the place of pests in a healthy world. Routledge, New York. https://doi.org/10.4324/9781003056034-3

Hawley W (1988) The biology of *Aedes albopictus*. J Am Mosq Control Assoc 1(Suppl):1–39

Hawley WA, Reiter P, Copeland RS, Pumpuni CB, Craig GB (1987) Aedes albopictus in North America: probable introduction in used tires from northern Asia. Science 236(4805):1114–1116. https://doi.org/10.1126/science.3576225. PMID: 3576225

Higa Y, Tsuda Y, Tuno N, Takagi M (2000) Tempo-spatial variation in feeding activity and density of Aedes albopictus (Diptera: Culicidae) at peridomestic habitat in Nagasaki, Japan. Med Entomol Zool 51(3):205–209. https://doi.org/10.7601/mez.51.205

Higa Y, Tsuda Y, Tuno N, Takagi M (2001) Preliminary field experiments on exophagy of *Aedes albopictus* (Diptera: Culicidae) in peridomestic habitat. Med Entomol Zool 52(2):105–116

Huang Y-M (1972) Contributions to the mosquito fauna of Southeast Asia. XIV. The subgenus *Stegomyia* of *Aedes* in Southeast Asia I – the *scutellaris* group of species. Contrib Am Entomol Inst 9(1):1–110

James SP (1902) Malaria in India. Sci Mem 2:160

Kitzmiller JB (1953) Mosquito genetics and cytogenetlcs. Rev Bras Malariol Doencastrop 5(4):285–359

Kitzmiller JB (1976) Genetics, cytogenetics, and evolution of mosquitoes. Adv Genet 18:315–433

Kyoko S, Isawa H, Hoshino K, Sasaki T, Roychoudhury S, Higa Y, Kasai S, Tsuda Y, Nishiumi I, Hisai N, Hamao S, Kobayashi M (2010) Host-feeding habits of *Culex pipiens* and *Aedes albopictus* (Diptera: Culicidae) collected at the urban and suburban residential areas of Japan. J Med Entomol 47(3):442–450. https://doi.org/10.1093/jmedent/47.3.442

Lequime S, Paul RE, Lambrechts L (2016) Determinants of arbovirus vertical transmission in mosquitoes. PLoS Pathog 12:e1005548. https://doi.org/10.1371/journal.ppat.1005548

Liston WG (1901) A year's experience of the habits of Anopheles in Ellichipur. The description of species found in Ellichipur during the year. Ind Med Gaz 36:361–366

Maciel-de-Freitas R, Neto RB, Gonçalves JM, Codeço CT, Lourenço-De-Oliveira R (2006) Movement of dengue vectors between the human modified environment and an urban forest in Rio de Janeiro. J Med Entomol 43:1112–1120

Mattingly PF (1957) Genetic aspects of the *Aedes aegypti* Problem. I.-Taxonomy and bionomics. Ann Trop Med Parasitol 51:392–408

Mori A, Oda T (1981) Studies on the egg diapause and overwintering of *Aedes albopictus* in Nagasaki. Trop Med 23:79–90

Nagpal BN, Sharma VP (1995) Indian anophelines. New Delhi, Oxford IBH, p 416

Niebylski ML, Craig GB (1994) Dispersal and survival of Aedes albopictus at a scrap tire yard in Missouri. J Am Mosq Control Assoc 10(3):339–343. PMID: 7807074

Niebylski ML, Savage HM, Nasci RS, Craig GB (1994) Blood hosts of Aedes albopictus in the United States. J Am Mosq Control Assoc 10(3):447–450. PMID: 7807094

Peyton EL, Harrison BA (1979) *Anopheles (Cellia) dirus*, a new species of the *Leucosphyrus* group from Thailand (Diptera: Culicidae). Mosq Syst 11:40–52

Ponlawat A, Harrington LC (2005) Blood feeding patterns of Aedes aegypti and Aedes albopictus in Thailand. J Med Entomol 42(5):844–849. https://doi.org/10.1093/jmedent/42.5.844. PMID: 16363170

Rai KS (1991) Structural and functional aspects of mosquito genomes. In: Eukaryotic chromosomes: structural and functional aspects. Narosa Publishing House, New Delhi, pp 52–57

Rai KS, Pashley DP, Munstermann LE (1982) Genetics of speciation in *Aedine* mosquitoes. In: Recent developments in the genetics of insect disease vectors. Oxford University Press, Oxford, pp 84–128. (a)

Rey JR, Nishimura N, Wagner B, Braks MA, O'Connell SM, Lounibos LP (2006) Habitat segregation of mosquito arbovirus vectors in south Florida. J Med Entomol 43(6):1134–1141

Rodenwaldt E (1925) Entomologische notities. III. Geneeskundig Tijdschrift Nederlandsch Indie Batavia 65:173–201

Rodhain FR, Rosen L (1997) Mosquito vectors and dengue virus-vector relationships. In: Dengue and dengue hemorrhagic fever. CAB International, New York, pp 45–60

Rosen L (1986) The natural history of Japanese encephalitis virus. Ann Rev Microbiol 40:395–414. https://doi.org/10.1146/annurev.mi.40.100186.002143. PMID: 2877613

Rudolf I, Betášová L, Blažejová H, Venclíková K, Straková P, Šebesta O, Mendel J, Bakonyi T, Schaffner F, Nowotny N et al (2017) West Nile virus in overwintering mosquitoes, central Europe. Parasit Vectors 10:452. https://doi.org/10.1186/s13071-017-2399-7

Savage HM, Niebylski ML, Smith GC, Mitchell CJ, Craig GB Jr (1993) Host-feeding patterns of Aedes albopictus (Diptera: Culicidae) at a temperate North American site. J Med Entomol 30(1):27–34. https://doi.org/10.1093/jmedent/30.1.27. PMID: 8433336

Schäfer M, et al (1997) Dispersal behavior of adult snow melt mosquitoes in the Upper Rhine Valley, Germany, PMID: 9221732

Service (1968) Observations on the ecology of some British mosquitoes. Bull Entomol Res 59:161–194. https://doi.org/10.1017/S000748530000314X

Suwonkerd W, Mongkalangoon P, Parbaripai A, Grieco J, Achee N, Roberts D, Chareonviriyaphap T (2006) The effect of host type on movement patterns of Aedes aegypti (Diptera: Culicidae) into and out of experimental huts in Thailand. J Vector Ecol 31(2):311–318

Takagi M, Tsuda Y, Suzuki A, Wada Y (1995a) Movement of individually marked *Aedes albopictus* females in Nagasaki, Japan. Trop Med 37:79–85

Takagi M, Tsuda Y, Wada Y (1995b) Movement and oviposition of released *Aedes albopictus* (Diptera: Culicidae) in Nagasaki, Japan. Jpn J Sanit Zool 46(2):131–138

Takagi M, Tsuda Y, Wada Y (1995c) Temporal and spatial distribution of released *Aedes albopictus* (Diptera: Culicidae) in Nagasaki, Japan. Jpn J Sanit Zool 46:223–228

Tanaka K, Mizusawa K, Saugstad ES (1979) A revision of the adult and larval mosquitoes of Japan (including the Ryukyu Archipelago and the Ogasawara islands) and Korea (Diptera, Culicidae). Contrib Am Entomol Inst 16:1–987

Thangamani S, Huang J, Hart CE, Guzman H, Tesh RB (2016) Vertical transmission of Zika virus in Aedes aegypti mosquitoes. Am J Trop Med Hyg 95(5):1169–1173. https://doi.org/10.4269/ajtmh.16-0448. Epub 2016 Aug 29. PMID: 27573623; PMCID: PMC5094235

Theobald FV (1901) A monograph of the culicidae or mosquitoes. William Clowes and Sons, London

Toma T, Miyagi I (1986) The mosquito fauna of the Ryukyu Archipelago with identification keys, pupal descriptions and notes on biology, medical importance and distribution. Mosq Syst 18(1):1–109

Tsuda Y, Kobayashi J, Nambanya S, Miyagi I, Toma T, Phompida S, Manivang K (2002) An ecological survey of dengue vector mosquitos in central Lao PDR. Southeast Asian J Trop Med Public Health 33(1):63–67. PMID: 12118462

Tsuda Y, Suwonkerd W, Chawprom S, Prajakwong S, Takagi M (2006) Different spatial distribution of *Aedes aegypti* and *Aedes albopictus* along an urban-rural gradient and the relating environmental factors examined in three villages in northern Thailand. J Am Mosq Control Assoc 22:222–228

Tsuzuki A, Vu TD, Higa Y, Nguyen TY, Takagi M (2009) High potential risk of dengue transmission during the hot-dry season in Nha Trang City, Vietnam. Acta Trop 111(3):325–329. https://doi.org/10.1016/j.actatropica.2009.05.010. Epub 2009 May 23. PMID: 19467217

Weaver SC, Reisen WK (2010) Present and future arboviral threats. Antivir Res 85(2):328–345. https://doi.org/10.1016/j.antiviral.2009.10.008. Epub 2009 Oct 24. PMID: 19857523; PMCID: PMC2815176

Weaver SC, Charlier C, Vasilakis N, Lecuit M, Zika C (2018) Other emerging vector-borne viral diseases. Annu Rev Med 69:395–408. https://doi.org/10.1146/annurev-med-050715-105122. Epub 2017 Aug 28. PMID: 28846489; PMCID: PMC6343128

Mosquito Host Interaction

6

Dipanwita Das, Pranesh Paul, Sampa Banerjee, Milita Roy, Shreya Brahma, Goutam K. Saha, and Gautam Aditya

Abstract

Blood acts as a source of metabolic energy in female mosquitoes and is also essential for viable egg production and higher reproductive fitness. In this context, the host-seeking behaviour of mosquitoes is essential for their survival as well as for the transmission of pathogens from one host to another. Although several mosquitoes show non-specific host preference, most of the disease-transmitting species are host specific and can be classified as opportunistic, mammophilic, ornithophilic, and specialized species. The mosquitoes search the host for blood-feeding through many potential cues, including exhaled carbon dioxide, skin odours, host's visual and heat signatures, and moisture. Among different cues, the mosquitoes perceive odours through the olfactory receptors within the hair-like sensilla situated on the antenna and maxillary palps. The volatile skin odours of the host, specifically produced by the normal microbiota present on the skin, are known to serve as one of the important cues for host finding by the mosquitoes. On the other side, the pathogens (e.g. flavivirus, *Plasmodium*) present in the infected host can promote host attractiveness to the mosquitoes by reshaping the spectrum of host volatile odorants. The information regarding the relationship between host microbiota, mosquito attraction, and

D. Das
Department of Zoology, Bagnan College, Bagnan, West Bengal, India

P. Paul · S. Banerjee · S. Brahma · G. K. Saha · G. Aditya (✉)
Department of Zoology, University of Calcutta, Kolkata, West Bengal, India
e-mail: gazoo@caluniv.ac.in

M. Roy
Department of Zoology, Bejoy Narayan Mahavidyalaya, Itachuna, West Bengal, India

© The Author(s), under exclusive license to Springer Nature Singapore Pte Ltd. 2024
Omkar (ed.), *Mosquitoes*, https://doi.org/10.1007/978-981-97-4163-2_6

pathogen transmission can be crucial for developing targeted strategies to mitigate the spread of mosquito-borne diseases.

Keywords

Flavivirus · Microbiota · Olfactory receptor · Ionotropic receptor · Gustatory receptor · *Plasmodium* · Skin odour

Learning Objectives
1. Diverse types of mosquitoes depending on the hosts.
2. Host-seeking behaviour in mosquito—extrinsic and intrinsic factors.
3. Sensory factors and cues for detecting hosts by mosquitoes.
4. Host skin microbiota and odour cues for mosquitoes.
5. Pathogens as mediators in host finding by mosquitoes.

6.1 Introduction

The host-seeking behaviour of disease vectors is essential for the transmission of pathogens from one host to another. Disease vectors, such as mosquitoes, ticks, and flies, actively seek out hosts to obtain blood meals, which is essential for their survival and reproduction. When a disease vector feeds on an infected host, it can acquire pathogens, such as viruses, bacteria, or parasites. Subsequently, when the insect feeds on a new host, it can transmit these pathogens, thereby spreading diseases to the uninfected individuals. To locate suitable hosts, the disease vectors use a variety of cues, including visual, olfactory, and thermal cues. For instance, the body louse, *Pediculus humanus humanus,* is receptive to host odours, temperature, humidity, light, and contact cues (Wigglesworth 1941) through the chemo-, mechano- and thermo-hygro receptors present on its antenna (Keilin and Nuttall 1930; Miller 1969; Steinbrecht 1994). Among the disease vectors, a large number of mosquito species are involved in the transmission of virus, protozoan, and nematode pathogens to a range of vertebrates, including human and domestic animals. Several infectious diseases, such as malaria, dengue, Japanese encephalitis, chikungunya, lymphatic filariasis, and many more, are transmitted by mosquitoes. Along with plant sugars, blood acts as a source of metabolic energy in mosquitoes and is essential for viable egg production and higher reproductive fitness (Takken and Verhulst 2013). Although most mosquito species show non-specific host preferences, most disease-transmitting species are host-specific (Takken and Verhulst 2013).

In mosquitoes, the feeding pattern including the blood meal is a crucial factor influencing several life history traits, like longevity and fecundity. Complying with the evolutionary norms, extended longevity and higher fecundity are favoured for maximizing the reproductive success and fitness at the individual and population levels. On a comparative scale, extended longevity will increase the chance of multiple blood meal episodes as well as the increased risk of transmission of the diseases. Besides, the reproductive success, in terms of the eggs oviposited, will be quite higher, and thus the fitness level will also be superior. A successful blood meal by a female mosquito ensures egg production with the desired quality while ensuring longevity. On the other hand, the quality and quantity of the blood meal will determine the longevity and fecundity of the mosquitoes. While the quantity of the blood sucked is dependent on the mosquito body size, which in turn is linked with the larval feeding activity and the developmental pattern, the quality of the blood is dependent on the hosts selected. The essential components of the blood vary with the hosts, and thus, the effective production of the eggs and longevity of the mosquito vary accordingly. Differences in the quantity of selected amino acids may change the course of fecundity in mosquitoes as observed for *Aedes aegypti* feeding on human and rodents (Harrington et al. 2001). An 'only blood' meal in contrast to 'blood and sucrose' meal in mosquitoes may bear different consequences for fecundity as well as longevity (Harrington et al. 2001; Braks et al. 2006). Thus, selection of the host and the type of meal that the female mosquitoes select determines the reproductive success and fitness at the individual and the population levels. Selection of the host is of prime importance for the successful accomplishment of the life history of mosquitoes.

Mosquitoes exhibit a diverse extent of host selection and specificity toward the host that includes amphibians, birds, and mammals, including humans. The blood-feeding activity of mosquitoes is guided by several intrinsic and extrinsic factors, irrespective of the species and the corresponding host. Apart from physiological state, including the completion of mating, the environmental condition, and availability of the cues from the hosts, guides the host-seeking behaviour for successful blood feeding. Host-seeking behaviour has been studied in several blood-feeding insects, like *Culicoides* midges (Raich et al. 1997; Ayllón et al. 2014), bed bugs (Liu and Liu 2015; Singh et al. 2015), and mosquitoes (Raji and DeGennaro 2017; Zhang et al. 2022; Coutinho-Abreu et al. 2022). Although species-specific variations have been observed in respect to the blood feeding in mosquitoes, the pattern of host seeking remain similar, which is dependent on the odour emanating from the host (Takken and Knols 1999; Cardé 2015; de Obaldia et al. 2022; Zhang et al. 2022), the carbon dioxide and the thermal conditions of the ambient environment, and the visual reception by the mosquitoes. In reality, the integration of multiple stimuli by the mosquito sensory system enables successful host detection and landing on the host for the blood meal (Raji and DeGennaro 2017). In the present narrative, a brief idea about the classification of the mosquitoes based on the host specificity, the factors affecting the host-seeking behaviour of mosquitoes, and the underlying sensory mechanisms is emphasized to highlight the mosquito-host interactions. An outline of the microbe-mediated modification of the odorant production for facilitating mosquito host-seeking behaviour is provided to understand the key issues in the host preferences and host choice in mosquitoes.

6.2 Classification of Mosquitoes According to the Host Preference

The mosquitoes exhibit a wide range of host preferences, which can significantly impact their behaviour, ecology, and role in disease transmission (Table 6.1). Host preference is the trait by which mosquitoes selectively target certain host species over others for blood feeding (Chaves et al. 2010; Takken and Verhulst 2013).

1. Opportunistic species: Many mosquito species are opportunistic feeders, which indicate that their host preference is primarily driven by factors such as host abundance, availability, and environmental conditions. These mosquitoes may feed on various hosts, including mammals, birds, reptiles, and even humans, depending on what is readily accessible in their habitat.
2. Mammophilic species: Some mosquito species display a strong preference for feeding on mammals. These mosquitoes may preferentially target mammals,

Table 6.1 The mosquito species and the hosts and the mechanisms defining the specificity for blood-feeding host

Mosquito species	Host species	Specificity	References
Aedes aegypti	Human	Olfactory cues, visual and heat stimuli	Cardé (2015), Dekker et al. (2005), Verhulst et al. (2018), Alberto et al. (2022)
Anopheles gambiae	Human	Olfactory cues, visual and heat stimuli	Cardé (2015), Dekker et al. (2005), Verhulst et al. (2018)
An. stephensi	Human	Olfactory cues, visual and heat stimuli	Alberto et al. (2022)
Culex quinquefasciatus	Human	Olfactory cues, visual and heat stimuli	Lacey and Cardé (2011), Verhulst et al. (2018), Alberto et al. (2022)
Cx. quinquefasciatus	Aves	Skin volatiles	Burkett-Cadena et al. (2008), Spanoudis et al. (2020)
Cx. territans	*Rana clamitans*, *Rana catesbeiana*, *Pseudacris crucifer*, *Hyla versicolor*, *Hyla cinerea*, reptile, mammal	Olfactory cues, vocalization	Bartlett-Healy et al. (2008), Burkett-Cadena et al. (2011, 2013), Ferguson and Smith (2012)
Cx. peccator	Amphibian, Aves, reptile, mammal	Vocalization (for amphibian)	Burkett-Cadena et al. (2008, 2011, 2013)
Cx. pipiens molestus	Aves	Skin volatiles	Spanoudis et al. (2020)
Uranotaenia lowii	Amphibian	Vocalization	Borkent and Belton (2006), Bartlett-Healy et al. (2008), Burkett-Cadena et al. (2011, 2013)

such as humans, livestock, or other mammals present in their environment. Examples of mammophilic mosquitoes include species within the genera *Aedes* and *Culex*, which are known vectors for diseases like dengue fever, Zika virus, and West Nile virus.

3. Ornithophilic species: Other mosquito species exhibit a preference for feeding on birds. These ornithophilic mosquitoes are often associated with avian habitats and may play a significant role in the transmission of diseases among bird populations. Mosquito species within the genera *Culiseta* and *Coquillettidia* are examples of ornithophilic mosquitoes that can transmit pathogens, such as avian malaria and West Nile virus (Njabo et al. 2009).
4. Specialized feeders: Some mosquito species are highly specialized and preferentially feed on specific vertebrate groups, such as reptiles or amphibians. For instance, *Culex territans*, *Uranotaenia lowii,* and *Deinocerites* spp. prefer amphibian hosts (Borkent and Belton 2006; Bartlett-Healy et al. 2008; Burkett-Cadena et al. 2008), and *Culex peccator* prefers to feed on reptiles (Burkett-Cadena et al. 2008). These mosquitoes may have evolved unique adaptations to target and feed on their preferred hosts.

6.3 Host-Seeking Behaviour of Mosquito

Mosquito host-seeking behaviour refers to the series of actions mosquitoes undertake to locate and feed on a suitable host for a blood meal. This behaviour is essential for the survival and reproduction of mosquitoes, as blood serves as a critical nutrient source, especially for female mosquitoes, which require blood for egg development. Hence, the female mosquito must find a human host for successful reproduction. Many potential cues are available to guide such navigation, including exhaled carbon dioxide, a range of skin odours, the host's visual and heat signatures, and moisture (Potter 2014; Cardé 2015; van Breugel et al. 2015; Raji and DeGennaro 2017; de Obaldia et al. 2022; Zhang et al. 2022). Recent work is shedding new light on how the mosquito integrates these factors in targeting a human host (Braks and Takken 1999; Bosch et al. 2000; Lacey and Cardé 2011; van Breugel et al. 2015; Hinze et al. 2021). Comparable to other insects, like *Culicoides* midges (Raich et al. 1997; Ayllón et al. 2014), and bed bugs (Liu and Liu 2015; Singh et al. 2015), the mosquitoes perceive odours through the olfactory receptors within the hair-like sensilla situated on the antenna and maxillary palps (Potter 2014; Cardé 2015; Raji and DeGennaro 2017; Coutinho-Abreu et al. 2022). After the odour molecules bind with the olfactory receptors and open the ion pore, the olfactory receptor neuron depolarizes and propagates an action potential. Although most olfactory receptor neurons express a single type of olfactory receptor, the polycistronic mRNA transcription of olfactory receptor genes can lead to the presence of multiple olfactory receptors in a single olfactory receptor neuron, which leads to the response of the olfactory receptor neuron toward a blend of multiple odour compounds (Wolff and Riffell 2018). Changes in the olfactory receptor gene sequence can alter the odour compound binding properties, which alter the sensitivity, tuning, and specificity for odour compounds within various mosquito species. Similarly, host preference can fluctuate across populations of same species due to the environment and allelic variation and differences in olfactory receptor gene expressions (Wolff and Riffell 2018). For example, the forest-dwelling *Aedes aegypti* show strong zoophilic behaviour, while the domestic subspecies exhibit high preference for humans (McBride et al. 2014).

Apart from the types of mosquitoes based on preferred host, the human-blood feeding mosquitoes can be broadly divided into exophilic and endophilic mosquitoes. The exophilic mosquito species tend to rest outdoors and are ecologically independent of the human and domestic environment, while the endophilic mosquitoes exhibit preference to rest in indoors within human dwellings. However, few mosquito species, such as *Anopheles gambiae*, were earlier known as endophilic mosquitoes but now seek human hosts equally in indoors and outdoors (Reddy et al. 2011). This finding indicates that the application of indoor mosquito repellent may result in adaptive shifts of endophilic mosquitoes to seek hosts outside of human dwellings.

The mosquitoes can learn and adapt their host-seeking behaviour according to the negative and positive experiences associated with blood meals in a specific location or under the presence of specific host cue (Takken and Verhulst 2013). For instance, *Anopheles gambiae* and *Culex quinquefasciatus* can learn to associate odours with rewarding blood meals (Tomberlin et al. 2006; Chilaka et al. 2012). Similarly, *Culex vishnui* exhibits an inclination to feed on the host species to which they were attracted earlier and obtained successful blood meal (Mwandawiro et al. 2000). However, such kind of behavioural adaptations were not exhibited by the offspring suggesting that the behavioural and physiological conditioning regarding the host preference is limited to the parents only.

6.4 Factors Affecting the Host-Seeking Behaviour

The process of successful reproduction in female mosquitoes involves finding a human host, and they employ various cues to navigate and locate suitable hosts for blood feeding. Here are some critical cues involved in this process:

1. Carbon dioxide (CO_2): Exhaled carbon dioxide is a primary attractant for mosquitoes. They are highly sensitive to changes in CO_2 levels, and this serves as a crucial long-range cue for them to locate potential hosts (Cardé 2015; Dekker et al. 2005).
2. Skin odours: Mosquitoes are attracted to a variety of skin odours produced by humans. These odours are influenced by factors, such as body temperature, sweat, and the presence of specific chemicals on the skin.
3. Visual and heat signatures: Mosquitoes use visual cues to orient themselves toward potential hosts. They are particularly attracted to movement, which helps them identify a living, warm-blooded host. The heat signature emitted by the host's body is another crucial cue.

4. Moisture: Mosquitoes are attracted to moisture and often thrive in humid environments. The presence of moisture, such as sweat on the skin, can be an additional cue for mosquitoes.

By relying on a combination of these cues, female mosquitoes can locate suitable hosts for blood feeding. For instance, in the absence of wind, the orientation of the mosquitoes may largely be determined by visual features or odour cues present in the environment. The mosquitoes find the host through the direction of odour gradient (from low in the air to high near the host) by a mechanism known as 'klinotaxis.' One of the most studied host odours is 1-octen-3-ol, which is a mushroom alcohol with an earthy smell present in the sweat and breath of humans as well as several mammal species. This specific odour compound acts synergistically with CO_2 levels. Interestingly, different species often act contrarily in the presence of host odours, such as 1-octen-3-ol. For example, *Anopheles gambiae* and *Aedes aegypti* show attraction behaviour in the presence of CO_2 and 1-octen-3-ol, whereas *Culex quinquefasciatus* shows repellent behaviour in the presence of the same (Wolff and Riffell 2018).

6.5 Role of Skin Microbiota in Host Finding of Mosquitoes

Earlier research (Schreck and James 1968) marked a significant milestone in understanding the role of the skin microbiota in attracting mosquitoes. Their study demonstrated that the volatile compounds produced by bacteria, specifically *Bacillus cereus*, were attractive to the *Aedes aegypti* mosquito. Since then, numerous studies have supported the idea that human skin bacteria produce volatiles that serve as cues for mosquitoes during host-seeking behaviour. These findings underscore the influence of skin microbiota on mosquito attraction and host selection (De Jong and Knols 1995; Braks et al. 2000; Verhulst et al. 2009, 2018).

The symbiotic microbial communities inhabiting the skin of mammals and birds play a crucial role in odour production through the generation of various volatile compounds. These compounds, such as linear alcohols, methyl ketones, and carboxylic acids, contribute to the distinct odour profiles associated with different species and individuals (Madigan et al. 2010). In humans, for instance, the odour from axillary glands has been linked to *Corynebacteria*, which produce volatile fatty acids (Leyden et al. 1981). Similarly, *Staphylococcus* spp., which is common in both humans and birds, can metabolize branched-chain amino acids into volatile and odorous short-chain amino acids (James et al. 2004). In birds, volatile compounds found in uropygial gland secretions have been associated with several genera of bacteria, including *Pseudomonas* and *Staphylococcus* (Whittaker et al. 2019). The composition and abundance of these microbial communities can vary not only between species but also among individuals within a species (Theis et al. 2012; Leclaire et al. 2014). As a result, each individual may have a unique odour profile influenced by their specific skin or feather microbiota. These olfactory cues are not only indicative of individual identity but also carry information about factors such as sex, age, and social interactions (Fierer et al. 2008; Theis et al. 2012; Leclaire et al. 2014). Understanding the role of skin and feather microbiota in odour production and chemical communication provides valuable insights into animal behaviour, social dynamics, and ecological interactions.

The unique composition of skin microbiota in each individual contributes to the generation of distinct odour profiles, which in turn drive mosquito host-seeking behaviour. Mosquitoes use these odour cues to locate potential hosts for blood feeding. By targeting the interactions between mosquitoes and the skin microbiota of their hosts, researchers may uncover new avenues for mitigating the transmission of diseases such as malaria, dengue fever, and Zika virus.

The intraspecific differences in attractiveness to mosquitoes observed in humans and likely in other animals highlight the significant role of microbiota in mediating mosquito host-seeking behaviour. Studies have identified specific bacterial species, such as *Pseudomonas aeruginosa* and *Staphylococcus epidermis*, as key determinants of attractiveness to mosquitoes (Verhulst et al. 2011). Research on the *Anopheles gambiae* species complex has shown that they are not attracted to the odour produced by *P. aeruginosa*, while the odour of *S. epidermis* is attractive to them. This suggests that the composition of skin microbiota can directly influence the preference of mosquitoes for certain individuals.

Furthermore, studies have revealed that individuals who are more attractive to mosquitoes tend to have less diverse microbiota but host a higher abundance of specific bacteria, such as *Staphylococcus* (Verhulst et al. 2011). Conversely, less attractive individuals have a higher abundance of bacteria, like *Pseudomonas* (Ruiz-López 2020). This pattern underscores the role of microbiota composition in determining individual attractiveness to mosquitoes. These intraspecific differences in attractiveness have significant implications for pathogen exposure and transmission dynamics. More attractive individuals bitten more frequently by mosquitoes have an increased likelihood of being infected by mosquito-borne pathogens, which can contribute to the amplification and transmission of pathogens within human populations.

6.6 Role of Pathogens in Host Finding of Mosquitoes

The volatile compounds are one of the key factors that manipulate the host preference and attractiveness of mosquitoes. Apart from the direct influence of normal microbiota, the pathogens present in the infected host can promote host attractiveness to the mosquitoes by reshaping the spectrum of host volatile odorants. For instance, malaria-infected mice often show increased skin volatile emission, which was observed to be more attractive to the mosquito vectors (De Moraes et al. 2014). Similarly, a *Plasmodium*-produced isoprenoid precursor along with the increasing release of volatiles by human RBCs positively affects the blood meal-seeking and feeding behaviours of mosquitoes (Emami et al. 2017). Comparable to the protozoan parasites, mosquito-transmitted flaviviruses (e.g. dengue and Zika virus) are reported to manipulate the skin microbiota and produce specific scents that in turn attract other mosquitoes (Zhang et al. 2022). In mice, the flavivirus infection results in the suppression of an essential antimicrobial protein, RELMα, which promotes the growth of acetophenone-producing bacteria in the host skin. The volatile compound acetophenone effectively stimulates mosquito olfaction for attractiveness to the host. Incidentally, the dengue-infected hosts emitted more acetophenone in comparison to the healthy people (Zhang et al. 2022).

The mosquito host-seeking behaviour, host specificity, and blood meal accomplishment are based on a foundation of coordinated efforts of olfaction, vision, and thermal reception modulated by the CO_2 and odour emission from the host (Fig. 6.1).

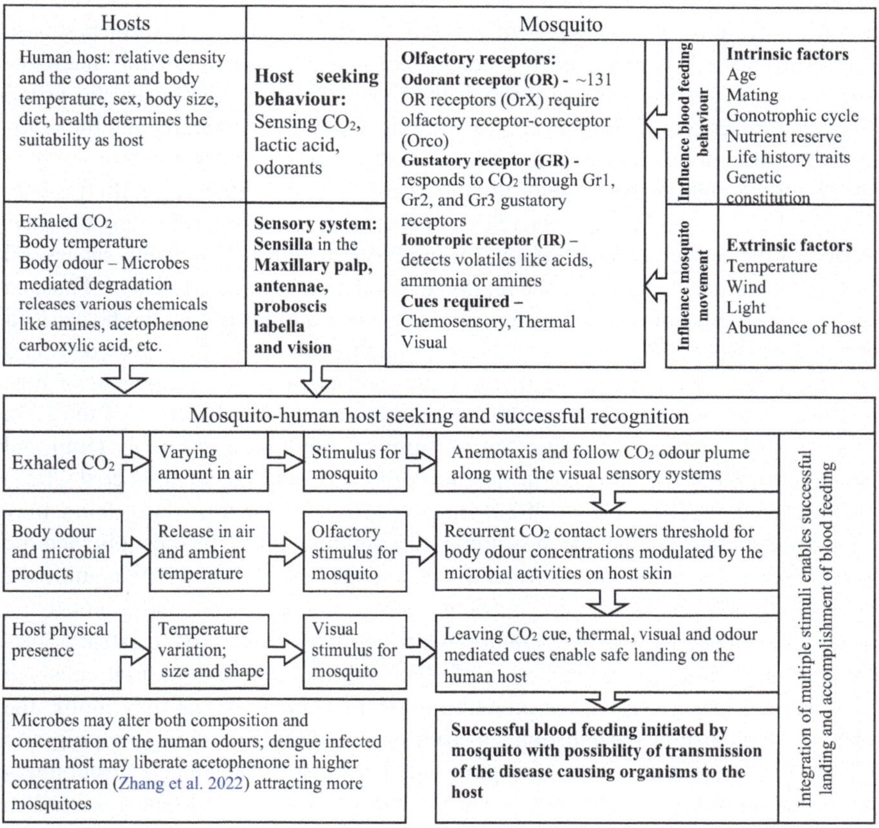

Fig. 6.1 The flow chart representing the various aspects of the mosquito host-seeking behaviour including the sensory systems and the corresponding cues. Apart from the genetic constitution of the mosquito individual, the sensilla associated with the olfactory, gustatory, and ionotropic receptors are significant in maintaining the host-seeking behaviour. Mutations related to the olfactory receptor coreceptor (*orco* receptor) may hinder host-seeking behaviour, but CO_2 and lactic acid-linked chemical cues may enable successful landing on the host body. The information mentioned above were collected from various sources (Bowen 1991; Takken and Knols 1999; Takken and Verhulst 2013; Potter 2014; Cardé 2015; van Breugel et al. 2015; Raji and DeGennaro 2017; Coutinho-Abreu et al. 2022; Zhang et al. 2022). As shown here, the integration of the three major stimuli in the form of host exhaled CO_2, host emanated odorants, and the ambient temperate by the sensory system of the mosquito enables landing on the human host successfully for a blood meal

The chemical and gustatory cues are received and processed by the sensilla in the antennae and maxillary palp, along with the vision and the taste senses through the legs and proboscis, in a concerted effort. At an initial state, the mosquitoes rely on the CO_2 levels in the air, followed by temperature sensing through vision and further aid from the odorants emitted from the hosts. An integration of multiple stimuli is adopted by the mosquitoes to find the hosts (Raji and DeGennaro 2017), which are received through several olfactory, gustatory, and ionotropic receptors. Among the olfactory receptors (OR receptors), around 131 (OrX) will require olfactory receptor-coreceptor (orco) to be functional. The response to the CO_2 in the air is mediated by the gustatory receptors (GR), and the odorants or volatiles emanating from the hosts are sensed through the ionotropic receptor. The response to the CO_2 stimuli is mediated by the capitate peg (cp) sensilla of the maxillary palp. Following the initial cue through the CO_2 exhaled by the hosts, the visual senses are in action that follows the CO_2 plume and the temperature, and the odorant of the host body enables safe landing on the host. Olfactory receptor neurons (ORNs) linked with expression of the orco gene extends between labella (proboscis) and the suboesophageal zone (SEZ) of the mosquito brain, reflecting the integration of the odour and taste stimuli during blood feeding. As a result, successful host finding and completion of the requisite blood meal are accomplished. Thus, an integration of multiple stimuli is adopted by the mosquitoes to find the hosts (Raji and DeGennaro 2017), which is received through several olfactory, gustatory, and ionotropic receptors. While the thermal cues and lactic acid (component of host skin odour) are not attractive on their own, but with the combination of the CO_2, either of these two cues enhance the mosquito attraction considerably (Raji and DeGennaro 2017). Certain chemicals, like carboxylic acid (de Obaldia et al. 2022) and acetophenone (Zhang et al. 2022) emanated from human host skin odours, may act as greater stimulant for the mosquitoes. The interaction of the mosquito and the human host is dependent on several intrinsic and the extrinsic factors that guide the host selection based on the age, nutrient reserve, mating status, and gonotrophic cycle of the mosquito. The temperature, wind, and seasonality are the extrinsic factors that may influence the detection of chemical, thermal, and odorant cues. An effort is therefore given for the preparation of the chemicals that may mimic the odorants and restrain the mosquitoes from host-seeking behaviour. As proposed (Coutinho-Abreu et al. 2022), mosquito repellents can be developed from human skin microbe modified odorants. A variant form of the skin odorant may fail to elicit any response from the mosquito, thereby preventing mosquito bites and disease transmission.

6.7 Conclusion

Understanding the host preferences of mosquito species is crucial for predicting their behaviour, identifying potential disease reservoirs, and implementing effective control strategies. Information about the relationship between host microbiota, mosquito attraction, and pathogen transmission is crucial for developing targeted strategies to mitigate the spread of mosquito-borne diseases. By manipulating the microbiota of individuals or producing repellents that interfere with mosquito attraction to specific bacterial odours, researchers may be able to reduce pathogen transmission and improve public health outcomes. Perceiving the ecological interactions between mosquitoes and their hosts, researchers can develop targeted approaches to mitigate the transmission of mosquito-borne diseases and protect both human and animal populations.

Points to Remember

1. Mosquitoes use a wide range of hosts as a source of blood meal; however, species-specific host specificity is observed. Many species share more than one host for blood meal. On this basis, the mosquitoes can be classified as zoophilic or anthropophilic. Despite host specificity at the broader level, certain species of mosquitoes may show a shift in the host preference depending on the seasons and the host availability.
2. Mosquito host-seeking behaviour is initiated depending on several endogenous/intrinsic and exogenous factors. Among the endogenous factors, age, parity state, nutrient reserve, mating accomplishment, and genetics guide the activation of the behaviour. The mating state and the hunger level of the mosquitoes determine the requirement of the blood meal and, thus, trigger the host-seeking behaviour. Appropriate ambient temperature, light, humidity, and host availability comprise the exogenous or extrinsic factors that modulate the recognition of the cues.
3. Mosquito sensory system comprising the olfactory receptor, gustatory receptor, and ionotropic receptor receives the stimuli through the sensilla located in the maxillary palp, proboscis, and antennae, apart from the sense of vision for the recognition of the host. The peripheral olfactory receptors perceive the chemical volatiles emanated from the hosts. Successful landing on the host body surface, however, requires the integration of multiple stimuli generated from the host and modulated by the extrinsic factors.
4. Mosquitoes follow the variations in the CO_2 concentrations of mammalian host exhaled air in the ambient air recognized by the olfactory receptors, which are reinforced by the vision and the temperature. Subsequently, the variations in the concentrations of the odour emanated from the host body enable the successful navigation of the mosquitoes to the host body. Mosquitoes locating amphibian hosts recognize the host's call for a successful landing.
5. Odorants comprising lactic acid, acetophenone, and carboxylic acid derived from the host attract mosquitoes. Microbes in the human skin may degrade the products, augmenting the production of the odorants and thereby facilitating the precise location of the host by mosquitoes.
6. A feasible option for mosquito control can be the development of chemicals that mimic the odorants emanated from the host but do not amplify into an adequate stimulus for attracting mosquitoes. Modification of the human skin microbes can also be adopted to reduce the production of the odorants that attract mosquitoes naturally.

Acknowledgments The authors acknowledge the Head, Department of Zoology, University of Calcutta, for providing the facilities in compiling this chapter, including the facilities of UGC DRS-SAP and DST-FIST, Government of India. DD acknowledges the Principal, Bagnan College, Bagnan, Howrah, and Department of Science and Technology and Biotechnology, Government of West Bengal, for the facilities provided, including the project sanction no. 208(Sanc.)/ST/P/S&T/1G-73/2017 dated 21.03.2018. PP acknowledges the financial support of the Council of Scientific and Industrial Research, Government of India [09/028(1115)/2019-EMR-1, dated 06/08/2019]. MR acknowledges the Principal, Bejoy Narayan Mahavidyalaya, Itachuna, Hooghly, for the facilities provided.

References

Alberto DAS, Rusch C, Zhan Y, Straw AD, Montell C, Riffell JA (2022) The olfactory gating of visual preferences to human skin and visible spectra in mosquitoes. Nat Commun 13:555

Ayllón T, Nijhof AM, Weiher W, Baer B, Allene X, Calusen PH (2014) Feeding behavior of *Culicoides* spp. (Diptera: Ceratopogonidae) on cattle and sheep in Northeast Germany. Parasit Vectors 7:34

Bartlett-Healy K, Crans W, Gaugler R (2008) Temporal and spatial synchrony of *Culex territans* (Diptera: Culicidae) with their amphibian hosts. J Med Entomol 45:1031–1038

Borkent A, Belton P (2006) Attraction of female *Uranotanenia lowii* (Diptera: Culicidae) to frog calls in Costa Rica. Can Entomol 138:91–94

Bosch OJ, Martin G, Jürgen B (2000) Contribution of fatty acids to olfactory host finding of female *Aedes aegypti*. Chem Senses 25(3):323–330

Bowen MF (1991) The sensory physiology of host-seeking behaviour in mosquitoes. Annu Rev Entomol 36:139–158

Braks MAH, Takken W (1999) Incubated human sweat but not fresh sweat attracts the malaria mosquito *Anopheles gambiae* sensu stricto. J Chem Ecol 25:663–672

Braks MAH, Scholte EJ, Takken W, Dekker T (2000) Microbial growth enhances the attractiveness of human sweat for the malaria mosquito, *Anopheles gambiae* sensu stricto (Diptera: Culicidae). Chemoecology 10:129–134

Braks MAH, Juliano SA, Lounibos LP (2006) Superior reproductive success on human blood without sugar is not limited to highly anthropophilic mosquito species. Med Vet Entomol 20(1):53–59

Burkett-Cadena ND, Graham SP, Hassan HK, Guyer C, Eubanks MD, Katholi CR, Unnasch TR (2008) Blood feeding patterns of potential arbovirus vectors of the genus *Culex* targeting ectothermic hosts. Am J Trop Med Hyg 79:809–815

Burkett-Cadena ND, McClure CJW, Ligon RA, Graham SP, Guyer C, Hill GE, Ditchkoff SS, Eubanks MD, Hassan HK, Unnasch TR (2011) Host reproductive phenology drives seasonal patterns of host use in mosquitoes. PLoS One 6:e17681

Burkett-Cadena ND, McClure CJW, Estep LK, Eubanks MD (2013) Hosts or habitats: what drives the spatial distribution of mosquitoes? Ecosphere 4(2):30

Cardé RT (2015) Multi-cue integration: how female mosquitoes locate a human host. Curr Biol 25:793–795

Chaves LF, Harrington LC, Keogh CL, Nguyen AM, Kitron UD (2010) Blood feeding patterns of mosquitoes: random or structured? Front Zool 7:3

Chilaka N, Perkins E, Tripet F (2012) Visual and olfactory associative learning in the malaria vector *Anopheles gambiae* sensu stricto. Malar J 11:27

Coutinho-Abreu IV, Riffell JA, Akbari OS (2022) Human attractive cues and mosquito host-seeking behavior. Trends Parasitol 38(3):246–264

De Jong R, Knols BG (1995) Selection of biting sites on man by two malaria mosquito species. Experientia 51:80–84

De Moraes CM, Stanczyk NM, Betz HS, Pulido H, Sim DG, Read AF, Mescher MC (2014) Malaria-induced changes in host odors enhance mosquito attraction. Proc Natl Acad Sci 111(30):11079–11084

de Obaldia ME, Morita T, Dedmon LC, Boehmier DJ, Jiang CS, Zeledon EV, Cross JR, Vosshall LB (2022) Differential mosquito attraction to humans is associated with skin-derived carboxylic acid levels. Cell 185:4009–4116

Dekker T, Geier M, Cardé RT (2005) Carbon dioxide instantly sensitizes female yellow fever mosquitoes to human skin odours. J Exp Biol 208(15):2963–2972

Emami SN, Lindberg BG, Hua S, Hill SR, Mozuraitis R, Lehmann P, Birgersson G, Borg-Karlson AK, Ignell R, Faye I (2017) A key malaria metabolite modulates vector blood seeking, feeding, and susceptibility to infection. Science 355(6329):1076–1080

Ferguson LV, Smith TG (2012) Reciprocal trophic interactions and transmission of blood parasites between mosquitoes and frogs. Insects 3:410–423

Fierer N, Hamady M, Lauber CL, Knight R (2008) The influence of sex, handedness, and washing on the diversity of hand surface bacteria. Proc Natl Acad Sci 105:17994–17999

Harrington LC, Edman JD, Scott TW (2001) Why do female *Aedes aegypti* (Diptera: Culicidae) preferentially and frequently on human blood? J Med Entomol 38(3):411–422

Hinze A, Lantz J, Hill SR, Ignell R (2021) Mosquito host seeking in 3D using a versatile climate controlled wind tunnel system. Front Behav Neurosci 15:643693

James AG, Hyliands D, Johnston H (2004) Generation of volatile fatty acids by axillary bacteria. Int J Cosmet Sci 26:149–156

Keilin D, Nuttall GHF (1930) Iconographic studies of *Pediculus humanus*. Parasitology 22:1–10

Lacey ES, Cardé RT (2011) Activation, orientation and landing of female *Culex quinquefasciatus* in response to carbon dioxide and odour from human feet: 3-D flight analysis in a wind tunnel. Med Vet Entomol 25:94–103

Leclaire S, Nielsen JF, Drea CM (2014) Bacterial communities in meerkat anal scent secretions vary with host sex, age, and group membership. Behav Ecol 25:996–1004

Leyden JJ, McGinley KJ, Hölzle E, Labows JN, Kligman AM (1981) The microbiology of the human axilla and its relationship to axillary odor. J Invest Dermatol 77:413–416

Liu F, Liu N (2015) Human odorant reception in the common bedbug, *Cimex lectularius*. Sci Rep 5:15558

Madigan MT, Martinko JM, Stahl DA, Clark DP (2010) Brock biology of microorganisms. Benjamin Cummings, San Francisco

McBride C, Baier F, Omondi A, Spitzer SA, Lutomiah J, Sang R, Ignell R, Vosshall LB (2014) Evolution of mosquito preference for humans linked to an odorant receptor. Nature 515:222–227

Miller JF (1969) Antennal tuft organs of *Pediculus humanus* Linn. and *Pthirus pubis* (Linn.) (Anoplura: Pediculidae). J N Y Entomol Soc 77:85–89

Mwandawiro C, Boots M, Tuno N, Suwonkerd W, Tsuda Y, Takagi M (2000) Heterogeneity in the host preference of Japanese encephalitis vectors in Chiang Mai, northern Thailand. Trans R Soc Trop Med Hyg 94:238–242

Njabo KY, Cornel AJ, Sehgal RN, Loiseau C, Buermann W, Harrigan RJ, Pollinger J, Valkiūnas G, Smith TB (2009) *Coquillettidia* (Culicidae, Diptera) mosquitoes are natural vectors of avian malaria in Africa. Malar J 8:193

Potter CJ (2014) Stop the biting: targeting a mosquito's sense of smell. Cell 156:878–881

Raich T, Jacobson M, Holbrook F, Babion R, Blair C, Beaty B (1997) *Culicoides variipennis* (Diptera: Ceratopogonidae) host selection in Colorado. J Med Entomol 34(2):247–249

Raji JI, DeGennaro M (2017) Genetic analysis of mosquito detections of humans. Curr Opin Insect Sci 20:34–38

Reddy MR, Overgaard HJ, Abaga S, Reddy VP, Caccone A, Kiszewski AE, Slotman MA (2011) Outdoor host seeking behaviour of *Anopheles gambiae* mosquitoes following initiation of malaria vector control on Bioko Island, Equatorial Guinea. Malar J 10:184

Ruiz-López MJ (2020) Mosquito behaviour and vertebrate microbiota interaction: implications for pathogen transmission. Front Microbiol 11:573371

Schreck CE, James J (1968) Broth cultures of bacteria that attract female mosquitoes. Mosq News 28(1):33–38
Singh N, Wang C, Cooper R (2015) Role of vision and mechanoreception in *Cimex lectularius* L. behaviour. PLoS One 10(3):e0118855
Spanoudis CG, Andreadis SS, Bray DP, Savopoulou-Soultani M, Ignell R (2020) Behavioural response of the house mosquitoes *Culex quinquefasciatus* and *Culex pipiens molestus* to avian odours and its reliance on carbon dioxide. Med Vet Entomol 34:129–137
Steinbrecht RA (1994) The tuft organs of the human body louse *Pediculus humanus corporis* – cryofixation study of a thermo-hygro sensitive sensillum. Tissue Cell 6:259–275
Takken W, Knols BGJ (1999) Odor-mediated behaviour of Afrotropical malaria mosquitoes. Annu Rev Entomol 44:131–157
Takken W, Verhulst NO (2013) Host preferences of blood-feeding mosquitoes. Annu Rev Entomol 58:433–453
Theis KR, Schmidt TM, Holekamp KE (2012) Evidence for a bacterial mechanism for group-specific social odors among hyenas. Sci Rep 2:615
Tomberlin J, Rains G, Allan S, Sanford M, Lewis W (2006) Associative learning of odor with food- or blood-meal by *Culex quinquefasciatus* Say (Diptera: Culicidae). Naturwissenschaften 93:551–556
van Breugel F, Riffell J, Fairhall A, Dickinson MH (2015) Mosquitoes use vision to associate odor plumes with thermal targets. Curr Biol 25(16):2123–2129
Verhulst NO, Beijleveld H, Knols BG, Takken W, Schraa G, Bouwmeester HJ, Smallegange RC (2009) Cultured skin microbiota attracts malaria mosquitoes. Malar J 8:302
Verhulst NO, Qiu YT, Beijleveld H, Maliepaard C, Knights D, Schulz S, Berg-Lyons D, Lauber CL, Verduijn W, Haasnoot GW, Mumm R, Bouwmeester HJ, Claas FH, Dicke M, van Loon JJ, Takken W, Knight R, Smallegange RC (2011) Composition of human skin microbiota affects attractiveness to malaria mosquitoes. PLoS One 6(12):e28991
Verhulst NO, Boulanger N, Spitzen J (2018) Impact of skin microbiome on attractiveness to arthropod vectors and pathogen transmission. In: Boulanger N (ed) Skin and arthropod vectors. Academic, Strasbourg, pp 55–81
Whittaker DJ, Slowinski SP, Greenberg JM, Alian O, Winters AD, Ahmad MM, Burrell MJE, Soini HA, Novotny MV, Ketterson ED, Theis KR (2019) Experimental evidence that symbiotic bacteria produce chemical cues in a songbird. J Exp Biol 222:jeb202978
Wigglesworth V (1941) The sensory physiology of the human louse *Pediculus humanus corporis* De Geer (Anoplura). Parasitology 32:67–109
Wolff GH, Riffell JA (2018) Olfaction, experience and neural mechanisms underlying mosquito host preference. J Exp Biol 221:157131
Zhang H, Zhu Y, Liu Z, Peng Y, Peng W, Tong L, Wang J, Liu Q, Wang P, Cheng GA (2022) Volatile from the skin microbiota of flavivirus-infected hosts promotes mosquito attractiveness. Cell 185(14):2510–2522

Host Finding, Recognition, Specificity, and Interaction with the Host

7

Tanvi Singh, Vaishali Saini, Pooja Rohilla, Sanjay Tevatiya, and Rajnikant Dixit

Abstract

The current chapter offers an in-depth exploration of the sensory world of malaria vectors. The first part of the chapter focuses on how these mosquitoes use a complex array of sensory information, integrating olfactory, visual, hygric, and thermal cues to efficiently locate their hosts. It delves into the mechanisms of host detection, emphasising the dominant role of olfaction and the contribution of other senses in guiding mosquitoes to their prey.

The second part examines the impact of genetic, ecological, and environmental factors on the evolution and adaptation of the host-seeking behavioural characteristics of these mosquitoes. It highlights how urbanization, environmental changes, and parasites influence mosquito host preferences, providing insights into the evolutionary aspects of host specificity. This comprehensive analysis not only deepens our understanding of mosquito behaviour but also underscores the implications for vector control and malaria prevention strategies.

T. Singh · V. Saini · P. Rohilla · R. Dixit (✉)
Department of Vector Genomics, ICMR-National Institute of Malaria Research, New Delhi, India

Academy of Scientific and Innovative Research, Ghaziabad, Uttar Pradesh, India
e-mail: rkdixit@icmr.gov.in

S. Tevatiya
Department of Vector Genomics, ICMR-National Institute of Malaria Research, New Delhi, India

© The Author(s), under exclusive license to Springer Nature Singapore Pte Ltd. 2024
Omkar (ed.), *Mosquitoes*, https://doi.org/10.1007/978-981-97-4163-2_7

Keywords
Malaria vectors · Olfactory cues · Host seeking · Host specificity · Mosquito behaviour · Ecological determinants · Genetic determinants · Mosquito host preferences · Vector control

Learning Objectives

1. To understand the sensory mechanisms in mosquito host-seeking behaviour to explore the evolutionary basis of mosquito behaviour: gain insights into the evolutionary adaptations
2. To investigate the role of host emissions in mosquito attraction to analyse genetic and ecological influences on host preferences
3. To understand mosquito host specificity
4. To comprehend the impact of evolution on mosquito behaviour
5. To discover the correlation between mosquito behaviour and survival
6. To assess the implications for human-mosquito interactions: contemplate the sophisticated and sometimes deadly interactions

7.1 Introduction

In the unfolding saga of mosquito ecology, the pursuit of hosts is not merely behaviour but an essential mechanism for survival, finely orchestrated by a symphony of sensory cues. This convoluted dance of stimuli, encompassing hygric, auditory, visual, olfactory, and thermos-receptive signals, is fundamental to the mosquito's ability to pinpoint and secure a blood meal, a process explored in depth by Wynne et al. (2020). It is a testament to the complexity of sensory physiology that defines these enigmatic behavioural properties of mosquito vectors. Before we venture deeper into this sensory labyrinth, it is crucial to uncover the mystery of the evolutionary backdrop that has given rise to such behaviours. Within the tapestry of mosquito evolution, a particular trait, known as an anautogeny, emerges prominently. This remarkable evolutionary adaptation necessitates that female mosquito takes part in blood meals to facilitate egg development (Clements 1992a, b). Thus, host seeking and blood feeding are inexorably woven into the fabric of the incredible sensory system empowering their existence, shaping their survival and reproductive success.

As denizens of the order Diptera, these mosquitoes boast a suite of specialized anatomical adaptations. Notable among these are their expansive compound eyes and halteres, the transformed hindwings that act as gyroscopic stabilizers, bestowing upon them an exceptional equilibrium and navigational prowess (McGavin 2003). These features are not mere evolutionary by-products but essential tools that enable mosquitoes to fulfil their ecological roles. To grasp the full narrative of mosquito-host interactions, we must journey back to the 1960s. It was then that Brown et al. (1966) first illuminated the pivotal role of host emissions in luring mosquitoes. While initial theories postulated that air-borne and visual cues were the primary beacons guiding mosquito navigation, later advancements could enable the dissection of these processes at molecular and genetic levels, revealing a far more complex picture.

In this chapter, we navigate the realm of malaria vectors with a lens focused on their host preferences. We traverse the genetic and ecological landscapes that sculpt these preferences, unveiling patterns etched by the relentless forces of evolution. We contemplate how mosquitoes fine-tune their host selection, influenced by an array of evolutionary pressures. The ensuing narrative weaves a tale of adaptation and survival, providing a comprehensive vista into the evolutionary determinants of mosquito host specificity. This is the prologue to a broader story, one that charts the course of mosquito evolution and behaviour through the eons, culminating in the sophisticated and sometimes deadly interactions these creatures have with their human hosts. Welcome to a journey through the evolutionary dynamics of blood feeding, an odyssey that explores the primal instincts driving the delicate balance of life and death in the world of mosquitoes.

7.2 Prevalent Malaria Vectors Across India

The subcontinent of India is a battleground for malaria, with six *Anopheles* species identified as primary transmitters of the disease: *An. baimaii*, *An. culicifacies*, *An. fluviatilis*, *An. minimus*, *An. stephensi*, and *An. sundaicus*, catalogued in Table 7.1. A distinguishing feature of *An. stephensi* is its assortment of ecological variants, as outlined by Thomas et al. (2017). In contrast, the remaining vectors constitute what are known as species complexes. While these vectors predominantly tend to seek human blood, described as anthropophilic behaviour, they are also known to adapt to zoophilic tendencies, feeding on animals in the absence of human hosts, a trait that highlights their opportunistic nature in host selection. This behavioural flexibility has been documented by Barik et al. (2009), as well as by Dev and Sharma (2013).

7.3 Host Recognition and Navigation in Mosquitoes: An Integrated Multisensory Approach

1. Host detection: Mosquitoes exhibit remarkable navigator abilities for host detection, necessitating a sophisticated integration of spatio-temporal cues encompassing olfactory, visual, hygric, and thermal information. This integration facilitates efficient and effective host localization, underscoring a complex biological interplay that drives mosquito behaviour.
2. Mechanisms of host detection: Mosquitoes adopt a multisensory approach to discern environmental cues, particularly those indicating potential hosts.

Table 7.1 Key characteristics of various *Anopheles* species, including sibling species within complexes, their ecological distribution, typical resting and breeding areas, host preferences, and peak biting times, essential for understanding malaria vector behaviour and informing control strategies

Vector species	Sibling species	Ecological distribution	Resting areas	Breeding areas	Host preference	Peak biting time
Anopheles culicifacies	*culicifacies* complex (A, B, C, D, E)	Rural plains	Indoor-cattle sheds, human dwellings	River-bed pools, rainwater collections, streams, rice fields, seepage water, borrow pits, irrigation channels	Zoophilic	22:00–24:00 h
Anopheles fluviatilis	*fluviatilis* complex (S, T, U, V)	Hilly forested	Human dwellings (indoors)	Seepage water streams with perceptible flow of water, river margins, irrigation channels, shallow wells, terraced rice fields along foothills	S-anthropophilic T, U-zoophilic	20:00–24:00 h
Anopheles dirus	*An. diruss.s.* (*species* A), *An. cracens* (species B), *An. scanloni* (species C), *An. Baimaii* (species D), *An. elegans* (species E), *An. anemophilous* (species F), *An. takasagoensis*	Northeastern forest regions	Exophilic tree trunks/creepers in forests	Forest including small transient pools, elephant footprints	Anthropophilic	21:00
Anopheles minimus	*An. minimus* (species A), *An. harrisoni* Harbach and Manguin (species C), *An. yaeyamaensis* Somboon and Harbach (species E)	Foothills of eastern and Northeastern region	Endophilic and endophagic	Perennial seepage water foothill streams, paddy fields	Anthropophilic	01:00–04:00 h

Anopheles stephensi	Three ecological variants: type form, intermediate form, variety 'mysorensis'	Urban, arid zones	Cattle sheds, barracks, poorly constructed houses	Freshwater ponds, stream beds, seepage canals, wells	Zoophilic	22:00–24:00 h
Anopheles sundaicus	*An. sundaicus An. epiroticus* Linton and Harbach (formerly species A) *An. sundaicus* (species D) *An. sundaicus* (species E)	Andaman and Nicobar Islands	Indoors and outdoors of human dwellings and cattle shed	Brackish and freshwater	Zoophilic	21:00–4:00 h

McMeniman et al. (2014) noted that chemosensory receptors, olfactory neurons, and sensory appendages are key components in this process. Sensilla on antennae ensure detection of CO_2 exhaled by humans from distances over 10 m, activating flight behaviour (Das De and Dixit 2020). Infrared radiation from body heat, along with sweat, moisture, and specific chemical compounds, like lactic acid and ammonia, provide distinctive cues for these heat-seeking insects. Visual stimuli, though secondary, enhance host visibility, particularly when individuals wear dark-coloured clothing.

3. Strategic sensory responsiveness: To minimise unnecessary exposure to hosts, mosquitoes demonstrate strategic responsiveness to sensory cues, balancing host-seeking with evasion of host defences (Coutinho-Abreu et al. 2022).
4. Host-seeking behaviour: Integrating multiple sensory cues—vectors employ an advanced process involving the amalgamation of various sensory modalities, yielding robust information about the location and quality of potential hosts. According to Takken et al. (2002), the vector's responses to these cues are equally influenced by internal factors, such as state, age, and past experiences, as well as external factors, like temperature, humidity, and time of day, while host defensiveness may also add unexpected complexity to their host-seeking behaviour.
5. Olfaction: The primary sensory modality—olfaction is the predominant mode in the initial stages of host-seeking activation. Wynne et al. (2020) emphasise that before engaging other senses, mosquitoes identify chemical signals. The olfactory system, comprising receptors on the dendrites of neurons within sensilla on antennae and maxillary palps, is highly conserved across insects (Wang et al. 2010). Odour molecules interact with odorant-binding proteins, facilitating their transport to receptors (Leal 2013; Das De et al. 2018). Olfactory receptor neurons (ORNs), expressing various receptors, convey signals to the central brain's antennal lobes, where they synapse with projection neurons to higher-order centres, like mushroom bodies and lateral horns, integral for learning, memory, and behaviour integration (Carey et al. 2010; Carraher et al. 2015; Das De and Dixit 2020).
6. Vision and its role in host detection: Vision in insect vectors is interlinked with olfactory and thermosensory behaviours. McMeniman et al. (2014) found that CO_2 presence not only serves as an indicator of vertebrate hosts but also regulates thermosensory behaviours, influencing olfactory responses. It is also observed that mosquitoes are more attracted to visual stimuli in the presence of CO_2, with subsequent cues, like heat and water vapour, affecting probing and biting behaviours (Zhou et al. 2018)
7. Thermosensation and hygrosensation: Crucial for host discrimination and biting decisions is the interplay between air, temperature, and water vapour (Takken and Verhulst 2013; Van Breugel et al. 2015). Variations in these cues across different hosts provide a rich sensory environment affecting mosquito behavioural properties.
8. Gustation in host interaction: Upon locating a host, mosquitoes employ gustatory receptors on their tarsae and labellum for pre-biting and biting behaviours

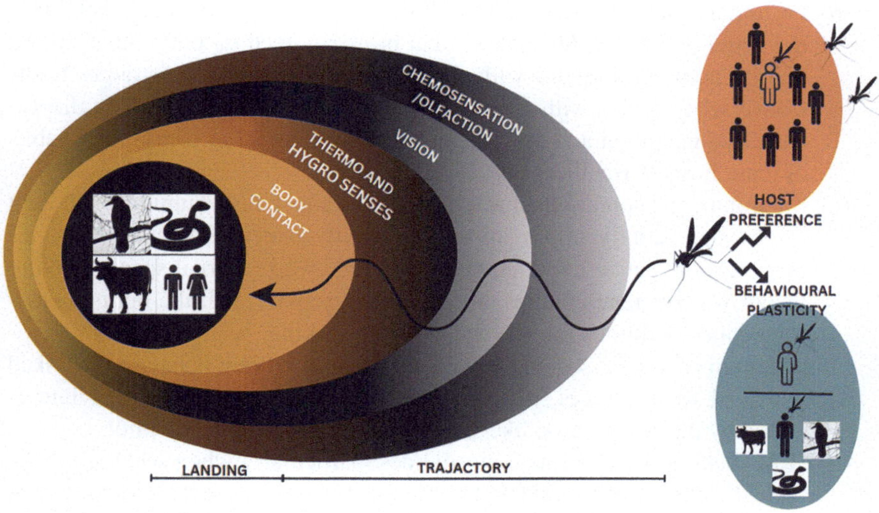

Fig. 7.1 An illustrative guide to the multisensory host-seeking process in mosquitoes, depicting the progression from initial landing to host preference, influenced by a combination of chemosensory, olfactory, visual, thermal, hygric, and body contact cues, culminating in behavioural plasticity that allows for adaptation to diverse human and animal hosts

(Sparks 2013). This gustatory system plays a vital role in the rapid acquisition of a blood meal, crucial for survival and reproduction.

This comprehensive overview depicted in Fig. 7.1 highlights the sophisticated multisensory system employed by mosquitoes in host detection and navigation, demonstrating the complexity of their ecological adaptations and behaviours.

7.4 Activity Patterns in Mosquito Species: Nocturnal vs. Diurnal Behaviours

1. Activity patterns: Mosquitoes display diverse activity patterns, varying significantly across species. Some species are predominantly nocturnal, being active during night hours, while others are diurnal, showing increased activity during daylight.
2. Factors influencing nocturnal and diurnal activities: The distinction between nocturnal and diurnal activity in mosquitoes is governed by several factors:
   (a) Species-specific biology: Different mosquito species exhibit variations in their physical characteristics, which correlate with their activity patterns. Notably, nocturnal and crepuscular species, such as *An. gambiae sensu stricto*, *An. stephensi*, and *C. pipiens molestus*, have larger facet diameters in their ommatidia, aiding in low-light vision. In contrast, diurnal species, like *Ae. aegypti*, possess smaller facets suitable for daylight conditions (Kawada et al. 2006).

(b) Preferred host types: The choice of host also plays a crucial role in determining activity patterns. Mosquitoes that primarily feed on birds often exhibit diurnal activity, aligning with the activity patterns of their avian hosts. Conversely, species with a preference for mammalian hosts might display increased nocturnal activity, adapting to the nocturnal or crepuscular habits of these hosts (Cribellier et al. 2022).

(c) Environmental conditions: Factors, such as temperature, humidity, and light levels, significantly impact mosquito activity patterns. Nocturnal species may be more active under cooler conditions, while diurnal species tend to favour warmer temperatures, aligning their activity with the optimal environmental conditions for survival and host-seeking.

3. Flight patterns and host-seeking behaviour: Mosquito flight behaviour is linked to their need for resources, including mating and feeding. This behaviour is influenced by internal factors, like satiety, mating status, and gravidity.
   (a) Host-seeking flight: Female mosquitoes, particularly when seeking a host, typically engage in upwind flight, orienting themselves toward the source of host odours and CO_2 emissions (Van Breugel et al. 2015). Upon detecting a plume of these cues, they adopt a strategic flight pattern known as 'casting flight', which involves a zigzagging motion into the odour plume, followed by a direct surge upwind. This tactic is designed to minimise the distance between the mosquito and the host, enhancing the efficiency of their host-seeking behaviour (Cardé and Willis 2008).
   (b) Navigational adaptations: The flight patterns of mosquitoes are sophisticated, enabling them to navigate complex environmental cues and locate hosts efficiently. These patterns reflect the coordination between the mosquito's sensory capabilities and their environmental context, demonstrating their evolutionary adaptations to diverse ecological niches (Das De et al. 2018).

7.5 Determinants of Host Preference

Understanding host preference in disease vectors, notably in mosquitoes, like *Anopheles*, reveals a dynamic reciprocity between their innate tendencies and environmental factors. While these vectors show a marked preference for certain mammalian hosts, their choices are not rigidly exclusive. They often feed on a range of available hosts, as evidenced by foundational studies of White et al. (1972), and further elaborated in the works of Takken and Verhulst (2013). Gillies' seminal experiment in 1964, involving *Anopheles* mosquitoes in an environment with both human and calf hosts, was pivotal in demonstrating that even the most anthropophilic vectors diversify their choices based on availability (Gillies 1964). Over the years, research has significantly deepened our understanding of host preference and biting behaviour. However, it is essential to recognize that controlled experimental settings may not fully capture the complexities of host selection that occur in natural environments. Laboratory observations often provide insights into intrinsic vector preferences but may not always align with mosquito behavioural activities occurring in the field conditions.

A crucial aspect in this context is the survival-driven adaptability of vectors, like *Anopheles* mosquitoes. Their survival and reproductive success are intimately linked to their proficiency in obtaining blood meals. This survival instinct prompts them to adjust their host choices based on what's available, as observed in *Anopheles gambiae sensu stricto* in São Tomé, West Africa, where they displayed a preference for dogs, an example of their versatile host selection (Sousa et al. 2001). Similarly, the work of Dumonteil et al. (2018) on triatomines underscores the importance of ecological associations in understanding vector behaviour and pathogen transmission cycles.

Building on these insights, our discussion now shifts toward the multifaceted determinants that shape vector behaviour and disease transmission dynamics. Extrinsic determinants, particularly ecological factors, play a pivotal role in vector distribution and host-vector interactions, influencing the transmission patterns of vector-borne diseases. Intrinsic determinants, including genetic variations within vector populations, underpin their adaptability and resistance mechanisms, affecting control strategies' effectiveness. Lastly, evolutionary determinants highlight the ongoing co-evolution between vectors, pathogens, and hosts, a process that continuously shapes the landscape of vector-borne diseases. This comprehensive approach allows us to understand the complexity of vector ecology and the challenges in managing vector-borne diseases.

7.6 Biting Choices: Delving Deep Into the Determinants of Vector Host Preference

As mentioned earlier broadly, the factors influencing host preference in disease vectors, like mosquitoes, are traditionally categorised into two distinct realms: intrinsic and extrinsic determinants (Takken and Verhulst 2013). However, there emerges a compelling need to consider a third, equally critical category: evolutionary determinants. This category, transcending the conventional binary classification, encapsulates the role of both intrinsic and extrinsic factors. Evolutionary determinants are shaped and honed over time by the complex interaction of internal genetic factors regulating the species-specific physiology and the external ecological-cum-environmental pressures. Their inclusion as a separate category, illustrated in Fig. 7.2, provides a more nuanced and comprehensive understanding of the driving forces behind host preference in these vectors, acknowledging the dynamic and adaptive nature of their behaviour through the lens of evolutionary biology.

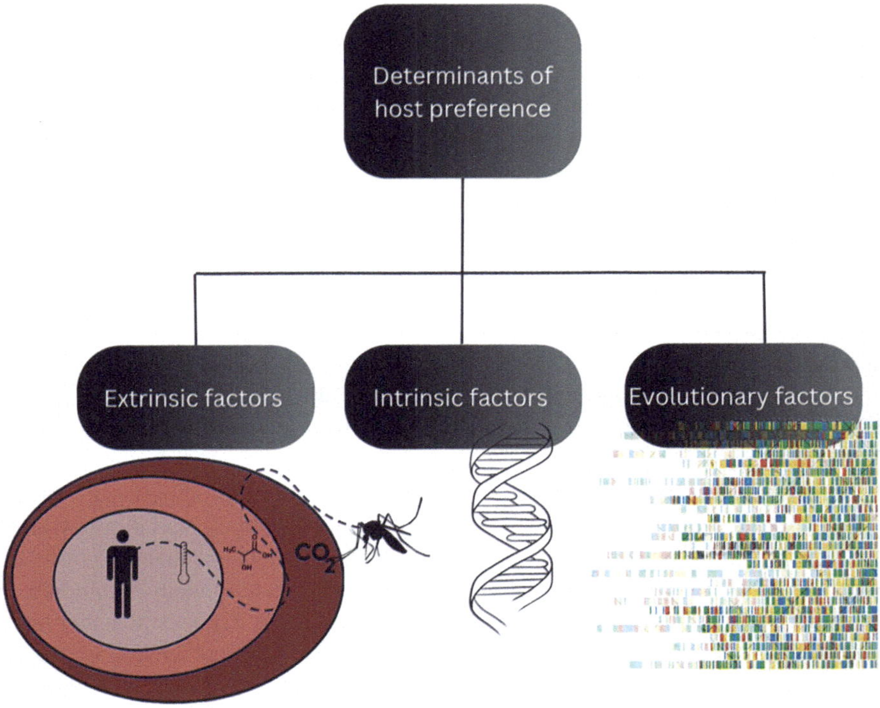

Fig. 7.2 Illustration of the multifaceted determinants of mosquito host preference, encompassing extrinsic factors, such as environmental cues, intrinsic factors rooted in the mosquito's biology, and the complex array of evolutionary factors that together drive the adaptive choices of these disease vectors in their natural habitats

7.7 Extrinsic Factors

Turning to extrinsic factors, the host preference of mosquitoes is influenced by a myriad of environmental and host-specific elements. Laboratory studies have revealed that controlled conditions can lead to the inactivation or loss of certain genetic traits over generations, potentially affecting natural feeding behaviours of specific host species (Franks et al. 2011; Lagisz et al. 2011). Conversely, field studies, which analyse the blood meal origins of field-collected specimens and mosquito behaviour in choice situations, offer deeper insights into natural host preferences (Garrett-Jones et al. 1980). These studies, however, may lean toward local host species abundance, introducing a potential bias.

In their natural environment, mosquitoes show an inclination for individual hosts within a species, an inclination likely shaped by variations in odorants and the sensitivity of their olfactory receptors (Qiu et al. 2006; Knols et al. 1995; Zwiebel et al. 2004). The availability of the preferred host, the mosquito's metabolic energy level, and adverse weather conditions are significant factors in host selection (Clements

1992a, b; Williams et al. 2010). Moreover, the role of skin bacteria in producing host-specific odorants adds to the complexity of mosquito-host interactions (Verhulst et al. 2011). Blood quality and host species also play crucial roles in influencing mosquito behaviour and reproductive fitness, further complicating the understanding of their host preferences (Langley et al. 1978; Moloo et al. 1988; Lyimo and Ferguson 2009; Takken et al. 2002). Collectively, these extrinsic factors form a complex network that shapes mosquito behaviour, underlining their role as disease vectors.

7.8 Ecological Determinants of Host Preference

Ecological factors play a profound role in shaping the behaviour and distribution of malaria vectors. These factors include an array of environmental conditions, such as deforestation, water bodies, temperature, vegetation, humidity, and urbanisation, all of which collectively influence vector ecology and, consequently, the dynamics of malaria transmission.

7.8.1 Deforestation

In the complicated phenomena of ecological change, deforestation emerges as a pivotal player in the narrative of malaria transmission, its influence modulated by the diversity of vector species within a given region. The retreat of the forest canopy ushers in a cascade of sunlight, casting its rays upon vector breeding grounds and resting sites once cloaked in arboreal shadow, consequently sculpting the habitats of these disease carriers. The shade-seeking proclivities of certain *Anopheles* species, such as *An. dirus*, *An. fluviatilis*, and *An. minimus*, have been well-documented, revealing a marked preference for the cool refuge provided by forested environments (Service 1991; Nazaroff 2011; Obsomer et al. 2007). These variations in vector populations and the incidence of malaria are not monolithic but instead are tied to the mosaic of human activity, including the patterns of agricultural development and the unique ecological characteristics of local environments.

7.8.2 Water Bodies

When we turn our gaze to the aqueous realms that serve as nurseries for the nascent stages of mosquito life, the role of water bodies in the saga of malaria transmission becomes even more pronounced. These aquatic cradles, vital for the development of mosquito larvae, are sensitive indicators of the health and intensity of malaria transmission. In the cooler, sheltered realms of forested landscapes, *An. fluviatilis* and *An. minimus* often select the flowing streams as their breeding sanctuaries (Lindsay 2004), while *An. dirus*, with a penchant for still waters, claims the small, clear pools nestled within Asian forests as its reproductive haven (Obsomer et al. 2007). The fates of these vectors, and hence the patterns of malaria transmission, are inexorably tied to the ebb and flow of these water sources. The alteration or diminution of these habitats through environmental change can precipitate significant shifts in the prevalence and spread of malaria, underscoring the delicate balance that exists between natural ecosystems and public health.

7.8.3 Vegetation

The presence of vegetation near to human settlements amplifies the population of malaria vectors, consequently elevating the transmission of malaria (Lowe 2018). These vectors typically exhibit a preference for tree canopy coverage and are known to seek refuge in tree holes (Yadav et al. 1997). The significance of flora and sugar availability as pivotal factors influencing vectorial capacity has been demonstrated. The accessibility of plant sugars has been linked to increased egg numbers and enhanced survival potential (Manda et al. 2007), extending beyond the age at which vectors become capable of transmitting malaria. Additionally, the deposition of leaves into larval habitats ensures the provision of sustainable microclimatic conditions and serves as a food source for larvae, creating favourable conditions for vectors, such as *An. dirus* in south-east Asia (Obsomer et al. 2007).

7.8.4 Temperature, Humidity, and Rainfall

The dynamics of mosquito population growth are profoundly influenced by the delicate balance between rainfall, and the combined processes of precipitation and transpiration, as elucidated by Koenraadt et al. (2004). This growth trajectory is notably accelerated as the dry season wanes, surging to its zenith following the advent of the rainy season, which heralds a significant uptick in humidity levels, a phenomenon meticulously documented by Jawara et al. (2008). In this ecological ballet, trees emerge as pivotal players, their transpiration activities enriching the atmosphere with moisture, thereby fostering a cooler climate and augmenting precipitation rates. This cascade of climatic conditions creates an idyllic haven for mosquito populations to flourish. The resultant humid milieu, coupled with the proliferation of aquatic habitats engendered by rainfall, not only bolsters the survival and expansion of these vector populations but also amplifies the potential for malaria transmission. The confluence of these environmental factors underscores the critical role of meteorological elements and vector ecology, paving the way for a deeper understanding of disease propagation mechanisms.

7.8.5 Urbanisation's Impact on Mosquito Populations and Disease Transmission

The pieces of evidence indicate that the process of urbanization has markedly intensified the density, larval development rates, and longevity of *Aedes albopictus* mosquitoes, thereby elevating their capacity as vectors and, by extension, enhancing the transmissibility of diseases they carry, such as dengue fever (Li et al. 2014). The transformation of landscapes through urban development precipitates the creation of artificial aquatic environments, serving as breeding sanctuaries for these mosquitoes. Particularly in urban expanses, these anthropogenic water bodies become critical for the survival of mosquito populations during prolonged and arid dry seasons, as they adapt to rely exclusively on water stored by humans for breeding (Rose et al. 2020). Furthermore, the burgeoning human populations characteristic of urbanised areas appear to influence mosquito host selection preferences significantly. In such environments, certain mosquito species exhibit a pronounced predilection for human hosts, a trend likely spurred by the sheer density and accessibility of humans. This shift toward anthropophily could substantially amplify the rates at which mosquito-borne illnesses are propagated among urban dwellers (Rose et al. 2020). This interconnection between urbanisation, mosquito ecology, and disease dynamics underscores the necessity of scrutinizing these relationships across diverse environmental contexts to elucidate the mechanisms underpinning disease transmission in an increasingly urbanised world.

7.9 Intrinsic Factors

Delving into the intrinsic aspects, the physiology and genetics of mosquitoes significantly dictate their host preferences. Post-emergence, both sexes initially seek nectar for the energy vital for flight, illustrating a key physiological response (Foster and Takken 2004). Notably, autogenous female mosquitoes experience a pivotal behavioural shift from sugar to blood feeding within 24–48 h after mating, aligning this change with their reproductive cycle (Foster 1995). The genetic basis of host preference is evident in species, like *An. quadriannulatus* and *An. arabiensis*, favouring cow odour, versus *An. gambiae.*, which is drawn to human volatiles (Dekker and Takken 1998; Pates et al. 2001). This genetic predisposition, however, can be overridden by urgent nutritional requirements, particularly related to reproduction, compelling mosquitoes to feed on non-preferred hosts. Additionally, while age does not significantly influence host preference, adaptive learning from past encounters can modify future host-seeking behaviours (Chilaka et al. 2012). These intrinsic elements critically influence the dynamic process of host selection in disease vectors.

7.10 Genetic Determinants of Mosquito Host Preference

Grasping the genetic underpinnings of mosquito host selection is crucial for devising strategies to curb vector-borne diseases effectively. The behaviour of mosquitoes, including their search for hosts, is a complex synergy of genetic, environmental, and physiological influences. These behaviours are critical as they directly impact the mosquitoes' ability to serve as vectors for disease pathogens. In this section, we explore the genetic factors that influence mosquitoes' preferences for certain hosts,

Table 7.2 Summary of the roles of various genetic factors in mosquito host preference: this table outlines how olfactory receptors (ORs), odorant-binding proteins (OBPs), sensory appendage proteins (SAPs), neuropeptides (neuropeptide Y and leucokinin), and the histamine-gated chloride channel (HR) gene contribute to the detection of host-specific cues, suppression of host-seeking behaviour post-feeding

Genetic factor	Role in host preference
Olfactory receptors (ORs)	Detect host-specific odours, crucial for host seeking
Odorant-binding proteins (OBPs)	Bind specific odours, mediating the detection of host cues
Sensory appendage proteins (SAPs)	Enhance sensitivity to host odours, found in leg sensilla
Neuropeptide Y and leucokinin	Suppress host-seeking post-blood meal, indicating gut-brain communication
Histamine-gated chloride channel (HR) gene	Affects immunity and microbial proliferation, indirectly influencing host preference

focusing on the roles of the olfactory system, salivary glands, and the neuroendocrine system, outlined in Table 7.2.

7.10.1 Olfactory System and Host Preference

The ability of mosquitoes to find their hosts is largely attributed to their sophisticated olfactory system. Mosquitoes rely on their antennae, maxillary palps, and labium, all of which are equipped with sensilla, to detect odours in their environment (Das De and Dixit 2020). Within these sensilla are various receptors, such as chemosensory proteins (SAP), odorant-binding proteins (OBP), and olfactory receptors (OR), which are crucial for initiating host seeking and feeding behaviours. Studies involving RNA-Seq analysis of the olfactory tissues of *An. culicifacies* both before and after a blood meal have demonstrated significant changes in the expression of genes related to olfaction. These findings suggest a genetic foundation underlying the mosquitoes' preferences for certain hosts (Das De et al. 2018).

7.10.2 Neuroendocrine System and Feeding Behaviour

The neuroendocrine system of mosquitoes, with the brain at its core, is instrumental in orchestrating vital functions, such as growth, development, and the nuanced behaviours associated with feeding. Within this system, neuropeptides, like neuropeptide Y and leucokinin, emerge as pivotal elements, particularly after the mosquito engages in a blood meal. The observable shifts in the expression of these neuropeptides hint at their integral role in dampening the mosquito's urge to seek further hosts, a mechanism driven by the attainment of satiety (Das De et al. 2022).

This adaptation is not merely about conserving energy but is a sophisticated survival strategy deeply embedded in the mosquito's genetic makeup. Once a mosquito has successfully fed, the activation of satiety signals serves as a biological cue to halt further host-seeking activities. This cessation is crucial for survival, as each additional attempt at feeding increases the mosquito's exposure to potential threats from hosts or environmental hazards. Consequently, it could be hypothesized that these satiety signals ensure that the mosquito minimizes its risk by avoiding unnecessary host-seeking behaviour.

In the context of seeking host preference, the role of the neuroendocrine system becomes even more pronounced. By modulating feeding behaviour based on internal signals of fullness, the system contributes significantly to the mosquito's selective approach to host seeking. This selectiveness might not be arbitrary but a reflection of an evolved preference that optimises survival chances by balancing the need for nourishment with the imperative of minimising danger.

Therefore, the dynamics of the gut-brain axis in mosquitoes are not just a fascinating aspect of their biology but a crucial determinant of their behavioural ecology (Das De et al. 2022). The signals of satiety that guide post-feeding behaviour may illuminate the complex interaction between nutritional intake and survival strategy, underscoring how deeply survival instincts are woven into the fabric of their host-seeking behaviour. This elegant balance between the need to feed and the need to evade risk underscores the mosquito's evolutionary success, revealing the profound impact of neuroendocrine mechanisms on host preference and the overarching survival strategies of these ubiquitous vectors.

7.10.3 Influence of *Plasmodium* on Anopheles Mosquito Host-Seeking Behaviour

The interaction between adult female *Anopheles* mosquitoes and the *Plasmodium* parasite is a critical component in the cycle of malaria transmission, with the act of blood feeding serving as a linchpin in the dynamics of *Anopheles-Plasmodium* interactions. The host-seeking behaviours of these mosquitoes are fundamental in dictating the efficiency of malaria spread. Emerging research suggests that the pathogenesis of malaria within human hosts, coupled with *Plasmodium* infections in the mosquito vectors, may have profound effects on the mosquitoes' behavioural and physiological profiles, thereby potentially amplifying the likelihood of pathogen transmission (Anderson et al. 1999). It is, however, crucial to bear in mind that several investigations into the *Anopheles-Plasmodium* nexus have been predicated on infection intensities within mosquitoes that may not truly mirror the complexities and nuances of their natural counterparts (Olliaro 2008). Such a caveat is essential for a nuanced interpretation of research outcomes, especially in the context of elucidating the mechanisms through which pathogens influence mosquito host-seeking behaviours.

7.11 Evolutionary Determinants of Host Preference in Mosquitoes: Bridging Dietary Specialism and Generalism

7.11.1 Evolutionary Dynamics of Host Preference in Mosquitoes

The evolutionary journey of host preference in mosquitoes, particularly in species, like *Anopheles*, unfolds a fascinating tale that firmly weaves together dietary specialisation and generalisation with the threads of environmental and genetic evolution. The conundrum of dietary specialisation in the natural world, choosing a limited array of resources amidst abundancies entanglement of fitness trade-offs derived from different food sources. Specialisation is theorised to emerge when focusing on a narrow range of resources maximises net energy gains. Conversely, a generalist approach is favoured in environments, where diverse resources offer comparable energetic benefits (Egas et al. 2004). In the world of mosquitoes, generalism likely gains traction in environments, where encountering a host is infrequent and the perils of waiting for an ideal host surpass the risks of imminent mortality. On the other side of the spectrum, specialisation is inclined to develop in environments abundant with preferred hosts, shaped by a blend of optimal foraging on accessible hosts and the historical patterns of host presence (Kelly and Thompson 2000; Page 2003). Contrast to the adaptive metabolic shift from sugar to blood is a unique evolutionary speciality in mosquitoes, though the reason behind is yet unclear. Data revealing the distinct nature of genes, such as plant-like transcripts (PLTs), likely a result of horizontal gene transfer (HGTs) evidence their evolutionary benefits in some local species (Sharma et al. 2015a, b) (Fig. 7.3).

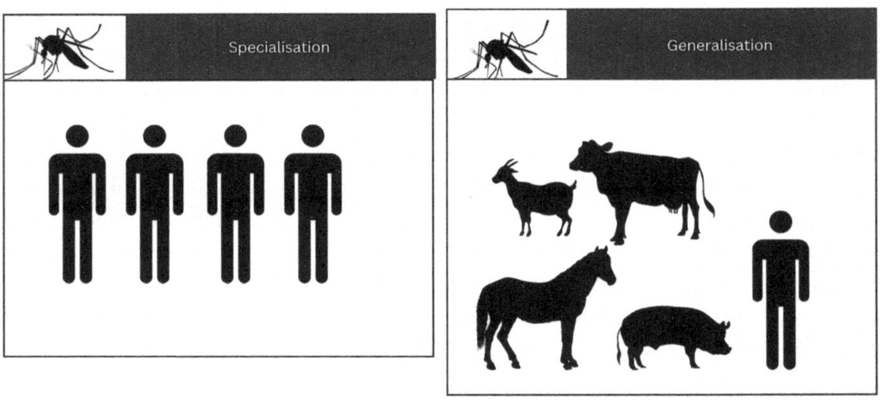

Fig. 7.3 Evolution of host preference in mosquitoes: (left) specialization is depicted where a mosquito species has evolved to preferentially feed on humans; (right) generalization is shown, illustrating a mosquito species with a diverse host range, feeding on various animals and humans, an adaptation driven by host availability

7.12 Genetic and Physiological Adaptations in Mosquito Host Selection

Parallel to these behavioural adaptations, the genetic and physiological evolution in species, such as *An. culicifacies*, has fine-tuned their olfactory system, pivotal for successful blood feeding. Post-prandial RNA-Seq analyses have divulged significant alterations in critical olfactory genes, indicative of sensory adaptations. This evolution is most apparent in the harmonised functioning of odorant-binding proteins and odorant receptors, fundamental to the host-seeking process. These molecular developments underscore the significant impact of intrinsic genetic and physiological factors in moulding the host preferences of these vectors (Das De et al. 2018).

Thus, the narrative of host preference in mosquitoes is a rich mosaic, crafted by environmental influences, genetic diversity, and physiological adaptations. It delineates the strategic manoeuvres of mosquitoes within their ecological realms, adjusting their feeding behaviours in response to host availability and the evolutionary forces at play. The delicate balance between generalist and specialist feeding strategies, a product of both current ecological realities and historical host distributions, eloquently illustrates the evolutionary forces that govern host preference in these pivotal disease vectors.

7.13 Conclusion

The quest for a host by *Anopheles* mosquitoes, critical vectors of malaria, is influenced by a myriad of factors that span ecological, extrinsic, intrinsic, and genetic dimensions. Ecological and extrinsic factors encompass the environment in which mosquitoes thrive, including climate conditions and the availability of breeding sites, which in turn affect their population dynamics and host-seeking behaviour. Intrinsic factors relate to the physiological state of the mosquito, such as energy needs and reproductive status, which drive the urgency of their search for blood meals. Genetic factors, however, provide the underlying blueprint that orchestrates this complex behaviour, integrating signals from the environment with the mosquito's internal state to guide their host-seeking actions. This ensures that mosquitoes not only find suitable hosts to sustain their life cycle but also navigate the delicate balance between feeding for survival and avoiding potential threats.

At the genetic core of this behaviour lies the mosquito's olfactory system, a sophisticated network of sensors and receptors that decode environmental cues to locate hosts. The antennae, maxillary palps, and labium, adorned with sensilla, house an array of chemosensory proteins (SAP), odorant-binding proteins (OBP), and olfactory receptors (OR) essential for this purpose. Genetic studies, particularly those involving RNA-Seq analysis of *An. culicifacies* olfactory tissues, reveal significant post-blood meal changes in gene expression related to olfaction, underscoring the genetic foundation of host preference (Das De et al. 2018). Furthermore, the neuroendocrine system, through neuropeptides, like neuropeptide Y and leucokinin, modulates the mosquito's feeding behaviour post-engagement, highlighting a genetically encoded mechanism that curtails further host-seeking post-satiety (Das De et al. 2022). This genetic machinery not only facilitates efficient host location but also embeds a survival strategy that minimises risk by modulating host-seeking intensity based on the mosquito's satiety status (Fig. 7.4).

Understanding the complex web of factors that govern mosquito host-seeking behaviour opens avenues for targeted vector control strategies. By unravelling the genetic underpinnings of olfactory-guided host preference and the neuroendocrine controls of satiety, we can develop interventions that disrupt these processes, potentially reducing the efficacy of malaria transmission. For instance, manipulating environmental cues or introducing genetic modifications to alter the expression of key olfactory or neuroendocrine genes could impair the mosquito's ability to locate hosts or increase the threshold for satiety, thereby reducing feeding frequency and the chance of pathogen transmission. Moreover, exploring how pathogens, like

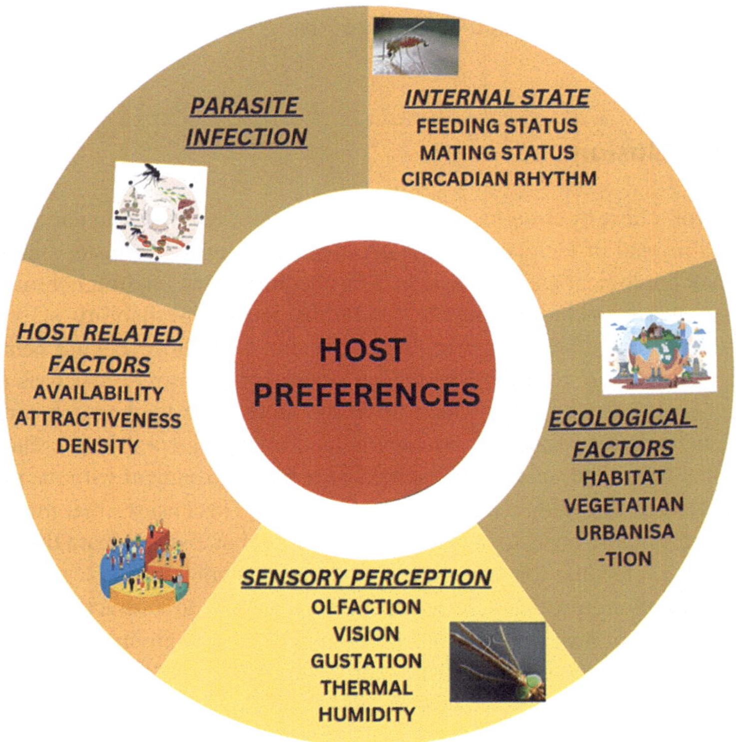

Fig. 7.4 Factors influencing mosquito host preferences: an infographic detailing the various elements such as host-related factors, the mosquito's internal state, ecological variables, and sensory perception mechanisms that collectively determine mosquito host-seeking behaviour and preferences

Plasmodium, influence mosquito behaviour offers additional targets for disrupting the cycle of malaria transmission. Integrating these insights with traditional vector control measures promises a more holistic and effective approach to combating malaria, highlighting the importance of a multifaceted strategy that considers the ecological, intrinsic, genetic, and pathogen-driven factors in mosquito host-seeking behaviour.

Points to Remember

1. Host-seeking behaviour: *Anopheles* mosquitoes exhibit host-seeking behaviour influenced by ecological conditions, such as climate and breeding site availability, and intrinsic factors, like energy needs and reproductive status, and genetic components that determine olfactory sensitivity.
2. Olfactory system: The olfactory system, crucial for detecting host odours, consists of antennae, maxillary palps, and labium, which contain sensilla with chemosensory proteins (SAP), odorant-binding proteins (OBP), and olfactory receptors (OR).
3. Genetic basis of host preference: Post-blood meal changes in gene expression within olfactory tissues of *An. culicifacies* suggest a genetic basis for host preference, as indicated by RNA-Seq analysis.
4. Neuroendocrine regulation: Neuropeptides, such as neuropeptide Y and leucokinin, in the neuroendocrine system influence feeding behaviour and host seeking after a blood meal, serving as a survival mechanism to minimise risk.
5. Satiety signals: The activation of satiety signals post-feeding helps conserve energy and reduces unnecessary exposure to potential threats, thus minimising the mosquito's risk of engaging in additional host-seeking behaviour.
6. Pathogen impact on behaviour: The interaction with pathogens, like *Plasmodium*, may alter mosquito behaviour and physiology, potentially enhancing pathogen transmission and affecting host-seeking behaviour.
7. Vector control strategies: Understanding the genetic and neuroendocrine factors in mosquito behaviour can lead to targeted vector control strategies, such as manipulating environmental cues or genetic modifications.
8. Multifaceted approach: A comprehensive vector management strategy should consider ecological, intrinsic, genetic, and pathogen-driven factors in mosquito host-seeking behaviour to effectively combat malaria transmission.
9. Implications for public health: Insights into mosquito host preference and feeding behaviour have direct implications for public health strategies, emphasising the need for integrated control measures.
10. Research considerations: Studies exploring the *Anopheles-Plasmodium* relationship often use unnaturally high infection intensities, which may not accurately reflect natural interactions; this must be considered when interpreting findings.

References

Anderson RA, Koella JC, Hurd H (1999) The effect of Plasmodium yoelii nigeriensis infection on the feeding persistence of Anopheles stephensi Liston throughout the sporogonic cycle. Proc Biol Sci 266:1729–1733. https://doi.org/10.1098/rspb.1999.0839

Barik TK, Sahu B, Swain V (2009) A review on Anopheles culicifacies: From bionomics to control with special reference to Indian subcontinent. Acta Tropica. 109:87–97

Brown AWA (1966) The Attraction of Mosquitoes to Hosts. JAMA. 196(3):249–252. https://doi.org/10.1001/jama.1966.03100160099028

Cardé RT, Willis MA (2008) Navigational strategies used by insects to find distant, wind-borne sources of odor. J Chem Ecol 34(7):854–866. https://doi.org/10.1007/s10886-008-9484-5

Carey AF, Wang G, Chih Ying S, Zwiebel LJ, Carlson JR (2010) Odorant reception in the malaria mosquito anopheles gambiae. Nature 464(7285):66–71. https://doi.org/10.1038/nature08834

Carraher C, Dalziel J, Jordan MD, Christie DL, Newcomb RD, Kralicek AV (2015) Towards an understanding of the structural basis for insect olfaction by odorant receptors. Insect Biochem Mol Biol 66:31–41. https://doi.org/10.1016/j.ibmb.2015.09.010

Chilaka N, Perkins E, Tripet F (2012) Visual and olfactory associative learning in the malaria vector Anopheles gambiae sensu stricto. Malar. J. 11:27

Clements AN (1992a) The biology of mosquitoes: development, nutrition and reproduction, vol 1. Chapman & Hall, London

Clements AN (1992b) The biology of mosquitoes. Chapman & Hall, London, p 509

Coutinho-Abreu IV, Riffell JA, Akbari OS (2022) Human attractive cues and mosquito host-seeking behavior. Trends Parasitol 38(3):246–264. https://doi.org/10.1016/j.pt.2021.09.012. Epub 2021 Oct 18. PMID: 34674963; PMCID: PMC10789295

Cribellier A, Straw AD, Spitzen J, Pieters RPM, van Leeuwen JL, Muijres FT (2022) Diurnal and nocturnal mosquitoes escape looming threats using distinct flight strategies. Curr Biol 32(6):1232–1246. https://doi.org/10.1016/j.cub.2022.01.036

Das De T, Dixit R (2020) Neuro-olfactory regulation and salivary actions: a coordinated event for successful blood-feeding behavior of mosquitoes. In: Dysfunction of olfactory system. IntechOpen, London. https://doi.org/10.5772/intechopen.90768

Das De T, Thomas T, Verma S, Singla D, Chauhan C, Srivastava V, Sharma P, Kumari S, Tevatiya S, Rani J, Hasija Y, Pandey KC, Dixit R (2018) A synergistic transcriptional regulation of olfactory genes drives blood-feeding associated complex behavioral responses in the mosquito Anopheles culicifacies. Front Physiol 9:577. https://doi.org/10.3389/fphys.2018.00577. PMID: 29875685; PMCID: PMC5974117

Das De T, Sharma P, Tevatiya S, Chauhan C, Kumari S, Yadav P, Singla D, Srivastava V, Rani J, Hasija Y, Pandey KC, Kajla M, Dixit R (2022) Bidirectional microbiome-gut-brain-axis communication influences metabolic switch-associated responses in the mosquito Anopheles culicifacies. Cells 11(11):1798. https://doi.org/10.3390/cells11111798. PMID: 35681493; PMCID: PMC9180301

Dekker T, Takken W (1998) Differential responses of mosquito sibling species Anopheles arabiensis and An. quadriannulatus to carbon dioxide, a man or a calf. Med Vet Entomol 12:136–140

Dev V, Sharma VP (2013) The dominant mosquito vectors of human malaria in India. In: Anopheles mosquitoes - new insights into malaria vectors. IntechOpen, London. https://doi.org/10.5772/55215

Dumonteil E, Ramirez-Sierra M-J, Pérez-Carrillo S, Teh-Poot C, Herrera C, Gourbière S et al (2018) Detailed ecological associations of triatomines revealed by metabarcoding and next-generation sequencing: implications for triatomine behaviour and Trypanosoma cruzi transmission cycles. Sci Rep 8:4140

Egas M et al (2004) Evolution restricts the coexistence of specialists and generalists: the role of trade-off structure. Am Nat 163:518–531

Foster WF (1995) Mosquito sugar feeding and reproductive energetics. Annu Rev Entomol 40:443–474

Foster WA, Takken W (2004) Nectar-related versus human-related volatiles: behavioral response and choice by female and male Anopheles gambiae (Diptera: Culicidae) between emergence and first feeding. Bull Entomol Res 94:145–157

Franks SJ, Pratt PD, Tsutsui ND (2011) The genetic consequences of a demographic bottleneck in an introduced biological control insect. Conserv. Genet. 12:201–211

Garrett-Jones C, Boreham PFL, Pant CP (1980) Feeding habits of anophelines (Diptera: Culicidae) in 1971–78, with reference to the human blood index: a review. Bull Entomol Res 70:165–185

Gillies MT (1964) Selection for host preference in Anopheles gambiae. Nature 203:852

Jawara M, Pinder M, Drakeley CJ, Nwakanma DC, Jallow E, Bogh C, Lindsay SW, Conway DJ (2008) Dry season ecology of anopheles gambiae complex mosquitoes in the Gambia. Malar J 7:1–9. https://doi.org/10.1186/1475-2875-7-156

Kawada H, Tatsuta H, Arikawa K, Takagi M. Comparative study on the relationship between photoperiodic host-seeking behavioral patterns and the eye parameters of mosquitoes. J Insect Physiol. 2006;52(1):67–5. https://doi.org/10.1016/j.jinsphys.2005.09.004.. Epub 2005 Oct 25.

Kelly DW, Thompson CE (2000) Epidemiology and optimal foraging: modelling the ideal free distribution of insect vectors. Parasitology 120:319–327

Knols BGJ, De Jong R, Takken W (1995) Differential attractiveness of isolated humans to mosquitoes in Tanzania. Trans R Soc Trop Med Hyg 89:604–606

Koenraadt CJ, Githeko AK, Takken W (2004) The effects of rainfall and evapotranspiration on the temporal dynamics of Anopheles gambiae s.s. and Anopheles arabiensis in a Kenyan village. Acta Trop 90:141–153

Lagisz M, Port G, Wolff K (2011) Living in a jar: genetic variation and differentiation among laboratory strains of the red flour beetle. J Appl Entomol 135:682–692

Langley PA, Pimley RW, Mews AR, Flood MET (1978) Effect of diet composition on feeding, digestion, and reproduction in Glossina morsitans. J Insect Physiol 24:233–238

Leal WS. 2013. Odorant reception in insects: roles of receptors, binding proteins, and degrading enzymes. Annual Review of Entomology 58:373–391. https://doi.org/10.1146/annurev-ento-120811-153635. PMID: 23020622

Li Y, Kamara F, Zhou G, Puthiyakunnon S, Li C, Liu Y, Zhou Y, Yao L, Yan G, Chen XG (2014) Urbanization increases aedes albopictus larval habitats and accelerates mosquito development and survivorship. PLoS Negl Trop Dis 8(11):e3301. https://doi.org/10.1371/journal.pntd.0003301

Lindsay S (2004) Environmental management for malaria control in the East Asia and Pacific (EAP) region. World Bank, Washington, pp 1–46

Lowe R (2018) The impact of global environmental change on vector-borne disease risk: a modelling study. Lancet Planetary Health 2:S1. https://doi.org/10.1016/s2542-5196(18)30086-x

Lyimo IN, Ferguson HM (2009) Ecological and evolutionary determinants of host species choice in mosquito vectors. Trends Parasitol 25:189–196

Manda H, Gouagna LC, Foster WA, Jackson RR, Beier JC, Githure JI, Hassanali A (2007) Effect of discriminative plant-sugar feeding on the survival and fecundity of anopheles gambiae. Malar J 6:1–11. https://doi.org/10.1186/1475-2875-6-113

McGavin GC (2003) Essential entomology: an order by order introduction. Oxford University Press, Oxford

McMeniman CJ, Corfas RA, Matthews BJ, Ritchie SA, Vosshall LB. Multimodal integration of carbon dioxide and other sensory cues drives mosquito attraction to humans. Cell. 2014;156(5):1060–1071. https://doi.org/10.1016/j.cell.2013.12.044. PMID: 24581501; PMCID: PMC4007582.

Moloo SK, Grootenhuis JG, Kar SK, Karstad L (1988) Survival and reproductive performance of female Glossina morsitansmorsitans when maintained on the blood of different species of wild mammals. Med Vet Entomol 2:347–350

Nazaroff WW (2011) Best paper awards: indoor air (2008-2010). Indoor Air 21(4):265–266. https://doi.org/10.1111/j.1600-0668.2011.00729.x

Obsomer V, Defourny P, Coosemans M (2007) The anopheles dirus complex: spatial distribution and environmental drivers. Malar J 6:1–16. https://doi.org/10.1186/1475-2875-6-26

Olliaro P (2008) Mortality associated with severe Plasmodium falciparum malaria increases with age. Clin Infect Dis 47:158–160

Page RDM (2003) Tangled trees: phylogeny, cospeciation, and coevolution. University of Chicago Press, Chicago

Pates HV, Takken W, Stuke K, Curtis CF (2001) Differential behaviour of Anopheles gambiae sensu stricto (Diptera: Culicidae) to human and cow odours in the laboratory. Bull Entomol Res 91:289–296

Qiu YT, Smallegange RC, Van Loon JJA, Ter Braak CJF, Takken W (2006) Interindividual variation in the attractiveness of human odours to the malaria mosquito Anopheles gambiae ss. Med Vet Entomol 20:280–287

Rose NH, Sylla M, Badolo A, Lutomiah J, Ayala D, Aribodor OB, Ibe N et al (2020) Climate and urbanization drive mosquito preference for humans. Curr Biol 30(18):3570–3579. https://doi.org/10.1016/j.cub.2020.06.092

Service (1991) Agricultural development and arthropod-borne diseases: a review. Rev Saúde Públ 25(3):165–178. https://doi.org/10.1590/S0034-89101991000300002

Sharma P, Das De T, Sharma S, Kumar Mishra A, Thomas T, Verma S, Kumari V, Lata S, Singh N, Valecha N, Chand Pandey K, Dixit R. Deep sequencing revealed molecular signature of horizontal gene transfer of plant like transcripts in the mosquito Anopheles culicifacies: an evolutionary puzzle. F1000Res. 2015a;4:1523. doi: https://doi.org/10.12688/f1000research.7534.1. PMID: 26998230; PMCID: PMC4786938.

Sharma P, Sharma S, Mishra AK, Thomas T, Das De T, Rohilla SL, Singh N, Pandey KC, Valecha N, Dixit R. Unraveling dual feeding associated molecular complexity of salivary glands in the mosquito Anopheles culicifacies. Biol Open. 2015b;4(8):1002–1015. https://doi.org/10.1242/bio.012294. PMID: 26163527; PMCID: PMC4542284.

Sousa CA, Pinto J, Almeida PG, Ferreira C, Do Rosário VE, Charlwood JD (2001) Dogs as a favored host choice of Anopheles gambiae sensu stricto (Diptera: Culicidae) of São Tomé, West Africa. J Med Entomol 38:122–125

Sparks JT, Vinyard BT, Dickens JC (2013) Gustatory receptor expression in the Labella and tarsi of aedes aegypti. Insect Biochemistry and Molecular Biology. 43:1161–1171. https://doi.org/10.1016/j.ibmb.2013.10.005

Takken W, Verhulst NO (2013) Host preferences of blood-feeding mosquitoes. Annu Rev Entomol 58:433–453. https://doi.org/10.1146/annurev-ento-120811-153618

Takken W, Stuke K, Klowden MJ (2002) Biological differences in reproductive strategy between the mosquito sibling species Anopheles gambiae sensu stricto and An. quadriannulatus. Entomol Exp Appl 103:83–89

Thomas S, Sangamithra Ravishankaran NA, Justin JA, Asokan A, Mathai MT, Valecha N, Montgomery J, Thomas MB, Eapen A (2017) Resting and feeding preferences of Anopheles stephensi in an urban setting, perennial for malaria. Malar J 16(1):1–7. https://doi.org/10.1186/s12936-017-1764-5

Van Breugel F, Riffell J, Fairhall A, Dickinson MH (2015) Mosquitoes use vision to associate odour plumes with thermal targets. Curr Biol 25(16):2123–2129. https://doi.org/10.1016/j.cub.2015.06.046

Verhulst NO, Qiu YT, Beijleveld H, Maliepaard C, Knights D et al (2011) Human skin microbiota affects attractiveness to malaria mosquitoes. PLoS ONE 6:e2899

Wang G et al (2010) Molecular basis of odor coding in the malaria vector mosquito Anopheles gambiae. Proc Natl Acad Sci USA 107(9):4418–4423. https://doi.org/10.1073/pnas.0913392107

White GB, Magayuka SA, Boreham PFL (1972) Comparative studies on sibling species of the Anopheles gambiae Giles complex (Dipt., Culicidae): bionomics and vectorial activity of species A and species B at Segera, Tanzania. Bull Entomol Res 62:295–317

Williams CR, Bader CA, Kearney MR, Ritchie SA, Russell RC. The extinction of dengue through natural vulnerability of its vectors. PLoS Negl Trop Dis. 2010;4(12):e922. https://doi.org/10.1371/journal.pntd.0000922. PMID: 21200424; PMCID: PMC3006136.

Wynne NE, Lorenzo MG, Vinauger C (2020) Mechanism and plasticity of vectors' host-seeking behavior. Curr Opin Insect Sci 40:1–5. https://doi.org/10.1016/j.cois.2020.02.001

Yadav RS, Sharma VP, Chand SK (1997) Mosquito breeding and resting in treeholes in a forest ecosystem in Orissa. Indian J Malariol 34:8–16

Zhou YH, Zhang ZW, Fu YF, Zhang GC, Yuan S (2018) Carbon dioxide, odorants, heat and visible cues affect wild mosquito landing in open spaces. Front Behav Neurosci 12:86. https://doi.org/10.3389/fnbeh.2018.00086. PMID: 29867387; PMCID: PMC5949359

Zwiebel LJ, Zwiebel LJ, Takken W (2004) Olfactory regulation of mosquito-host interactions. Insect Biochem Mol Biol 34:645–652

Behaviour of Mosquitoes

8

K. John Ravindran and Alex Eapen

Abstract

Mosquitoes exhibit varied behavioural trends in their lifetime. The behaviour varies in each species and is a manifestation of the innate and adaptive nature of the species. Each behavioural trait contributes greatly to the survival and establishment of the population in its habitat. Prodigality in production and successful survival of mosquitoes that can support pathogens or parasites result in the transmission of disease in a vulnerable population. The study of mosquito behaviour particularly the vectors contributes to the knowledge on effective control of vector population and to prevent disease transmission. Mosquito behaviour, beginning from the immature stages to adult emergence, swarming, mating, feeding, resting and oviposition, has been intensively studied in various disease vectors in different endemic areas of the world. In the present chapter, some of the interesting behaviours of mosquitoes are discussed. An understanding of these will be helpful from an academic point of view and provide insight into the fundamentals of vector control.

Keywords

Feeding · Swarming · Mating · Biting · Resting · Oviposition · Vector · Mosquitoes

K. John Ravindran (✉) · A. Eapen
ICMR – National Institute of Malaria Research, Field Unit, ICMR-NIE Campus, Chennai, India

© The Author(s), under exclusive license to Springer Nature Singapore Pte Ltd. 2024

Omkar (ed.), *Mosquitoes*, https://doi.org/10.1007/978-981-97-4163-2_8

Learning Objectives

1. To understand various behaviours of different mosquito species, like feeding behaviour, reproductive behaviour, oviposition behaviour, etc.

8.1 Introduction

Mosquito behaviour is an outcome of interactions between innate and non-innate environmental factors in the region of their prevalence. Each mosquito species possesses specific innate attributes that cause differences in their behaviour. Variations in morphology, genetic factors and physiological processes determine the differences in their behaviour and encompass a wide range of activities, but not limited to feeding, reproduction, resting, aestivation or diapause and migration that contribute to their survival. In addition, the innate factors also determine their ability to sustain and transmit various diseases, like malaria, filaria, Japanese encephalitis, dengue, chikungunya, yellow fever, Zika, etc., affecting mankind. Nevertheless, all mosquito species do not support and transmit these diseases. It is, therefore, pertinent to incriminate a mosquito species as a vector. Extensive research work has been carried out on vectors of malaria in different parts of the world and substantial knowledge of mosquito behaviour obtained. In this chapter, resources to understand mosquito behaviour have been considered from studies on malaria vectors in India with added information on mosquitoes (vector/non-vectors) from other places. An understanding of the behaviour of vector mosquito species is of interest both from the academic point of view and for the control of mosquito-borne diseases (Fig. 8.1).

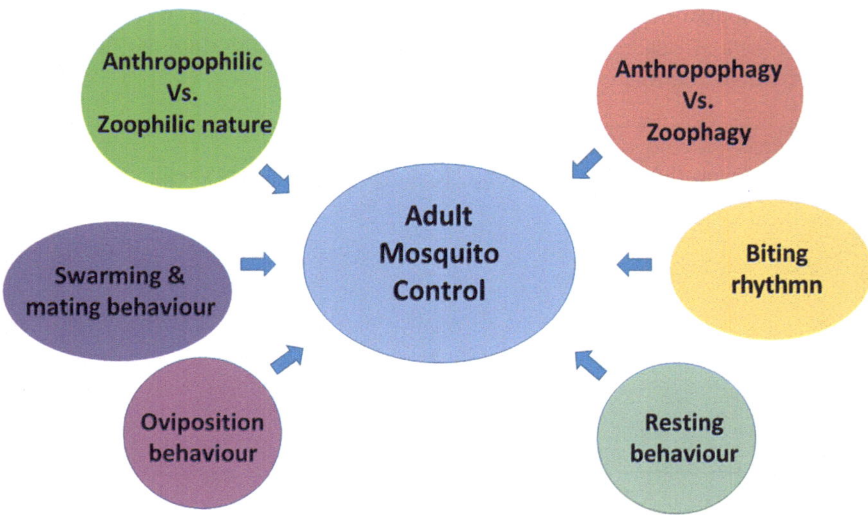

Fig. 8.1 A schematic diagram on the bionomics of mosquitoes essential for successful vector control

8.2 The Immature Mosquito

8.2.1 Behaviour and Activity

Mosquitoes have four stages in life cycle (egg, larva, pupa and adult) and occupy two different niches in their lifetime. The immatures are aquatic (egg, larva [I–IV instars], and pupa), and the adults are terrestrial and free-flying. Immatures of various mosquito species have evolved to adapt to variations in aquatic habitats which may be due to factors, viz., light, temperature, dissolved gases, dissolved inorganic salts, pH and flow of water (Bickley 1972), with extremely low or high temperatures affecting their development and survival (Agyekum et al. 2021). Active growth and development are seen in the immature phases, with foraging being the main activity of the larval instars. Recently, microbial RNA in the aquatic habitat has been reported to be a foraging attractant, and chemokinesis to navigate chemical gradients (Lutz et al. 2019). The anal papillae of mosquito larvae play an important role in xenobiotic detoxification including ammonia excretion and osmoregulation when encountering higher salinity in the aquatic habitat (Durant et al. 2021). Larvae are usually found to be aggregated or dispersed in a habitat, and this depends on the availability of food resources and predators. In rice fields, larvae were observed to be aggregated in the presence of *Azolla* and *Lemna* plants but were found dispersed in the presence of the predator (*Anisops* sp.) (Milita et al. 2019). Mosquito larvae are metapneustic (breathe through a single pair of posterior spiracles) (Clements 1992) and are found to frequent the water surface for breathing, since they depend on atmospheric oxygen. All mosquitoes are holometabolous, and insect hormones, like juvenile hormones together with ecdysteroids, orchestrate the expression of larval-specific genes, each time the insect moults to cause morphogenetic changes in different instars (Adam 2009). Pupae are non-feeding stages and tend to be very agile frequenting the water surface for breathing.

8.2.2 Feeding Behaviour

Mosquito larvae are filter feeders and have two alternative feeding mechanisms, viz., filter-feeding on microorganisms at the water surface, and scraping biofilms from underwater vegetation and rocks (Roberts 2014). They are primarily aquatic detritivores (Clements 1992) and feed on other microorganisms (bacteria or protozoa), unicellular and filamentous algae, and small metazoans (i.e. crustaceans) present in the aquatic habitat (Merritt et al. 1992). Larvae of the non-haematophagous mosquito belonging to the genus *Toxorhynchites* are carnivores and prey on the immatures and eggs of other mosquitoes present in the habitat. Many other species that are detritivores also exhibit a cannibalistic nature when challenged with food shortage, as reported in malaria vectors, viz., *Anopheles culicifacies* Giles (Reisen et al. 1982), and in *An. stephensi* Liston (Reisen and Emory 1976) in the Indian subcontinent, and *An. messeae* Fallerons (Gordeev and Troshkov 1990), *An. pharoensis* Theobald (Gillies and DeMeillon 1968), *An. punctipennis* Say (Hinman 1934), *An.*

maculatus Theobald (Senior-White 1926), and *An. gambiae* (s.l.) Giles (Koenraadt and Takken 2003) in laboratory-based observations elsewhere.

Food availability including its quality in the aquatic habitat, which is required for the growth and development of larvae, greatly determines the body size of the adult mosquito, its longevity and biting frequency and affects their competence for pathogen transmission (Yan et al. 2021). At higher food concentrations, the rate of larval growth was found to be fast (Levi et al. 2014; Gunathilaka et al. 2019), and their development with fitness was strongly contingent on both diet and microbial community composition in the gut (Martinson and Strand 2021). Padmanabha et al. (2011) reported that both thermal and nutrient conditions together affect development rather than each independently. Anthropogenic stressors, like insecticides and fertilizers, are known to impact larval behaviour and alter the microbial communities in freshwater ecosystems (Widenfalk et al. 2008), resulting in the change of quality and quantity of food available for mosquito larvae. Further, predator cues, like kairomones, are likely to cause changes in the larval feeding behaviour (Roberts 2014) and affect growth and development.

8.3 Adult Mosquito

8.3.1 Swarming and Mating Behaviour

Dipterans, including mosquitoes, are known to exhibit swarming behaviour, an aggregation of male and female mosquitoes as a part of their mating behaviour, which is unique to each species (Downes 1969; Yuval 2006). Aggregation pheromones are responsible for this behaviour and a five-compound blend of volatiles, acetoin, sulcatone, octanal, nonanal and decanal, have been identified as an aggregation pheromone among *An. gambiae* and *An. arabiensis* Patton, which also is said to have increased mating in *An. funestus* Giles, *An. coluzzii* Coetzee and Wilkerson and *An. merus* Donitz mosquitoes too (Mozūraitis et al. 2020). The role of male sex pheromones in attracting females into a swarm has not yet been identified (Gendrin 2017). However, in mating bioassays, heptacosane, a contact sex pheromone promoted mating activity in *Aedes aegypti* Linnaeus (Wang et al. 2023).

Swarming takes place over landmarks called swarm markers, which are generally the dark-light contrast areas that are attractive to both sexes (Howell and Knols 2009). Visual cues play a vital role in the selection of swarm markers (Yuval 2006; Diabate and Tripet 2015). Mosquito species differ in the choice of swarm markers. *An. gambiae* was found to swarm over galleries, manure, waste piles (Sawadogo et al. 2014) and in the bare ground (Diabate et al. 2003) whereas *An. coluzzii* over wood stocks, bare earth, sheds, wells, manure stacks, wastes and grass (Diabate and Tripet 2015; Sawadogo et al. 2014; Gueye et al. 2023). Factors, like spatial swarm segregation, variations in time of swarm formation, ground markers, height, flight one and other temporal variations, act as a pre-mating reproductive isolation mechanism among mosquitoes (Baeshen 2022). Though females of other mosquito species can be found in a monospecific swarm (Diabate et al. 2003; Poda et al. 2022), assortative mating mediated most probably by close-range mate recognition cues is observed (Dabire et al. 2013).

Swarming usually takes place during dusk (Charlwood and Jones 1980; Clements 1999; Charlwood et al. 2003; Achinko et al. 2016; Zawada et al. 2018; Kaindoa et al. 2019), and daylight and clock genes regulate the circadian rhythmicity of swarming and mating activity (Wang et al. 2021). A swarm may contain ten or fewer, to thousands or more, of male mosquitoes, and the swarm size may increase rapidly in a few minutes (Clements 1999; Sawadogo et al. 2014). The number of individuals may however remain the same over time (Charlwood et al. 2003). Individual conspecific females fly into the swarm to acquire a mate and leave the swarm *in copula* (Sawadogo et al. 2014; Diabate and Tripet 2015). Male mosquitoes mate on sexual maturity and after their terminalia become inverted, which may take 24 h post-emergence (Charlwood and Jones 1979; Clements 1999). After insemination, a male hormone transferred to the female during copulation causes a post-mating switch in the female, incapacitating further fertilization by other males (Gabrieli et al. 2014). Vector control through genetic control strategies generally targets swarming and mating mosquitoes.

8.3.2 Feeding Behaviour

Male mosquitoes are purely phytophagous, feeding on sugars found in nectars, fruit juices, exudates from plant tissues and plant-derived sugars from insects, like honeydew (Foster 1995; Gary Jr and Foster 2004; Impoinvil et al. 2004; Müller et al. 2011). Plant-based sugars are also energy sources for female mosquitoes a few days after emergence for flight and anemotactic behaviour (Foster 1995). Female mosquitoes are generally haematophagous, and their mouthparts constitute a tubular proboscis made of a bundle of stylets called a fascicle and a lower lip, the labium that helps in piercing and sucking blood from hosts (Kong and Wu 2010). However, some mosquitoes, particularly all species of the genera *Toxorhynchites* and *Malaya*, and possibly a few species of *Topomyia*, and *Maorigoeldia* feed exclusively on plant-derived sugars (Service 1993a; Reeves et al. 2018).

Metabolic energy and proteins obtained from blood are essential for the production of eggs (Clements 1992). Such mosquitoes that depend on blood for egg production are called an autogenous and others that are not hematophagous as autogenous mosquitoes (Clements 1963, 1992; Briegel 2003; Attardo et al. 2005). But, mosquitoes, like *Ae. taeniorhynchus* Wiedemann and *Culex pipiens molestus* Forskål, are capable of producing eggs without blood feeding for first oviposition but do require blood for subsequent ones and are referred to as facultatively autogenous. Mosquitoes of all species of the genera *Toxorhynchites* and *Malaya*, and possibly a few species of *Topomyia*, and *Maorigoeldia* rely on plant-derived sugars and are said to be obligately autogenous (O'Meara 1985; Clements 1992). It has been reported that multiple factors including blood meal-dependent release of ovary ecdysteroidogenic hormone and insulin-like peptide 3 from the brain prevent autogenous and facilitate anautogenous reproduction (Gulia-Nuss et al. 2015).

8.3.2.1 Host Selection

Mosquitoes feed on the blood of all vertebrates, i.e. mammals, aves, reptiles, amphibians and pisces (Tempelis 1975; Tamashiro et al. 2011; Takken and Verhulst 2013; Miyake et al. 2019). The mosquito species, *Uranotaenia sapphirina* Osten Sacken, has been reported to feed on invertebrates, namely, annelids (Reeves et al. 2018). Mosquitoes exhibit host preference and have been defined as 'the choice of a particular vertebrate host as a food source, rather than other host species equally available' (Boreham and Garrett-Jones 1973). Mosquito preference to host varies among different species, within species and among the sibling species. Among the same host species, mosquitoes exhibit a preference for a host, which is likely caused by the natural variation in odorants between individuals and demonstrates the sensitivity of the odorant receptors to semiochemicals (Zwiebel and Takken 2004). For example, lactic acid, an excretory product of humans, is an important cue in the host selection process of the anthropophilic mosquitoes, *Ae. aegypti* and *An. gambiae* (s.s.) (Dekker et al. 2002; Smallegange et al. 2011). Carbon dioxide exhaled by all vertebrates is also a general cue for all hematophagous arthropods, causing activation and attraction to hosts (Gillies 1980; Mboera and Takken 1997; Russell 2004). In addition to the odour, colour, heat and temperature of the host body are major cues for detecting preferential hosts by the mosquitoes. Host choice is reported to depend on the innate host preference of the mosquito species, and on the behavioural characteristics, which may be driven by selection and therefore have a genetic background (Takken and Verhulst 2013). The genetic control of host preferences was confirmed in the mosquitoes, *Ae. (Stegomyia) simpsoni* Theobald and *Ae. aegypti* (Mukwaya 1977).

8.3.2.1.1 Anthropophagy Vs. Zoophagy

Some mosquitoes are primarily anthropophagous preferring human blood feed, while others are zoophagic showing an inclination toward animal blood. However, in the wild, mosquitoes exhibit a high degree of plasticity toward host selection, and one of the strongest drivers is the relative abundance of the different host species (Russell et al. 2016). Vector mosquitoes feed alternatively on different blood sources during their life span (Waite et al. 2017; Chaumeau et al. 2019), and in the event of the non-availability of a preferential blood source (Reeves et al. 2018), they have been found to feed on the available host in the area of their prevalence. In the Indian subcontinent, malaria vectors, viz., *An. dirus* Peyton and Harrison (Dutta et al. 1996) and *An. fluviatilis* James (Nanda et al. 2000, 2012; Sahu et al. 2017), have been reported as anthropophagous while *An. culicifacies* (Nanda et al. 2000; Sahu et al. 2017; Baharia et al. 2022) and *An. sundaicus* Rodenwaldt (Kumari et al. 1993) as zoophagous mosquitoes. It is the preference of vectors to exploit human hosts wherever they find that makes them an efficient vector than the devoutly zoophagic vectors (Killeen et al. 2017).

8.3.2.1.2 Anthropophily Vs. Zoophily

Besides, a preference to be in the vicinity of humans or animals is noted among the mosquitoes and is termed anthropophilic or zoophilic, respectively. *An. baimaii*

Sallum and Peyton (Obsomer et al. 2007; Dev et al. 2015; Subbarao et al. 2019), *An. dirus* (Prakash et al. 1997; Nandi et al. 2000; Obsomer et al. 2007), *An. fluviatilis* (sibling species S) (Nanda et al. 1996), and *An. minimus* Theobald (Dev 1996; Dev et al. 2015) are considered anthropophilic, whereas *An. culicifacies* (except the sibling species E) (Rao 1984; Barik et al. 2009; Sharma and Dev 2015; Subbarao et al. 2019) and *An. stephensi* (Thomas et al. 2017) are primarily zoophilic. The anthropophilic or zoophilic nature of mosquitoes is determined by human landing collections of mosquitoes and is subject to variations by the presence of the availability of humans or cattle at the feeding time. The anthropophilic and anthropophagy nature of a vector is to be determined to ascertain the host preferences of mosquitoes to undertake appropriate vector control measures to prevent transmission of vector-borne diseases.

8.3.2.2 Feeding Site: Endophagy Vs. Exophagy

Mosquitoes also exhibit their tendency to either feed within enclosed structures, like houses, cattle sheds, etc., (indoor) or in the open (outdoor), and those that preferentially feed indoors are termed endophagic and those outdoors as exophagic. Some mosquitoes, however, readily feed both indoors and outdoors. The vector mosquito *An. minimus* is reported to be endophagic in Sonapur, Assam (Dev 1996) and in Tripura (Dev et al. 2015), and the same species showed both endophagic and exophagic behaviour in Jhargaon, Assam (Dev et al. 2001). *An. fluviatilis* and *An. baimaii* also exhibited endophagic and exophagic behaviour in Jeypore Hills, Orissa (Gunasekaran et al. 1994) and Tripura (Dev et al. 2015), respectively. Dev et al. (2015) also reported *An. maculatus* and *An. niveps* Theobald/*An. philippinensis* Ludlow to be exophagic, and the latter being exclusively exophagic. From the outcome of various observations, it is evident that variations in indoor or outdoor feeding preferences do exist among and within species. Both environmental and anthropogenic factors can influence this behaviour (Russell et al. 2011; Ravishankaran et al. 2022). In addition, human interventions by indoor residual spray or by usage of long-lasting impregnated nets (LLINs) were reported to alter indoor host-seeking preferences in vector mosquitoes (Asale et al. 2017).

8.3.2.3 Biting Rhythm

Different mosquito species have evolved distinct circadian timing of behaviours according to their temporal/ecological niche (Baik et al. 2020). Mosquito species can be nocturnal, diurnal or crepuscular in activity, particularly in biting or feeding time. Mosquitoes belonging to the genera *Aedes* are generally diurnal, and those of *Anopheles* and *Culex* are nocturnal. They exhibit a period of low and peak biting activity at various time periods, and this varies among species, in regions and is of immense importance to strategize vector control. In India, in the state of Goa, the peak biting activity of the malaria vectors, *An. stephensi* and *An. Fluviatilis*, was during the early morning hours (between 03:00 and 06:00 h) with *An. fluviatilis* showing a bimodal peak, one from 21:00 to 23:00 h and a higher peak from 03:00 to 06:00 h (Korgaonkar et al. 2012). A similar study from the same region, however, showed variation in the peak biting activity of *An. stephensi* and was observed between 21:00 and 24:00 h (Kumar et al. 1995). *An. culicifacies* did not show any trend in biting activity but showed maximum activity before 02:00 h after the monsoon (Korgaonkar et al. 2012). *An. dirus*, a vector in Assam, was predominantly a midnight biter, and peak biting time was between 21:00 and 03:00 h (Prakash et al. 1997). In Thailand, two peak biting times, one immediately after dusk (18:00 to 21:00 h) and a mild one before dawn (03:00 to 06:00 h), were noted among *An. minimus* (Chareonviriyaphap et al. 2003). Among *An. dirus* (s.l.), the peak biting time indoors and outdoors were 20:00 to 21:00 h in Western Thailand (Sungvornyothin et al. 2009), and 22:00 to 01:00 and 22:00 to 24:00 h in Bangladesh (Rosenberg and Maheswary 1982). The peak biting time of *An. sundaicus* was between 21:00 and 01:00 h in both indoor and outdoor observations in Indonesia (Syahrani et al. 2022).

In the African region, an unimodal peak biting activity between 19:00 and 20:00 h was observed among *An. ziemanni* Grünberg in Ethiopia (Kenea et al. 2016) and between 18:00 and 21:00 h in both indoor and outdoor settings in *An. gambiae* (s.l.), *An. coustani* (s.l.) Laveran and *An. pharoensis* (Taye et al. 2016). A bimodal or two peak biting activities between 22:00 and 23:00 h and from 02:00 to 04:00 h was observed in *An. funestus* in Zimbabwe (Sande et al. 2016), and from 20:00 to 21:00 h, and from 24:00 to 01:00 h (indoors) in *An. arabiensis* in Ethiopia (Kenea et al. 2016). Multiple peak biting activity was also observed in the predawn period in *An. arabiensis* in South Africa (Braack et al. 1994). Though *Anopheles* mosquitoes are nocturnal, there are, however, reports of substantial mosquito biting activity during the daytime among vector mosquitoes, like *An. gambiae*, *An. coluzzii*, *An. funestus* and *An. pharoensis*, which substantially bites during the daytime inside human dwellings (Sangbakembi-Ngounou et al. 2022).

Among the Culicines, the peak biting activity of *Cx. quinquefasciatus* Say, *Cx. vishnui* Theobald and *Cx. gelidus* Theobald was between 03:00 and 06:00 h in Goa. *Cx. gelidus* showed a trimodal biting pattern, the first peak from 20:00 to 21:00 h, the second from 23:00 to 24:00 h and the third much higher at 05:00 to 06:00 h during the pre-monsoon season (Korgaonkar et al. 2012). In the metropolitan area of Chennai, the peak biting activity of *Cx. quinquefasciatus* was reported between 12:00 and 03:00 h (Ravindran et al. 2002). In tea gardens in Assam, India, *Cx. quinquefasciatus* showed two peaks of biting activity, the first from 19:00 to 20:00 h and the second peak between 22:00 to 23:00 h (Mahanta et al. 1999). Another mosquito, *Armigeres subalbatus* Coquillett, which preferred to feed outdoors, exhibited bimodal peak, the higher and the lower peak biting times being 06:00 to 07:00 and 18:00 to 19:00 h in Nagamalai near Madurai, India (Selvaraj and Chandrashekaran 1980). *Cx. quinquefasciatus* is a known vector of bancroftian filariasis, whereas *Cx. vishnui* and *Cx. gelidus* have been reported to be responsible for the transmission of Japanese encephalitis in India. *Ar. subalbatus* is a vector of zoonotic *Brugia pahangi* Buckley and Edeson filariasis in Thailand and Malaysia (Muslim et al. 2013; Bhumiratana et al. 2023).

From the preceding examples, it can be seen that different vector species, each exposed to varied field conditions, exhibited variation in peak biting time. The biting rhythm is influenced by local weather conditions, human sleeping behaviour, light and lunar cycle and causes variations in biting rhythm. Mosquito abundance and mobility are generally affected by heavy rainfall and drop in temperature and strong winds prior to rainfall (Gillies and Wilkes 1981; Sharp 1983; Costantini et al. 1996) resulting in seasonal changes in their biting rhythm. In addition, human resting or sleeping behaviour, sleep time and usage of personal protection measures including repellents can alter the biting time. Further, illumination indoors also affect biting rhythm. It was observed that pulses of artificial light at night suppressed (photic inhibition) *An. gambiae*, a nocturnal mosquito's biting behaviour (Das and Dimopoulos 2008; Sheppard et al. 2017). On the other hand, the biting period of the diurnally active anthropophilic *Ae. aegypti* mosquito on human hosts was found to be extended on exposure to artificial light at night (Rund et al. 2020). Under moonlight, it was observed that there was an initial peak of biting activity of *An. farauti* Laveran after sunset, which was later dependent on the phase of the moon during the latter part of the night (Charlwood et al. 1986). Intervention measures, like indoor residual spray (IRS) and usage of LLINs can bring about a change in the biting rhythm of mosquitoes. It was speculated long ago that the medium- to long-term use of insecticide-treated nets would result in the nocturnal mosquitoes shifting their biting habits toward daytime (Ferreira et al. 2017).

8.3.3 Resting Behaviour

One of the most important behaviours of mosquitoes studied for intervention purposes is their resting behaviour. Mosquitoes generally prefer dark and humid places to rest and exhibit plasticity in the selection of resting places/habitats (Fig. 8.2). Freshly emerged mosquitoes are usually found nearby to the habitats from where they have emerged. Unfed and blood-fed mosquitoes tend to rest in places further away from the place of emergence. It is well known that blood-fed mosquitoes usually rest until the development and maturity of eggs. Those species that prefer to rest mainly indoors are termed endophilic and those outdoors as exophilic mosquitoes. Among the vector mosquitoes in India, *An. culicifacies* (Rao 1984; Singh et al. 1999; Dhiman et al. 2005), *An. fluviatilis* (Rao 1984; Nanda et al. 2000) and *An. minimus* (Dev 1996; Dev et al. 2015; Dev and Manguin 2020; Khan et al. 2023) were endophilic, while *An. baimaii* (Subbarao et al. 2019; Khan et al. 2023) and *An. stephensi* (Sumodan et al. 2004; Thomas et al. 2017) were exophilic. It is reported that resting behaviour appears to be relatively plastic with considerable potential for variation between and within species (Paaijmans and Thomas 2011). As an example, *An. minimus* was reported to exhibit exophilic nature in Thailand (Ismail et al. 1973), Cambodia, Laos and Vietnam (Trung et al. 2005), and the same species also showed endophilic behaviour in the plains of northern Thailand (Nustsathapana et al. 1986) and in the hilly forested part of northern Vietnam (Trung et al. 2005). Similarly, in the sub-Saharan regions, the primary malaria vectors, *An. gambiae* (s.s.) and *An. funestus*, were endophilic (Aniedu 1993; Faye et al. 1997; Githeko et al. 1997; Rajaonarivelo et al. 2004), and *An. arabiensis* was exophilic (Randrianasolo and Coluzzi 1987; Fontenille et al. 1990; Tirados et al. 2006). Nonetheless, reports indicate *An. gambiae* (s.s.) and *An. funestus* to be exophilic

Fig. 8.2 Some preferential resting places of adult mosquitoes

(Fontenille et al. 1990; Bockarie et al. 1994) and *An. arabiensis* to be endophilic (Dukeen and Omer 1986; Ameneshewa and Service 1996; Faye et al. 1997). Owing to the heterogeneity in vector resting behaviour, therefore, an understanding of this behaviour at the ecotype or regional level becomes imperative in planning an effective vector control strategy. It should also be noted that resting site preference can get altered during interventions with IRS, particularly when insecticides with exito-repellency effect are used (Boreham and Garrett-Jones 1973; Mnzava et al. 1995). This is in agreement with the report of Asale et al. (2017) when they noted that the use of LLINs and IRS impacted the resting habits of *An. arabiensis*.

8.3.4 Oviposition Behaviour

As stated earlier, immature stages of mosquitoes being aquatic, an appropriate habitat for gravid mosquitoes to oviposit is imperative for the survival and propagation of mosquitoes. Many mosquito species oviposit during the crepuscular periods (Day 2016). Warm, humid and low wind velocity conditions favour mosquito flight for oviposition (Bidlingmayer 1974). Gravid females generally use sensory cues (visual, olfactory and tactile) to seek an appropriate oviposition site (Bentley and Day 1989; Clements 1999). The visual cues are more prominent in the twilight hours and on moonlit days. In one study, the dark phase of a lunar cycle was observed to extend the oviposition cycle of *An. farauti* (Birley and Charlwood 1989). In addition to the visual cues, the olfactory cues are significant as mosquito-breeding habitats are said to contain a complex mixture of biologically derived chemicals that influence oviposition choices of gravid mosquitoes (Kramer and Mulla 1979; Bentley and Day 1989; Du and Millar 1999; Ponnusamy et al. 2008). Mosquitoes have well-developed olfactory sensors, which help in seeking a preferential aquatic habitat for oviposition. The antennae and maxillary palp and, recently, the proboscis too have been reported to be involved in the reception of chemical cues from the environment (Maekawa et al. 2011; Choo et al. 2015). An understanding of the various factors associated with oviposition helps to develop effective, species-specific surveillance and mosquito control protocols (Day 2016). It is well known that Anophelines generally prefer clean fresh water sources for laying eggs, whereas many Culicines are tolerant to organically polluted and foul-smelling waters.

Mosquitoes lay eggs singly and are deposited either individually, as rafts or as clusters. They may be deposited directly on the water, or the wet substrate nearby or even on dry land (Bernáth et al. 2008; Day 2016). Mosquitoes belonging to the genera *Anopheles* and *Aedes* deposit eggs singly but those of the genus *Armigeres*, *Culex*, *Culiseta*, *Coquillettidia* and *Uranotaenia* as rafts. Some species in the genus *Mansonia*, including *Ma. annulifera* Theobald, *Ma. uniformis* Theobald and *Ma. indiana* Edwards lay eggs as clusters under the water surface, which are attached to leaves, stems and roots of aquatic plants (Krishnamoorthy et al. 1994). The biology behind this egg-laying nature including the egg-brooding behaviour as reported in *Ar. flavus* Leicesteria (Miyagi et al. 2005) presents a wide-ranging adaptive oviposition behaviour that determines the establishment and distribution of mosquito species in a geographical region. A variety of breeding habitats exists in different ecotypes, and the mosquitoes prevalent therein exploit those habitats, which are most appropriate to their oviposition habit and behaviour. Elimination or reduction of a preferable habitat in an area can cause a considerable reduction in the density of mosquitoes that prefer to breed in them. An in-depth knowledge of the preferential vector breeding or oviposition habitats is therefore essential to undertake effective vector control through anti-larval measures.

8.4 Conclusion

To conclude, mosquito-borne diseases are a scourge to mankind. Vector mosquito control is a major and significant component to check the build-up of vector population to prevent transmission of vector-borne diseases. An understanding of the vector mosquito behaviour and its bionomics particularly at the local level is essential to carry out effective vector control activities. Various manuals on the techniques to be adopted and to quantify information are available (Service 1993b; World Health Organization 1995; Benedict 2007; Wirtz et al. 2007; Williams and Pinto 2012). Global warming can cause various changes in the ecology of a region, which is expected to result in a change in the behaviour and bionomics of vector mosquitoes. A constant vigil and update on the knowledge of vector mosquito behaviour including its bionomics is therefore indispensable for effective control of vector population and in the prevention of transmission of mosquito-borne diseases.

Points to Remember

1. Mosquitoes have a four-stage life cycle (egg, larva, pupa and adult) and occupy two different niches in their lifetime. Immatures are aquatic and adults are free-flying.
2. Mosquito larvae generally are detritivores, but few species are carnivores. Nutrients, thermal conditions and the presence of predator cues affect immature growth and development.
3. Swarming behaviour, an aggregation of male and female adult mosquitoes for selection of mates is unique to each species. Factors, like spatial swarm segregation, variations in time of swarm formation, ground markers, height, flight one and other temporal variations are pre-mating reproductive isolation mechanisms among different species of mosquitoes.
4. Male and female mosquitoes both feed on plant sugars. Females of non-haematophagous mosquitoes feed exclusively on plant sugars and haematophagous mosquitoes on the blood of hosts.
5. Haematophagous mosquitoes require blood feed for egg production, and reproduction is termed anautogenous reproduction. Autogenous reproduction is observed in non-haematophagous mosquitoes, which produce eggs without blood feed. Some are obligately autogenous depending only on plant-based sugars, and others are facultatively autogenous requiring sugars during the first feed, and thereafter blood feed for subsequent egg production.

6. Mosquitoes feed on the blood of all vertebrates, and some species have been reported to feed on invertebrates too. Host preference is however specific to each species. Odour, colour, heat and temperature of the host body are major cues for detecting preferential hosts. In addition, the genetic component has been reported to play a role in host preference.
7. Mosquitoes exhibit a preference to feed either on human or non-human hosts, the former being termed anthropophagy and the latter zoophagy, both determined by identifying the blood fed by the mosquitoes. Mosquitoes that generally tend to be in the vicinity of human hosts, which is assessed by mosquito landing collections on hosts commonly termed anthropophilic nature also indicate the preference of mosquitoes for human blood. Preference to feed on the human host is an important behaviour to be assessed for effective control measures.
8. Mosquitoes feed on hosts indoors or outdoors with distinct preference either to feed indoors or outdoors. Indoor and outdoor feeding preferences are termed endophilic or exophilic behaviour. The choice varies among species, and both environmental conditions and anthropogenic factors alter this feeding behaviour.
9. Mosquitoes exhibit nocturnal, diurnal or crepuscular activity when considering the time of feeding. The peak biting time varies among different species, geographical regions and seasons. It is generally influenced by local weather conditions, human sleeping behaviour, light and the lunar cycle. Human intervention measure by the use of indoor residual spray also alters their biting rhythm.
10. Mosquitoes prefer to rest in dark and humid places. Endophilic mosquitoes prefer to rest indoors and exophilic mosquitoes outdoors. Resting behaviour is said to be relatively plastic with considerable potential for variation between and within species. Resting site preference can get altered due to intervention activities using insecticides.
11. Mosquitoes oviposit eggs on water or in substratum near an aquatic habitat. Visual and olfactory cues are vital to search and seek an appropriate habitat for oviposition. Eggs are laid singly and deposited as individual eggs, as rafts or as clusters depending on the species. Biologically derived chemicals can influence oviposition choices of gravid mosquitoes. Determination of preferential oviposition sites of the target vector mosquito is essential for effective immature vector mosquito control.

References

Achinko D, Thailayil J, Paton D, Mireji PO, Talesa V, Masiga D, Catteruccia F (2016) Swarming and mating activity of *Anopheles gambiae* mosquitoes in semi-field enclosures. Med Vet Entomol 30(1):14–20

Adam ME (2009) Juvenile Hormones. In: Resh VH, Cardé RT (eds) Encyclopedia of insects, 2nd edn. Academic Press, San Diego, pp 541–546

Agyekum TP, Botwe PK, Arko-Mensah J, Issah I, Acquah AA, Hogarh JN, Dwomoh D, Robins TG, Fobil JN (2021) A systematic review of the effects of temperature on *Anopheles* mosquito development and survival: Implications for malaria control in a future warmer climate. Int J Environ Res Public Health 18:7255

Ameneshewa B, Service (1996) Resting habits of *Anopheles arabiensis* in the Awash river valley of Ethiopia. Ann Trop Med Parasitol 90(5):515–521

Aniedu I (1993) Biting activity and resting habits of malaria vectors in Baringo district, Kenya. Anzeiger Schädlingskunde für Pflanzenschutz Umweltschutz 66:72–76

Asale A, Duchateau L, Devleesschauwer B, Huisman G, Yewhalaw D (2017) Zooprophylaxis as a control strategy for malaria caused by the vector *Anopheles arabiensis* (Diptera: Culicidae): a systematic review. Infect Dis Poverty 6:160

Attardo GM, Hansen IA, Raikhel AS (2005) Nutritional regulation of vitellogenesis in mosquitoes: implications for anautogeny. Insect Biochem Mol Biol 35(7):661–675

Baeshen R (2022) Swarming behaviour in *Anopheles gambiae* (sensu lato): Current knowledge and future outlook. J Med Entomol 59(1):56–66

Baharia RK, Kaur J, Sindhania A, Sharma G, Srivastava HC, Pant CS, Rahi M, Raghavendra K, Singh OP (2022) Bionomics of *Anophelesculicifacies* Sensu Lato in two malaria endemic districts of central Gujarat, India. J Arthropod Borne Dis 16(2):108–123

Baik LS, Nave C, Au DD, Guda T, Chevez JA, Ray A, Holmes TC (2020) Circadian regulation of light-evoked attraction and avoidance behaviours in daytime-versus nighttime-biting mosquitoes. Curr Biol 30(16):3252–3259

Barik TK, Sahu B, Swain V (2009) A review on *Anopheles culicifacies*: from bionomics to control with special reference to Indian subcontinent. Acta Trop 109(2):87–97

Benedict MQ (2007) Methods in *Anopheles* research. Malaria Research and Reference Reagent Resource Center (MR4)

Bentley MD, Day JF (1989) Chemical ecology and behavioural aspects of mosquito oviposition. Annu Rev Entomol 34:401–421

Bernáth B, Horváth G, Gál J, Fekete G, Meyer-Rochow VB (2008) Polarized light and oviposition site selection in the yellow fever mosquito: no evidence for positive polarotaxis in *Aedes aegypti*. Vis Res 48:1449–1455

Bhumiratana A, Nunthawarasilp P, Intarapuk A, Pimnon S, Ritthison W (2023) Emergence of zoonotic *Brugia pahangi* parasite in Thailand. Vet World 16(4):752–765

Bickley WE (1972) Some aspects of the behaviour of mosquito larvae (Diptera: Culicidae). J Wash Acad Sci 62:261–264

Bidlingmayer WL (1974) The influence of environmental factors and physiological stage on flight patterns of mosquitoes taken in the vehicle aspirator and truck, suction, bait and New Jersey light traps. J Med Entomol 11(2):119–146

Birley MH, Charlwood JD (1989) The effect of moonlight and other factors on the oviposition cycle of malaria vectors in Madang, Papua New Guinea. Ann Trop Med Parasitol 83(4):415–422

Bockarie MJ, Service MW, Barnish G, Maude GH, Greenwod BM (1994) Malaria in a rural area of Sierra Leone. III. Vector ecology and disease transmission. Ann Trop Med Parasitol 88(3):251–262

Boreham PFL, Garrett-Jones C (1973) Prevalence of mixed blood meals and double feeding in a malaria vector (*Anopheles sacharovi Favre*). Bull World Health Organ 48(5):605–614

Braack LE, Coetzee M, Hunt RH, Biggs H, Cornel A, Gericke A (1994) Biting pattern and host-seeking behaviour of *Anopheles arabiensis* (Diptera: Culicidae) in northeastern South Africa. J Med Entomol 31(3):333–339

Briegel H (2003) Physiological bases of mosquito ecology. J Vector Ecol 28(1):1–11

Chareonviriyaphap T, Prabaripai A, Bangs MJ, Aum-Aung B (2003) Seasonal abundance and blood feeding activity of *Anopheles minimus* Theobald (Diptera: Culicidae) in Thailand. J Med Entomol 40(6):876–881

Charlwood JD, Jones MDR (1979) Mating behaviour in the mosquito, *Anopheles gambiae* s.l. I: close range and contact behaviour. Physiol Entomol 4(2):111–120

Charlwood JD, Jones MDR (1980) Mating in the mosquito, *Anopheles gambiae* s.l. II: swarming behaviour. Physiol Entomol 5(4):315–320

Charlwood J, Paru R, Dagoro H, Lagog M (1986) Influence of moonlight and gonotrophic age on biting activity of *Anopheles farauti* (Diptera: Culicidae) from Papua New Guinea. J Med Entomol 23(2):132–135

Charlwood J, Thompson R, Madsen H (2003) Observations on the swarming and mating behaviour of *Anopheles funestus* from southern Mozambique. Malar J 2:2

Chaumeau V, Fustec B, Hsel SN, Montazeau C, Nyo SN, Metaane S, Sawasdichai S, Kittiphanakun P, Phatharakokordbun P, Kwansomboon N, Andolina C, Cerqueir D, Chareonviriyaphap T, Nosten FH, Corbel V (2019) Entomological determinants of malaria transmission in Kayin state, Eastern Myanmar: a 24-month longitudinal study in four villages. Wellcome Open Res 3:109

Choo YM, Buss GK, Tan K, Leal WS (2015) Multitasking roles of mosquito labrum in oviposition and blood feeding. Front Physiol 6:306

Clements AN (1963) The physiology of mosquitoes, vol 17, 1st edn. Pergamon Press, Oxford

Clements AN (1992) The biology of mosquitoes. Volume 1: development, nutrition and reproduction. Cambridge University Press, London

Clements, A.N. (1999). The biology of mosquitoes. Volume 2: sensory reception and behaviour. CABI Publishing, Wallingford

Costantini C, Li SG, Torre DA, Sagnon N, Coluzzi M, Taylor CE (1996) Density, survival and dispersal of *Anopheles gambiae* complex mosquitoes in a west African Sudan savanna village. Med Vet Entomol 10(3):203–219

Dabire KR, Sawadodgo S, Diabate A, Toe KH, Kengne P, Ouari A, Costantini C, Gouagna C, Simard F, Baldet T, Lehmann T, Gibson G (2013) Assortative mating in mixed swarms of the mosquito *Anopheles gambiae* s.s. M and S molecular forms, in Burkina Faso, West Africa. Med Vet Entomol 27(3):298–312

Das S, Dimopoulos G (2008) Molecular analysis of photic inhibition of blood-feeding in *Anopheles gambiae*. BMC Physiol 8:23

Day JF (2016) Mosquito oviposition behaviour and vector control. Insects 7:65

Dekker T, Steib B, Cardé RT, Geier M (2002) L-lactic acid: a human-signifying host cue for the anthropophilic mosquito *Anopheles gambiae*. Med Vet Entomol 16(1):91–98

Dev V (1996) *Anopheles minimus*: its bionomics and role in the transmission of malaria in Assam, India. Bull World Health Organ 74(1):61–66

Dev V, Manguin S (2020) Bionomics and control of *Anopheles* (Cellia) *minimus* in the context of disappearing malaria in India. In: Dev V (ed) Vector biology and control. An update for malaria elimination initiative in India. The National Academy of Sciences, Allahabad, pp 33–42

Dev V, Ansari MA, Hira CR, Barman K (2001) An outbreak of *Plasmodium falciparum* malaria due to *Anopheles minimus* in central Assam, India. Indian J Malariol 38(1-2):32–38

Dev V, Adak T, Singh OP, Nanda N, Baidya BK (2015) Malaria transmission in Tripura: disease distribution & determinants. Indian J Med Res 142(Suppl 1):S12–S22

Dhiman RC, Shahi B, Sharma SN, Nanda N, Khargiwarkar VN, Subbarao SK (2005) Persistence of malaria transmission in a tribal area in Maharashtra, India. Curr Sci 88(3):475–478

Diabate A, Tripet F (2015) Targeting male mosquito mating behaviour for malaria control. Parasit Vectors 8:347

Diabate A, Baldet T, Brengues C, Kengne P, Dabire KR, Simard F, Chandre F, Hougard JM, Hemingway J, Ouedraogo JB, Fontenille D (2003) Natural swarming behaviour of the molecular M form of *Anopheles gambiae*. Trans R Soc Trop Med Hyg 97(6):713–716

Downes JA (1969) The swarming and mating flight of Diptera. Annu Rev Entomol 14(1):271–298

Du YJ, Millar JG (1999) Electroantennogram and oviposition bioassay responses of *Culex quinquefasciatus* and *Culex tarsalis* (Diptera: Culicidae) to chemicals in odors from Bermuda grass infusions. J Med Entomol 36(2):158–166

Dukeen MYH, Omer SM (1986) Ecology of the malaria vector *Anopheles arabiensis* Patton (Diptera: Culicidae) by the Nile in northern Sudan. Bull Entomol Res 76(3):451–467

Durant AC, Guardian GE, Kolosov D, Donini A (2021) The transcriptome of anal papillae of *Aedes aegypti* reveals their importance in xenobiotic detoxification and adds significant knowledge on ion, water and ammonia transport mechanisms. J Insect Physiol 132:104269

Dutta P, Bhattacharyya DR, Khan SA, Sharma CK, Mahanta J (1996) Feeding patterns of *Anopheles dirus*, the major vector of forest malaria in northeast India. Southeast Asian J Trop Med Public Health 27(2):378–381

Faye O, Konate L, Mouchet J, Fontenille D, Sy N, Hebrard G, Herve JP (1997) Indoor resting by outdoor biting females of *Anopheles gambiae* complex (Diptera: Culicidae) in the Sahel of northern Senegal. J Med Entomol 34(3):285–289

Ferreira CP, Lyra SP, Azevedo F, Greenhalgh D, Massad E (2017) Modelling the impact of the long-term use of insecticide-treated bed nets on *Anopheles* mosquito biting time. Malar J 16:373

Fontenille D, Lepers JP, Campbell GH, Coluzzi M, Rakotoarivony I, Coulanges P (1990) Malaria transmission and vector biology in Manarintsoa, high plateaux of Madagascar. Am J Trop Med Hyg 43(2):107–115

Foster WA (1995) Mosquito sugar feeding and reproductive energetics. Annu Rev Entomol 40:443–474

Gabrieli P, Kakani EG, Mitchell SN, Mameli E, Want EJ, Anton MA, Serrao A, Baldini F, Catteruccia F (2014) Sexual transfer of the steroid hormone 20E induces the postmating switch in *Anopheles gambiae*. Proc Natl Acad Sci USA 111(46):16353–16358

Gary RE Jr, Foster WA (2004) *Anopheles gambiae* feeding and survival on honeydew and extra-floral nectar of peridomestic plants. Med Vet Entomol 18(2):102–107

Gendrin MA (2017) Swiss army knife to cut malaria transmission. Cell Host Microbe 22(5):577–579

Gillies MT (1980) The role of carbon dioxide in host-finding by mosquitoes (Diptera: Culicidae): a review. Bull Entomol Res 70:525–532

Gillies MT, DeMeillon B (1968) The Anophelinae of Africa South of the Sahara. Publication no. 54. South African Institute for Medical Research, Johannesburg

Gillies MT, Wilkes TJ (1981) Field experiments with a wind tunnel on the flight speed of some West African mosquitoes (Diptera: Culicidae). Bull Entomol Res 71(1):65–70

Githeko A, Service M, Mbogo C, Atieli F (1997) Resting behaviour, ecology and genetics of malaria vectors in large scale agricultural areas of Western Kenya. Parassitologia 38(3):481–489

Gordeev MI, Troshkov NI (1990) Inversion polymorphism of the malaria mosquito *Anopheles messeae*. IX. Cannibalism in larvae as a selection factor. Genetika 26:1597–1603

Gueye OK, Niang A, Faye MB, Dia AK, Ahmed AA, Sy O, Tripet F, Konaté L, Dia I, Wondji CS, Gaye O, Faye O, Diabate A, Niang EHA (2023) Characterization of the swarming behaviour of *Anopheles coluzzii* and *Anopheles gambiae* (Diptera: Culicidae) populations in a hybrid zone of Senegal. J Med Entomol 60(6):1278-1287 https://doi.org/10.1093/jme/tjad120 .

Gulia-Nuss M, Elliot A, Brown MR, Strand MR (2015) Multiple factors contribute to anautogenous reproduction by the mosquito *Aedes aegypti*. J Insect Physiol 82:8–16

Gunasekaran K, Sadanandane C, Parida SK, Sahu SS, Patra KP, Jambulingam P (1994) Observations on nocturnal activity and man biting habits of malaria vectors, *Anopheles fluviatilis*, *An. annularis* and *An. culicifacies* in the hill tracts of Koraput District, Orissa, India. Southeast Asian J Trop Med Public Health 25(1):187–195

Gunathilaka N, Upulika H, Udayanga L, Amarasinghe D (2019) Effect of larval nutritional regimes on morphometry and vectorial capacity of *Aedes aegypti* for dengue transmission. Biomed Res Int 2019:3607342

Hinman EH (1934) Predators of the Culicidae (Mosquitoes). I: the predators of larvae and pupae exclusive of fish. J Trop Med Hyg 37:129–134

Howell PI, Knols BG (2009) Male mating biology. Malar J 8(Suppl 2):S8

Impoinvil DE, Kongere JO, Foster WA, Njiru BN, Killeen GF, Githure JI, Beier JC, Hassanali A, Knols BGJ (2004) Feeding and survival of the malaria vector Anopheles gambiae on plants growing in Kenya. Med Vet Entomol 18(2):108–115

Ismail IAH, Notananda V, Schepens J (1973) Studies on malaria and responses of *Anopheles balabacensis balabacensis* and *Anopheles minimus* to DDT residual spraying in Thailand. Part I: pre-spraying observations. World Health Organization, Geneva

Kaindoa EW, Ngowo HS, Limwagu AJ, Tchouakui M, Hape E, Abbasi S, Kihonda J, Mmbando AS, Njalambaha RM, Mkandawile G, Bwanary H, Coetzee M, Okumu FO (2019) Swarms of the malaria vector *Anopheles funestus* in Tanzania. Malar J 18:29

Kenea O, Balkew M, Tekie H, Gebre-Michael T, Deressa W, Loha E, Lindtjørn B, Overgaard HJ (2016) Human-biting activities of *Anopheles* species in south-central Ethiopia. Parasit Vectors 9:527

Khan SA, Sarmah B, Bhattacharyya DR, Raghavendra K, Rahi M, Dutta P (2023) A study on the bionomics of primary malaria vectors *Anopheles minimus* and *Anopheles baimaii* in some states of north east region of India. J Trop Med. https://doi.org/10.1155/2023/9946754

Killeen GF, Kiware SS, Okumu FO, Sinka ME, Moyes CL, Massey NC, Gething PW, Marshall JM, Chaccour CJ, Tusting LS (2017) Going beyond personal protection against mosquito bites to eliminate malaria transmission: population suppression of malaria vectors that exploit both human and animal blood. BMJ Glob Health 2:e000198

Koenraadt CJM, Takken W (2003) Cannibalism and predation among larvae of the *Anopheles gambiae* complex. Med Vet Entomol 17(1):61–66

Kong XQ, Wu CW (2010) Mosquito proboscis: an elegant biomicroelectromechanical system. Phys Rev E 82:011910

Korgaonkar NS, Kumar A, Yadav RS, Kabadi D, Dash AP (2012) Mosquito biting activity on humans & detection of *Plasmodium falciparum* infection in *Anopheles stephensi* in Goa, India. Indian J Med Res 135(1):120–126

Kramer WL, Mulla MS (1979) Oviposition attractants and repellents of mosquitoes: oviposition responses of *Culex* mosquitoes to organic infusions. Environ Entomol 8:1111–1117

Krishnamoorthy K, Rajendran G, Panicker KN (1994) Aquatic vegetation and their natural hospitability to the immatures of Mansonia mosquitoes, the vectors of Brugia malayi in Shertallai, Kerala, India. Southeast Asian J Trop Med Public Health 25:760–765

Kumar A, Thavaselvam D, Sharma VP (1995) Biting behaviour of disease vectors in Goa. J Parasit Dis 19:73–76

Kumari R, Joshi H, Giri A, Sharma VP (1993) Feeding preferences of *Anopheles sundaicus* in Car Nicobar island. Indian J Malariol 30(4):201–206

Levi T, Ben-Dov E, Shahi P, Borovsky D, Zaritsky A (2014) Growth and development of *Aedes aegypti* larvae at limiting food concentrations. Acta Trop 133:42–44

Lutz EK, Grewal TS, Riffell JA (2019) Computational and experimental insights into the chemosensory navigation of *Aedes aegypti* mosquito larvae. Proc R Soc B Biol Sci 286:20191495

Maekawa E, Aonuma H, Nelson B, Yoshimura A, Tokunaga F, Fukumoto S, Kanuka H (2011) The role of proboscis of the malaria vector mosquito *Anopheles stephensi* in host-seeking behaviour. Parasit Vectors 4:10

Mahanta B, Handique R, Dutta P, Narain K, Mahanta J (1999) Temporal variations in biting density and rhythm of *Culex quinquefasciatus* in tea agro-ecosystem of Assam, India. Southeast Asian J Trop Med Public Health 30(4):804–809

Martinson VG, Strand MR (2021) Diet–microbiota interactions alter mosquito development. Front Microbiol 12:650743

Mboera LEG, Takken W (1997) Carbon dioxide chemotropism in mosquitoes (Diptera: Culicidae) and its potential in vector surveillance and management programmes. Med Vet Entomol 7:355–368

Merritt RW, Dadd RH, Walker ED (1992) Feeding behaviour, natural food, and nutritional relationships of larval mosquitoes. Annu Rev Entomol 37:349–374

Milita R, Malini K, Soumendranath C, Gautam A (2019) Distribution of mosquito larvae in rice field habitats: a spatial scale analysis in semi-field condition. Eur J Ecol 5(1):38–48

Miyagi I, Toma T, Okazawa T, Mogi M, Hashim R (2005) Female *Armigeres* (Leicesteria) *flavus* holding an egg raft with her hind legs. J Am Mosq Control Assoc 21:466–468

Miyake T, Aihara N, Maeda K, Shinzato C, Koyanagi R, Kobayashi H, Yamahira K (2019) Bloodmeal host identification with inferences to feeding habits of a fish-fed mosquito, *Aedes baisasi*. Sci Rep 9:4002

Mnzava AE, Rwegoshora RT, Wilkes TJ, Tanner M, Curtis CF (1995) *Anopheles arabiensis* and *An. gambiae* chromosomal inversion polymorphism, feeding and resting behaviour in relation to insecticide house-spraying in Tanzania. Med Vet Entomol 9(3):316–324

Mozūraitis R, Melika H, Zawada JW, Szymczak J, Pålsson K, Vaishnovi S, Inna B, Friedländer MR, Koekemoer LL, Kevin BJ, Anna-Karin B-K, Noushin ES (2020) Male swarming aggregation pheromones increase female attraction and mating success among multiple African malaria vector mosquito species. Nat Ecol Evol 4(10):1395–1401

Mukwaya LG (1977) Genetic control of feeding preferences in the mosquitoes *Aedes* (*Stegomyia*) *simpsoni* and *aegypti*. Physiol Entomol 2:133–145

Müller GC, Xue RD, Schlein Y (2011) Differential attraction of *Aedes albopictus* in the field to flowers, fruits and honeydew. Acta Trop 118:45–49

Muslim A, Fong MY, Mahmud R, Lau YL, Sivanandam S (2013) Armigeres subalbatus incriminated as a vector of zoonotic Brugia pahangi filariasis in suburban Kuala Lumpur, Peninsular Malaysia. Parasit Vectors 6:219

Nanda N, Joshi H, Subbarao SK, Yadav RS, Shukla RP, Dua VK, Sharma VP (1996) *Anopheles fluviatilis* complex: host feeding patterns of species S, T, and U. J Am Mosq Control Assoc 12(1):147–149

Nanda N, Yadav RS, Subbarao SK, Joshi H, Sharma VP (2000) Studies on *Anopheles fluviatilis* and *Anopheles culicifacies* sibling species in relation to malaria in forested hilly and deforested riverine ecosystems in northern Orissa, India. J Am Mosq Control Assoc 16(3):199–205

Nanda N, Bhatt RM, Sharma SN, Rana PK, Kar NP, Sharma A, Adak T (2012) Prevalence and incrimination of *Anopheles fluviatilis* species S (Diptera: Culicidae) in a malaria endemic forest area of Chhattisgarh state, central India. Parasit Vectors 5:215

Nandi J, Kaul SM, Sharma SN, Lal S (2000) Anthropophily of anophelines in duars of West Bengal and other regions of India. J Commun Dis 32(2):95–99

Nustsathapana S, Sawasdiwongphorn P, Chitprarop U, Cullen JR (1986) The behaviour of *Anopheles minimus* Theobald (Diptera: Culicidae) subjected to differing levels of DDT selection pressure in northern Thailand. Bull Entomol Res 76(2):303–312

O'Meara GF (1985) Gonotrophic Interactions in mosquitoes: kicking the blood-feeding habit. Fla Entomol 68(1):122–133

Obsomer V, Defourny P, Coosemans M (2007) The *Anopheles dirus* complex: spatial distribution and environmental drivers. Malar J 6:26

Paaijmans KP, Thomas MB (2011) The influence of mosquito resting behaviour and associated microclimate for malaria risk. Malar J 10:183

Padmanabha H, Bolker B, Lord CC, Rubio C, Lounibos LP (2011) Food availability alters the effects of larval temperature on *Aedes aegypti* growth. J Med Entomol 48(5):974–984

Poda SB, Buatois B, Lapeyre B, Dormont L, Diabaté A, Gnankiné O, Dabiré RK, Roux O (2022) No evidence for long-range male sex pheromones in two malaria mosquitoes. Nat Ecol Evol 6(11):1676–1686

Ponnusamy L, Xu N, Nojima S, Wesson DM, Schal C, Apperson CS (2008) Identification of bacteria and bacteria-associated chemical cues that mediate oviposition site preferences by *Aedes aegypti*. Proc Natl Acad Sci USA 105(27):9262–9267

Prakash A, Bhattacharyya D, Mohapatra P, Mahanta J (1997) Indoor biting behaviour of *Anopheles dirus* Peyton and Harrison, 1979 in upper Assam, India. Mosquito Borne Dis Bull 14(1-2):31–37

Rajaonarivelo V, Le Goff G, Cot M, Brutus L (2004) *Anopheles* and malaria transmission in Ambohimena, a village in the Occidental fringe of Madagascar Highlands. Parasite 11(1):75–82

Randrianasolo BOR, Coluzzi M (1987) Genetical investigations on zoophilic and exophilic *Anopheles arabiensis* from Antananarivo area (Madagascar). Parassitologia 29(1):93–97

Rao TR (1984) The Anophelines of India. Indian Council of Medical Research, New Delhi

Ravindran J, Eapen A, Kar I (2002) Evaluation of repellent action of neem oil against the filarial vector, *Culex quinquefasciatus* (Diptera: Culicidae). Indian J Malariol 39(1-2):13–17

Ravishankaran S, Asokan A, Justin NAJA, Thomas S, Joshua V, Mathai MT, Eapen A (2022) Does the roof type of a house influence the presence of adult *Anopheles stephensi*, urban malaria vector? Evidence from a few slum settings in Chennai, India. Parasitol Res 121(1):105–114

Reeves LE, Holderman CJ, Blosser EM, Gillett-Kaufman JL, Kawahara AY, Kaufman PE, Burkett-Cadena ND (2018) Identification of *Uranotaenia sapphirina* as a specialist of annelids broadens known mosquito host use patterns. Commun Biol 1:92

Reisen WK, Emory RW (1976) Cannibalism in *Anopheles stephensi* Liston. Mosq News 36:198–200

Reisen WK, Azra K, Mahmood F (1982) *Anopheles culicifacies* (Diptera: Culicidae): horizontal and vertical estimates of immature development and survivorship in rural Punjab province, Pakistan. J Med Entomol 19:413–422

Roberts D (2014) Mosquito larvae change their feeding behaviour in response to kairomones from some predators. J Med Entomol 51(2):368–374
Rosenberg R, Maheswary NP (1982) Forest malaria in Bangladesh. II. Transmission by *Anopheles dirus*. Am J Trop Med Hyg 31(2):183–191
Rund SSC, Labb LF, Benefiel OM, Duffield GE (2020) Artificial light at night increases *Aedes aegypti* mosquito biting behaviour with Implications for arboviral disease transmission. Am J Trop Med Hyg 103(6):2450–2452
Russell RC (2004) The relative attractiveness of carbon dioxide and octenol in CDC-and EVS-type light traps for sampling the mosquitoes *Aedes aegypti* (L.), *Aedes polynesiensis* Marks, and *Culex quinquefasciatus* Say in Moorea, French Polynesia. J Vector Ecol 29(2):309–314
Russell TL, Govella NJ, Azizi S, Drakeley CJ, Kachur SP, Killeen GF (2011) Increased proportions of outdoor feeding among residual malaria vector populations following increased use of insecticide-treated nets in rural Tanzania. Malar J 10:80
Russell TL, Beebe NW, Bugoro H, Apairamo A, Cooper RD, Collins FH, Lobo NF, Burkot TR (2016) Determinants of host feeding success by *Anopheles farauti*. Malar J 15:152
Sahu SS, Gunasekaran K, Krishnamoorthy N, Vanamail P, Mathivanan A, Manonmani A, Jambulingam P (2017) Bionomics of *Anopheles fluviatilis* and *Anopheles culicifacies* (Diptera: Culicidae) in relation to malaria transmission in East-Central India. J Med Entomol 54(4):821–830
Sande S, Zimba M, Chinwada P, Masendu HT, Makuwaza A (2016) Biting behaviour of *Anopheles funestus* populations in Mutare and Mutasa districts, Manicaland province, Zimbabwe: implications for the malaria control programme. J Vector Borne Dis 53(2):118–126
Sangbakembi-Ngounou C, Costantini C, Longo-Pendy NM, Ngoagouni C, Akone-Ella O, Rahola N, Cornelie S, Kengne P, Nakouné ER, Komas NP, Ayala D (2022) Diurnal biting of malaria mosquitoes in the central African Republic indicates residual transmission may be "out of control". Proc Natl Acad Sci USA 119(21):e2104282119
Sawadogo PS, Namountougou M, Toé KH, Rouamba J, Maïga H, Ouédraogo KR, Baldet T, Gouagna LC, Kengne P, Simard F, Costantini C, Gibson G, Diabaté A, Lees RS, Gilles JR, Dabiré KR (2014) Swarming behaviour in natural populations of *Anopheles gambiae* and *An. coluzzii*: review of 4 years survey in rural areas of sympatry, Burkina Faso (West Africa). Acta Trop 132(Suppl):S42–S52
Selvaraj PR, Chandrashekaran MK (1980) Rhythms in the biting behaviour of a mosquito *Armigeres subalbatus*. Oecologia 47(1):89–95
Senior-White R (1926) Physical factors in mosquito ecology. Bull Entomol Res 16:187–248
Service MW (1993a) Mosquitoes (culicidae). In: Lane RP, Crosskey RW (eds) Medical insects and arachnids. Chapman and Hall, London, pp 120–240
Service MW (1993b) Mosquito ecology: field sampling methods, 2nd edn. Springer, Dordrecht, p 987
Sharma VP, Dev V (2015) Biology & control of *Anopheles culicifacies* Giles 1901. Indian J Med Res 141(5):525–536
Sharp BL (1983) *Anopheles merus* (Donitz) its biting cycle in relation to environmental parameters. J Entomol Soc South Afr 46:367–374
Sheppard AD, Rund SSC, George GF, Clark E, Acri DJ, Duffield GE (2017) Light manipulation of mosquito behaviour: acute and sustained photic suppression of biting activity in the *Anopheles gambiae* malaria mosquito. Parasit Vectors 10:255
Singh N, Mishra A, Chand S, Sharma V (1999) Population dynamics of *Anopheles culicifacies* and malaria in the tribal area of central India. J Am Mosq Control Assoc 15(3):283–290
Smallegange RC, Verhulst NO, Takken W (2011) Sweaty skin: an invitation to bite? Trends Parasitol 27(4):143–148
Subbarao SK, Nanda N, Rahi M, Raghavendra K (2019) Biology and bionomics of malaria vectors in India: existing information and what more needs to be known for strategizing elimination of malaria. Malar J 18:396
Sumodan PK, Kumar A, Yadav RS (2004) Resting behaviour and malaria vector incrimination of *Anopheles stephensi* in Goa, India. J Am Mosq Control Assoc 20(3):317–318

Sungvornyothin S, Kongmee M, Muenvorn V, Polsomboon S, Bangs MJ, Prabaripai A, Tantakom S, Chareonviriyaphap T (2009) Seasonal abundance and blood feeding activity of *Anopheles dirus* sensu lato in western Thailand. J Am Mosq Control Assoc 25(4):425–430
Syahrani L, Permana DH, Syafruddin D, Zubaidah S, Asih PBS, Rozi IE, Hidayati APN, Kosasih S, Dewayanti FK, Rachmawati N, Risandi R, Bangs MJ, Bøgh C, Davidson J, Hendershot A, Burton T, Grieco JP, Achee NL, Lobo NF (2022) An inventory of human night-biting mosquitoes and their bionomics in Sumba, Indonesia. PLoS Negl Trop Dis 16(3):e0010316
Takken W, Verhulst NO (2013) Host preferences of blood-feeding mosquitoes. Annu Rev Entomol 58:433–453
Tamashiro M, Toma T, Mannen K, Higa Y, Miyagi I (2011) Bloodmeal identification and feeding habits of mosquitoes (Diptera: Culicidae) collected at five islands in the Ryukyu Archipelago, Japan. Med Entomol Zool 62(1):53–70
Taye B, Lelisa K, Emana D, Asale A, Yewhalaw D (2016) Seasonal dynamics, longevity, and biting activity of Anopheline mosquitoes in Southwestern Ethiopia. J Insect Sci 16(1):6
Tempelis CH (1975) Host-feeding patterns of mosquitoes, with a review of advances in analysis of blood meals by serology 2. J Med Entomol 11(6):635–653
Thomas S, Ravishankaran S, Justin NAJA, Asokan A, Mathai MT, Valecha N, Montgomery J, Thomas MB, Eapen A (2017) Resting and feeding preferences of *Anopheles stephensi* in an urban setting, perennial for malaria. Malar J 16:111
Tirados I, Costantini C, Gibson G, Torr SJ (2006) Blood-feeding behaviour of the malarial mosquito *Anopheles arabiensis*: implications for vector control. Med Vet Entomol 20(4):425–437
Trung HD, Bortel WV, Sochantha T, Keokenchanh K, Briët OJT, Coosemans M (2005) Behavioural heterogeneity of *Anopheles* species in ecologically different localities in Southeast Asia: a challenge for vector control. Trop Med Int Health 10(3):251–262
Waite JL, Swain S, Lynch PA, Sharma SK, Haque MA, Montgomery J, Thomas MB (2017) Increasing the potential for malaria elimination by targeting zoophilic vectors. Sci Rep 7:40551
Wang G, Vega-Rodríguez J, Diabate A, Liu J, Cui C, Nignan C, Dong L, Li F, Ouedrago CO, Bandaogo AM, Sawadogo PS, Maiga H, Alves E, Silva TL, Pascini TV, Wang S, Jacobs-Lorena M (2021) Clock genes and environmental cues coordinate *Anopheles* pheromone synthesis, swarming, and mating. Science 371(6527):411–415
Wang LM, Li N, Zhang M, Tang Q, Lu H, Zhou Q, Niu J, Xiao L, Peng Z, Zhang C, Liu M, Wang D, Deng S (2023) The sex pheromone heptacosane enhances the mating competitiveness of sterile *Aedes aegypti* males. Parasit Vectors 16:102
Widenfalk A, Bertilsson S, Sundh I, Goedkoop W (2008) Effects of pesticides on community composition and activity of sediment microbes–responses at various levels of microbial community organization. Environ Pollut 152(3):576–584
Williams J, Pinto J (2012) Training manual on malaria entomology for entomology and vector control technicians (basic level). United States Agency for International Development, Washington
Wirtz R, Avery M, Benedict M, Sutcliffe A (2007) *Plasmodium* sporozoite ELISA. Methods in Anopheles research. The Malaria Research and Reference Reagent Resource Center, pp 1–8
World Health Organization (1995) Manual on practical entomology in malaria. World Health Organization, Geneva
Yan J, Kibech R, Stone CM (2021) Differential effects of larval and adult nutrition on female survival, fecundity, and size of the yellow fever mosquito, *Aedes aegypti*. Front Zool 18(1):10
Yuval B (2006) Mating systems of blood-feeding flies. Annu Rev Entomol 51(1):413–440
Zawada JW, Dahan-Moss YL, Muleba M, Dabire RK, Maïga H, Venter N, Davies C, Hunt RH, Coetzee M, Koekemoer LL (2018) Molecular and physiological analysis of *Anopheles funestus* swarms in Nchelenge, Zambia. Malar J 17:49
Zwiebel LJ, Takken W (2004) Olfactory regulation of mosquito-host interactions. Insect Biochem Mol Biol 34(7):645–652

Abiotic and Biotic Interactions of Mosquitoes

9

Gaurav Kumar, Shweta Pasi, Jaspreet Kaur, and Himmat Singh

Abstract

Mosquitoes, being cold-blooded organisms, lack the ability to generate internal body heat and, as a result, rely on the surrounding temperature. The temporal and spatial distribution of mosquitoes is intricately tied to abiotic factors, such as temperature, humidity, and rainfall. In tropical and subtropical regions, these small creatures play a crucial role in the transmission of pathogens, such as the malaria parasite, dengue, and the chikungunya virus. The dynamic interplay of abiotic and biotic interactions creates a unique ecosystem that fosters the thriving of mosquitoes.

The lifespan of mosquitoes spans approximately 2–4 weeks, a duration influenced by temperature and humidity, besides other factors. Progressing through larval and pupal stages, they complete their cycle within a specific number of days, and finally emerge as adults. The overall life characteristics, including longevity, mating, and survival, are significantly contingent on both abiotic and biotic factors. As global temperatures rise, there is a pressing need for thorough investigations into the effects of these changes on the diversity of mosquitoes, as it has far-reaching implications for public health and disease transmission dynamics.

G. Kumar (✉) · J. Kaur · H. Singh
ICMR-National Institute of Malaria Research, New Delhi, India

S. Pasi
ICMR-National Institute of Occupational Health, Ahmedabad, Gujrat, India

© The Author(s), under exclusive license to Springer Nature Singapore Pte Ltd. 2024

Omkar (ed.), *Mosquitoes*, https://doi.org/10.1007/978-981-97-4163-2_9

Keywords
Mosquitoes · Microclimate · Temperature · Humidity · Rainfall · Prey-predator interaction

Learning Objectives

1. Understand the roles of abiotic and biotic components in the life cycle of mosquitoes, recognizing their indispensable influence on mosquito development and survival.
2. Identify the key abiotic factors, including temperature, humidity, and rainfall, and their role in creating breeding habitats and regulating mosquito physiology throughout various life stages.
3. Examine the ecological impact of predators feeding on mosquito larvae in the aquatic microenvironment, and comprehend how this natural regulation influences mosquito abundance and population dynamics.
4. Explore the intricate relationships between mosquitoes and both pathogens and commensals, understanding their role in regulating mosquito reproduction and transforming them into vectors of infectious diseases.
5. Recognize the pivotal role of mosquitoes as intermediate hosts for *Plasmodium* spp. and arboviruses, and understand their contribution to the transmission of vector-borne diseases, such as malaria, chikungunya, and dengue.

9.1 Introduction

Mosquitoes, members of the Culicidae family, are among the most pervasive and consequential insects on earth. With over 3500 species identified to date, these small, winged insects exhibit a remarkable ability to adapt to diverse ecosystems, playing a crucial role in shaping the dynamics of the natural world. However, their impact on global health is equally profound, as mosquitoes serve as vectors for a myriad of infectious diseases, affecting millions of people annually. They are notorious for their ability to transmit diseases, such as malaria, dengue fever, and Zika virus. However, their ecological interactions go beyond their impact on human health. Mosquitoes are an integral part of various ecosystems, participating in intricate abiotic and biotic interactions that shape their populations, behavior, and distribution. We delve into the intricate web of ecological interactions that govern mosquito life cycles, population dynamics, and the resulting consequences for ecosystems and public health. Recognizing the urgency of addressing mosquito-borne diseases, this chapter seeks to unravel the ecological intricacies of mosquitoes as a foundation for informed strategies in disease management (Table 9.1).

Table 9.1 Common vector-borne diseases and their mosquito vectors

Disease	Reservoir of infection	Mosquito vector
Malaria	Human	Anopheline mosquitoes (9 species: *Anopheles culicifacies*, *An. stephensi*, *An. minimus*, *An. fluviatilis*, *An. sundaicus*, *An. dirus*, *An. annularis*, *An. Varuna*, and *An. philippinensis*)
Filariasis	Human	Culex mosquitoes (*Cx. quinquefasciatus*, *Mansonia*)
Kala-azar	Human	Phlebotomine sandfly (*P. argentipes*)
Dengue fever	Human	Aedes mosquitoes (*Ae. aegypti*, *Ae. albopictus*)
Chikungunya	Human	Aedes mosquitoes (*Ae. aegypti*, *Ae. albopictus*)
Japanese Encephalitis	Ardeid birds and animals, especially pigs	Culex mosquitoes (*Cx. tritaeniorhynchus*, *Cx. vishnui* group)

9.2 Abiotic Interactions

Abiotic factors refer to the nonliving components of an ecosystem that influence the survival and behavior of organisms. Mosquitoes are cold-blooded creatures, i.e., unable to regulate their body heat and rely on ambient temperature for their growth and survival. Thus, they are highly sensitive to abiotic factors, such as temperature, humidity, and rainfall. Different species exhibit varying climate preferences, thereby determining their varying geographical distribution. Warmer temperatures generally accelerate the development of mosquito larvae, while cooler temperatures extend their life cycle. Changes in climate patterns cannot only influence mosquito abundance but also alter disease transmission dynamics.

9.2.1 Temperature

The life cycle of mosquitoes is completed in water and air. There are four distinct stages in the life cycle of mosquito: egg, larva, pupa, and adult. The first three stages develop in water, while the adults are terrestrial. Among them, only the female feeds on blood and lays eggs during its reproductive cycle, also known as gonotrophic cycle. Mosquitoes are poikilothermic, and in consequence, their body temperature and metabolic processes are governed by environmental temperature. During their lifetime, mosquitoes are exposed to both water and air temperature. So it is the most important abiotic factor in their life cycle. Temperature profoundly affects their development, survival, reproductive ability, and vector competence (Hardy et al. 1983).

During its development from egg to pupal stage, exposure of mosquito to temperature fluctuations occurs within water. Here, temperature is shown to significantly affect egg viability and hatching time. Temperature may differentially affect physiological processes even within species. For example, eggs of *Ae. albopictus*, a temperate mosquito, can withstand lower temperatures better than its tropical counterpart *Ae. aegypti*, which has led to the geographical expansion of the former to both tropical and temperature regions (Hanson and Craig 1994). Cold temperatures and photoperiods are also known to trigger overwintering in *An. albopictus* mosquitoes with the production of diapausing eggs (Thomas et al. 2012). The temperature affects hatching time more than the hatching rate in *Anophelines*. Eggs of *An. arabiensis* are shown to hatch faster at 27 °C than at 22 °C with most of the eggs hatched after 2 days, irrespective of the water temperature. The mean hatching rate of eggs of *An. sawadwongporni* and *An. dirus* incubated at 23 and 30 °C was significantly similar (Phasomkusolsil et al. 2011). However, temperatures around 42 °C can dramatically affect egg hatchability and result in low mean hatching counts with no hatchability at extremely prolonged exposure to high temperatures (greater than 2 days) (Impoinvil et al. 2007). After egg hatching, the survival, development of immature stages, and adult emergence are critically dependent on temperature. The optimal survival temperature for immature stages of *Ae. aegypti* is seen around 26 °C, while development is compromised below 16 °C and above 38 °C (Carrington et al. 2013). Generally, with increase in temperature, the development time of immature stages decreases up to a critical thermal threshold that is essential for survival. For instance, the time taken by *Ae. aegypti* to develop from egg to adult ranges from 7.2 days at 35 °C, while it increases to 39.7 days at 15 °C (Tun-Lin et al. 2000). Further, thermal stress experienced by mosquito during its immature stages greatly impacts its adult life traits, such as survival, body size, and fecundity (Ezeakacha and Yee 2019; Mohammed and Chadee 2011; Dodson et al. 2012).

The impact of temperature on mosquito physiology can differ significantly between species, even within the same complex in *Anophelines*. Larval stages of *An. arabiensis* can survive higher temperature than aquatic stages of *An. funestus* or *An. quadriannulatus* (Lyons et al. 2012). Higher temperatures significantly reduce development rate and survival of immature stages of *An. stephensi* (Paaijmans et al. 2013). In general, increasing temperatures reduce larval size, hatching rate, and pupation time of the larval stages of *Anopheles* mosquitoes. Exposure of adult *Anophelines* to higher temperatures reduces their body size, length of gonotrophic cycle, fecundity, and longevity. Temperature changes are also known to affect insecticide resistance in mosquitoes. *Anopheles arabiensis* mosquitoes reared at high temperatures exhibited resistance against pyrethroids (Oliver and Brooke 2017). High temperatures are shown to enhance the expression of nitric oxide synthase (NOS) and reduce insecticide toxicity (Murdock et al. 2014a, b). The effect of temperature on flight activity and olfaction is also reported for mosquitoes. Studies conducted using mosquito mill on *Ae. aegypti* have shown 21 °C to be optimal temperature for flight and also that maximum flight speed occurred between temperature range of 27–32 °C (Rowley and Graham 1968). The frequency of flight increases with temperature with frequency of 367 beat/s reported at 18 °C, which increases to 427 beat/s at 25 °C (Christophers 1960). For every 1 °C increase in temperature, an increase of 8–13 Hz in wing beat frequency is seen for *Ae. aegypti* (Villarreal et al. 2017). Temperature is also a modulator of olfactory behavior and biting rate in mosquitoes (Lahondère et al. 2023; Shapiro et al. 2017). Biting rates of *An. stephensi* are almost doubled with increase in temperature from 21 to 32 °C (Shapiro et al. 2017) (Table 9.1).

Hence, temperature shifts from cool to warm, aligning with global warming trends, have the potential to significantly broaden the geographical range of mosquitoes. This acceleration in their life cycle and enhanced reproductive potential may result in the expansion of vector-borne diseases, including malaria and dengue (Githeko et al. 2000; Hales et al. 2002; Patz et al. 1998).

9.2.2 Humidity

Humidity is defined by the amount of water vapor in the air. It can be articulated either in absolute terms as the mass of water vapor per unit volume of air or in relative terms as the ratio of actual vapor pressure to saturation vapor pressure at ambient pressure and temperature. Relative humidity (RH) is the most commonly used meteorological parameter to define the atmospheric conditions. RH varies with temperature even at constant amounts of water in the air and changes diurnally. It reaches maximum at the coolest time of night and minimum at the warmest time of the day. After temperature, humidity is the most important abiotic factor in the life cycle of mosquitoes, which determines the risk of desiccation in mosquitoes and affects several physiological processes. Humidity has important implications on mosquitoes' ability of hydro-regulation. Mosquitoes, like other insects, regulate water balance by changing the permeability of their cuticle, regulating production of excreta, changes in activities, such as flight. Mosquitoes due to their blood-feeding behavior are also faced by overhydration in addition to the risk of dehydration when environmental conditions are unfavorable. The effect of RH is most pronounced on the adult stage of mosquitoes, and changes in RH can either enhance or buffer the effects of temperature on adult physiology. By promoting longevity in mosquitoes at ambient temperatures, humidity not only governs mosquito fitness and abundance but also influences indirectly the transmission of mosquito borne diseases. Increases in RH are experimentally shown to affect survival, reproduction, and desiccation tolerance in mosquitoes in a positive manner. However, it may negatively impact biting rates, which are seen to increase in drier conditions. Exposure of *An. pharoensis*, *Ae. aegypti*, *Ae. taeniorhynchus*, *Ae. vexans*, and *Ae. trivittatus* to RH variations in the range of 34–95% led to increased egg production vis-à-vis drier conditions (Canyon et al. 1999; Gaaboub et al. 1971; Mcgaughey and Knight 1967). Increased RH also promotes longevity in female *Ae. aegypti* and *Ae. albopictus* in comparison to males (Schmidt et al. 2018). Mosquitoes are able to survive longest in combinations of low temperature and high humidity. RH is also seen to modulate flight and host-seeking activity across a wide range of mosquito species and genera (Wright and Knight 1966; Rudolfs 1925; Rowley and Graham 1968; Dow and Gerrish 1970). Cooler and humid macroclimates are also the preferred places of refuge by newly emerged mosquitoes *An. gambiae*, *An. stephensi*, *Ae. Aegypti*, and *Cx. pipiens* when they have no access to water or sugar in microclimate studies (Kessler and Guerin 2008). Thus, variations in RH and its interaction with temperature may significantly impact the thermal performance of mosquitoes.

9.2.3 Rainfall and Breeding Dynamics

Development of mosquito from egg to pupa occurs in water. Therefore, the aquatic stages of mosquito require water sources, which may be permanent or temporary for development into an adult. Mosquitoes are capable of breeding in nearly all types of aquatic habitats ranging from large permanent pools or streams, rain pools, small water collection in leaves and even polluted waters. The effect of rainfall is primarily in the creation and persistence of breeding habitats for mosquitoes where adults emerge from immature stages. Also, precipitation increases relative humidity in the environment, which favors adult survival and promotes longevity. Rainfall patterns influence mosquito populations by affecting the availability and persistence of breeding sites (Fig. 9.1). Mosquitoes lay their eggs in various aquatic habitats, including stagnant water bodies, such as puddles, ponds, and marshes. Environmental conditions, like water quality, temperature, and vegetation cover, play a crucial role in determining the suitability of a breeding site. Human activities that modify landscapes, leading to the creation of new water sources/wetlands, can inadvertently promote mosquito proliferation. Heavy rainfall can create temporary breeding habitats, promoting rapid population growth. However, excessive rainfall can be detrimental and flush away immature stages, disrupting the mosquito life cycle (Koenraadt and Harrington 2008; Seidahmed and Eltahir 2016; Dieng et al. 2012).

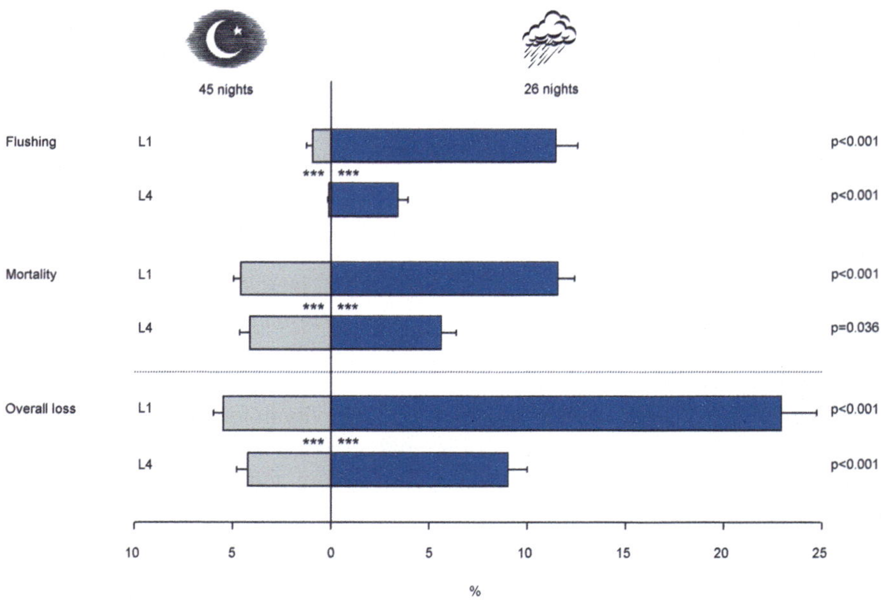

Fig. 9.1 Losses of *Anopheles gambiae* larvae during nights with and without rainfall. (Percentages of L1 and L4 larvae of *An. gambiae* that were flushed away or died and the overall loss during nights without rainfall (on the left) and nights with rainfall (on the right). The asterisks (***) indicate the level of significance between L1 and L4 larvae (p <0.001). The error bars indicate the standard error of the mean). (**Source:** https://doi.org/10.1371/journal.pone.0001146.g002)

With excessive rains, breeding sites overflow and get flushed away when water levels exceed the breeding sites' drainage threshold. In such cases, the developing larvae and pupae are washed away. Through laboratory studies simulating the effects of rainfall on breeding habitats, it has been seen that heavy downpour of water not only washes away larvae but also enhances their mortality (Koenraadt and Harrington 2008; Dieng et al. 2012). Intensity of rainfall, type of breeding habitat, and larval age are important factors that determine the extent of effect that flushing may have (Koenraadt and Harrington 2008; Dieng et al. 2012). The flushing effect may also vary with mosquito genera. For example, in a study simulating flushing effect of rainfall on container breeders *Ae. aegypti* and *Cx. pipiens*, it was seen that immature stages *aegypti* were able to withstand the flushing effect of rain better than larvae and pupae of *Cx. pipiens* even at lower water temperatures and also for longer durations (Koenraadt and Harrington 2008). In tropics, seasons occasionally have fluctuating rains, and studies have shown densities of different mosquito species to closely follow rainfall patterns, thus establishing a relationship between mosquito density and rainfall. However, it is difficult to draw a relationship between rainfall and mosquito density in case of mosquitoes, which breed in artificial aquatic habitats, such as man-made containers. Nevertheless, understanding the intricate relationships between rainfall and mosquito populations is crucial for effective vector control strategies.

9.3 Biotic Interactions

Biotic interactions involve the relationships between living organisms within an ecosystem. Biotic factors have the potential to influence spatial and temporal distribution of vector species and thereby influencing disease distribution. During biotic interactions, mutualism leads to enhance the mosquito distribution, whereas predation aids to restrict its population and thereby useful for disease control.

9.3.1 Predator-Prey Relationships

Mosquitoes play a crucial role in complex food webs, functioning as both predators and prey. The larvae of mosquitoes serve as a vital food source for various aquatic organisms, including insects, amphibians, and fish. In their adult stage, mosquitoes become essential prey for many animals, such as birds, bats, and spiders. These predator-prey interactions play a key role in regulating mosquito populations, creating cascading effects throughout the ecosystem. The dynamics of predator-prey relationships among mosquitoes are influenced by habitat structure and interactions among multiple predators. Research has shown that mosquito larvae are consumed by wild-caught diving beetles, swimming beetles, newts, damselfly larvae, and dragonfly larvae. Laboratory experiments have observed live mosquito larvae being fed upon by amphipods, ditch shrimp, fish, water boatmen, and beetle larvae. Furthermore, a negative correlation was identified between the distributions of ditch shrimp (Palaemonetes) and fish and the four mosquito species, suggesting an intricate interplay between predator and prey populations that influences mosquito distribution.

9.3.2 Parasitism and Mutualism

Mosquitoes engage in various parasitic and mutualistic relationships. Parasitic nematodes, for example, infect mosquito larvae and alter their development. Some mosquito species are known to harbor microorganisms, like *Wolbachia* bacterium, which can manipulate mosquito reproduction and impact disease transmission. Additionally, mosquitoes participate in pollination by feeding on nectar, forming mutualistic interactions with flowering plants. Parasite-infected mosquito has an advantage over the noninfected ones as their sense of smell to seek host and capability to reproduce is enhanced and thereby aiding in disease transmission (Carr et al. 2021). Interactions among pathogens and bacteria that colonize mosquito gut or symbionts, such as *Wolbachia*, may potentially alter disease vector competence. Experimental studies in laboratory have revealed that *Wolbachia* may or may not alter malaria parasite development in mosquitoes depending upon the temperature (Murdock et al. 2014a, b). Further, mosquito species have also been reported to be affected by the presence of green filamentous algae and vegetation cover, and their presence is positively associated with larval density (Mouatassem et al. 2020). Xeno-monitoring studies have identified various pathogenic alpha-proteobacteria, including *Anaplasma* spp., *Ehrlichia* spp., *Candidatus*, *Neoehrlichia*, *Bartonella* spp., and *Rickettsia* spp., in adult mosquitoes (Guo et al. 2016; Krajacich et al. 2018). These findings underscore the complexity of the relationship mosquitoes maintain with various organisms and highlight the potential implications for disease transmission dynamics.

9.3.3 Disease Transmission

Perhaps the most significant biotic interaction involving mosquitoes is their role as disease vectors. Many pathogens, including malaria parasites and arboviruses, rely on mosquitoes to complete their life cycles. Mosquitoes by blood feeding can acquire pathogens or parasites from one vertebrate host and transmit them to another, provided that the mosquito's ecology and physiology are conducive to the transmission process. One of the most well-known diseases transmitted by mosquitoes is malaria, caused by *Plasmodium* parasites and resulting in severe illness and fatalities, particularly in tropical regions. Additionally, mosquitoes are vectors for arboviruses, such as the dengue virus, Zika virus, and chikungunya virus. These diseases can cause a range of symptoms from mild fever to severe complications, and they have become global concerns due to the expanding geographical range of mosquito species capable of transmitting them.

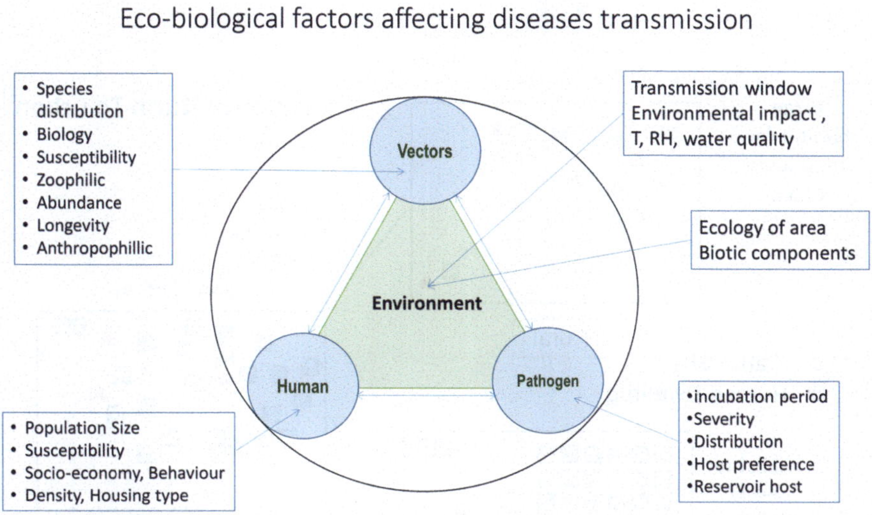

Fig. 9.2 Epidemiological triad of vector-borne diseases

The transmission cycle of these diseases typically involves mosquitoes acting as intermediate hosts. When a mosquito feeds on an infected host, it acquires the pathogen and can subsequently transmit it to other individuals during subsequent blood meals. Mosquito-borne diseases have a significant impact on public health, necessitating efforts in vector control, disease surveillance, and research for effective prevention and treatment strategies. The interaction between pathogens, mosquitoes, and vertebrate hosts creates a complex transmission dynamic that influences disease prevalence and distribution (Fig. 9.2). Understanding these interactions is essential for designing effective disease control strategies.

9.3.4 Altered Ecosystems and Human Influence

Human activities can substantially modify mosquito ecosystems (Fig. 9.3). Urbanization, deforestation, and the construction of artificial water bodies can create new breeding sites and alter mosquito populations. These changes can lead to increased disease transmission risks in certain areas. Moreover, factors, such as global increase in temperature, population movement, and increasing trade between countries, have expanded the horizons of disease vectors. Additionally, the use of pesticides to control mosquitoes can have unintended consequences, affecting non-target species and disrupting the balance of ecosystems.

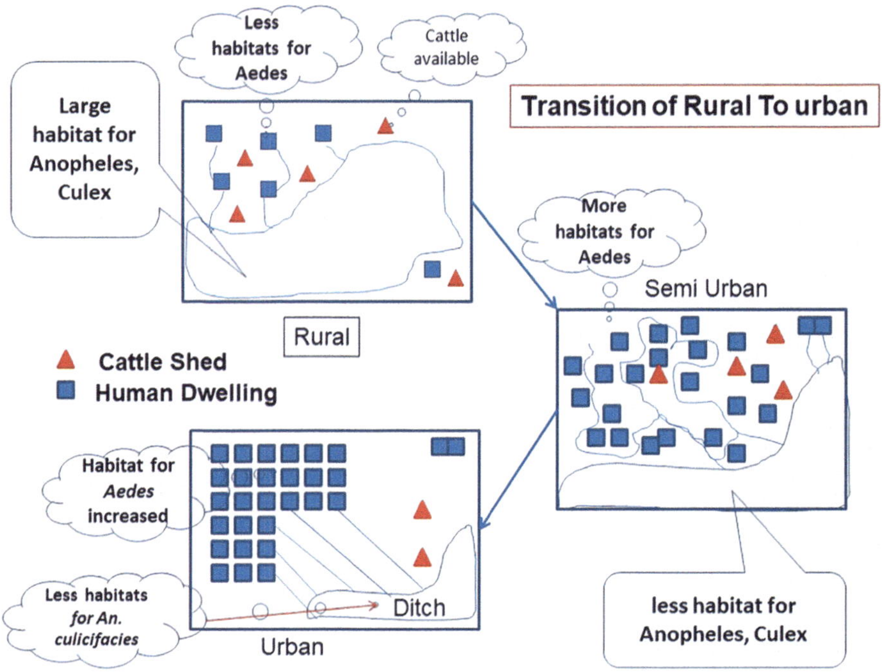

Fig. 9.3 Effect of urbanization on mosquito's habitation

9.3.5 Integrated Pest Management

Recognizing the intricate ecological interactions involving mosquitoes, integrated pest management (IPM) strategies have gained importance. IPM involves a multidisciplinary approach that combines ecological, biological, chemical, and cultural methods to control mosquito populations while minimizing ecological disruption. Strategies may include introducing mosquito predators, manipulating breeding sites, and utilizing targeted pesticide applications.

9.4 Conclusion

It is essential to unravel the adaptive potential of mosquitoes in view of change in climatic conditions. With the rise in temperature due to global warming, these short-lived ectotherm species have evolved themselves and are dispersed to distant places. The ecological interactions of mosquitoes are multifaceted, involving intricate relationships with both abiotic and biotic elements in their environment. To effectively manage mosquito populations and mitigate disease transmission, it is crucial to consider these interactions within the broader context of ecosystems. Future research should continue to explore these connections, leading to innovative strategies that promote ecological balance while safeguarding human health. Understanding the interplay between mosquitoes and their environment can ultimately pave the way for more sustainable and comprehensive mosquito management practices. As we navigate the intricate landscape of abiotic and biotic interactions of mosquitoes, it becomes evident that their story is one of complexity and nuance. Embracing this complexity is not only a scientific endeavor but a call to action. It beckons us to move beyond the conventional narrative of mosquitoes as mere pests, urging us to comprehend their ecological significance and the pivotal role they play in global health.

This chapter sets the stage for an in-depth exploration into the world of mosquitoes. It is an invitation to researchers, public health professionals, and environmentalists to collaborate in understanding, managing, and ultimately mitigating the impact of mosquitoes on both ecosystems and human societies. As we embark on this journey, we treat the fine line between fascination and urgency, driven by the conviction that by unraveling the intricacies of mosquito interactions, to pave the way for a healthier, harmonious coexistence with these enigmatic insects. In particular, we would highlight that both abiotic and biotic interactions may combine to promote the persistence of pathogens and mosquitoes.

Points to Remember

1. Abiotic and biotic components of the environment play indispensable roles in the life cycle of mosquitoes.
2. Beginning from creation of breeding habitats to regulating several aspects of mosquito physiology temperature, humidity, and rainfall constitute key abiotic factors that influence all the life stages of mosquitoes.
3. Predators feed on mosquito larvae in the aquatic microenvironment and naturally regulate mosquito populations eventually affecting abundance and population dynamics.
4. Mosquitoes harbor both pathogens and commensals and share intricate relationships with them. While some regulate reproduction in mosquitoes, others transform them into important vectors of various infectious diseases.
5. Mosquitoes serve as intermediate hosts for *Plasmodium* spp. and arboviruses and thus play key role in transmission of vector borne diseases, such as malaria, chikungunya, dengue, etc.
6. Human activities, such as deforestation and urbanization, have substantially modified mosquito ecosystems, which have dramatically changed the geographic distribution of mosquitoes and mosquito-borne diseases.
7. Mosquito control tools in the form of integrated pest management are also important manipulators of mosquito populations and in turn disease transmission dynamics.

Acknowledgments We would like to thank the Director, ICMR-National Institute of Malaria Research, New Delhi, for their support provided in writing the chapter.

References

Canyon DV, Hii JLK, Müller R (1999) Adaptation of *Aedes aegypti* (Diptera: Culicidae) oviposition behavior in response to humidity and diet. J Insect Physiol 45(10):959–964

Carr AL, Rinker DC, Dong Y, Dimopoulos G, Zwiebel LJ (2021) Transcriptome profiles of *Anopheles gambiae* harboring natural low-level Plasmodium infection reveal adaptive advantages for the mosquito. Sci Rep 11(1):22578

Carrington LB, Armijos MV, Lambrechts L, Barker CM, Scott TW (2013) Effects of fluctuating daily temperatures at critical thermal extremes on *Aedes aegypti* life-history traits. PLoS One 8(3):e58824

Christophers SR (1960) Aedes aegypti: the yellow fever mosquito. Cambridge University Press, Cambridge. 750 pp

Dieng H, Rahman GS, Abu Hassan A, Che Salmah MR, Satho T, Miake F, Boots M, Sazaly A (2012) The effects of simulated rainfall on immature population dynamics of *Aedes albopictus* and female oviposition. Int J Biometeorol 56:113–120

Dodson BL, Kramer LD, Rasgon JL (2012) Effects of larval rearing temperature on immature development and West Nile virus vector competence of *Culex tarsalis*. Parasit Vectors 5:1–6

Dow RP, Gerrish GM (1970) Day-to-day change in relative humidity and the activity of *Culex nigripalpus* (Diptera: Culicidae). Ann Entomol Soc Am 63(4):995–999

Ezeakacha NF, Yee DA (2019) The role of temperature in affecting carry-over effects and larval competition in the globally invasive mosquito *Aedes albopictus*. Parasit Vectors 12:1–11

Gaaboub IA, El-Sawaf SK, El-Latif MA (1971) Effect of different relative humidities and temperatures on egg-production and longevity of adults of *Anopheles (Myzomyia) pharoensis* Theob. 1. Z Angew Entomol 67(1–4):88–94

Githeko AK, Lindsay SW, Confalonieri UE, Patz JA (2000) Climate change and vector-borne diseases: a regional analysis. Bull World Health Organ 78(9):1136–1147

Guo WP, Tian JH, Lin XD, Ni XB, Chen XP, Liao Y, Yang SY, Dumler JS, Holmes EC, Zhang YZ (2016) Extensive genetic diversity of Rickettsiales bacteria in multiple mosquito species. Sci Rep 6(1):38770

Hales S, De Wet N, Maindonald J, Woodward A (2002) Potential effect of population and climate changes on global distribution of dengue fever: an empirical model. Lancet 360(9336):830–834

Hanson SM, Craig GB Jr (1994) Cold acclimation, diapause, and geographic origin affect cold hardiness in eggs of *Aedes albopictus* (Diptera: Culicidae). J Med Entomol 31(2):192–201

Hardy JL, Houk EJ, Kramer LD, Reeves WC (1983) Intrinsic factors affecting vector competence of mosquitoes for arboviruses. Annu Rev Entomol 28(1):229–262

Impoinvil DE, Cardenas GA, Gihture JI, Mbogo CM, Beier JC (2007) Constant temperature and time period effects on *Anopheles gambiae* egg hatching. J Am Mosq Control Assoc 23(2):124

Kessler S, Guerin PM (2008) Responses of *Anopheles gambiae*, *Anopheles stephensi*, *Aedes aegypti*, and *Culex pipiens* mosquitoes (Diptera: Culicidae) to cool and humid refugium conditions. J Vector Ecol 33(1):145–149

Koenraadt CJM, Harrington LC (2008) Flushing effect of rain on container-inhabiting mosquitoes *Aedes aegypti* and *Culex pipiens* (Diptera: Culicidae). J Med Entomol 45(1):28–35

Krajacich BJ, Huestis DL, Dao A, Yaro AS, Diallo M, Krishna A, Xu J, Lehmann T (2018) Investigation of the seasonal microbiome of Anopheles coluzzii mosquitoes in Mali. PLoS One 13(3):e0194899

Lahondère C, Vinauger C, Liaw JE, Tobin KK, Joiner JM, Riffell JA (2023) Effect of temperature on mosquito olfaction. Integr Comp Biol 63(2):356–367

Lyons CL, Coetzee M, Terblanche JS, Chown SL (2012) Thermal limits of wild and laboratory strains of two African malaria vector species, *Anopheles arabiensis* and *Anopheles funestus*. Malar J 11:1–14

Mcgaughey WH, Knight KL (1967) Preoviposition activity of the black salt-marsh mosquito, *Aedes taeniorhynchus* (Diptera: Culicidae). Ann Entomol Soc Am 60(1):107–115

Mohammed A, Chadee DD (2011) Effects of different temperature regimens on the development of *Aedes aegypti* (L.) (Diptera: Culicidae) mosquitoes. Acta Trop 119(1):38–43

Mouatassem TF, El Ouali Lalami A, Faraj C, Rais N, Guemmouh R (2020) Study of abiotic and biotic parameters affecting the abundance of mosquito larvae (Diptera: Culicidae) in the region of Fez (Morocco). Int J Zool 2020:1–7

Murdock CC, Blanford S, Hughes GL, Rasgon JL, Thomas MB (2014a) Temperature alters Plasmodium blocking by *Wolbachia*. Sci Rep 4(1):3932

Murdock CC, Blanford S, Luckhart S, Thomas MB (2014b) Ambient temperature and dietary supplementation interact to shape mosquito vector competence for malaria. J Insect Physiol 67:37–44

Oliver SV, Brooke BD (2017) The effect of elevated temperatures on the life history and insecticide resistance phenotype of the major malaria vector *Anopheles arabiensis* (Diptera: Culicidae). Malar J 16:1–13

Paaijmans KP, Heinig RL, Seliga RA, Blanford JI, Blanford S, Murdock CC, Thomas MB (2013) Temperature variation makes ectotherms more sensitive to climate change. Glob Chang Biol 19(8):2373–2380

Patz JA, Martens WJ, Focks DA, Jetten TH (1998) Dengue fever epidemic potential as projected by general circulation models of global climate change. Environ Health Perspect 106(3):147–153

Phasomkusolsil S, Lerdthusnee K, Khuntirat B, Kongtak W, Pantuwatana K, Murphy JR (2011) Effect of temperature on laboratory reared *Anopheles dirus* Peyton and Harrison and *Anopheles sawadwongporni* Rattanarithikul and green. Southeast Asian J Trop Med Public Health 42(1):63

Rowley WA, Graham CL (1968) The effect of temperature and relative humidity on the flight performance of female *Aedes aegypti*. J Insect Physiol 14(9):1251–1257

Rudolfs W (1925) Relation between temperature, humidity and activity of house mosquitoes. J N Y Entomol Soc 33(3):163–169

Schmidt CA, Comeau G, Monaghan AJ, Williamson DJ, Ernst KC (2018) Effects of desiccation stress on adult female longevity in *Aedes aegypti* and *Ae. Albopictus* (Diptera: Culicidae): results of a systematic review and pooled survival analysis. Parasit Vectors 11:1–21

Seidahmed OM, Eltahir EA (2016) A sequence of flushing and drying of breeding habitats of *Aedes aegypti* (L.) prior to the low dengue season in Singapore. PLoS Negl Trop Dis 10(7):e0004842

Shapiro LL, Whitehead SA, Thomas MB (2017) Quantifying the effects of temperature on mosquito and parasite traits that determine the transmission potential of human malaria. PLoS Biol 15(10):e2003489

Thomas SM, Obermayr U, Fischer D, Kreyling J, Beierkuhnlein C (2012) Low-temperature threshold for egg survival of a post-diapause and non-diapause European aedine strain, *Aedes albopictus* (Diptera: Culicidae). Parasit Vectors 5:1–7

Tun-Lin W, Burkot TR, Kay BH (2000) Effects of temperature and larval diet on development rates and survival of the dengue vector *Aedes aegypti* in north Queensland, Australia. Med Vet Entomol 14(1):31–37

Villarreal SM, Winokur O, Harrington L (2017) The impact of temperature and body size on fundamental flight tone variation in the mosquito vector *Aedes aegypti* (Diptera: Culicidae): implications for acoustic lures. J Med Entomol 54(5):1116–1121

Wright RE, Knight KL (1966) Effect of environmental factors on biting activity of *Aedes vexans* (Meigen) and *Aedes trivittatus* (Coquillett). Mosq News 26(4):565–578

Mosquitoes as Disease Vectors and Underlying Factors for Transmission

10

Ramesh C. Dhiman and Poonam Singh

Abstract

Mosquito-borne diseases pose a considerable public health problem globally as well as in India. Malaria, dengue, chikungunya, Zika, Japanese encephalitis and lymphatic filariasis are major mosquito-borne diseases in India. There are several species of mosquito vectors, but only a few are considered as vectors of diseases. Various reasons, like longevity, impact of climatic conditions, role of peritrophic membrane, presence/absence of microbiota in vectors, vectorial capacity, etc., have been put forth. As there is conflicting information on the climatic thresholds of transmission of malaria, dengue, chikungunya and Zika, the thresholds need to be backed up with reported data of a particular disease for public health implications of the projected data. Still a few species, like *An. subpictus*, are not considered as vector of malaria in India. Under what conditions the secondary vectors of malaria, like *An. philippinensis*, *An. varuna* and *An. annularis*, can become primary vectors, is not known. The penetration of peritrophic membrane by *Plasmodium* to the gut epithelial layer of *Anopheles* mosquito is facilitated by the chitinases secreted by *Plasmodium*, opening scope for development of chitinases inhibitor transgenic *Anopheles* vector, which can serve as transmission blocking vaccine. Studies have provided evidence that there is higher development of oocysts and sporozoites of *Plasmodium* in pyrethroid-resistant *An. gambiae* and *An. funestus*, while a few studies found no difference warranting further research. The presence of a particular species of microbiota inhibit the development of malaria parasite in mosquito, thus providing a way for transmission blocking. The quantification of repeated bites by mosquito vector for resulting into elephantiasis is not well elucidated. Filariasis is transmitted by many genera of mosquitoes, and the transmission of Zika virus is also not restricted to *Aedes* only. In the wake of elimination target of many diseases, like malaria and

R. C. Dhiman (✉) · P. Singh
Formerly at ICMR-National Institute of Malaria Research, New Delhi, India

© The Author(s), under exclusive license to Springer Nature Singapore Pte Ltd. 2024

Omkar (ed.), *Mosquitoes*, https://doi.org/10.1007/978-981-97-4163-2_10

lymphatic filariasis, transmitted by mosquitoes, the role of a few mosquito vectors still needs to be elucidated. The present communication highlights a few thought-provoking issues warranting research.

Keywords

Malaria · Dengue · Transmission · Mosquito borne · Climatic factors · Vectors

Learning Objectives

1. The criteria of incriminating a vector-borne disease vector.
2. The components of vectorial competence affecting transmission in an area.
3. Thresholds of temperature and relative humidity for major vector-borne diseases and the impact of climate change on projected scenario of transmission windows of malaria.
4. The role of chitinases released by *Plasmodium* in penetration of peritrophic membrane so as to reach the gut epithelial cells of mosquito.
5. Role of microbiota in transmission of *Plasmodium.*
6. Transmission blocking vaccines in malaria.
7. Resistant strains of *Anopheles* vectors facilitate better development of oocysts, sporozoites in *Anopheles* vector.
8. Transmission routes of Zika virus.
9. Why man is a dead-end host for Japanese encephalitis.

10.1 Introduction

Vector-borne diseases (VBD) represent a significant threat to public health worldwide, caused by various pathogens, such as viruses, bacteria and parasites. These pathogens are transmitted to humans and animals primarily through the bites of infected vectors, predominantly arthropods, like mosquitoes, ticks, fleas and sandflies. The prevalence of vector-borne diseases is particularly pronounced in tropical and subtropical regions, where environmental conditions are conducive to the proliferation of these vectors.

Within the vast array of mosquito species, numbering approximately 35,000 identified species worldwide not all serve as vectors for disease transmission. Various factors determine whether a mosquito species qualifies as a vector for a particular disease. These factors can include the mosquito's biology, behaviour and its ability to harbour and transmit specific pathogens.

In India, several mosquito-borne diseases pose significant health challenges. Among them most notable are malaria, dengue fever, chikungunya, filariasis and Japanese encephalitis. As per the World Health Organization (WHO), India reported 65.7% of malaria cases in southeast Asia region in the year 2022. Understanding the intricate mechanisms by which these diseases are transmitted is crucial for devising effective control measures and advancing research efforts.

In the following discussion, we aim to delve into the mechanisms underlying the transmission of some common mosquito-borne diseases. By shedding light on these mechanisms, we hope to provide valuable insights that can inform further research endeavours and facilitate the development of more targeted and efficient vector control strategies.

10.2 Vectors of Major Mosquito-Borne Diseases

10.2.1 Malaria

Not all types of mosquitoes are capable of transmitting malarial parasites to humans. Among the numerous species of mosquitoes identified, only a portion of those belonging to the genus *Anopheles* act as vectors. Of 58 species of anopheline mosquitoes, only 6 species, namely, *An. culicifacies*, *An. stephensi*, *An. fluviatilis*, *An. baimii* and *An. epiroticus* (formerly *An. sundaicus*), are major vectors of malaria and 3 species, namely, *An. philippinensis*, *An. varuna* and *An. Annularis*, are vectors of secondary importance in India.

Kumar et al. (2022) have highlighted the need for studies on the bionomics of primary and secondary vectors of malaria in India, as the data on a number of vector features are either missing or not completely defined. District level mapping of malaria vectors carried out by Singh et al. (2017) shows the missing information in terms of presence or absence of vectors. Additionally, there are approx. 183 districts in the country where not even a single survey has ever been conducted.

Plasmodium falciparum and *P. vivax* are the two major human malaria species in India, with *P. malariae* being reported from the eastern part of the country, and *P. ovale* is also present but rare. Both primary malaria parasites occur in varying proportions and can also result in mixed infections. Like the vectors, malaria parasites development is also temperature-dependent process. Co-infection of *P. falciparum* and *P. vivax* leads to the complicated malaria diagnosis, which further complicates the treatment process.

First, let us define the criteria based on which the vectorial status is given to any vector. Based on empirical knowledge, the arthropod/insect species should fulfil the five conditions are termed as vector:

1. It should be distributed in the geographic area.
2. It should feed on human beings.
3. It should be found naturally infected with the pathogen of concern.
4. It should acquire infection when artificially fed in laboratory.
5. A laboratory-infected mosquito should be able to transmit the infection to healthy person.

Based on the above-mentioned conditions, we can evaluate the vectorial status of some anopheline species, which have been found infected with malaria parasite in nature but are not considered as vector.

10.2.1.1 Why all the Anopheline Mosquitoes Are Not Vectors of Malaria?

Some *Anopheles* species either do not feed on humans or are not susceptible to human malaria parasites or have lifespans too brief for the parasites to develop fully. A few vector species, which are abundant, have longer lifespans, frequently bite humans, and tend to live in close proximity to human populations which are a major concern. Their role in spreading malaria largely hinges on the availability of suitable environments for both larval growth and adult survival, as well as their ability to feed on humans. Additionally, the transmission of malaria is significantly influenced by human behaviours that increase contact between hosts and vectors.

An. subpictus is widely distributed in India (Singh et al. 2014) and has been reported naturally infected with *P. falciparum* and *P. vivax* (Kumar et al. 2016) but is still not regarded as vector of malaria in India. Analysis of geographic areas where *An. subpictus* has been found naturally infected reveals that the coastal regions (more reports from east coast and primarily Goa from the west coast) are humid, which support longevity of *An. subpictus*. Further, this species has also been reported as a secondary vector of malaria in Sri Lanka (Gunathilaka et al. 2019), which is mainly humid.

The circumstantial evidence indicates that in non-coastal areas, the longevity of *An. subpictus* is not sufficient to transmit malaria. May be *An. subpictus* fulfils the first three criteria of becoming a vector, except for the lack of experimental evidence supporting its ability to successfully transmit the infection to a human host. Furthermore, laboratory work may demonstrate transmission under controlled temperature and relative humidity (RH), which is not met in natural conditions. Therefore, experimental work to determine the transmission potential of *An. subpictus* under varying RH conditions is necessary to establish the vectorial status as a malaria vector in India.

In addition to *An. subpictus*, the authors found *Anopheles aconitus*, *Anopheles annularis*, *Anopheles barbirostris*, *Anopheles nigerrimus*, *Anopheles pallidus*, *Anopheles subpictus*, *Anopheles tessellatus*, *Anopheles vagus* and *Anopheles varuna* naturally infected with malaria parasites, but they are still not considered as vectors of malaria in Sri Lanka (Gunathilaka et al. 2015).

Different geographic areas have different levels of endemicity of malaria in spite of the presence of vector species. There are several biological parameters, like density, feeding rate on human beings, longevity, etc., which affect the capacity of vector species in a given area. The components of vector competence are given below which determine high, moderate or low intensity of malaria transmission in any geographic area.

10.2.1.2 Vector Competence

The vector competence is defined as a component of vectorial capacity about the vector's ability to transmit the infection to human and is governed by genetic factors also. The competence of a mosquito to become a vector depends upon several factors, like suitable breeding habitats, density, feeding preference on animals or human beings and suitable climatic conditions (Temperature, rainfall and RH) (Lewis et al. 2023).

The vectorial capacity of mosquito vector can be defined by the below given formula:

$$\mathrm{VC} = ma2pn - \ln(p)$$

where the components of vectorial capacity (VC) are (a) extrinsic incubation period (EIP) (developmental time of malaria parasite in mosquito) in days (n); the ratio of mosquito to human beings (m), survival of mosquitoes through 1 day (p) and human biting rates (a).

The EIP depends on temperature and RH. At 16 °C temperature, it may take 55 days for completion of 1 cycle of malaria parasite, while at 25 °C, the development of parasite is completed in 10 days. Similarly, the ratio of mosquito to human beings depends on the density of vector species. If the breeding and resting habitats are suitable and there is no scarcity of source of blood meal, high density of vectors is observed. The suitable resting places are also important for high density of vectors. For example, high density of *An. culicifacies* is found in riverine areas with rocky riverbed pools, and the houses are made of mud material with thatched roofs and have plenty of cattle population. The high density of *An. stephensi* mosquito is encountered at construction sites, which provide ideal breeding grounds in curing tanks, underground cement tanks and overhead tanks. At construction sites, there is less availability of animals; therefore, this species prefers to feed on human beings. Thus, the components of VC are dependent on favourable breeding sites, availability of blood meal, suitable temperature and RH for development and survival of mosquito vector. If the biological and climatic requirements are not fulfilled for longer duration sufficient enough to provide required density of vectors and survival for completion of EIP, the rate of transmission may remain very low. The components of vectorial capacity reflect the transmission suitability for a shorter time, longer time or unsuitability reflecting low, high or non-endemicity of malaria or any other VBD.

The survival of mosquitoes through 1 day (p) also depends on the suitable temperature and RH to mosquitoes. These parameters are critical for transmission to another human being. The purpose of indoor residual spray (IRS) for vector control is to reduce the longevity of mosquitoes, so that they do not survive to complete the EIP, thus curtailing the transmission.

Human biting rate (a) or anthropophagic index depends on the availability of human. The lower the human biting rate, the more is the requirement of density of vector and vice versa. *An. fluviatilis* and *An. stephensi* have predilection toward human beings; therefore, low density of these vectors is sufficient to cause transmission. On the other hand, the *An. culicifacies* is zoophagic to the extent of 98%; therefore, very high density of *An. culicifacies* is required for transmission.

Over the years, knowledge has been generated on various components of vector competence, like role of mosquito immunity (Kumar et al. 2018), role of microbiota, physical barriers (Franz et al. 2015) and role of the genetics (Beerntsen et al.

2000). Additionally, Kramer and Ciota (2015) have studied the overall contribution of vector competence to the vectorial capacity of vectors.

The climatic parameters affect almost all aspects of biology of mosquito vectors affecting vectorial capacity and transmission in an area. The details of climatic thresholds and their role in projected scenarios of malaria in India by the year 2030s are detailed here.

10.2.1.3 Climatic Thresholds Affecting Transmission

The role of climatic thresholds is very crucial for ensuing indigenous transmission of any vector-borne disease. Thresholds of temperature and RH in respect of major VBDs in India are given in Table 10.1.

In recent times, wealth of data has been generated on the thresholds of transmission of malaria. As per Craig et al. (1999), 18–32 °C is the suitable range of transmission, while Mordecai et al. (2013) pointed out that the upper threshold for malaria transmission is 28 °C (but lower than 32 °C). Further, Shapiro et al. (2017) observed that the suitable range of transmission of malaria is 17–34 °C and the optimum temperature is 29 °C. The knowledge of climatic thresholds of malaria transmission has been useful in assessing the impacts of climate change. Based on the threshold 18–32 °C temperature and >55% RH, the projected scenario of malaria at district level in India by the year 2030s has been provided by Sarkar et al. (2019). The authors found that the Himalayan areas are projected to have new foci of transmission of malaria while extended windows of transmission in north-eastern parts of India. A few regions were projected to experience reduction in transmission months by the 2030s. Reduction in transmission windows in a few areas is likely to lead toward unstable malaria and may become prone to malaria outbreaks (Fig. 10.1).

10.2.1.4 Role of Peritrophic Membrane in Transmission

The peritrophic membrane serves as a physical barrier, which separates the blood bolus from the midgut epithelium. It can prevent direct contact between ingested pathogens and the insect's gut epithelial cells, providing a first line of defence against pathogens. It may impact the ability of the arthropod to acquire, maintain and transmit pathogens during blood feeding. The structure of peritrophic

Table 10.1 Climatic thresholds of major vector-borne diseases

S. no.	VBDs	Temperature (°C)	Relative humidity	References
1.	Malaria	18–32	55–80%	Craig et al. (1999)
2.	Dengue	13–35	50–90%	Watts et al. (1987)
3.	Chikungunya	18–34	50–90%	Mordecai et al. (2017)
4.	Zika	22.7–34.7		Tesla et al. (2018)
5.	Japanese encephalitis	20–32	>50%	Based on occurrence of cases in Gorakhpur (UP) (India)
6.	Kala Azar	7–37	–	Napier (1931)
7.	Kyasanur Forest Disease	26–31	–	Based on occurrence of cases in Wayanad (Kerala) and Goa (India)

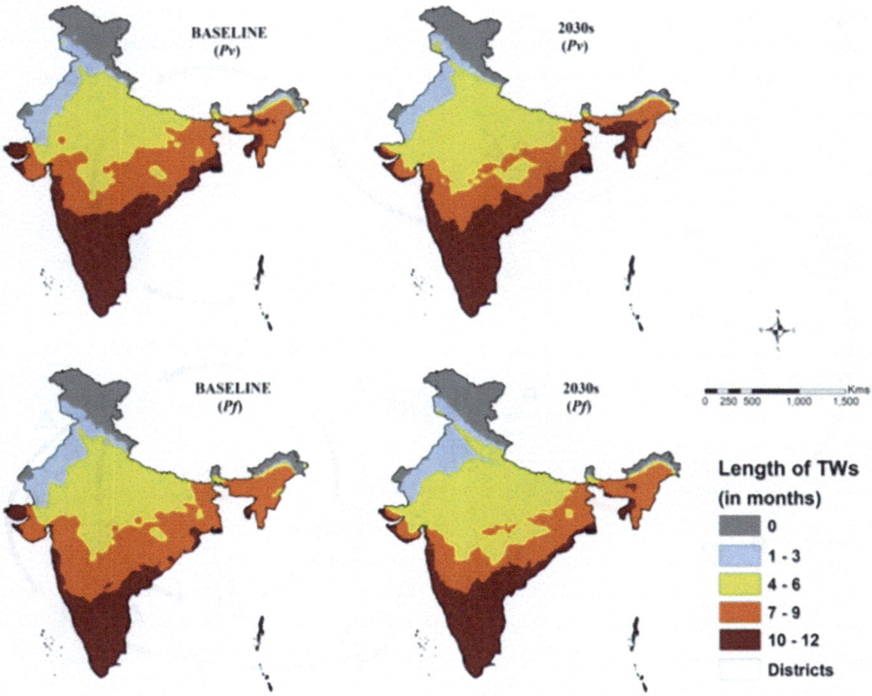

Fig. 10.1 Composite climate suitability map for malaria transmission for the baseline and projected 2030s. For *P. vivax*, new foci of transmission are visible in parts of Jammu and Kashmir, Uttarakhand and Himachal Pradesh. Increase in the number of months of transmission windows in north-eastern states and reduction in parts of Orissa and Gujarat states are also seen. In the case of *P. falciparum*, few foci in Jammu and Kashmir and Uttarakhand; an increase in transmission months in north-eastern states and reduction in parts of Orissa and Gujarat states are visible (Source: Sarkar et al. 2019)

membrane (PM) is composed of chitin and proteins that lines the midgut or hindgut of these organisms. PM plays a crucial role in the transmission of pathogens in vectors, like mosquitoes. Erlandsona et al. (2019) have reviewed the role of peritrophic matrix in insect pathogen interactions.

The role of PM in transmission of pathogens is more important in disease, like malaria, wherein the malaria parasite has to penetrate the epithelial wall of the gut to develop into ookinete (Fig. 10.2). *Plasmodium* chitinases belonging to the glycoside hydrolase-18 (GH18) family are secreted inside the mosquito midgut, during the ookinete stage of the parasite. Chitinases mediate the penetration of parasite through the PM, facilitating access to the gut epithelial layer. Langer and Vinetz (2001) described *Plasmodium* chitinases with special emphasis on chitinases from *P. falciparum* and *P. vivax*, the representative examples of the short and long forms of this protein.

Thus, chitinase enzyme is a candidate target for blocking malaria transmission. Future research lies in development of chitinase-inhibitor-producing transgenic

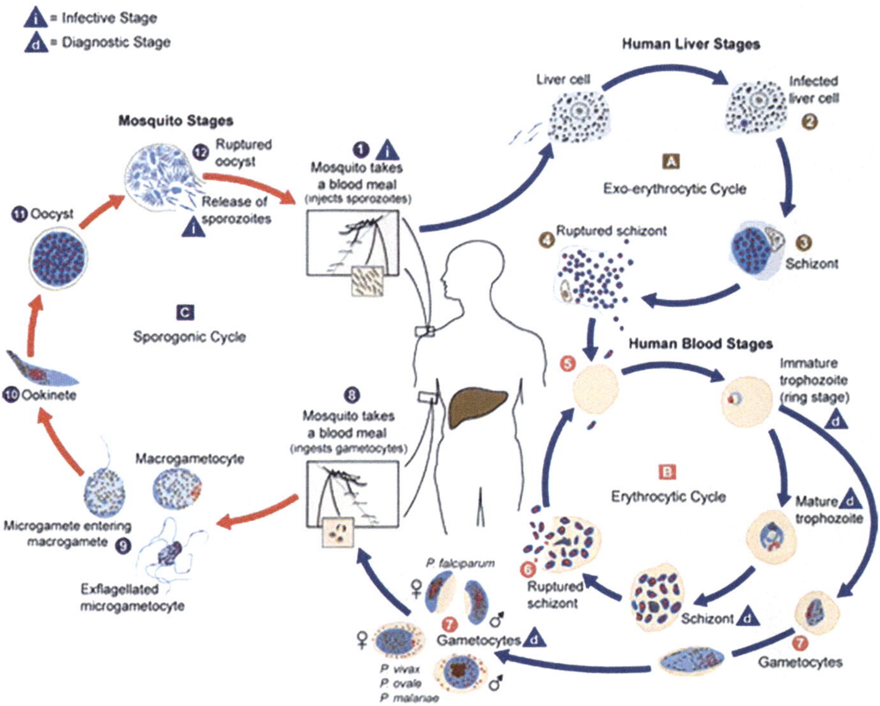

Fig. 10.2 Transmission cycle of malaria parasite. (Source: CDC)

mosquitoes. Understanding the role of these amino acids in *Plasmodium* chitinases will help us elucidate the mechanism of catalysis and the mode of inhibition, which will be the key for identification of potent inhibitors or antibodies demonstrating transmission blocking activity (Viswanath et al. 2021).

10.2.1.5 Role of Microbiota

In view of increasing risk of VBDs in areas where they were not previously distributed, this has spurred both research and public health sectors to hasten the development of innovative vector control/transmission methods to mitigate the future risk of mosquito-borne diseases. Among others, one such strategy is symbiotic control, which involves identifying beneficial microbes, like bacteria, yeast, etc., capable of spreading within vector populations. Microbiota are the living microorganisms found in a defined environment, such as oral and gut microbiota. Many microorganisms, including bacteria, yeasts and viruses, coexist in various parts of the human body (gut, skin, lung, oral cavity and vagina) (Hou et al. 2022). The gut bacteria serve several functions, such as fermentation of food, protection against pathogens, stimulating immune response and vitamin production. Generally, the gut microbiota is composed of six phyla including Actinobacteria, Bacteroidetes, Firmicutes, Lactobacillus, Streptococci and Enterobacteria, of which Firmicutes and Bacteroidetes are the major types. The most studied fungi (gut mycobiont) are

Candida, *Saccharomyces*, *Malassezia* and *Cladosporium*. In addition to bacteria and fungi, the human gut microbiota also contains viruses, phages and archaea, mainly *M. smithii*. The human microbiota not only protects the host from external pathogens by producing antimicrobial substances but also serve as a significant component in the development of intestinal mucosa and immune system.

The role of microbiota has been recently studied in the context of transmission of malaria through mosquito vectors as well as in the clinical outcome of malaria in human host. Recent evidence shows that microbiota of both mosquito and human host play important roles in transmission, progression and clearance of *Plasmodium* infection. *Plasmodium* can also cause significant alterations in the human host and mosquito gut microbiota, affecting the clinical outcome of malaria. When mosquitoes ingest both male and female gametocytes of *Plasmodium*, exflagellation occurs from male gametocyte, which fuses with female gametocyte to form zygote, which is called as ookinite (Fig. 10.2), which traverses the midgut epithelium. Subsequently, they mature into oocysts, releasing sporozoites. These sporozoites migrate into the circulatory system and eventually reach the salivary glands (Frischknecht and Matuschewski 2017). The presence of some microbiota in the gut of *Anopheles* vector is regarded as transmission blocking, which is still a matter of research for determining specific microbiota.

The microbiota interferes with *Plasmodium* colonization by stimulation of the mosquito immune response, and production of metabolites directly impairing parasite survival in addition, the number of ookinetes is reduced in microbiota-dependent *Anopheles* suggesting microbiota-induced immunity. Enterobacter Esp_Z is a Gram-negative bacterium isolated from the gut of *An. arabiensis* mosquitoes, which was found to reduce *P. falciparum* ookinete, oocyst and sporozoite loads in *An. gambiae* via production of reactive oxygen species and/or inhibition of the oxidative defence system of the parasite.

Wickerhamomyces anomalus, a yeast isolated from the midgut of *An. stephensi* inhibited *P. berghei* ookinetes in vitro by producing a killing toxin with β-1,3-glucanase activity.

The microbiota is also involved in the synthesis of the peritrophic matrix, a layer composed of chitin and proteins, which surrounds the midgut epithelium after blood feeding and protects the mosquito from the dissemination of midgut bacteria into the body cavity. *Plasmodium* ookinetes secrete a chitinase required to cross the peritrophic matrix, indicating that this barrier, if not impenetrable for malaria parasites, exerts at least some selection pressure on them. From this point of view, it may be a colonization resistance mechanism induced by the microbiota against *Plasmodium*.

Modifications of microbiota may be an important potential strategy in blocking *Plasmodium* transmission in vectors and in the diagnosis, treatment and prevention of malaria in humans in the future (Omondi and Caner 2022).

10.2.1.6 Transmission Blocking Vaccines (TBV)

Currently, three approaches are being adopted for the development of malaria vaccines to combat malaria transmission. First approach is the development of pre-erythrocytic vaccines targeting the sporozoites and liver stages of the malaria parasite life cycle (Fig. 10.2). Second strategy is to target the asexual blood stages of malaria parasite, and the third is by utilizing vector midgut antigens and other sexual stages of parasite development inside the mosquito.

While significant progress has been made in developing malaria vaccines, currently, RTS, S/AS01 vaccine, targeting *Plasmodium falciparum*, has been endorsed by the WHO for widespread use, primarily among children in sub-Saharan Africa. This vaccine shows modest efficacy. In 2021, WHO also recommended R21/Matrix-M vaccine for wider use.

The development of transmission-blocking vaccines (TBVs), which aim to prevent parasite transmission from humans to mosquitoes, is crucial for malaria elimination efforts. Despite some TBVs advancing to clinical development, challenges

Table 10.2 Discovery of malaria transmission-blocking vaccine antigens with years, target parasite, developmental stage, gene and expression system

Antigen[a]	Year[b]	Target parasite[c]	Developmental stage[d]	Discovery[e]	Expression system[f]
Pre-genomic era					
Pfs25	1988	*Pf*	Zygote/ookinete	Gene	–
Pfs25	1994	*Pf*	Zygote/ookinete	TRA	Yeast
Pfs28	1997	*Pf*	Zygote/ookinete	Gene/TRA	Yeast
Pfs48/45	1993	*Pf*	Gametocyte/gamete	Gene	–
Pfs48/45	2008	*Pf*	Gametocyte/gamete	TRA	Bacteria
Pfs230	1993	*Pf*	Gametocyte/gamete	Gene	–
Pfs230	1995	*Pf*	Gametocyte/gamete	TRA	Bacteria
Pvs25 and Pvs28	1998	*Pv*	Zygote/ookinete	Gene	–
Pvs25 and Pvs28	2000	*Pv*	Zygote/ookinete	TRA	Yeast
Post-genomic era					
HAP2/GCS1	2008	*Pb*	Gamete	Gene	–
HAP2/GCS1	2009	*Pb*	Gamete	TRA	Bacteria
HAP2/GCS1	2013	*Pf*	Gamete	TRA	WGCFS
HAP2/GCS1	2017	*Pb, Pf*	Gamete	TRA	Peptide
HAP2/GCS1	2020	*Pv*	Gamete	TRA	Baculovirus
Pvs230	2012	*Pv*	Gametocyte/gamete	TRA	DNA
Pvs48/45, Pvs47	2015	*Pv*	Gametocyte/gamete	TRA	DNA, bacteria
Pfs47	2010	*Pf*	Gametocyte/gamete	Gene	–
Pfs47	2018	*Pf*	Gametocyte/gamete	TRA	Bacteria
AnAPN1	2014	*Pf, Pv*	*Anopheles* midgut	Gene/TRA	*Drosophila* S2

Table 10.2 (continued)

Antigen[a]	Year[b]	Target parasite[c]	Developmental stage[d]	Discovery[e]	Expression system[f]
PbPSOP12	2015	*Pb*	Gamete—ookinete	TRA	BDES
PbPH	2016	*Pb*	Gamete—ookinete	TRA	Bacteria
PbPSOP7, 25 and 26PbPSOP25	20162017	*Pb*	Ookinete	TRA	Bacteria
Pb51	2017	*Pb*	Gametocyte—ookinete	TRA	Bacteria
Pbg37	2018	*Pb*	Gametocyte—zygote	TRA	Bacteria
PyMiGS	20182020	*Py*	Gametocyte/gamete	TRA	WGCFS
Pb22	2021	*Pb*	Gamete—ookinete	TRA	Bacteria

[a] Antigen, abbreviated names of TBV antigens
[b] Year, year of publication
[c] Target parasite, Pf, *Plasmodium falciparum*; Pv, *P. vivax*; Pb, *P. berghei*; Py, *P. yoelii*
[d] Developmental stage, parasite developmental stage(s) in which target antigen is expressed
[e] Discovery, gene, target gene discovered; TRA, antigens-specific antibodies with confirmed transmission reducing/blocking activity identified
[f] Expression system indicates the platform used to express the antigen as either in yeast cells, bacteria, wheat germ cell-free system (WGCFS), *Drosophila* S2 cells, baculovirus-vectored protein expression system or a synthetic peptide (peptide). Alternatively, DNA vaccine used as the antigen (DNA). BDES indicates target antigen was expressed in baculovirus dual expression system (Source: Takashima et al. 2021)

such as low immunogenicity or high reactogenicity in humans persist. Therefore, there is a pressing need for identification of new innovative approaches to expedite TBV research and development. Takashima et al. (2021) have reviewed the whole process of identification of TBV candidates and highlighted the need of discovery of new proteins and the list of identified TBV candidates is given in (Table 10.2).

10.2.1.7 Insecticide Resistance and Infectivity of Malaria

In insecticide-resistant malaria vectors, growth and development of sporozoites or oocysts are reported to be high as compared to susceptible ones (Table 10.3).

A study carried out by Ndiath et al. (2014) with an aim to determine the impact of difference in genotypes at the *kdr* locus toward the susceptibility of *Plasmodium* infection revealed that the existence of the resistance allele at the *kdr* locus not only heightens vulnerability to *Plasmodium* during the oocyst stage but also during the sporozoite stage in non-genetically modified wild mosquitoes. These findings carry substantial implications and warrant consideration in the formulation of malaria control strategies.

Table 10.3 Studies evaluating the impact of insecticide resistance on malaria vectors susceptibility to *P. falciparum* infection

Mosquito species	Type of experimen	Type performed	Prevalence of infection	Intensity	
An. gambiae[a]	*kdr*-resistant strain compared to susceptible strain	Natural	Higher sporozoite infection prevalence in resistant strain[s]		
An. funestuss.s[a]	*GSTe2*-resistant genotypes compared to susceptible genotypes	Natural	No difference in oocysts infection prevalence[ns] Higher sporozoite infection prevalence in resistant strain[s]		
An. gambiae s.l.[a]	*ace-1*-resistant strain compared to susceptible strain	Natural	Higher oocyst infection prevalence in resistant strain[s]		
An. gambiae s.l.[a]	*kdr* resistant strain compared to susceptible strain	Natural	No difference in oocysts infection prevalence[ns]		]
An. gambiae[a]	*kdr*-resistant strain compared to susceptible strain	Experimental	Higher oocyst infection prevalence in resistant strain[s] Higher sporozoite infection prevalence in resistant strain[s]	Higher oocyst load in the resistant strain[s] Higher sporozoite infection load in the resistant strain[s]	]
An. coluzzii[a]	*kdr*-resistant strain compared to susceptible strain	Experimental	Higher oocyst prevalence in the resistant strain[s] Higher sporozoite infection prevalence in resistant strain[s]	Higher oocysts load in the resistant strain[s] Higher sporozoite load prevalence in the resistant strain[s]	
An. funestuss.s[a]	*GSTe2*-resistant genotypes compared to susceptible genotypes	Experimental	Lower oocyst infection prevalence in homozygous resistant genotypes[s]	Higher oocysts load in homozygous and heterozygous resistant genotypes[s]	
An. gambiae[b]	*ace-1*-resistant strain compared to susceptible strain	Experimental	Higher oocyst prevalence infection in resistant strain [s]No difference in sporozoite infection prevalence[ns]	No difference in oocyst and sporozoite infection load[ns]	
An. gambiae[b]	*kdr*-resistant strain compared to susceptible strain	Experimental	Higher oocyst infection prevalence in resistant strain[s] Higher sporozoite infection prevalence In resistant strain[s]	Lower oocyst and sporozoite infection load in resistant strain[s]	

An. gambiae[b]	*ace-1*-resistant strain compared to susceptible strain		Higher oocyst infection prevalence in resistant strain[s]		
An. gambiae[b]	*kdr*-resistant strain compared to susceptible strain	Experimental	Higher oocyst infection prevalence in resistant strain[s]		
An. gambiae[b]	*kdr*-resistant strain compared to *kdr*-resistant strain exposed to insecticides		Lower oocyst infection prevalence in resistant strain[s]	No difference in oocyst infection load[ns]	
An. gambiae[b]	*ace-1*-resistant strain compared to *ace-1*-resistant strain exposed to insecticides	Experimental	Lower oocyst infection prevalence in resistant strain[s]	Lower oocyst infection load in resistant strain[s]	
An. gambiae[a]	*kdr*-resistant strain compared to *kdr*-resistant strain exposed to insecticide-treated nets	Experimental	Lower oocyst infection prevalence in resistant strain[s]	Lower oocyst infection load in resistant strain[s]	

a, field strain; b, laboratory strain; ns, nonsignificant; s, significant; α, insecticide exposure to confirm resistance status probably postinfection; β, insecticide exposure to confirm resistance status prior to infection (Source: Suh et al. 2023); kdr, knockdown resistant

Kabula et al. (2016) studied deltamethrin-resistant strain (22.3%) of *An. gambiae* and *An. arabiensis*, 96.5% of which were *An. gambiae*. In both the species, *Vgsc*-1014S mutation was detected. A higher sporozoite infection was found in resistant strains. The *P. falciparum* sporozoite infection rate was 4.2%. A significant association has been reported between the presence of sporozoites and *Vgsc*-L1014S genotype in *A. gambiae* sensu.*stricto*. These findings indicate that resistant strains of *Anopheles* may impact malaria transmission.

Genetic sequencing of glutathione *S*-transferase metabolic resistance gene (GSTe2)-based study carried out by Tchouakui et al. (2019) showed that vectorial capacity of *An. funestus*-resistant mosquitoes is increasing. It could be matter of concern in the areas where rise in GSTe2-based metabolic resistance to insecticides is being reported as it may lead to increased vectorial capacity of vectors.

Collins et al. (2019) investigated the relationship between pyrethroid and carbamate resistance with parity, *P. falciparum* infection and molecular insecticide resistance mechanisms in *Anopheles gambiae* sensu lato. These findings offer a comprehensive assessment of current levels of insecticide resistance and the underlying target site mutations in Maferinyah, Guinea, an area characterized by high malaria transmission rates. Local mosquito populations exhibited strong resistance to pyrethroids (alpha-cypermethrin, deltamethrin and permethrin), correlated with high occurrences of the L1014F *kdr*. The restoration of mosquito susceptibility following pre-exposure to piperonyl butoxide (PBO) suggests an increased activity of detoxification enzymes contributing to pyrethroid resistance in this area, demanding further characterization. Malaria infection did not correlate with pyrethroid resistance but was associated with bendiocarb resistance. Generally, resistant vectors are younger than their susceptible counterparts. Further studies are desired to explore the impact of insecticide resistance on vector fitness, encompassing mosquito fecundity, egg viability, hatchability and parasite development following an infected blood meal.

The review of current understanding of factors influencing vectorial competence in insecticide-resistant vectors remains inadequate to conclusively determine the epidemiological significance of these vectors compared to their susceptible counterparts (Suh et al. 2023). Further research is required to address significant knowledge gaps that could aid in better understanding of malaria epidemiology, particularly in settings where the selection and proliferation of insecticide-resistant vectors are ongoing. However, it is advisable to periodically determine insecticide resistance in malaria vectors and adopt appropriate insecticide for indoor residual spray to eliminate resistant strains.

10.2.2 Dengue Fever

The role of climatic thresholds operates in arboviral diseases as well. The distribution maps of dengue, chikungunya and Zika reflect that the diseases are not prevalent in colder areas in Himalayan region. Kakarla et al. (2019) found that transmission of dengue in India occurs between 17 and 34 °C. Thus, the climatic thresholds given in Table 10.1 for dengue, chikungunya and Zika mention minimum and maximum thresholds of temperature and RH indicating the possibility of transmission; however, at increasing temperatures, the infection, dissemination and transmission of these viruses have been found to be higher (Delrieu et al. 2023).

As regards the role of peritrophic membrane (PM), Franz et al. (2015) reviewed the barriers to arbovirus infection in mosquitoes and mentioned that PM is one postulated barrier to infection. Experimental evidence for barriers of infection is much stronger for midgut infection (entry into and replication within midgut epithelial cells) and midgut escape, with the basal lamina underlying the midgut epithelium being a major impediment to virus expulsion from the midgut and subsequent systemic infection of the arthropod host.

Lehane (1997) postulated that the mosquito, which produces a type II PM, is difficult to infect with viruses, whereas adult mosquitoes producing a type I PM are often efficient vectors. It was proposed that this distinction is related to the delay, sometimes up to several hours, in producing a robust PM following the uptake of a blood meal.

It has also been observed that the presence of bacterial microbiota in insect guts can impact the efficiency of pathogen infections in these hosts (Rodgers et al. 2017). In a recent paper, Wu et al. (2019) demonstrated that the presence of *Serratia marcescens* in the gut of adult *Aedes aegypti* facilitates the acquisition of dengue virus. They also demonstrated that *S. marcescens* secretes a protein referred to as Sm enhancin that, like baculovirus enhancin, cleaves and digests midgut mucins. This suggests that, in mosquitoes, it is probable that midgut epithelial membrane-bound mucins, but not the PM itself, play a role in protecting the midgut from arbovirus infections (Wu et al. 2019).

10.2.3 Transmission of Zika Virus (ZKV)

Indigenous cases of Zika virus transmitted by *Aedes aegypti* have recently been reported from India (Singh et al. 2019). *Aedes aegypti* and *Ae. albopictus* are the main vectors. The climatic thresholds of Zika are wider than dengue and chikungunya. Gregory et al. (2017) reviewed the modes of transmission of Zika, which reported several non-vector routes, like sexual transmission, intrauterine and intrapartum, blood transfusion, laboratory exposure and organ and tissue transplantation. In addition to *Aedes aegypti* and *Ae. albopictus*, *Culex* mosquitoes have also been implicated as vector of Zika virus. Guo et al. (2016) could infect mice with ZKV providing evidence that Cx. p. quinquefasciatus could transmit ZKV.

10.2.4 Japanese Encephalitis

Japanese encephalitis is caused by a flavivirus, JEV, which belongs to Flavivirdae family and transmitted mainly by *Culex vishnui* group of mosquitoes. In addition to mosquitoes of genus *Culex*, JE is also transmitted by *Aedes* in Andaman and Nicobar

Islands. It shows that the JEV crosses the barrier of the genus of mosquitoes as well as in becoming vector.

The pigs and cattle egrets are the reservoirs of the JEV, and pigs serve as amplifier host. The glaring aspect is that man is dead-end host, which means the vectors cannot pick up JEV from infected human as the level of viraemia is very low. In pigs, the viraemia remains from day 1 to 5 days of infection. It has been reported that the viraemia levels of about 10^4 infectious units per mL is sufficient to transmit the virus to mosquitoes (Scherer et al. 1959). Depending on the study, viraemia was detected between 1 and 5 days of infection in pigs and lasted from 3 to 5 days (Carey et al. 1969; Williams et al. 2001). Another study concluded that the higher the concentration of inoculated virus, the earlier the virus was detected (as early as 1 day). When detected, virus titres reached 2.6 log LD_{50}/0.03 mL of blood in weanling mice, with LD_{50} being 50% of the lethal dose (Scherer et al. 1959). In addition to pigs and egret birds, there may be other animals, which may serve as reservoirs of the JEV, and this is a matter of research.

The reason of inadequate viraemia in infected human is early presence of antibodies to bring down the level of virus load. May be under immuno-compromised conditions, viraemia level may sometimes become sufficient enough to make the vector mosquitoes infective, which is a matter of further research.

10.2.5 Lymphatic Filariasis

Lymphatic filariasis (LF) is caused by helminthic parasite of the genus *Wuchereria bancrofti*, *Brugia malayi* and *B. timori*. Several species of the mosquitoes belonging to *Anopheles*, *Mansonia*, *Culex* and *Aedes* genera are reported as vectors of human LF. *Culex quinquefasciatus*, *Culex pipiens molestus*, *Culex pipiens pipiens*, *An. sinensis*, *An. gambiae*, *An. melas*, *An. meras* and *An. maculatus* are natural vectors of periodic *W. bancrofti*. In Andaman and Nicobar Islands (India), diurnal subperiodic *W. bancrofti* is transmitted by *Downsiomyia nivea*, *Aedes oceanicus*, *Ae. polynesiensis* and *Ae. pseudoscutellaris* (Shriram et al. 2015). It indicates that *W. bancrofti* and *B. malayi* cross all the physiological/molecular barriers in the gut of mosquitoes to become vectors of LF. It is still to prove the vectorial status of *Aedes scutellaris* in Andaman and Nicobar Islands.

Another interesting aspect in the transmission of lymphatic filariasis leading to elephantiasis is the repeated bites of vectors required for resulting into the elephantiasis syndrome. There is a need to quantify the repeated bites required by infected mosquito vector and/or duration of stay in endemic areas to turn into a case of elephantiasis.

10.3 Conclusion

In conclusion, this chapter highlights the complex landscape of mosquito-borne diseases in India, underscoring the multifaceted challenges they pose to public health. The prevalence of diseases such as malaria, dengue, chikungunya, Zika, Japanese encephalitis and lymphatic filariasis underscores the urgent need for effective control measures. While several mosquito species serve as vectors, the factors influencing their vectorial capacity are diverse and not yet fully understood.

There should be emphasis on the importance of data-driven approaches in determining climatic thresholds for disease transmission and the need for further research to elucidate the conditions under which secondary vectors may become primary vectors. Moreover, the potential development of transmission-blocking vaccines targeting chitinases in mosquito vectors presents a promising avenue for disease control.

Insights into the role of microbiota in inhibiting malaria parasite development in mosquitoes offer new strategies for transmission control, while the complex transmission dynamics of diseases like filariasis and Zika underscore the necessity of a comprehensive approach to vector management.

As efforts to eliminate mosquito-borne diseases progress, it is crucial to continue investigating the roles of different mosquito species and their interactions with pathogens. By addressing the knowledge gaps highlighted in this chapter, researchers can contribute to the development of more effective prevention and control strategies, ultimately reducing the burden of mosquito-borne diseases on public health in India and beyond.

Points to Remember

1. Any insect vector should fulfil five conditions to be labelled as a vector of disease.
2. The components of vectorial capacity of a vector depend on extrinsic incubation period in days (n): the ratio of mosquito to human beings (m), survival of mosquitoes through 1 day (p) and human biting rates (a).
3. Climatic thresholds are critical in ensuing transmission of a particular vector-borne disease (VBD) in an area. Upper parts of Himalayan region do not have indigenous transmission of malaria as the temperature is <16 °C temperature.
4. Projected scenario of malaria by 2030s indicate opening of new foci of malaria transmission in Himalayan region of India in view of projected increase in temperature.
5. Peritrophic membrane (PM) also plays important role in any insect to act as vector of a particular disease. *Plasmodium* chitinases belonging to the GH18 family are secreted inside the mosquito midgut, during the ookinete stage of the parasite. Chitinases mediate the penetration of parasite through the peritrophic membrane, facilitating access to the gut epithelial layer.
6. The presence of microbiota in *Anopheles* mosquitoes as well as human beings affect malaria transmission.

7. The development of transmission-blocking vaccines, which aim to prevent parasite transmission from humans to mosquitoes is crucial for malaria elimination efforts. Despite some transmission-blocking vaccines advancing to clinical development, challenges, such as low immunogenicity or high reactogenicity in humans, persist, necessitating the need for identification of new innovative approaches to expedite transmission-blocking vaccine research and development.
8. Non-vector routes, like sexual transmission, intrauterine and intrapartum, blood transfusion, laboratory exposure and organ and tissue transplantation, are also reported as modes of transmission of Zika virus. In addition to *Aedes aegypti* and *Ae. albopictus*, *Culex* mosquitoes have also been implicated as vector of Zika virus.
9. There is need to explore several other reservoirs of JE virus than egret birds. May be under immuno-compromised conditions, viraemia level may sometimes become sufficient enough to make the vector mosquitoes infective.
10. *W. bancrofti* and *B. malayi* cross all the physiological/molecular barriers in the gut of mosquitoes to become vectors of LF as *Anopheles*, *Culex*, *Mansonia* and *Aedes* genera are vectors of filariasis. The vector status of *Aedes scutellaris* in Andaman and Nicobar Islands still needs to be proved.
11. Insecticide-resistant Anopheline vectors of malaria are by and large more infective to malaria parasite.

References

Beerntsen BT, James AA, Christensen BM (2000) Genetics of mosquito vector competence. Microbiol Mol Biol Rev 64:115–137. https://doi.org/10.1128/MMBR.64.1.115-137.2000

Carey DE, Reuben R, Myers RM (1969) Japanese encephalitis studies in Vellore, South India. V. Experimental infection and transmission. Indian J Med Res 57(2):282–289

Collins E, Vaselli NM, Sylla M, Beavogui AH, Orsborne J, Lawrence G, Ryan E, Wiegand RE, Irish SR, Walker T, Messenger LA (2019) The relationship between insecticide resistance, mosquito age and malaria prevalence in Anopheles gambiae s.l. from Guinea. Sci Rep 9:8846

Craig MH, Snow RW, le Sueur D (1999) A climate-based distribution model of malaria transmission in sub-Saharan Africa. Parasitol Today 15(3):105–111. https://doi.org/10.1016/s0169-4758(99)01396-4

Delrieu M, Martinet JP, O'Connor O, Viennet E, Menkes C, Burtet-Sarramegna V, Frentiu FD, Dupont-Rouzeyrol M (2023) Temperature and transmission of chikungunya, dengue, and Zika viruses: a systematic review of experimental studies on Aedes aegypti and Aedes albopictus. Curr Res Parasitol Vector Borne Dis 4:100139. https://doi.org/10.1016/j.crpvbd.2023.100139

Erlandsona MA, Toprakc U, Hegedusa DD (2019) Role of the peritrophic matrix in insect-pathogen interactions. J Insect Physiol 117:103894. https://doi.org/10.1016/j.jinsphys.2019.103894

Franz AW, Kantor AM, Passarelli AL, Clem RJ (2015) Tissue barriers to arbovirus infection in mosquitoes. Viruses 7(7):3741–3767. https://doi.org/10.3390/v7072795

Frischknecht F, Matuschewski K (2017) Plasmodium sporozoite biology. Cold Spring Harb Perspect Med 7:a025478. https://doi.org/10.1101/cshperspect.a025478

Gregory CJ, Oduyebo T, Brault AC, Brooks JT, Chung KW, Hills S, Kuehnert MJ, Mead P, Meaney-Delman D, Rabe I, Staples E, Petersen LR (2017) Modes of transmission of Zika virus. J Infect Dis 216(Suppl 10):S875–S883. https://doi.org/10.1093/infdis/jix396

Gunathilaka N, Hapugoda M, Abeyewickreme W, Wickremasinghe R (2015) Entomological investigations on malaria vectors in some war-torn areas in the Trincomalee district of Sri Lanka after settlement of 30-year civil disturbance. Malar Res Treat 2015:367635. https://doi.org/10.1155/2015/367635

Gunathilaka N, Hapugoda M, Wickremasinghe R, Abeyewickreme W (2019) A comprehensive analysis on abundance, distribution, and bionomics of potential malaria vectors in Mannar District of Sri Lanka. Malar Res Treat 2019:1650180. https://doi.org/10.1155/2019/1650180

Guo XX, Li CX, Deng YQ, Xing D, Liu QM, Wu Q, Sun AJ, Dong YD, Cao WC, Qin CF, Zhao TY (2016) Culex pipiens quinquefasciatus: a potential vector to transmit Zika virus. Emerg Microbes Infect 5(9):e102. https://doi.org/10.1038/emi.2016.102

Kabula B, Tungu P, Rippon EJ, Steen K, Kisinza W, Magesa S, Mosha F, Donnelly MJ (2016) A significant association between deltamethrin resistance, *Plasmodium falciparum* infection and theVgsc 1014S resistance mutation in Anopheles gambiae highlights the epidemiological importance of resistance markers. Malar J 2016(15):289

Hou K, Wu ZX, Chen XY, Wang JQ, Zhang D, Xiao C, Zhu D, Koya JB, Wei L, Li J, Chen ZS. Microbiota in health and diseases. Signal Transduct Target Ther. 2022;7(1):135. https://doi.org/10.1038/s41392-022-00974-4

Kakarla SG, Mopuri R, Mutheneni SR, Bhimala KR, Kumaraswamy S, Kadiri MR, Gouda KC, Upadhyayula SM (2019) Temperature dependent transmission potential model for chikungunya in India. Sci Total Environ 647:66–74. https://doi.org/10.1016/j.scitotenv.2018.07.461

Kramer LD, Ciota AT (2015) Dissecting vectorial capacity for mosquito-borne viruses. Curr Opin Virol 15:112–118. https://doi.org/10.1016/j.coviro.2015.10.003

Kumar A, Hosmani R, Jadhav S, de Sousa T, Mohanty A, Naik M, Shettigar A, Kale S, Valecha N, Chery L, Rathod PK (2016) *Anopheles subpictus* carry human malaria parasites in an urban area of Western India and may facilitate perennial malaria transmission. Malar J 15:124. https://doi.org/10.1186/s12936-016-1177-x

Kumar A, Srivastava P, Sirisena P, Dubey SK, Kumar R, Shrinet J, Sunil S (2018) Mosquito innate immunity. Insects 9:95. https://doi.org/10.3390/insects9030095

Kumar G, Gupta SK, Rahi M, Sharma A (2022) Challenges in understanding the bionomics of Indian malaria vectors. Am J Trop Med Hyg 107(5):1005–1014. https://doi.org/10.4269/ajtmh.22-0137

Langer RC, Vinetz JM (2001) Plasmodium ookinete-secreted chitinase and parasite penetration of the mosquito peritrophic matrix. Trends Parasitol 17(6):269–272. https://doi.org/10.1016/s1471-4922(01)01918-3

Lehane MJ (1997) Peritrophic matrix structure and function. Annu Rev Entomol 42:525–550. https://doi.org/10.1146/annurev.ento.42.1.525

Lewis J, Gallichotte EN, Randall J, Glass A, Foy BD et al (2023) Intrinsic factors driving mosquito vector competence and viral evolution: a review. Front Cell Infect Microbiol 13:1330600. https://doi.org/10.3389/fcimb.2023.1330600

Mordecai EA, Paaijmans KP, Johnson LR, Balzer C, Ben-Horin T, Moor ED, McNally A, Pawar S, Ryan SJ, Smith TC, Lafferty KD (2013) Optimal temperature for malaria transmission is dramatically lower than previously predicted. Ecol Lett 16(1):22–30. https://doi.org/10.1111/ele.12015

Mordecai EA, Cohen JM, Evans MV, Gudapati P, Johnson LR, Lippi CA, Miazgowicz K, Murdock CC, Rohr JR, Ryan SJ, Savage V, Shocket MS, Ibarra AS, Thomas MB, Weikel DP (2017) Detecting the impact of temperature on transmission of Zika, dengue, and chikungunya using mechanistic models. PLoS Negl Trop Dis 11(4):e0005568. https://doi.org/10.1371/journal.pntd.0005568

Napier LE (1931) An epidemiological consideration of the transmission of Kala-azar in India. Indian Med Res Memoirs 25:219–265

Ndiath MO, Cailleau A, Diedhiou SM, Gaye A, Boudin C, Vincent Richard V, Trape J (2014) Effects of the kdr resistance mutation on the susceptibility of wild Anopheles gambiae populations to Plasmodium falciparum: a hindrance for vector control. Malar J 2014(3):340

Omondi ZN, Caner A (2022) An overview on the impact of microbiota on malaria transmission and severity: plasmodium–vector–host axis. Acta Parasitol 67:1471–1486

Rodgers FH, Gendrin M, Wyer CAS, Christophides GK (2017) Microbiota-induced peritrophic matrix regulates midgut homeostasis and prevents systemic infection of malaria vector mosquitoes. PLoS Pathog 13(5):e1006391. https://doi.org/10.1371/journal.ppat.1006391

Sarkar S, Gangare V, Singh P, Dhiman RC (2019) Shift in potential malaria transmission areas in India, using the fuzzy-based climate suitability malaria transmission (FCSMT) model under changing climatic conditions. Int J Environ Res Public Health 16(3474):1–16

Scherer WF, Moyer JT, Izumi T (1959) Immunologic studies of Japanese encephalitis virus in Japan. V. Maternal antibodies, antibody responses and viremia following infection of swine. J Immunol 83:620–626

Shapiro LLM, Whitehead SA, Thomas MB (2017) Quantifying the effects of temperature on mosquito and parasite traits that determine the transmission potential of human malaria. PLoS Biol 15(10):e2003489. https://doi.org/10.1371/journal.pbio.2003489

Shriram AN, Krishnamoorthy K, Vijayachari P (2015) Diurnally subperiodic filariasis among the Nicobarese of Nicobar district - epidemiology, vector dynamics & prospects of elimination. Indian J Med Res 141(5):598–607. https://doi.org/10.4103/0971-5916.159537

Singh RK, Kumar G, Mittal PK, Dhiman RC (2014) Bionomics and vector potential of *Anopheles subpictus* as a malaria vector in India: an overview. Int J Mosquito Res 1(1):29–37

Singh P, Lingala MA, Sarkar S, Dhiman RC (2017) Mapping of malaria vectors at district level in India: changing scenario and identified gaps. Vector Borne Zoonotic Dis 17(2):91–98. https://doi.org/10.1089/vbz.2016.2018

Singh H, Singh OP, Akhtar N, Sharma G, Sindhania A, Gupta N, Valecha N (2019) First report on the transmission of Zika virus by Aedes (Stegomyia) aegypti (L.) (Diptera: Culicidae) during the 2018 Zika outbreak in India. Acta Trop 199:105114. https://doi.org/10.1016/j.actatropica.2019.105114

Suh PF, Elanga-Ndille E, Tchouakui M, Sandeu MM, Tagne D, Wondji C, Ndo C (2023) Impact of insecticide resistance on malaria vector competence: a literature review. Malar J 22(1):19. https://doi.org/10.1186/s12936-023-04444-2

Takashima E, Tachibana M, Morita M, Nagaoka H, Kanoi BN et al (2021) Identification of novel malaria transmission-blocking vaccine candidates. Front Cell Infect Microbiol 11:805482. https://doi.org/10.3389/fcimb.2021.805482

Tchouakui M, Chiang MC, Ndo C, Kuicheu CK, Amvongo-Adjia N, Wondji MJ, Tchoupo M, Kusimo MO, Riveron JM, Wondji CS (2019) A marker of glutathione S-transferase-mediated resistance to insecticides is associated with higher Plasmodium infection in the African malaria vector Anopheles funestus. Sci Rep 9(1):5772

Tesla B, Demakovsky LR, Mordecai EA, Ryan SJ, Bonds MH, Ngonghala CN, Brindley MA, Murdock CC (2018) Temperature drives Zika virus transmission: evidence from empirical and mathematical models. Proc Biol Sci 285(1884):20180795. https://doi.org/10.1098/rspb.2018.0795

Viswanath VK, Gore ST, Valiyaparambil A, Mukherjee S, Lakshminarasimhan A (2021) Plasmodium chitinases: revisiting a target of transmission-blockade against malaria. Protein Sci 30(8):1493–1501. https://doi.org/10.1002/pro.4095

Watts DM, Burke DS, Harrison BA, Whitmire RE, Nisalak A (1987) Effect of temperature on the vector efficiency of Aedes aegypti for dengue 2 virus. Am J Trop Med Hyg 36(1):143–152. https://doi.org/10.4269/ajtmh.1987.36.143

Williams DT, Daniels PW, Lunt RA, Wang LF, Newberry KM, Mackenzie JS (2001) Experimental infections of pigs with Japanese encephalitis virus and closely related Australian flaviviruses. Am J Trop Med Hyg 65(4):379–387. https://doi.org/10.4269/ajtmh.2001.65.379

Wu P, Sun P, Nie K, Zhu Y, Shi M (2019) A gut commensal bacterium promotes mosquito permissiveness to arboviruses. Cell Host Microbe 25(1):101–112.e5. https://doi.org/10.1016/j.chom.2018.11.004

Mosquito-Borne Human Diseases

11

Goutam Chandra and Indranil Bhattacharjee

Abstract

Mosquitoes are the transmitters of several deadly diseases. Protozoan parasites and viruses are the main causes of vector-borne diseases. Most diseases spread by mosquitoes are found in tropical and subtropical regions of the world. The Global Vector Control Response (GVCR) 2017–2030 recommends the strategic control of vector-borne disease outbreaks. It is proposed that improved infrastructure, monitoring systems, and technical capacity are necessary to prevent vector-borne diseases and maintain human health. Mosquito-borne diseases pose the greatest threat to public health worldwide and are typically associated with severe economic hardships in the countries they infect. Vector-borne diseases have been on the rise and have reemerged in the last 40 years due to a variety of causes including population growth, migration, urbanization, globalization, lack of public health infrastructure, awareness, etc. We attempt to focus mostly on causes, symptoms, diagnosis, therapy, and control in this book chapter. With a general awareness, the key challenges related to significant vector-borne diseases may be addressed.

Keywords

Mosquito-borne diseases · Cause · Symptoms · Diagnosis

G. Chandra (✉)
Department of Zoology, The University of Burdwan, Burdwan, West Bengal, India

I. Bhattacharjee
Mosquito, Microbiology and Nanotechnology Research Units, Parasitology Laboratory, Department of Zoology, The University of Burdwan, Purba Bardhaman, West Bengal, India

Department of Zoology, Dr. Bhupendra Nath Dutta Smriti Mahavidyalaya, Hatgobindapur, Purba Bardhaman, West Bengal, India

© The Author(s), under exclusive license to Springer Nature Singapore Pte Ltd. 2024

Omkar (ed.), *Mosquitoes*, https://doi.org/10.1007/978-981-97-4163-2_11

Learning Objectives

1. Determine the key disease-causing vectors.
2. Sort the fundamentals of vector biology.
3. Understand how disease is spread via vectors.
4. Sort the diseases that each vector is capable of spreading.
5. Describe the fundamental pathophysiology of vector-borne diseases.

11.1 Introduction

Mosquito-borne infections are highly prevalent in over 150 countries worldwide. Mosquito-borne infections affect more than 500 million people globally, accounting for around 1 million fatalities. Nearly 40 million people in India contact diseases that are spread by mosquitoes annually. Mosquitoes are the primary carriers of a wide range of illnesses, including dengue, malaria, Japanese encephalitis, yellow fever, West Nile fever, chikungunya, Zika virus, lymphatic filariasis, etc. Many illnesses have the potential to be lethal if the symptoms are not correctly detected timely. In India, the three most common illnesses carried by mosquitoes are malaria, dengue, and filariasis. *Anopheles*, *Aedes*, and *Culex* mosquitoes are known to carry a number of dangerous diseases. Management of mosquito population is a better way to reduce the burden of mosquito-borne illnesses. The three largest mosquito genera found throughout the world are *Anopheles*, *Aedes*, and *Culex*.

11.2 *Anopheles* and Anopheline Borne Diseases

11.2.1 *Anopheles*

Out of about 425 known species of *Anopheles*, only 91 are known malaria vectors. In India, 58 species of *Anopheles* are recorded, out of which 6 are considered as primary vectors and 4 as secondary vectors. The primary vectors include *Anopheles culicifacies*, *An. stephensi*, *An. fluviatilis*, *An. dirus*, *An. minimus,* and *An. sundaicus,* whereas *An. annularis*, *An. philippiensis* (*nivipes*), *An. varuna,* and *An. jeyporiensis* are considered as secondary vectors (Rao 1984) though afterward *An. subpictus* has been incriminated as vector of malaria in some places of India (Chatterjee and Chandra 2000; Chandra et al. 2010). Among the primary vectors, *An. culicifacies* has been established as a major vector of malaria in rural and peri-urban areas of India (Subbarao et al. 1988).

11.3 The Mosquito Vectors

Table 11.1 shows the vectors of malaria of India, Pakistan, and Bangladesh and their distribution and breeding sites. The principal vectors are those species of mosquitoes that are mainly responsible for transmitting malaria in a situation, sometimes alternating seasonally. Secondary vectors are capable of maintaining malaria only at a reduced level in the absence of principal vectors.

Human malaria is caused by infection with sporozoan of the genus *Plasmodium* transmitted in nature by the bite of infective female anopheline mosquitoes.

The name malaria is derived from two Italian words *mala* and *aria* meaning bad air. It is also called *paludism*. Latin *palus* means a marsh. It was formerly believed that the disease was spread by miasma or mist, arising in marshes. Malaria still remains a very important disease of man. Malaria killed 800,000 people in India in 1953 out of nearly 75 million affected. Total loss of income due to malaria amounted to 500 million dollars per year (National Framework for Malaria Elimination in India, 2016–2030 n.d.).

Because of the National Malaria Control Programme and, subsequently, National Malaria Eradication Programme, hopes became alive about the eradication of

Table 11.1 Vectors of malaria in India, Pakistan, and Bangladesh and their distribution and breeding sites

Sl no.	Species	Distribution	Breeding sites
1.	*An. stephensi*	In cities like Calcutta, Mumbai, Delhi, Chennai, etc.	Small collection of clean water, overhead cisterns, wells, small water reservoirs, springs, marsh, and collection of rainwater
2.	*An. culicifacies*	Madhya Pradesh, U.P., Bihar, adjacent parts of West Bengal, Rajasthan, Punjab, Kashmir, Nainital area, West Pakistan	Irrigation channels, river beds, pools, ponds, lakes, burrow pits, gardens, wells
3.	*An. philippinensis*	West Bengal, Assam, North Bihar, Bangladesh	Tanks, burrow pits, and rice fields
4.	*An. fluviatilis*	Nainital, Nepal Terai area, Manipur, Assam, Bangladesh, Western Ghats of Mysore and Kerala	Slowly running perennial stream, channels, rice fields, betel ground trenches
5.	*An. varuna*	Calcutta area, Jeypore hills, Singhbhum hills, Satpura range	Stagnant freshwater: ponds, ditches
6.	*An. sundaicus*	Chilka lake area, coastal Orissa, Vishakhapatnam, Ganjam, Chennai area, coastal parts of West Bengal and Bangladesh	Brackish water
7.	*An. minimus*	Assam Dooars, West Bengal, the foot hills of the Himalayas in U.P. and South India	Slowly running perennial streams under shade
8.	*An. dirus*	Assam, Arunachal, Nagaland, Meghalaya, Mizoram, Tamil Nadu, Tripura, West Bengal, Karnataka, Andaman Islands, Bangladesh, etc.	Forest pools and nullahs in partially shaded places
9.	*An. maculatus*	Assam (Shillong). North West Himalayas, South India	Clear water, hill stream, exposed to sunlight
10.	*An. annularis*	West Bengal, Orissa	Stagnant water weedy tanks, rice fields, swamps, lakes, burrow pits

Table 11.2 Recorded year wise account of death due to malaria in India

Year	**1974**	**1975**	**1976**	**1977**	**1978**	**1979**	**1980**	**1981**	**1982**	**1983**	**1984**	**1985**	**1986**
Death	*3*	*99*	*59*	*55*	*74*	*198*	*207*	*170*	*187*	*239*	*247*	*213*	*323*
Year	**1987**	**1988**	**1989**	**1990**	**1991**	**1992**	**1993**	**1994**	**1995**	**1996**	**1997**	**1998**	**1999**
Death	*188*	*209*	*268*	*353*	*421*	*422*	*354*	*1122*	*1152*	*1110*	*653*	*853*	*879*
Year	**2000**	**2001**	**2002**	**2003**	**2004**	**2005**	**2006**	**2007**	**2008**	**2009**	**2010**	**2011**	**2012**
Death	*932*	*1005*	*973*	*1006*	*949*	*963*	*1707*	*1311*	*1055*	*1144*	*1018*	*754*	*519*
Year	**2013**	**2014**	**2015**	**2016**	**2017**	**2018**	**2019**	**2020**	**2021**	**2022**			
Death	*440*	*562*	*384*	*331*	*194*	*96*	*77*	*93*	*90*	*83*			

Table 11.3 Geographical distribution

P. vivax	*P. falciparum*	*P. malariae*	*P. ovale*
Tropical and temperate zones, Siberia, Africa (does not occur among the Negrocs of W. Africa), Iran, Bangladesh, Pakistan, many parts of India, Myanmar, Laos, Thailand, Malaysia. Indonesia, Vietnam, the Philippines and New Guinea, Central and S. Americas	Eastern Mediterranean and the Middle East, India, Sri Lanka, Bangladesh, Pakistan, Nepal. Afghanistan, Myanmar, Laos. Thailand, Cambodia, Malaysia, Indonesia, Portuguese Timor, Central and S. Americas, Caribbean region	Tropical and temperate climates; very wide but patchy distribution in Central and East Africa, India, Sri Lanka, Andaman and Maldives Islands, Pakistan, Myanmar, Malaysia, Far East, Caribbean region	East Africa, West Africa, Philippines, sporadic cases reported from India, Mauritius, Israel, Thailand, Malaysia, New Guinea, and S. America

malaria from the Indian subcontinent. But cases began to rise from 1969. The number of malaria cases rose to 350,000 in 1969 from 100,000 in 1965. 1,930,273 and 3,167,658 cases were detected in 1973 and 1974, respectively. In 1975, 1976, and 1977, about 5.1, 6.4, and 4.7 million people, respectively, suffered from malaria. In India, year wise number of malaria deaths is given in Table 11.2:

Species of malaria parasites that infect man are as follows:

1. *Plasmodium vivax.*
2. *P. falciparum.*
3. *P. ovale.*
4. *P. malariae.*

There are races and strains in *P. vivax* and *P. falciparum* according to which the clinical picture, geographical distribution, and immunological responses vary (Table 11.3).

In India, *P. vivax* constituted 65–69%, whereas *P. falciparum* constituted 25–30% of total malaria cases. *P. malariae* represented only 1% of cases. Distribution of *P. ovale* was sporadic. Recent data show that the proportion of *P. falciparum* to the total cases has increased and reached up to 40% (Lal et al. 1998; WHO 2011).

11.4 Habitat

Malaria parasites invade the parenchyma cells of the liver and, after passing through a developmental phase, attack and reside inside the red blood corpuscles.

11.5 Mode of Infection

Through the bite of infective female anopheline mosquito, harboring sporozoites in the salivary glands enter into the human blood. During the act of biting, the mosquito at first injects saliva through the punctured wound when sporozoites are introduced into the blood.

11.5.1 Other Modes of Infection

1. Sporozoite induced malaria. By injecting emulsion of salivary glands of female anopheline mosquitoes containing sporozoites.
2. Transfusion malaria. When infected people are used as donors, malaria occurs after blood transfusion.
3. Transplacental transmission. Fetus in utero may be infected from mother. The placenta is usually an effective barrier. But congenital infection has been documented without demonstrable damage to the placenta before delivery.
4. Malaria in drug addicts. By using the same syringe from an infected person, others may be infected.

11.6 Incubation Period in Humans

The period between the bite of an infective mosquito and the first attack of paroxysm in *P. vivax*, *P. falciparum*, and *P. ovale* is 10–14 days. In *P. malariae*, it varies between 18 days and 6 weeks.

11.7 Clinical Manifestations

A series of febrile paroxysms follows in three stages: (1) the cold stage (20 min to 1 h), (2) the hot stage (1–4 h), and (3) the sweating stage (2–3 h). The paroxysm recurs 1–2 days in case of *P. falciparum*, 3 days in *P. malariae*, and 2 days in *P. vivax* and *P. ovale*. This classical picture may not always be present. Gradually, anemia develops. There is enlargement of the spleen.

11.8 Disease Biology

11.8.1 Life Cycle of Malaria Parasites of Man

For completing the life cycle of malaria parasites, two quite different hosts are necessary: one, the man (intrinsic phase or intermediate host) in which asexual reproduction takes place and, the other, the insect vector (extrinsic phase or definitive host) in which sexual reproduction occurs.

11.8.2 The Cycle in the Mosquito

A susceptible *Anopheles* mosquito ingests from a malaria patient mature microgametocyte, macrogametocytes, and asexual stages of the parasites. In the stomach of the mosquito, asexual stages of the parasites are destroyed during the act of digestion, but microgametocytes and macrogametocytes come out of their respective erythrocyte coverings.

Each microgametocyte matures by exflagellation, the nucleus divides, and within 10 or 15 min 8 long flagellated bodies (microgametes) are formed. They swim actively in search of a macrogamete. Little change occurs in the macrogametocyte (except that the *P. falciparum* crescent becomes rounded). Increase in size occurs, and the macrogamete is developed where the nucleus shifts to the surface.

One microgamete penetrates the nucleus of the macrogamete (fertilization), forming the zygote. After about 20 min, the zygote moves by producing a pseudopod, taking fusiform shape, known as the ookinete or the motile vermicule. The ookinete grows in size, passes the barrier of peritrophic membrane, and then by penetrating the brush border enters an epithelial cell by liquefying the cell membrane. The ookinete traverses the epithelial cell and positions itself between the outer region of the epithelium and the elastic membrane covering it, becomes spherical, secretes a thin cyst wall, and grows into an oocyst, which is 50 μm or more in size, projecting into the hemocoele. During this journey, the ookinete destroys a number of epithelial cells following time bomb theory (Han et al. 2000). Meiotic and mitotic divisions of the nucleus take place, and numerous haploid sporozoites are formed. The number of oocysts on the stomach wall varies from a few to more than a hundred. The oocyst bursts, liberating the sporozoites into the hemocoele. A mature oocyst may contain about 10,000 sporozoites. Experimental infection of *An. balabacensis* with *P. falciparum* revealed 5–300 oocysts on the stomach wall and 326,000 sporozoites in the salivary glands of the individual mosquito (Li et al. 1981). Only zygote, ookinete, and oocysts are diploid; the rest of the life cycle stages are haploid.

They disperse throughout the body of the mosquito (except the ovaries). On reaching the salivary glands (their special predilection site), they penetrate through the acinar cells and accumulate in the acinar ducts. When an infective mosquito injects the saliva during a blood meal, the sporozoites are introduced into the bloodstream of the man.

The oocysts of various species of malaria parasites are distinguishable by the number and character of the pigment granules (Shute and Maryon 1952).

11.8.3 The Cycle in Man

The sporozoites disappear from the bloodstream within 45 min of the bite. Most of them invade the parenchyma cells of the liver, where preerythrocytic schizogony occurs. Time taken from infection until the invasion of the blood by parasites (prepatent period) in different species is given below:

P. falciparum—5.5–7 days.
P. vivax—6–8 days.
P. ovale—9 days.
P. malariae—12–16 days.

During this period, no clinical symptom and no pathological damage take place. The blood remains sterile. In the liver cell, the sporozoite enlarges and becomes rounded or oval (schizont), and the cytoplasm divides forming thousands of merozoites. No pigment is present. The liver cells containing merozoites rupture after 6–16 days. Merozoites reach the circulation and enter the red blood cells. Preerythrocytic schizogony terminates when preerythrocytic merozoites have invaded the red blood cells.

In *P. vivax* and *P. ovale* malaria, some of the schizonts remain dormant in the liver cells. The resting or dormant stages are known as hypnozoites (Krotoski et al. 1982). They awake weeks or months later and cause the relapse (Krotoski et al. 1982). This stage does not exist in *P. falciparum* and *P. malariae*.

11.8.4 Blood Invasion

Preerythrocytic merozoites enter the erythrocytes where they grow at the expense of the host cells. That the merozoite becomes intracellular has been made clear by use of electron microscope. In the red cell, the merozoite is vacuolated (parasitophorous vacuole), ring-shaped, more or less amoeboid, and uninucleate.

It is now called *trophozoite*. It feeds on hemoglobin which it does not completely metabolize. Proteolysis of hemoglobin occurs in the digestive food vacuole of the parasite. Amino acids are released and taken up by the growing parasite for the synthesis of protein. Heme is liberated and oxidized to toxic ferric form. Spontaneous polymerization takes place and an inert crystalline substance, hemozoin, is formed. Thus, the toxicity is avoided by the parasite. Hemozoin is a brown or black insoluble pigment.

The trophozoite grows and transforms into a schizont. Nucleus of mature schizont give birth to erythrocytic merozoites. The red cell ruptures and merozoites flow into the bloodstream. Schizont matures and contains various numbers of merozoites dependent on *Plasmodium* species. For instance, one mature *P. falciparum* schizont contains 8–32 merozoites, with 16 being the usual number, 12–24 merozoites for one mature *P. vivax*, 4–16 for *P. ovale*, and 6–12 for *P. malariae* (Liu et al. 2024). Many are destroyed by the host's forces of immunity, but others invade the red cells in which a new cycle of erythrocytic schizogony takes place.

After two or three erythrocytic generations, the phenomenon of gametogenesis begins. Some of the intracellular merozoites do not become schizonts but develop into male or female *microgametocytes* or *macrogametocytes*, respectively.

11.9 Animal Reservoir

P. vivax, *P. ovale*, and *P. falciparum* have no animal reservoir. *P. malariae* is present in chimpanzees at low levels in Africa. Its immunological characteristics are the same as of human strain; *P. rodhaini* of chimpanzees is actually the synonym of *P. malariae*. Human transmission from chimpanzee should occur, though it has not been documented. The vector of chimpanzee is not known.

P. falciparum and *P. vivax* commonly infect humans, leading to high morbidity and mortality. Only humans were considered the primary host for *Plasmodium* species including *P. falciparum*, *P. vivax*, *P. malariae*, and *P. ovale* before molecular diagnostics development. However, studies over the past two decades reported that the parasites originated from animals. Specifically, *P. falciparum* was traced back to gorillas (Liu et al. 2010) and chimpanzees (Duval et al. 2010; Krief et al. 2010), while *P. vivax* was associated with African apes (Liu et al. 2014). Other sources stated were the origination of *P. malariae* from chimpanzees (Duval et al. 2010) and *P. knowlesi* from monkeys (Knowlesi 1935; Zhang et al. 2016), while *P. ovale* found in humans and chimpanzees had genetic similarities (Duval et al. 2009). Factors, such as habitat loss among primates and human encroachment into forested areas, are believed to have facilitated the transmission of these parasites (Davidson et al. 2019).

11.10 Simian Malaria

Important plasmodia of simian malaria are *P. schwetzi* (of African chimpanzees and gorilla), *P. cylesi* (of Malayan gibbons), *P. cynomolgi* (of oriental monkeys, *Macaca* spp., and *Presbytis*), *P. conomolgibastianelli* (do), *P. simium* (of black monkeys of Brazil, *Alowatta* spp.), *P. inui* (of oriental monkeys, *Macaca* spp.), *P. brasilianum* (of New World monkeys, 16 spp. of monkeys), *P. shortti* (of Indian and Sri Lankan monkeys), *P. knowlesi* (of Oriental monkeys, *Macaca irus*), etc.

Natural infection of *P. cynomolgi*, *P. knowlesi*, and *P. simium* in humans has been reported. The possibility of establishing a virulent simian strain in course of evolution in man in the future through an appropriate vector should be kept in mind.

11.11 Reservoir of Infection

A person who harbors the gametocytes in the peripheral blood serves as the reservoir of infection. (a) There must be both male and female gametocytes. (b) The gametocytes must be matured. Immature gametocytes do not develop further in the mosquito. (c) The gametocytes must remain viable. Drug treatment usually kills the gametocytes of *vivax* and *ovale* malaria, but not that of the *falciparum* malaria. (d) The gametocytes must be sufficient in number to infect a mosquito. To infect a mosquito, peripheral blood must contain at least 1 gametocyte of *P. falciparum* to 200 leucocytes, 1 of *P. vivax* to 1000 leucocytes, and 1 of *P. malariae* to 830 leucocytes.

P. falciparum gametocytes live in the blood for 30–60 days or longer (maximum 120 days). *P. vivax* gametocytes do not live for more than a week (Gebru et al. 2017).

In *P. vivax* infection, gametocytes appear in the peripheral blood (Battle et al. 2019) on the 16th day of infection (first day of fever, i.e., 4 days after the appearance of asexual parasites in the peripheral blood). *P. falciparum* gametocytes in peripheral blood will appear on the 21st day of infection (after 9 days of fever, i.e., about 10 days after the appearance of asexual parasites in peripheral blood). The gametocytes grow fully in 4 days and circulate in the blood for some time; the maximum number is reached about 4–6 days after the peak density of asexual parasites. In some highly endemic areas, the maximum production of gametocytes occurs in children up to fifth year.

Variations of the quality of gametocytes have been observed. Some gametocyte carriers are poor infectors, and others good infectors.

In *P. vivax* and *P. ovale*, relapses may occur for up to 3 years. *P. malariae* infection may persist for 30 years or more.

In the absence of reinfection, recrudescence of *falciparum* malaria usually disappears completely in 6–8 months. A few species of *Anopheles* like *An. gambiae*, *An. Funestus*, etc. transmit filariasis (caused by *Wuchereria bancrofti*) in West Africa and *An. punctulatus*, *An. Koliensis*, etc. in Papua New Guinea.

11.12 *Aedes* Mosquito-Borne Diseases

11.12.1 *Aedes*

The genus *Aedes*, comprising approximately 888 species, includes mosquitoes that may spread some of the deadliest diseases, including dengue, chikungunya, Zika fever, yellow fever, and lymphatic filariasis. *Aedes* mosquitoes are marked in black and white on their bodies and legs. *Aedes* mosquitoes are principally day biters. Two prevalent species of *Aedes* mosquitoes, *Aedes aegypti* (L.) and *Aedes albopictus* (Skuse), carry the bulk of diseases spread by *Aedes* mosquitoes.

Artificial or natural water bodies, such as discarded tires, flower pots, clogged roof gutters, and residential water storage containers, are common *Aedes* mosquito breeding grounds. *Aedes* mosquitoes deposit their eggs in plastic household water containers. *Ae. albopictus*, popularly known as the Skuse, breeds in areas with abundant vegetation and also uses natural habitats, such as tree holes, bamboo holes, etc.

11.12.1.1 Dengue

11.12.1.1.1 Dengue and Dengue Hemorrhagic Fever (DHF)

11.12.1.1.1.1 Geographical Distribution

Dengue occurs in Southeast Asia (types 1–4), the Pacific (types 1–3), East and West Africa (types 1–4), Caribbean (types 1–4), and the Americas (types 1–4). DHF occurs in Southeast Asia, Cuba, the Caribbean, Fiji, Venezuela, and Brazil. It has not been recorded from Africa also (Holmes and Twiddy 2003).

Dengue virus is a member of the flavivirus group (Gubler 1998). It is an RNA virus, 17–25 μm in diameter (Sobhia et al. 2019). There are four serotypes (1, 2, 3, and 4) all of which can be involved in both classical dengue and dengue hemorrhagic fever. In October 2013, Dengue-5 has been reported, which follows the sylvatic cycle (Mustafa et al. 2015). DENV-5 has been circulating among nonhuman primates in the forests of Southeast Asia for centuries before jumping the human barrier. The virus has been maintained in macaques with a spillover into humans, resulting in the Sarawak outbreak. Phylogenetic evaluation revealed that DENV-5 is genetically similar to the other four serotypes, thereby suggesting to a common ancestral origin. There is no indication of the presence of DENV-5 in India. It may be assumed that as human DENV circulates in abundance in India, it may be providing cross-immunity against sylvatic dengue by competitive exclusion.

11.12.1.1.2 Disease Biology

11.12.1.1.2.1 Classical Dengue

Symptoms.

1. Fever up to 105 °C.
2. Headache.
3. Sever back pain around the eyes and joint pain.
4. Skin rash.

Warning symptoms

1. Vomiting.
2. Loose motion.
3. Abdominal pain.
4. Pleural effusion
5. Leucopenia.
6. Thrombocytopenia.
7. Respiratory trouble.

11.12.1.1.2.2 Dengue Hemorrhagic Fever (DHF)

Some patients with dengue fever go on to develop dengue hemorrhagic fever (DHF), a severe and sometimes fatal form of the disease. Around the time the fever begins to subside (usually 3–7 days after symptom onset), the patient may develop warning signs of severe disease. Warning signs include severe abdominal pain, persistent vomiting, marked change in temperature (from fever to hypothermia), hemorrhagic manifestations, or change in mental status (irritability, confusion, or obtundation). The patient also may have early signs of shock, including restlessness, cold clammy skin, rapid weak pulse, and narrowing of the pulse pressure (systolic blood pressure-diastolic blood pressure). Patients with dengue fever should be told to return to the hospital if they develop any of these signs. DHF is currently defined by the following four World Health Organization (WHO) criteria:

1. Fever or recent history of fever lasting 2–7 days.
2. Any hemorrhagic manifestation.
3. Thrombocytopenia (platelet count of <100,000/mm^3).
4. Evidence of increased vascular permeability.

Dengue hemorrhagic fever (DHF) has been classified into three stages:

1. **Febrile stage.** The manifestations of illness begin with high fever (39–41 °C) lasting 2–7 days in most cases, accompanied by flushed face, headache, anorexia, epigastrium pain or tenderness at the right costal margin, enlarged tender liver, vomiting, and myalgia. Hemorrhagic diathesis is manifested by fine petechial hemorrhage or epistaxis and hematemesis, and tourniquet test is positive in most cases.
2. **Critical stage.** After 2–7 days, fever subsides with plasma leakage and hemoconcentration. In severe cases, the patient becomes restless and has cold, clammy skin; pulse is rapid and feeble, with narrowing pulse pressure (<20 mmHg). Prolonged shock is often complicated by severe bleeding.
3. **Convalescent stage.** Fluid is reabsorbed into the circulation. Patients have a good appetite, increased urine output, relative bradycardia accompanied by confluent petechial rash (convalescent rash), and itching.

11.12.1.1.2.3 Grading of the Severity of DHF (Technical guide for diagnosis, treatment, surveillance, prevention and control of dengue haemorrhagic fever 1975) (Table 11.4)

11.12.1.1.2.4 Dengue Shock Syndrome (DSS)

1. Dengue shock syndrome (DSS) (Simmons et al. 2012) is defined as any case that meets the criteria for DHF and has evidence of circulatory failure manifested by
   (a) Rapid, weak pulse and narrow pulse pressure (≤20 mm Hg).
   (b) Hypotension for age, restlessness, and cold, clammy skin. Patients with dengue can rapidly progress into DSS, which, if not treated correctly, can lead to severe complications and death.

Table 11.4 Grading of severity of dengue hemorrhagic fever

DHF		Grade	I	Fever, nonspecific constitutional symptoms. The only hemorrhagic manifestation is a positive tourniquet test
		Grade	II	In addition to above manifestations, there is spontaneous bleeding from the skin and/or other hemorrhages
	DSS	Grade	III	Circulatory failure, rapid and weak pulse, narrowing of pulse pressure (20 mmHg or less), or hypotension with the presence of cold clammy skin and restlessness
		Grade	IV	Profound shock with undetectable blood pressure and pulse

2. Fatality rates among patients with DSS can be 10% or higher but, with early recognition and treatment, can be less than 1%. DHF and DSS can occur in both children and adults.

11.12.1.1.2.5 The Human Phase

Bone marrow (BM) megakaryocytes are known to be highly permissible to DENV replication and myelosuppression; and a reduction in the mass of BM cells is one of the hallmarks of DENV infection (Clark et al. 2012; Noisakran et al. 2012; Sridharan et al. 2013; Vogt et al. 2019). Independent of disease severity, DENV infection causes thrombocytopenia to a lesser or greater extent with severe patients exhibiting acute drop in platelet level along with leakage of fluid from the blood vessels (Martina et al. 2009). Multiple mechanisms have been suggested to contribute to thrombocytopenia including attenuated platelet biogenesis and increased platelet decay through either direct virus interaction or upon binding of antiplatelet antibodies (Fang et al. 2013; de Azeredo et al. 2015; Simon et al. 2015; Ojha et al. 2017, 2019).

11.12.1.1.2.6 The Mosquito Phase

Incubation period in human varies from 4 to 10 days. The most important vector is *Ae. aegypti*, *Ae. albopictus*, and *Ae. scutellaris*. The mosquito becomes infected only during the first 3 days of the patient's illness. Incubation period in mosquito varies from 8 to 11 days.

The virus is found in all the tissues of the infective mosquito. Infection takes place when the mosquito introduces saliva into the skin. Once infected, the mosquito remains infected for its lifetime. Transovarian transmission does take place through which virus gets its passage from infected female mosquito to its eggs through reproductive system and consequently to the next generation mosquito in the absence of human reservoirs (Gubler 1998).

11.12.1.1.3 Epidemiology of Dengue and DHF

Dengue is found mainly in Asia and Africa, although it has been reported from the USA (Philadelphia) as well. The first recorded epidemic of dengue occurred in Egypt in 1779, and in 1780, dengue outbreak was reported from Indonesia and the USA (Schlesinger 1977). In Asia and Africa, all four dengue serotypes are present. In the tropical region, about 1.5 billion people are at risk of infection with dengue (Yacoub and Wills 2014).

Though the transmission occurs throughout the year, the peak transmission season is the rainy season. The disease is transmitted from human to human by the bites of mosquitoes. If the mosquito bites a viremic patient, it can transmit the virus as a result of interrupted feeding or after an extrinsic incubation period of 8–10 days.

In endemic areas, young children become infected with multiple serotypes of dengue (Teo et al. 2009). Adults are usually immune and classical dengue fever is rare among them. In these endemic areas mainly, children may suffer from mild dengue and DHF. Dengue is mainly endemic in metropolitan cities in India. Rural outbreaks have also been demonstrated from Maharashtra and West Bengal (Ilkal et al. 1991; Bhattacharyya et al. 1997).

DHF was first recognized in the Philippines in 1954. It is now widespread in different countries of Asia. More than 3 million people were affected with DHF with about 50,000 deaths between 1956 and 1990 in 12 Asian countries, Cuba, Venezuela, and Pacific Islands. Young children (2–15 years of age) were mostly affected. Peak transmission occurred in the rainy season.

During 1963–1964, about 100,000 people of Calcutta suffered from DHF of which 200 died (Sarkar et al. 1967). Outbreaks were reported from Calcutta and Visakhapatnam (Gupta 1981). It again broke out in Calcutta in 1990, affecting 60 children, of which 12 succumbed (Hati 1991). Delhi was also invaded in 1996, with 500 deaths (Anuradha et al. 1996). This sudden appearance and disappearance here and there in India have not been properly explained.

There is a strong association between DHF and secondary dengue infection. When a second infection occurs with a dengue virus, which differs from the first one in serotype, activation of complement and clotting cascade ultimately produces DHF (Halstead 1981). Two other theories are virulent strain of dengue virus (Rosen 1977) and genetic difference in the hosts (Guzman et al. 1990). The theory of secondary infection and presence of virulent strain are widely accepted.

11.12.1.1.4 Reservoir

Humans seem to be the principal reservoir host. In Northern Queensland, Australia, flying foxes (*Pteropus*) and migratory shore birds act as reservoirs. In Malaysia and Africa, monkey acts as the jungle reservoir (Pavri 1964).

11.12.1.2 Chikungunya

This belongs to group A arbovirus (CHIK) (alphavirus). The virus was isolated in Africa in 1952–1953. Chikungunya is a local word; the root verb *kungunyala* means "drying up" or contorted. This fever occurs in many parts of Africa and Asia (Weaver and Lecuit 2015). The incubation period is 2–4 days. The clinical symptoms are characterized by high fever and severe pain in the trunk and joints. Usually the disease does not appear to be dangerous. The virus was first isolated in India during the outbreak of dengue hemorrhagic fever in Calcutta in 1963 (Wang et al. 2000). The epidemic continued up to the early part of 1965. Although the virus was isolated in Calcutta in 1963, examination of sera collected in 1960 from Calcutta revealed antibodies to chikungunya virus, indicating that this virus or an antigenically related virus was active in this area many years before the outbreak of hemorrhagic fever

(Burt et al. 2012). Subsequently the virus was detected in Madras and Vellore in South India (Rao et al. 1965; Myers et al. 1965; Carey et al. 1969). In one case, both dengue and chikungunya viruses were found (Myers and Carey 1967; Carey et al. 1969). In previous years, Asia and the USA were the areas most affected by chikungunya. 2019 had 62,000 instances in India, while a prolonged outbreak in Pakistan resulted in 8387 cases. In the Caribbean and the USA, 185,000 cases were reported.

The disease is biphasic. After 1–4 days, the fever subsides. Then there is an afebrile period of 3 days. The fever then returns. A maculopapular rash appears on the trunk and extensor surfaces of the limbs which is different from dengue rash. After another 3–6 days, complete recovery occurs with remission of the fever, but second phase may be sometimes complicated. In some parts of Asia, mild hemorrhagic feature has been observed, but no shock.

Vectors: *Aedes aegypti*, *Ae. africanus*, *Cx. quinquefasciatus*, *Mansonia* spp. (Africa), *Ae. aegypti*, *Cx. quinquefasciatus*, *Cx. tritaeniorhynchus*, and *Cx. gelidus* (Thailand). *Ae. aegypti* is the well-known vector in India. *Ae. albopictus* also can act as an experimental vector. The incubation period in mosquito is about 14 days.

11.12.1.2.1 Disease Biology (Table 11.5)

11.12.1.3 Zika Fever

The flavivirus known as the Zika virus (ZIKV) is the source of Zika fever in the western hemisphere. In monkeys, it was initially found in Uganda in the 1940s. There have been reports of Zika virus epidemics in Asia, the Pacific, Africa, and the USA (Musso et al. 2014). To date, 86 countries and territories have reported discovering proof of mosquito-borne Zika virus infection (Zika Virus Symptoms, Diagnosis, & Treatment 2016). Only three indigenous Zika cases have been reported from Bapunagar area, Gujarat, India (Zika Virus Infection 2017). ZIKV transmission is comparable to that of the two most deadly arbovirus illnesses,

Table 11.5 Disease biology of chikungunya

Symptoms	Diagnostic methods	Treatment
Symptoms of chikungunya include severe joint pain and fever. Chikungunya-related joint pain can range in severity and duration from mild to severe. Additional symptoms include rash, fatigue, headaches, nausea, joint swelling, and muscle discomfort	As part of serological testing for the illness, ELISA is used to confirm IgM and IgG anti-chikungunya antibodies	While fatalities from the virus are rare, its symptoms can be severe and debilitating. Joint pain may persist for months, although most individuals recover from infections in a week or less. One year after their infection was treated, 20% of individuals still have chronic joint pain. There are no specific drugs that can be used to treat chikungunya. To put it plainly, experts recommend getting plenty of rest and water. Certain drugs, such as acetaminophen, ibuprofen, and naproxen, can help lower temperature and stiffness in the joints

dengue and chikungunya. The Zika virus is carried by the *Aedes* mosquito species, which has the potential to infect humans by biting. *Aedes africanus*, a mosquito that lives in trees, and mosquitoes with wide distributions in the tropics and subtropics, *Aedes aegypti* (L.) and *Aedes albopictus* (Skuse), respectively, were among the mosquito species from which ZIKV was identified in Africa and Asia. Additionally, the Zika virus can be passed from mother to fetus during pregnancy through sexual contact, blood and blood product transfusions, and organ transplants.

11.12.1.3.1 Disease Biology (Table 11.6)

11.12.1.4 Yellow Fever

11.12.1.4.1 Geographical Distribution

It mainly occurs in Africa and South America. In Ethiopia during 1960–1962 and in Senegal during 1965–1966, many people died of yellow fever (Staples and Monath 2008). In Trinidad, in 1978–1979, there were 18 cases and 8 deaths. Many tourists visiting West Africa contracted yellow fever and succumbed.

11.12.1.4.2 Clinical Features

In some, the infection is inapparent. Immunity develops in them. Incubation period in humans is 5–6 days.

In clinically recognizable form, the onset is sudden, fever rises up to 39.4–40 °C, and chill, rigor, headache, body ache, nausea, and vomiting are present. The patient is dehydrated. The tongue is coated. Jaundice appears. Temperature comes down during the next few days. In mild cases, recovery takes place. In severe cases, the liver and kidney are grossly damaged, there is albuminuria, jaundice deepens, and bleeding occurs from the eyes, nose, mouth, bladder, etc. There is shock. Meningitis or encephalitis may occur. Death is due to shock, uraemia, hemorrhage, or myocarditis. Death may occur as early as the third day. Overall mortality rate is 5–10% or more. The virus (Group B arbovirus—*Flavivirus fabricius*) circulates in the blood up to 107 h after the onset or possibly more.

Table 11.6 Disease biology of Zika

Symptoms	Diagnostic methods	Treatment
Many of the infected individuals donot show any symptoms of the Zika virus at all. Most of the symptoms, such as fever, rash, conjunctivitis, discomfort in the joints and muscles, malaise, and headache, frequently last between 2 and 7 days. Infection in pregnant women may give birth to microcephalic offspring	When symptoms start to show up less than 7 days later, those with viral RNA or DNA are tested for using macromolecule testing (NAT). When interpreting negative NAT findings, care should be taken since, 7 days after symptoms appear, viraemia rapidly declines and remains unnoticed	Individuals with the infection should take paracetamol for pain and fever, drink plenty of water, and rest a lot. Using aspirin or other nonsteroidal anti-inflammatory drugs (NSAIDs) increases a person's risk of bleeding if they are not under medical care. Nevertheless, there is no vaccine to guard against the infection

11.12.1.4.3 Epidemiology

In 1973 in South America, 207 people were attacked out of which 146 died, and in Africa, out of 8 patients, 5 died. Limited outbreaks occur in forest areas (Delaporte 1991; Mutebi and Barrett 2002).

It is not present in Asia, though potential vectors, especially *Aedes aegypti*, abound. If it could be introduced in Asia, the result would have been disastrous (McNeill 2002). To prevent its introduction, the strictest control of immigrants, aircrafts, and ships is maintained.

Yellow fever is primarily a sylvatic disease. The virus is found in certain jungle monkeys and other reservoir hosts. The infection was detected in chimpanzees in Congo. In eastern Columbia, marsupials probably serve as important mammalian hosts. In South America, day-biting *Haemagogus spegazzinii*, *H. leucocelaneus*, and *Sabethes chloropterus* are vectors, and monkey → mosquito → monkey cycle is going on. Virus has been detected in *A. fulvus* in Brazil. In West Africa, the situation is alike; the mosquitoes concerned are night-biting *Aedes*, *Culex*, *Ochlerotatus*, *Sabethinus*, and *Anopheles*. If a man enters the jungle, he becomes accidentally infected. He comes back home and, if *A. aegypti* mosquitoes are present, may start the urban yellow fever with man → mosquito → man cycle.

In East Africa, the situation is different. Man rarely acquires the infection in the jungle. In the jungle, the enzootic cycle is going on. The mosquito involved is *A. africanus*. The infected monkeys invade human habitations for cultivated food, such as bananas. *A. simpsoni* and possibly other semidomestic day-biting mosquitoes (*A. vittatus*, *A. metallicus*, *A. taylori*) are infected and they, in turn, infect humans. The virus remains in the mosquito during its whole life and can be recovered for a period of several hours after its death. Mosquito here, in the broad sense, acts as the true reservoir.

Urban yellow fever begins with man → mosquito → man cycle, where *A. aegypti* persists. Urban yellow fever is epidemic yellow fever. Man becomes infected through mosquito bite; with its droplets of saliva, the virus is introduced. Occasionally human infection may occur by "swatting" an infected mosquito on the skin. All ages and both sexes are susceptible. Hunters, woodcutters, etc. may contract yellow fever and bring the infection from the forest and initiate the urban cycle.

Incubation period The virus is available to the mosquito 3–6 days of incubation period prior to the onset of symptoms and during the first 3–4 days of illness. The mosquito becomes infective within 10–12 days of taking the viremic blood. It can retain the infection throughout its life (maximum period recorded 168 days after infective meal). Incubation period in man is 3–6 days.

No transovarian transmission takes place. The larvae exposed to liquid containing the virus develop into infective adults.

Table 11.7 Disease biology of yellow fever

Symptoms	Diagnostic methods	Treatment
The majority of those infected with yellow fever have extremely minor symptoms, if any at all. Yellow fever takes 3–6 days to incubate, and symptoms appear 3–6 days after a mosquito bite. Humans are not capable of spreading the virus. By biting, only infected mosquitoes may spread the illness to people. Common signs of yellow fever include fever, sluggish heartbeat, jaundice, congestion in the face, albuminuria, and bleeding or hemorrhaging	Different yellow fever diagnostic techniques are used depending on the illness stage. PCR may even be able to identify the virus in the blood or urine in its early stages	Yellow fever has no particular therapy at this time. The goal of the supportive treatment is to reduce the disease's symptoms, such as pain and fever. Depending on how severe the illness is, supportive measures should be taken, such as the following: 1. Administering oxygen 2. Administering fluids intravenously to treat dehydration 3. In situations where there is a circulatory collapse, use medicine to raise blood pressure 4. Blood product transfusions in cases of severe bleeding 5. Antibiotics for bacterial infections that come back later 6. Dialysis in cases of renal failure 7. When a patient has respiratory failure, mechanical ventilation is used 8. Due to the higher risk, stay away from aspirin and nonsteroidal anti-inflammatory drug (NSAID) medications

11.12.1.4.4 Disease Biology (Table 11.7)

11.13 *Culex* and *Culex* Mosquito-Borne Diseases

11.13.1 *Culex*

The genus *Culex* has a large number of species of mosquitoes that can spread serious illnesses to people, animals, and birds, including avian malaria, filariasis, Japanese encephalitis, the West Nile virus, etc. There are 715 species in the genus, 12 of which are vectors. These mosquitoes are the most prevalent type found in several major American cities and are found across the planet, with the exception of the most northern regions of the temperate zone. Adult *Culex* mosquitoes are brown to dark brown in color, and they lack the *Aedes* mosquito's distinguishing white band on their abdomen. The maxillary palps' length is less than half that of the female *Culex*'s proboscis.

The way that *Culex* mosquitoes deposit their eggs is another important feature that sets them apart from other mosquito species. *Culex* females lay their eggs in a cluster, known as an egg raft.

11.13.1.1 Japanese Encephalitis (JE)

The causative organism is a group B arbovirus. "Virus of encephalitis of Japan" was proposed by Itsuma Takagi, and "type B" was added to differentiate it from encephalitis lethargica of von Economo (type A) (Kaneko and Aoki 1928; Takagi 1925).

11.13.1.1.1 Geographical Distribution

It occurs in countries like Russia (Siberia), India, Japan, Guam, the Philippines, and Indonesia.

11.13.1.1.2 Clinical Features

Incubation period is 4–14 days. Encephalitis develops with sudden onset of severe headache, nausea, vomiting, and high fever. Cerebral and meningeal involvement occur. There is retracted head, arms and knees bent, and shoulders pressed toward the chest, which are very characteristic of the disease (Campbell et al. 2011). Coma and convulsion develop in fatal cases. Death occurs within 10 days.

11.13.1.1.3 Epidemiology

In India, during late 1955, six cases of JE were encountered in Madras State. The first record of large-scale epidemic of Japanese encephalitis in India was in June 1973 affecting the State of West Bengal, covering 1000 square miles where about 700 clinical cases occurred with an overall mortality of 40–50% (Gupta et al. 2010). JE virus was isolated from two human brain tissues collected postmortem and from mosquito pools of *Cx. vishnui*, *An. hyrcanus*, and *An. barbirostris* (Chakraborty et al. 1974).

The epidemic of encephalitis in West Bengal died down around November 1973. The incidence of this disease reappeared in 1976, 1977, and 1978, indicating the activity of JE. Since 1978, epidemics of JE occurred in many parts of India invading more than 30 states and union territories, causing colossal calamity. Year wise attack and death due to JE in India and West Bengal are given in Table 11.8.

In Japan, JE virus is the only group B arbovirus present, whereas in India, dengue, another group B arbovirus, sharing its antigen with JE virus is fairly common. It is expected that the presence of group B antibody in the sera of high percentage of population in this region, although mostly due to dengue virus infection, will have some impact on the epidemiology of the disease.

The vectors are *Cx. tritaeniorhynchus*, *Cx. pipiens pallens*, *Cx. gelidus*, *Cx. fuscocephalus*, *Cx. pseudovishnui*, *Cx. quinquefasciatus = (Cx. P. fatigans)*, *Cx. bitaeniorhynchus*, *Cx. vishnui*, *Ae. japonicus*, *Ae. togoi*, *An. hyrcanus*, and *Ae. barbirostris* (Hati 1976). It is considered that *Cx. tritaeniorhynchus* is responsible for transmitting the disease among man during the epidemics, but other vectors are also involved elsewhere. In West Bengal, the virus was isolated thrice from *Cx. vishnui* (Hati 1990). So, this is an important vector in West Bengal.

In North Asia and Japan, *Cx. tritaeniorhynchus* acts as a main vector. *Cx. annulirostris* and *Cx. gelidus* are vectors in Guam and Malaysia, respectively. Virus Research Centre, Pune, isolated four strains of JE from mosquitoes at Bankura, two

Table 11.8 Disease biology of Japanese encephalitis

Symptoms	Diagnostic methods	Treatment
The majority of JE patients either have moderate symptoms like headache and fever or no symptoms at all. Patients with severe infections may experience paralysis, tremors, convulsions (particularly in youngsters), headache, high temperature, stiff neck, altered mental status, and coma	Several test kinds are employed to identify the JE pathogen: • **Hematologic tests.** To identify the antibodies in the blood • **Lumbar puncture.** It is also used to examine the spinal fluid and CSF for the presence of antibodies • **Scan.** A scan is necessary if brain encephalitis is suspected • **Computerized tomography (CT) scan.** To provide a clear picture of the inside of the body for encephalitis patients; a series of X-rays of the body taken at slightly various angles are combined	While there are currently no known therapies for JE, hospitalization is usually necessary for intensive monitoring and supportive care. The illness is treated symptomatically. Some symptoms may be relieved by rest, drinks, and the use of medications and painkillers to lower fever

from *An. hyrcanus* and one each from *Cx. epidesmus* and *Cx. bitaeniorhynchus* (Banerjee et al. 1979).

JE virus multiplies in the body of the mosquito vector. Up till now, there is no evidence that mosquitoes die owing to virus infection. It is about 8 days on an average for the virus to develop an infective stage in the mosquito. Mosquito, once infected, remains so for the rest of its life. The question of transovarian transmission appears to have remained unsettled. It is interesting to get the virus from three species of mosquitoes in West Bengal within a very short span of time, which may be correlated to the peak period of the epidemic. It should be remembered that the virus of JE is maintained in mosquitoes and hosts other than man, the latter becoming accidentally involved. *An. barbirostris* and *An. hyrcanus* are mainly zoophilic, and they can maintain pig-mosquito-pig cycle (Hati 1976). The virus is maintained in a cycle between mosquitoes and vertebrate hosts, primarily pigs and wading birds (also referred to as amplifying hosts or natural reservoirs). Humans are incidental or dead-end hosts, because they usually do not develop high enough concentrations of JE virus in their bloodstreams to infect feeding mosquitoes. JE virus transmission occurs primarily in rural agricultural areas, often associated with rice production and flooding irrigation. In some areas of Asia, these conditions can occur near urban centers. In temperate areas of Asia, JE virus transmission is seasonal. Human disease usually peaks in the summer and fall. In the subtropics and tropics, transmission can occur year-round, often with a peak during the rainy season.

11.13.1.1.4 Reservoirs

The virus has been isolated from pigs, night herons, egrets, etc. In these animals, the virus increases and these animals serve to infect more of vectors. Hence, they are known as amplifiers. Horse infection may occur at the time of pig infection, which has economic significance, but it is not significant epidemiologically, because horses experience a low and transient blood titer.

11.13.1.1.5 Disease Biology (Table 11.8)

11.13.1.2 West Nile Virus

In the West Nile area of Uganda, a woman was the first person to contract the West Nile virus (WNV) in 1937. It was discovered in the Nile delta region in 1953 in birds such as Columbiformes and crows. WNV was not thought to be harmful to birds before to 1997, but at that time, a more virulent strain was found in Israel that killed a variety of bird species and showed symptoms of encephalitis and paralysis (Carson et al. 2006). It is a member of the family Flaviviridae, which is related to the flavivirus genus and the Japanese encephalitis antigenic complex. Typically, WNV is found throughout West Asia, Africa, Europe, the Middle East, North America, and India (Khatun and Chatterjee 2017). Virus activity has been reported in southern, central, and western India. WNV has been isolated in India from *Culex vishnui* mosquitoes in Andhra Pradesh and Tamil Nadu, from *Cx. quinquefasciatus* mosquitoes in Maharashtra, and from humans in Karnataka State (Paramasivan et al. 2003). Japanese encephalitis virus (JEV) has caused sporadic epidemics in Assam since 1976. Studies conducted during 2000–2002 in Assam showed that 187 (53.7%) of 348 persons with acute encephalitis syndrome were infected with JEV. The primary vectors of WNV are widely thought to be *Culex* mosquitoes, particularly *Cx. pipiens*. The virus spreads from birds to mosquitoes. Both horses and humans serve as inadvertent hosts. Most often, bites from infected mosquitoes result in human infection. After feeding on infected birds, mosquitoes absorb the virus (Jozan et al. 2003), which stays in their bodies for a few days and spreads throughout their blood. Eventually, the virus was discovered inside the salivary glands of the mosquito. The virus may potentially infect people and other animals through mosquito bites, where it may multiply and perhaps result in disease.

11.13.1.2.1 Disease Biology (Table 11.9)

11.13.1.3 Filariases

Filariasis is a morbid condition. The nematode worms, which cause filariasis in humans, are (1) *Wuchereria bancrofti*, (2) *Brugia malayi*, and (3) *Brugia timori*. The adult worms live in the lymphatics and produce live embryos (microfilariae), which invade the bloodstream. The embryos do not develop further in the blood.

11.13.1.3.1 *Wuchereria bancrofti*

(Bancroftian filariasis)

(Superfamily Filarioidea, genus *Wuchereria*)

W. bancrofti occurs in two forms: periodic and subperiodic. In periodic form, more microfilariae are found in the peripheral blood by night (peak hours 10 p.m. to 2 a.m.). The subperiodic form microfilariae are found in the blood for the whole

Table 11.9 Disease biology of West Nile virus

Symptoms	Diagnostic methods	Treatment
Twenty percent or more of those infected with West Nile fever will have symptoms such as fever, headache, lethargy, aches and pains throughout the body, nausea, vomiting, skin rash, and enlarged lymph nodes	Different techniques are employed to identify West Nile virus. By analyzing two specimens that were taken 1 week apart, ELISA is used to determine IgG antibody seroconversion. A different diagnostic method is IgM antibody capture ELISA	Since West Nile fever is caused by a virus, there is no known treatment for it. However, we may use painkillers like aspirin or ibuprofen to treat West Nile virus symptoms including headaches and muscular aches. A doctor should be seen if there are any significant symptoms, such as brain swelling, or other severe symptoms. It is recommended to use intravenous fluids and medicines to reduce the risk of infection

24 h, more during daytime in diurnally subperiodic and more during nighttime in nocturnally subperiodic form.

In Asia, (a) periodic (nocturnal) form is prevalent, which is carried chiefly by the mosquito, *Culex quinquefasciatus*, and (b) a diurnal subperiodic form is found in the South Pacific and in the Andaman and Nicobar Islands. In the South Pacific, the vector is *Aedes polynesiensis* and other species of *Aedes*; (c) subperiodic (nocturnal) form has been reported from Thailand, where the vector is *Aedes harinasutai*, a mosquito belonging to *Aedes niveus* (nocturnal group); and (d) there is also a rural form in Malaysia and Indonesia, transmitted by anopheline and other mosquitoes.

In India, periodic nocturnal form occurs.

11.13.1.3.2 Disease Biology

11.13.1.3.2.1 Life Cycle

1. ***In man (definitive host).***

   Adult worms lodge in the lymphatic system of humans (lower limbs, groin, upper limbs, etc.). Life span is 18 years or more. The adult is long, white, and threadlike (so it is called filaria), both ends are tapering, and the anterior part ends with a small rounded swelling. The male measures 2.5–4 cm × 0.1 mm; tail end is curved. Female measures 8–10 cm × 0.3 mm; tail end is narrow and pointed.

   The two ovaries are placed posteriorly. From these, two oviducts originate and end in two paired uteri, which ultimately fuse. The vagina is short, which opens at the vulva, 0.8 mm from the anterior end. Females are ovoviviparous, i.e., they lay well-developed embryonated eggs, known as microfilariae, measuring 250 and 6–7 μm (Fig. 11.1). Each microfilaria remains inside a sheath. The sheath is formed by the prolongation of the egg membrane. They are found to invade peripheral circulation at night.

   The sheath is hyaline in nature, much longer (350 μm) than the larva. The larva moves forward and backward within the sheath. A microfilaria can live for 3–6 months. The anterior end of microfilaria is rounded. Its posterior end is pointed.

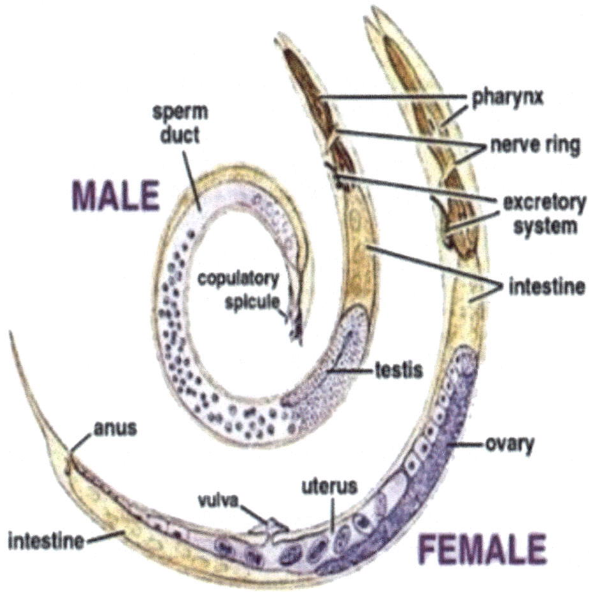

Fig. 11.1 Male and female *Wuchereria bancrofti*. (Courtesy: Vandana Kumari)

The cuticle of the larva is composed of subcuticular cells. The central axis of the body extends from the head to the tail and is filled in with the granules, known as somatic cells or nuclei. There is no nucleus at the extreme anterior end, and the area is known as cephalic space. The tail-up is also devoid of any granules. By breaking up at different places, they form (a) nerve ring, (b) anterior V spot (rudimentary excretory system), (c) oral pore, (d) G cells (Fig. 11.2) (genital cells G 2-3 and 4 anterior to anal pore and G-1 is situated further in front), and Innenkorper of Fulleborn or central body of Manson (rudimentary alimentary canal) extending from anterior V spot to G-1.

2. ***In mosquitoes (intermediate host).***

A mosquito imbibes microfilariae with its blood meal. In its stomach, the microfilariae shed their sheaths within 2–4 h. They penetrate the proventriculus and cardiac portion of the stomach apparently with the anterior hooks and reach the thoracic muscle within 12 h. Microfilariae further develop inside the thorax of the mosquito. They do not undergo multiplication (i.e., one microfilaria gives rise to one third-stage larva only). During the next 2 days, each microfilaria metamorphoses into a short sausage-like structure, measuring 124–250 μm × 10–17 μm. This is the first stage larva, bearing a rudimentary digestive tract. In 3–7 days' time, the larva grows, sheds its cuticle, and measures 220–330 μm × 15–30 μm (the second-stage larva). The tail has atrophied to a mere stump. The digestive tract has become well differentiated. Between the 10th and the 11th day, a second ecdysis takes place; this infective third-stage filiform elongated worm measures 1.4–2 mm × 18–23 μm (Fig. 11.3).

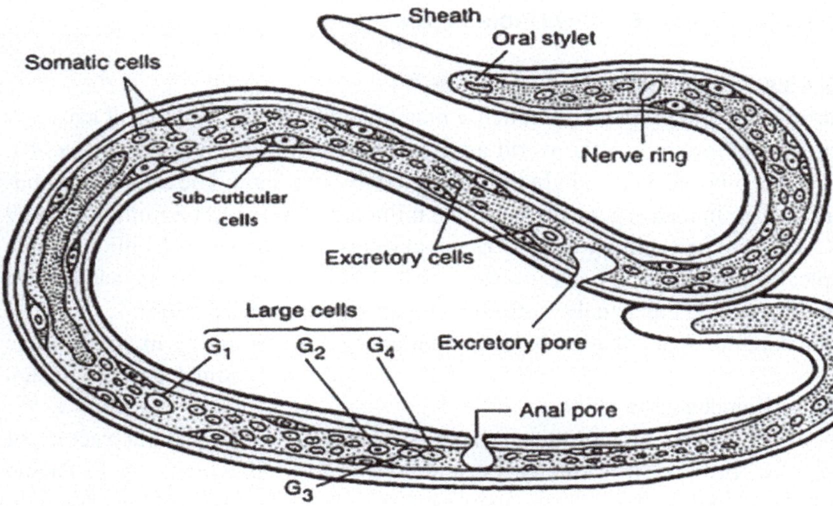

Fig. 11.2 G spots of *Wuchereria* larvae. (Courtesy: Anita Kumari)

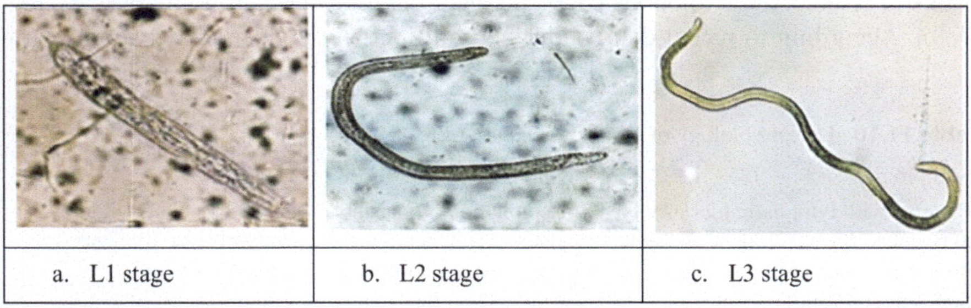

Fig. 11.3 Different stages of larvae in mosquito. (**a**) L1 stage, (**b**) L2 stage, (**c**) L3 stage

The infective larvae migrate toward the proboscis. During the next feed on humans, the mosquito bends the proboscis, and the infective larvae escape through the extreme tip of the labium. It was observed that under condition of massive accumulation of larvae, rupture of the wall of the labium might take place .

The third-stage infective larvae, deposited on the human skin, enter through the puncture wound. Some workers believe that these larvae may penetrate through the skin on their own. For the complete development of microfilaria in the mosquito, 10–12 days are required. This depends on temperature, humidity, and species of mosquito. There will be three equal bubble-like caudal papillae in the infective larva of *W. bancrofti*. In *B. malayi*, all three papillae are poorly developed. Time taken from the entry of the third-stage infective larvae into the skin until the microfilariae first appear in the peripheral blood (i.e., the biological incubation period) is 1 year (minimum recorded time is 82 days). Females begin to liberate microfilariae, which invade the bloodstream. The cycle is thus repeated.

11.13.1.3.3 Disease Biology (Table 11.10)

11.13.1.3.4 Epidemiology

Filariasis is a world problem of immense magnitude. It is estimated that at least 250 million people throughout the world are infected with *W. bancrofti* and *B. malayi*. As a mosquito-borne disease in India, it is next only to malaria. The estimated population at risk in India has gone up from 69.2 million in 1960 to 121.8 million in 1969 (WHO 1974) to 286 million in 1991 (Annual Report 1991). About 1 billion people in tropical and subtropical countries are at risk of *W. bancrofti* infection, and at least 200 million people are actually suffering from the disease (Sasa 1968).

Ramchandran states that in India, the number of people living in areas where *W. bancrofti* is endemic is 120 million and about 5 million is estimated to be affected clinically (Ramchandran 1970).

About 80 species of mosquitoes have been reported to be natural vectors of *W. bancrofti*. In India, it is chiefly transmitted by *Cx. quinquefasciatus*. In Pacific Islands, *Aedes polynesiensis* is the main vector. In rural East Africa, the important vectors are *Anopheles gambiae* A and B and *A. funestus*.

In India, bancroftian filariasis is distributed mainly along the coastal parts and banks of rivers. Autochthonous cases are also seen in Rajasthan, Punjab, U.P., and Delhi. According to recent data, about 420 million people in India are living with

Table 11.10 Disease biology of *Wuchereria bancrofti*

Symptoms	Diagnostic methods	Treatment
Patients with lymphatic filariasis experience fevers, chills, shaking, body pains, and enlarged lymph nodes in addition to acute inflammation of the lymphatic vessels (lymphangitis). In the afflicted regions (arms or legs), excessive fluid buildup may occur (edema); however, this accumulation usually goes away once the other symptoms disappear. Acute genital inflammation caused by this virus most commonly affects men and causes discomfort and swelling in the testes (orchitis), sperm tract (funiculitis), and sperm ducts (epididymitis). There might be discomfort and unusual swelling in the scrotum called hydrocele. Extreme case may cause elephantiasis of any or many body parts	Lymphatic filariasis not always ends in clinical symptoms; if someone is infected, one can be detected by also biopsy. In most parts of the Earth, the parasites have a "nocturnal periodicity" that restricts their appearance within the blood to only from 10 p.m. to 2 a.m. Therefore, between 10 p.m. and 2 a.m. blood extraction for laboratory examination when microfilaria is present in the peripheral blood is trusted time	Preventive chemotherapy for lymphatic filariasis consists of yearly administration of a combination dosage of two medications to all individuals at risk: 400 mg of albendazole with 150–200 μg/kg of ivermectin or 6 mg/kg of diethylcarbamazine citrate (DEC). Ideally, MDA and albendazole (400 mg) should be administered twice a year to stop the spread of lymphatic filariasis

the risk of filaria infection, and about 44 million people are actually affected by filariasis (Training Manual 1949).

New cases of filariasis are observed throughout the year, but the most favorable period of transmission is during the monsoon. Lymphatic filariasis is now found in Australia, North America, and Japan. In China, the prevalence is greatly decreased due to effective control programs.

The zoonotic aspect is not important. *Wuchereria* type microfilaria has however been reported from the Potto, *Perodicticus potto*. Experimentally almost complete development was obtained in macaques.

11.13.1.3.4.1 *Brugia malayi*

(Superfamily Filarioidea, genus *Brugia*)

There are two strains of *B. malayi*: (a) nocturnally periodic, found in India, Malaysia, etc., reported only from man, transmitted mainly by *Mansonia* spp., wherein in some areas, *Anopheles* and *Aedes* mosquitoes may act as vectors, and (b) nocturnally subperiodic found in the Philippines. The nematode is also commonly found in wild monkeys, and the vectors are *Mansonia* spp.

11.13.1.3.5 Distribution in India

The infection has been reported from Orissa, Madhya Pradesh, Tamil Nadu, Kerala, West Bengal, and Assam. It is rural in distribution. According to a study carried out some years ago, the rate of *Brugian filariasis* in the rural Kerala was estimated to be between 3.5% and 13.4% of the population (Technical Report 1984).

11.13.1.3.6 Vectors in India

Ma. annulifera, *Ma. indiana*, *Ma. uniformis*, and *An. barbirostris* are well-known vectors of *Brugian filariasis*.

11.13.1.3.7 Timor Microfilaria

This species is found in Timor Island, Indonesia. In stained films, they resemble *B. malayi* in some respects but differ in overall length, in cephalic space ratio, and in failure of the sheaths to stain with Giemsa. It is nocturnally periodic, not found in cats, dogs, and monkeys. Adult worms have not yet been discovered.

This is known as *Brugia timori* (Fig. 11.4), and *Anopheles barbirostris* has been implicated as its vector (Table 11.11).

Other diseases transmitted by different species of mosquitoes are eastern equine encephalitis (EEE), western equine encephalitis (WEE), Venezuelan equine encephalitis (VEE), Rift Valley fever (RVF), O'nyong-nyong virus (ONNV), Saint Louis encephalitis, Ťahyňa virus (TAHV), Ilheus virus, Murray Valley encephalitis virus, Ganjam virus (GANV), Itaqui virus, etc.

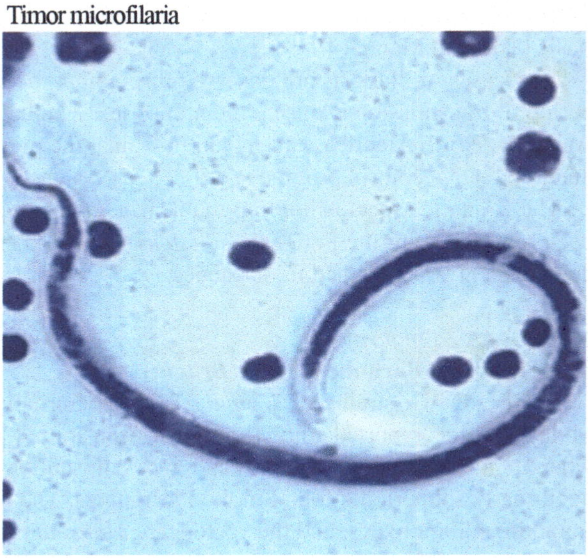

Fig. 11.4 *Brugia timori.* (Courtesy: Maria Rebollo Polo)

Table 11.11 Differentiating features between microfilaria of *W. bancrofti* and *B. malayi*

W. bancrofti	*B. malayi*
1. Measurement: 290 μm × 6–7 μm	230 μm × 6 μm
2. The curve: smooth	Secondary kinks are present
3. Stylet one at the anterior end	Presence of double stylets
4. The cephalic space: as long as broad The body is studded with somatic nuclei, the distribution of which has got species specificity	The cephalic space is longer
5. Nuclei discrete	Blurred
6. The tail tip is free from nuclei	Two distinct terminal nuclei are present

11.14 Conclusion

Mosquitoes are regarded as one of the world's most dangerous creatures due to their capacity to transmit a variety of deadly diseases. Each year, around 700 million people get sick from diseases spread by mosquitoes, which causes over 1 million fatalities. A rising number of scientists are concerned about the recent sharp increase in the rate of infection spread, since they believe that global warming may cause the global prevalence of mosquito-borne diseases to explode. The Zika virus, West Nile virus, chikungunya virus infections, dengue fever, and malaria are some of the most prevalent illnesses that are carried by mosquito bites and are disseminated globally. For the most part, there is no medication or vaccination to prevent these illnesses.

Points to Remember

1. Mosquito-borne diseases are caused by the bite of an infected mosquito.
2. The transmission of diseases by viruses, such as dengue, West Nile and other arboviruses, yellow fever, Zika, and chikungunya, as well as parasites, like malaria, is known as mosquito-borne disease transmission.
3. *Aedes* mosquitoes are the primary vector for dengue, which is the most common virus infection globally.
4. The majority of viral infections do not cause any symptoms or only cause moderate or transient disease; yet, serious or protracted illness can happen, and certain viruses can even be deadly.
5. The etiology of a mosquito-borne disease may be inferred from many factors such as mosquito exposure, living in or visiting endemic regions, travel dates and destinations, pathogen incubation periods, and vaccination history.
6. It's critical to identify potentially fatal illnesses as soon as possible.
7. Supportive measures are part of the treatment.
8. There's no antiviral medication that works.
9. The danger of infections carried by mosquitoes can be reduced by taking precautions against them.
10. Furthermore, vaccinations against Japanese encephalitis [JE], dengue, and yellow fever are efficient means of preventing illness.

References

Annual Report (1991) National Malaria Eradication Programme, Delhi, p 56

Anuradha S, Singh NP, Rizvi SN, Agarwal SK, Gur R, Mathur MD (1996) The 1996 outbreak of dengue hemorrhagic fever in Delhi, India. Southeast Asian J Trop Med Public Health 29(3):503–506

Banerjee K, Mahadev PV, Ilkal MA, Mishra AC, Dhanda V, Modi GB, Geevarghese G, Kaul HN, Shetty PS, George PJ (1979) Isolation of Japanese encephalitis virus from mosquitoes collected in Bankura District (West Bengal) during October 1974 to December 1975. Indian J Med Res 69:201–205

Battle KE, Lucas TC, Nguyen M, Howes RE, Nandi AK, Twohig KA et al (2019) Mapping the global endemicity and clinical burden of Plasmodium vivax, 2000-17: a spatial and temporal modelling study. Lancet 394(10195):332–343

Bhattacharyya N, Neogi DK, Hati AK et al (1997) An outbreak of dengue fever in a rural area of West Bengal. Ind J Med Microbiol 15:139–141

Burt FJ, Rolph MS, Rulli NE, Mahalingam S, Heise MT (2012) Chikungunya: a re-emerging virus. Lancet 379(9816):662–671

Campbell GL, Hills SL, Fischer M, Jacobson JA, Hoke CH, Hombach JM et al (2011) Estimated global incidence of Japanese encephalitis: a systematic review. Bull World Health Organ 89(10):766–774

Carey DM et al (1969) The 1964 chikungunya epidemic at Vellore, South India, including observations of concurrent dengue. Trans R Soc Trop Med Hyg 63:434

Carson PJ, Konewko P, Wold KS et al (2006) Long-term clinical and neuropsychological outcomes of West Nile virus infection. Clin Infect Dis 43(6):723–730

Chakraborty SK, Sarkar JK, Chakravarty MK, Mukherjee MK, Mukherjee KK, Das BC, Hati AK (1974) The first epidemic of Japanese encephalitis studied in India-virological studies. Ind J Med Res 63:77

Chandra G, Bhattacharjee I, Chatterjee SN (2010) A review on *Anopheles subpictus* Grassi—a biological vector. Acta Trop 115(1–2):142–154

Chatterjee SN, Chandra G (2000) Role of *Anopheles subpictus* as a primary vector of malaria in an area in India. Jpn J Trop Med Hyg 28(3):177–181

Clark KB, Noisakran S, Onlamoon N, Hsiao HM, Roback J, Villinger F et al (2012) Multiploid CD61+ cells are the pre-dominant cell lineage infected during acute dengue virus infection in bone marrow. PLoS One 7:e52902. https://doi.org/10.1371/journal.pone.0052902

Davidson G, Chua T, Cook A et al (2019) Defining the ecological and evolutionary drivers of *Plasmodium knowlesi* transmission within a multi-scale framework. Malar J 18:66

de Azeredo EL, Monteiro RQ, de-Oliveira Pinto LM (2015) Thrombocytopenia in dengue: interrelationship between virus and the imbalance between coagulation and fibrinolysis and inflammatory mediators. Mediat Inflamm 2015:313842. https://doi.org/10.1155/2015/313842

Delaporte F (1991) The history of yellow fever: an essay on the birth of tropical medicine. MIT Press, Cambridge, pp 89–90

Duval L, Nerrienet E, Rousset D et al (2009) Chimpanzee malaria parasites related to *Plasmodium ovale* in Africa. PLoS One 4:e5520

Duval L, Fourment M, Nerrienet E et al (2010) African apes as reservoirs of *Plasmodium falciparum* and the origin and diversification of the Laverania subgenus. Proc Natl Acad Sci 107(23):10561–10566

Fang S, Wu Y, Wu N, Zhang J, An J (2013) Recent advances in DENV receptors. Sci World J 2013:684690. https://doi.org/10.1155/2013/684690

Gebru T, Lalremruata A, Kremsner PG, Mordmüller B, Held J (2017) Life-span of in vitro differentiated *Plasmodium falciparum* gametocytes. Malar J 16:330. https://doi.org/10.1186/s12936-017-1986-6

Gubler DJ (1998) Dengue and dengue hemorrhagic fever. Clin Microbiol Rev 11(3):480–496

Gupta NP (1981) Dengue haemorrhagic fever. Bull Cal Sch Trop Med 29:101–106

Gupta N, Lomash V, Rao PVL (2010) Expression profile of Japanese encephalitis virus induced neuroinflammation and its implication in disease severity. J Clin Virol 49(1):04–10

Guzman MG et al (1990) Dengue haemorrhagic fever in Cuba, 1981; a retrospective seroepidemiologic study. Am J Trop Med Hyg 42:179–184

Halstead SB (1981) The Alexander D. Langmuir Lecture. The pathogenesis of dengue. Am J Trop Med Hyg 114:632–648

Han YS, Thompson J, Kafatos FC, Barillas-Mury C (2000) Molecular interaction between *Anopheles stephensi* midgut cells and *Plasmodium berghei*: the time bomb theory of ookinete invasion of mosquitoes. EMBO J 19:6030–6040. https://doi.org/10.1093/emboj/19.22.6030

Hati AK (1976) Vectors of Japanese encephalitis. Indian J Public Health 20:149

Hati AK (1990) Epidemiology of Japanese encephalitis in West Bengal. Bull Cal STM 38:38–41

Hati AK (1991) Dengue and dengue haemorrhagic fever. Commun Health 2:10–11

Holmes EC, Twiddy SS (2003) The origin, emergence and evolutionary genetics of dengue virus. Infect Genet Evol 3:19–28

Ilkal MA et al (1991) Entomological investigations during outbreaks of dengue fever in certain villages of Maharashtra State. Ind J Med Res A 93:174–178

Jozan M, Evans R, McLean R, Hall R, Tangredi B, Reed L et al (2003) Detection of West Nile virus infection in birds in the United States by blocking ELISA and immunohistochemistry. Vector Borne Zoonotic Dis 3(3):99–110

Kaneko R, Aoki Y (1928) Ergebnisse der Encephalitis epidemica in Japan. Erg Inn Med Und Der Kinderheilkd 34:342–456

Khatun T, Chatterjee S (2017) Emergence of West Nile Virus in West Bengal, India: a new report. Trans R Soc Trop Med Hyg 111(4):178–184

Knowlesi R (1935) Monkey malaria. Br Med J 3907:1020

Krief S, Escalante A, Pacheco M et al (2010) On the diversity of malaria parasites in African apes and the origin of *Plasmodium falciparum* from Bonobos. PloS Pathog 6:e1000765

Krotoski WA, Garnham PCC et al (1982) Observation on early and late post sporozoite tissue stages in primary malaria 1. Discovery of a new latent from of Plasmodium cynomolgi (the hypnozoite) and failure to detect hepatic forms within the first 24 hours after infection. Am J Trop Hyg 31:24–35

Lal S, Dhillon GPS, Sonal GS, Rao R, Sita B (1998) Country scenario malaria and its control in India. National Malaria Eradication Programme (Enhanced Malaria Control Project), vol 110054. Directorate General of Health Services, Ministry of Health and Family Welfare, Delhi, p 98

Li ZY et al (1981) Experimental infection of A. balabacensis with P. falciparum in vitro. Chinese J Microbiol Immunol 1:362

Liu W, Li Y, Learn GH et al (2010) Origin of the human parasite *Plasmodium falciparum* in gorillas. Nature 467(7314):420–425

Liu W, Li Y, Shaw KS et al (2014) African origin of the malaria parasite *Plasmodium* vivax. Nat Commun 5:3346

Liu T, Zhu F, Tan N, Chen S, Xu W (2024) *Plasmodium* (Chapter 151). In: Molecular medical microbiology, 3rd edn. Academic Press, New York, pp 3005–3029

Martina BE, Koraka P, Osterhaus AD (2009) Dengue virus pathogenesis: an integrated view. Clin Microbiol Rev 22:564–581. https://doi.org/10.1128/CMR.00035-09

McNeill JR (2002) Yellow fever and geopolitics: environment, epidemics, and the struggles for empire in the American tropics, 1650-1900. Hist Now 8(2):10–16

Musso D, Nhan T, Robin E, Roche C, Bierlaire D, Zisou K (2014) Potential for Zika virus transmission through blood transfusion demonstrated during an outbreak in French Polynesia, November 2013 to February 2014. Euro Surveill 1(14):20761

Mustafa MS, Rasotgi V, Jain S, Gupta V (2015) Discovery of fifth serotype of dengue virus (DENV-5): a new public health dilemma in dengue control. Med J Arm Forces India 71(1):67–70

Mutebi JP, Barrett AD (2002) The epidemiology of yellow fever in Africa. Microbes Infect 4(14):1459–1468

Myers RM, Carey DM (1967) Concurrent isolation from a patient of two arboviruses, chikungunya and dengue type II. Science 157:1307–1308

Myers RM et al (1965) The 1964 epidemic of dengue-like fever in South India, isolation of chikungunya virus from human sera and from mosquitoes. Ibid 53:425

National Framework for Malaria Elimination in India, NVBDCP, 2016-2030 (n.d.)

Noisakran S, Onlamoon N, Hsiao HM, Clark KB, Villinger F, Ansari AA et al (2012) Infection of bone marrow cells by dengue virus *in vivo*. Exp Hematol 40:250–259.e4. https://doi.org/10.1016/j.exphem.2011.11.011

Ojha A, Nandi D, Batra H, Singhal R, Annarapu GK, Bhattacharyya S et al (2017) Platelet activation determines the severity of thrombocytopenia in dengue infection. Sci Rep 7:41697. https://doi.org/10.1038/srep41697

Ojha A, Bhasym A, Mukherjee S, Annarapu GK, Bhakuni T, Akbar I et al (2019) Platelet factor 4 promotes rapid replication and propagation of Dengue and Japanese encephalitis viruses. EBioMedicine 39:332–347. https://doi.org/10.1016/j.ebiom.2018.11.049

Paramasivan R, Mishra AC, Mourya DT (2003) West Nile virus: the Indian scenario. Indian J Med Res 118:101–108

Pavri KM (1964) Presence of chikungunya antibodies in human sera collected from Calcutta and Jamshedpur before 1963. Ind J Med Res 52:698

Ramchandran CP (1970) Guide to methods and techniques in filariasis investigation. Bulletin No. 15. Institute for Medical Research, Kuala Lumpur

Rao TR (1984) The *Anopheles* of India. Malaria Research Centre (ICMR), Delhi

Rao et al (1965) Preliminary isolation of chikungunya virus of dengue-like illness in Madras city. Ind J Med Res 53:689

Rosen L (1977) The Emperors New Clothes revisited, or reflections on the pathogenesis of dengue haemorrhagic fever. Am J Trop Med Hyg 23:337–343

Sarkar JK, Chatterjee SN, Chakrobarty SK (1967) Three year study of mosquito-borne haemorrhagic fever in Calcutta. Trans R Soc Trop Med Hyg 61:725–735

Sasa M (1968) Review 5th international congress on tropical medicine and malaria, p 106
Schlesinger RW (1977) Dengue viruses. Springer, New York, pp 90–91
Shute PG, Maryon M (1952) A study of human malaria oocysts as an aid to species diagnosis. Trans R Soc Trop Med Hyg 46:275
Simmons CP, Farrar MD, Chau NVV, Wills B (2012) Dengue. N Engl J Med 366:1423–1432
Simon AY, Sutherland MR, Pryzdial EL (2015) Dengue virus binding and replication by platelets. Blood 126:378–385. https://doi.org/10.1182/blood-2014-09-598029
Sobhia ME, Ghosh K, Singh A et al (2019) A multi-perspective review on dengue research. Curr Drug Targets (Review) 20(15):1550–1562
Sridharan A, Chen Q, Tang KF, Ooi EE, Hibberd ML, Chen J (2013) Inhibition of megakaryocyte development in the bone marrow underlies dengue virus-induced thrombocytopenia in humanized mice. J Virol 87:11648–11658. https://doi.org/10.1128/JVI.01156-13
Staples JE, Monath TP (2008) Yellow fever: 100 years of discovery. J Am Med Assoc 300(8):960–962
Subbarao SK, Vasantha K, Raghuvendra K, Sharma VP, Sharma GK (1988) *Anopheles culicifacies*: sibling species composition and its relationship with incidence. J Am Mosq Control Assoc 4:29–33
Takagi I (1925) On the etiology of encephalitis that occurred epidemically in Japan in 1924. Jpn Med World 5:147. (In Japanese)
Technical guide for diagnosis, treatment, surveillance, prevention and control of dengue haemorrhagic fever (1975) Technical Advisor Committee on DHF for the SE Asian and Eastern Pacific Regions. WHO
Technical Report Series (1984) Lymphatic filariasis. No. 702. WHO, Geneva
Teo D, Ng LC, Lam S (2009) Is dengue a threat to the blood supply. Transfus Med 19(2):66–77
Training Course, National Health Programmes for Chief Medical Officers/District Officers (1949) Background document, vol I. Department of Community Health Administration and Family Welfare, Munirka
Vogt MB, Lahon A, Arya RP, Spencer Clinton JL, Rico-Hesse R (2019) Dengue viruses infect human megakaryocytes, with probable clinical consequences. PLoS Negl Trop Dis 13:e0007837. https://doi.org/10.1371/journal.pntd.0007837
Wang E, Ni H, Xu R, Barrett AD, Watowich SJ, Gubler DJ et al (2000) Evolutionary relationships of endemic/epidemic and sylvatic dengue viruses. J Virol 74:3227–3234
Weaver SC, Lecuit M (2015) Chikungunya virus and the global spread of a mosquito-borne disease. N Engl J Med 372(13):1231–1239
WHO Tech Rep Ser (1974) No. 542
World Health Organization (2011) Current states of Malaria
Yacoub S, Wills B (2014) Predicting outcome from dengue. BMC Med 12(1):147
Zhang X, Kadir KA, Quintanilla-Zariñan LF et al (2016) Distribution and prevalence of malaria parasites among long-tailed macaques (*Macaca fascicularis*) in regional populations across Southeast Asia. Malar J 15(1):450
Zika Virus Infection - India (2017) Disease outbreak news. http://www.who.int/csr/don/26-may-2017-zika-ind/en/. Accessed 2 Jun 2017
Zika Virus Symptoms, Diagnosis, & Treatment (2016) Centers for Disease Control and Prevention, Atlanta. Accessed 4 Mar 2016

Mosquito-Borne Veterinary Diseases

12

Ankita Sarkar, Paramita Banerjee,
and Abhijit Mazumdar

Abstract

Mosquito-borne diseases (MBDs) significantly threaten human and animal health. The causative pathogens, i.e., viruses, bacteria, protozoans, and nematodes, of these diseases are vectored by different mosquito species, mainly, *Anopheles*, *Culex*, and *Aedes*. These soft-bodied minute dipterans are implicated in the transmission of some deadly viruses, namely, dengue, chikungunya, yellow fever, and Zika virus. Besides that, protozoan and filarial parasites also cause some other essential vector-borne diseases, such as malaria and lymphatic filariasis in animals and humans. Tropical and subtropical regions around the world are significant targets of mosquito-borne diseases. The Global Vector Control Response (GVCR) 2017–2030 has suggested to develop strategic management plan for the vector-borne diseases and their outbreaks with immediate effect. Here, it attempts to give an overview of the vector species, causes, and symptoms of various medico-veterinary vital diseases transmitted by these mosquitoes. A general understanding can address the critical issues regarding significant vector-borne diseases.

Keywords

Mosquito · Vector · Disease · Virus · Protozoa · Animals

A. Sarkar · P. Banerjee · A. Mazumdar (✉)
Entomology Research Unit, Department of Zoology, The University of Burdwan,
Burdwan, West Bengal, India

© The Author(s), under exclusive license to Springer Nature Singapore Pte Ltd. 2024
Omkar (ed.), *Mosquitoes*, https://doi.org/10.1007/978-981-97-4163-2_12

Learning Objectives

1. This chapter addresses the zoonotic transmission of deadly pathogens and parasites by three important genera of mosquitoes (*Anopheles*, *Culex*, and *Aedes*) to animals.
2. The host range of these mosquitoes and sylvatic transmission of various important veterinary pathogens (viruses, bacteria, protozoa, fungi, and nematodes) have been discussed.
3. The disease epidemiology and symptoms of various mosquito-borne diseases (MBDs) have been elaborated.
4. Control strategies of these vectors have been discussed to some extent.

12.1 Introduction

Mosquitoes (Diptera: Culicidae) are nematoceran flies belonging to the Infraorder Culicomorpha and superfamily Culicoidea, consisting of 50 genera, more or less 42 subgenera, and more than 3541 species (Tyagi et al. 2015). Family Culicidae is categorized into three subfamilies, i.e. Anophelinae, Culicinae, and Toxorhynchitinae (Mullen and Durden 2018). They are ubiquitously distributed throughout the tropical and temperate regions, at 3500 to 1250 m above sea level, except Antarctica and a few islands. They not only pose the deadliest threat to humans but also to several animals. They achieve their prime importance as they serve as vectors of various pathogens of public health and medico-veterinary important fatal diseases. Their worldwide geographic expansion is associated with an increasing frequency of outbreaks of these diseases.

12.2 Mosquito Vectors and Disease Epidemiology

Among 3541 species, 404 mosquitoes were recorded from India (Tyagi et al. 2015). About 100 are recognized as vectors of pathogens of humans and other vertebrates (Pavela 2015). These species belong to three main genera of Culicidae, namely, *Anopheles*, *Aedes*, and *Culex* (Pohlit et al. 2011). Bites of these mosquitoes transmit pathogens of fatal diseases, such as malaria, dengue, yellow fever, West Nile, and filariasis (Sanfelice 2022). These diseases cause high morbidity and mortality, mainly in developing countries where they represent a significant economic burden. These diseases are transmitted by the above three most abundant and common medico-veterinary important genera of mosquitoes.

12.2.1 *Anopheles*

They are known as marsh mosquitoes; more than 500 species have been recorded worldwide. Over 30 species of it led to public health problems (Escobar et al. 2020).

These mosquitoes are responsible for malaria transmission (Dieme et al. 2017); in 2022, the global tally of malaria cases reached 249 million based on the World Malaria Report (2023). This genus does not only contain the vectors of human-pathogenic plasmodia, but there is evidence that they may transmit other veterinary-linked pathogens, e.g., viruses and nematodes. *Anopheles* mosquitoes also transmit the nematode *Wuchereria bancrofti*, the causing agent of filariasis in the tropics (Manguin et al. 2010); in addition, the primary vectors of O'nyong-nyong virus (ONNV) are *Anopheles gambiae* and *An. funestus*, which causes fever and polyarthritis in Africa (Brault et al. 2004). Generally, anophelines exhibit a wide range of host preferences, including humans, cattle, sheep, horses, pigs, dogs, cats, birds, and even reptiles (Bruce-Chwatt et al. 1966; Gunathilaka et al. 2016). According to Adugna et al. (2021), blood meal analysis reports suggested that the hosts of this mosquito included not only humans but also bovines, sheep, donkeys, mules, chickens, and dogs. *Anopheles stephensi* successfully feed on the hemolymph of larvae of two common moth species, *Manduca sexta* and *Heliothis subflexa* (George et al. 2014). Six *Anopheles* species namely, *Anopheles baimaii*, *An. culicifacies*, *An. fluviatilis*, *An. minimus*, *An. stephensi*, and *An. sundaicus*, are implicated as primary vectors transmitting malaria in different ecogeographical regions of India (Subbarao et al. 2019). The most prominent species is *An. gambiae*, well-known for vectoring malarial plasmodia in Africa; *An. freeborni* is also known as vector in North America, while *An. albimanus*, *An. pseudopunctipennis*, *An. darlingi*, *An. vestitipennis* and *An. punctimacula* are implicated as vectors in Mesoamerica (Escobar et al. 2020).

12.2.2 *Culex*

These are known as the house mosquito and recently have 817 valid species divided into 28 subgenera (Reis et al. 2023). They have a cosmopolitan distribution, being most abundant in countries with a tropical climate. *Culex quinquefasciatus* showed a greater preference for a blood meal in galliform birds, such as roosters, hens, and turkeys, as well as passerines, such as songbirds, pintails, and sparrows, and columbiform birds, such as doves and turtle doves. Besides these, blood meal analysis reports also detected positive *Culex* spp. for snakes, alligators, monitor lizards, and crocodile (Mendoza-Roldan et al. 2021). These feeding patterns are related to the transmission cycles of several arboviruses, as these vertebrates act as amplifiers and as a source of blood meals for invertebrates, which in turn act as vectors and reservoirs, transmitting the virus to other susceptible hosts. This genus contains several species that are vectors of West Nile virus, chikungunya virus, Zika virus, dengue virus (DENV2, 3, 4), Oropouche virus, St. Louis encephalitis, Rift Valley fever, and lymphatic filariasis transmitted by *Culex pipiens*, *Cx. tarsalis*, *Cx. mimeticus*, *Cx. quinquefasciatus*, *Cx. coronator*, *Cx. declarator*, *Cx. spissipes*, and *Cx. portesi* to human (Reis et al. 2023). Besides humans, *Culex* species transmit other pathogens including dog heartworm, West Nile virus, and eastern equine encephalitis to animals.

12.2.3 *Aedes*

The genus *Aedes* (tiger mosquito) has more than 950 species; some species cause serious biting nuisances and transmit vector-borne diseases to humans and other animals (Rogers 2023). The distribution of different species of these mosquitoes is temperate and tropical habitats worldwide. Some species have spread beyond their native ranges by human activity or changing environmental influences (Ryan et al. 2019). *Aedes aegypti* and *Ae. albopictus* have been widely distributed. Both species spread infectious diseases, like dengue, chikungunya, and Zika fever (Kraemer et al. 2015). Therefore, the crucial arboviruses transmitted by *Ae. aegypti* and *Ae. albopictus* are chikungunya virus (CHIKV), equine encephalitis virus, dengue virus (DENV), yellow fever virus (YFV), Zika virus (ZIKAV), Japanese encephalitis virus (JEV), West Nile virus (WNV), Kunjin virus, and Rift Valley fever (Murphy et al. 2012; Kaavya et al. 2022). Recently, it was reported that these vector species also transmit many microorganisms, such as bacteria, viruses, and parasites (Gao et al. 2020; Huang et al. 2020; Villegas et al. 2018). It is found in suburban and rural areas' habitats with a wide range of hosts, including livestock, humans, amphibians, reptiles, and birds (Paupy et al. 2009; Kraemer et al. 2015). *Aedes albopictus* and *Ae. aegypti* are reportedly fed on squamata reptiles and associated with the zoonotic Zika virus transmission (Mendoza-Roldan et al. 2021).

12.3 Pathogen Transmission of Medico-Veterinary Importance

Mosquito-borne diseases are threats even in the twenty-first century and thus have become a focal point for researchers and the world community. The World Health Organization recorded that vector-borne diseases alone contribute to 17% of all infectious diseases combined and are responsible for 700,000 deaths worldwide annually. Many dengue and chikungunya viruses occur in many Asian countries, and more than 3.9 billion people in over 129 countries are at risk of contacting dengue, with an estimated 96 million symptomatic cases and an estimated 40,000 deaths every year. Several other virus species were reported from mosquitoes back in the twentieth century, and an estimated 100 of those virus species were able to cause infection (Gould et al. 2017).

12.3.1 Virus

12.3.1.1 Chikungunya Virus (CHIKV)

This is a positive-sense, single-stranded RNA virus belonging to the family Togaviridae and the genus *Alphavirus*. The first isolation occurred in 1954 in Tanzania (Vu et al. 2017). This virus was reported in Africa and spread over Asia (Powers and Logue 2007). Its enzootic transmission cycle mainly occurs between *Aedes* mosquitoes as the primary vectors and nonhuman primates as reservoirs and amplification hosts (Reis et al. 2023). In Africa, this virus affects nonhuman primates, i.e., vervet monkeys and baboons by *Aedes africanus*, *Ae. opok*, and *Ae. taylori*. Sylvatic CHIKV has been recorded and this virus has been isolated from African green monkeys (*Chlorocebus sabaeus*), patas monkeys (*Erythrocebus patas*), guinea baboons (*Papio papio*), guenons (*Cercopithecus aethiops*), and bushbaby (*Galago senegalensis*) in Senegal; mandrills (*Mandrillus sphinx*) in Gabon; red-tail monkeys (*Cercopithecus ascanius schmidti*) in Uganda; and African green monkeys (*Cercopithecus aethiops*) and Chacma baboons (*Papio ursinus*) in South Africa and Zimbabwe (McIntosh 1970; McCrae et al. 1971; Diallo et al. 1999; Chevillon et al. 2008; Pruetz et al. 2010; Kading et al. 2013). *Aedes aegypti* and *Ae. albopictus* is the vector of this virus that causes arthralgic illness of fever, rashes, and joint pains (Wimalasiri-Yapa et al. 2019). Detection of this virus was also done from pools of *Culex quinquefasciatus* by da Silva Neves et al. (2022).

12.3.1.2 Dengue Virus (DENV)

This single-stranded RNA virus causes this infection and belongs to the family Flaviviridae and genus *Flavivirus*. *Aedes aegypti* has long been recognized as a vector of this virus, causing major fever epidemics in the America and Southeast Asia (Jansen and Beebe 2010; Stanaway et al. 2016). All four serotypes have been isolated from field-collected *Ae. aegypti* (Gratz 2004). According to Brady et al. (2012), at least 128 countries, including 36 previously declared dengue-free countries with an estimated population of 3.97 billion, are at risk of a dengue outbreak. Its spread in some countries, such as India, Pakistan, and Bangladesh, is associated with urbanization and population growth (Monath 2007). Dengue fever was first reported in Kolkata (India) (Karamchandani 1946). The symptoms of this disease are fever, myalgia, arthralgia, rash, nausea, vomiting, and bleeding (Jelinek et al. 1997). Recently, Thongyuan and Kittayapong (2017) and Gwee et al. (2021) detected dengue positivity in pigs, nonhuman primates, marsupials, birds, bats, horses, bovids (cattle, sheep, goats), rodents, domestic dogs, and other small animals (shrew, sloths, acouchi, agouti, porcupine, armadillo, and kangaroo). They suggested that animals may play a role in both the sylvatic and urban endemic cycles as potential secondary hosts.

12.3.1.3 Eastern Equine Encephalitis Virus (EEEV)

Eastern equine encephalitis virus (EEEV) is a non-segmented RNA arbovirus (family Togaviridae, genus *Alphavirus*) responsible for several epidemics and epizootics in horses and various birds (Komar and Spielman 1994). This virus was first isolated in 1931 in Delaware (USA) from the brains of horses exhibiting symptoms of incoordination, fever, hyperirritability, convulsions, and eventually death (Hanson 1957; Sah et al. 2023). Based on the CDC, the total number of fatal cases of this virus between the period of 2011 and 2020 was 47, giving a mortality rate of 43%. Its natural cycle includes passerine birds as the primary amplification hosts, and mainly by the *Culiseta melanura*, ornithophilic mosquitoes serve as vectors (Komar and Spielman 1994). According to Armstrong and Andreadis (2010), other mosquito species include *Aedes vexans*, *Coquillettidia perturbans*, *Ochlerotatus canadensis*, and *Oc. sollicitans* that feed opportunistically on mammals that have been incriminated as bridge vectors between viremic birds, humans, and horses. Besides this, Oliver et al. (2018) also detected this virus in other mosquitoes, such as *Anopheles punctipennis*, *An. quadrimaculatus*, *Psorophora ferox*, *Culex salinarius*, and *Cx. pipiens-restuans* groups.

12.3.1.4 Oropouche Virus (OROV)

This is a negative-sense, single-stranded RNA arbovirus member of the Simbu group of orthobunyaviruses. This virus was first isolated in 1955 in Trinidad from a human blood sample (Pinheiro and Travassos 1981). The sylvatic cycle involves sloths, monkeys, and *Coquillettidia venezuelensis*. The virus was first isolated in 1960 from the blood of a sloth, *Bradypus tridactylus*, captured in a forest area of Brazil in 1960 and also from a pool of *Ochlerotatus serratus* mosquitoes caught near the same area (Pinheiro et al. 2004). Gibrail et al. (2016) detected seropositive nonhuman primates (*Cebus libidinosus*, *Alouatta caraya*) for this virus. However, McGregor et al. (2021) detected this virus from *Culex quinquefasciatus* and *Cx. tarsalis*, and de Mendonça et al. (2021) also identified this virus from two anthropophilic and urban species of mosquitoes, *Aedes aegypti* and *Ae. albopictus*. This disease is not fatal, but patients suffer from abrupt onset of fever, headache, anorexia, arthralgias, and dizziness.

12.3.1.5 Rift Valley Fever (RVF)

Aedes aegypti and *Ae. albopictus* transmit this single-stranded RNA virus (RVFV) (Moutailler et al. 2008). This virus affects many animals, including humans, and has severely impacted public health and the economy, especially in African countries (Brustolin et al. 2017).

12.3.1.6 St. Louis Encephalitis Virus (SLEV)

This is a member of the Japanese encephalitis virus complex and is a flavivirus, a single-stranded positive-sense RNA virus. Its first isolation in Brazil occurred in 1960 from a mosquito pool of *Sabethes belisarioi* (Heinen et al. 2015). *Culex* spp. particularly *Cx. pipiens*, *Cx. quinquefasciatus*, *Cx. tarsalis*, and *Cx. nigripalpus* are the primary vectors, and wild birds (passerine and columbiform) serve as amplifying hosts (Reisen 2003). The birds, including pigeons, bluejays, robins, and house sparrows, are in urban and rural environments. The virus was isolated from *Cx. declarator*, *Cx. coronator*, *Cx. spissipes*, and *Cx. quinquefasciatus* (Pinheiro 1982; Heinen et al. 2015; Reis et al. 2023). Most cases are asymptomatic, and some have flulike symptoms, such as fatigue, headaches, nausea, vomiting, body aches, encephalitis of the brain, dizziness, agitation, confusion, tremors, or coma.

12.3.1.7 West Nile Virus (WNV)

West Nile virus (WNV) is a zoonotic flavivirus transmitted by mosquito vectors (Rappole and Hubálek 2003). This *Flavivirus* affects the nervous system and was first isolated in 1937 from a human sample taken in the West Nile district of Uganda. It causes disease in horses, humans, and several species of birds. Besides that, Marra et al. (2004) showed that this virus infects at least 225 species of birds and other animals, including horses, cattle, llamas, alligators, cats, dogs, wolves, and sheep. Mosquitoes of the *Culex* genus are mainly responsible for transmitting the virus, and wild birds are the primary amplification hosts. The virus was isolated from a pool of *Culex quinquefasciatus* and exhibited a significantly high susceptibility to WNV infection. Hubálek and Halouzka (1999) isolated this virus from *Mimomyia* spp. from Africa. According to Boreham et al. (1975), these mosquitoes occasionally feed on humans and other mammals. These mosquitoes have a preference for feeding on amphibians, as *Mimomyia elegans* landing on an adult Australian tree frog, *Litoria fallax*, and subsequently feeding on their nostrils (Van Beurden 1980; Gould and Valdez 2024). Besides that, *Aedes* species feed on the blood of birds, and mammals can transmit WNV to them, including humans (Veronesi et al. 2018).

12.3.1.8 Yellow Fever Virus (YFV)

This virus belongs to the Flaviviridae and is a single-stranded positive-sense RNA (Beasley et al. 2015). Yellow fever virus causes massive mortality in New World monkeys and around 30% in humans (Amarela 2017). This virus is transmitted by mosquitoes, such as *Aedes aegypti* and *Ae. albopictus* (Aitken et al. 1979). The symptoms of infection of this virus lead to bleeding, shock, liver function decay, and jaundice in infected individuals. In severe cases, acute hepatic failure occurs rapidly (Staples et al. 2010). The transmission of the virus involves an urban cycle among humans and nonhuman primates by *Aedes* spp. (Butler 2016).

12.3.1.9 Zika Virus

This *Flavivirus* has a single-stranded positive-sense RNA genome (Brown et al. 2016). Zika virus, named after a forest region in Uganda (Africa), was first detected in a monkey in Central Africa. The Zika virus is transmitted in the tropical and subtropical regions by *Aedes aegypti*. Earlier studies also isolated this virus from *Anopheles africanus*, *An. coustani*, *An. gambiae*, *Aedes albopictus*, *Ae. africanus*, *Ae. aegypti*, *Ae. dalzieli*, *Ae. fowleri*, *Ae. furcifer*, *Ae. luteocephalus*, *Ae. vittatus*, *Ae. neoafricanus*, *Ae. metallicus*, and *Ae. minutus* (Althouse et al. 2015). This virus was detected in wild orangutans, semi-captive orangutans, ducks, goats, cows, horses, bats, rats, carabaos (water buffalo), and various monkeys (Olson et al. 1983; Vorou 2016). Since this virus invaded Europe, sporadic cases of Zika virus infections have occurred in Southeast Asian countries including Thailand, Cambodia, Indonesia, Malaysia, and the Philippines (Song et al. 2017). Human cases occurred in countries like Australia, Italy, Japan, Norway, etc. (Song et al. 2017; Possas 2016; Ali et al. 2017; Waddell and Greig 2016). At the beginning of the year 2015, many cases of Zika virus infections were detected in Brazil which increased in numbers up to 1.3 million cases, including a vast number of at least 4000 instances of formations of microcephalus in newborn babies (Hennessey et al. 2016). Ali et al. (2017) documented that typical symptoms are rash, headache, muscle pain, inflammation of the eyes, and fever.

12.3.1.10 Japanese Encephalitis Virus (JEV)

This is a single-stranded RNA flavivirus, and the first clinical case of JEV was recorded in 1871 in Japan, involving >6000 cases (Erlanger et al. 2009). The detection of this virus is in *Culex pipiens* and birds from Europe (Platonov et al. 2012; Ravanini et al. 2012; Zeller 2012). This zoonotic pathogen is mainly found in East and Southeast Asia (Auerswald et al. 2021). Pigs and birds, particularly those belonging to the family Ardeidae (e.g., cattle egrets, pond herons), are natural hosts (Ladreyt et al. 2019). This virus infects a wide array of animals, including equines, birds, dogs, bats, and snakes, and these serve as dead-end hosts (Mansfield et al. 2017). *Culex tritaeniorhynchus* and *Cx. gelidus* are primary vectors, while *Ae. japonicus*, *Cx. fuscocephala*, and *Cx. vishnui* are secondary vectors (Oliveira et al. 2018; Auerswald et al. 2021). Besides, Auerswald et al. (2021) enlisted the data regarding the isolation and detection of this virus from the following field-caught mosquitoes: *Anopheles subpictus*, *An. sinensis*, *An. vagus*, *An. barbirostris*, *An. hyrcanus*, *An. pallidus*, *An. annularis*, *Culex epidesmus*, *Cx. infula*, *Cx. whitmorei*, *Cx. rubithoracis*, *Aedes butleri*, *Ae. curtipes*, *Ae. lineatopennis*, and *Ae. vexans nocturnus* reported from India, Sri Lanka, China, Taiwan, Malaysia, and Indonesia. Symptomatic patients develop a high fever, headache, disorientation, coma, tremors, cerebral inflammation, movement disorders, neurologic deficits, and seizures.

12.3.2 Bacteria

Francisella tularensis, the causative agent of tularemia disease, was reported from mosquitoes, i.e., *Aedes* and *Culex* spp. (Abdellahoum et al. 2020). The clinical manifestations included skin lesions in the legs combined with inguinal lymphadenopathy or lesions in arms, face, or neck with axillary or cervical lymphadenopathy, fever, and body aches. *Aeromonas hydrophila*, a pathogenic bacteria, was identified from *Anopheles gambiae* (Lindh et al. 2005). Many other pathogenic bacteria are identified from the salivary glands and gut of mosquito species. Some of the bacteria associated with the different organs of *Aedes* are *Acetobacter*, *Burkholderia*, *Cupriavidus*, *Elizabethkingia*, *Pantoea*, *Serratia*, and *Sphingomonas* in the salivary gland; *Asaia*, *Bacillus*, *Chryseobacterium*, *Chromobacterium*, *Enterobacter*, *Enterococcus*, *Klebsiella*, *Kluyvera*, *Leucobacter*, *Pantoea*, *Pseudomonas*, *Serratia*, and *Sphingomonas* in the midgut; and *Pseudomonas*, *Acinetobacter*, *Cupriavidus*, *Ochrobacterium*, *Stenotrophomonas*, and *Wolbachia* in the reproductive organs (Mancini et al. 2018; Díaz et al. 2021).

12.3.3 Protozoa

12.3.3.1 Malaria

Malaria is one of the most important mosquito-borne diseases. According to the WHO, more than 228 million cases occurred worldwide in 2018. Yanmanee et al. (2023) recorded that the salivary glands of *Anopheles latens* harbored *P. knowlesi*,

P. inui, *P. fieldi*, *P. coatneyi*, and *P. hylobati*, and an unnamed *Plasmodium* species is known to infect both long-tailed and pig-tailed macaques, and the salivary glands of *An. introlatus* possessed *P. cynomolgi*, *P. inui*, and *P. hylobati*, with the coexistence of *P. knowlesi* and *P. inui*. An avian malaria parasite, *P. juxtanucleare*, has been identified in the salivary glands of *An. latens*. Three other mosquitoes, *An. latens*, *An. introlatus*, and *An. macarthuri*, were recognized as avian malaria species. The salivary gland sample of *An. maculatus* contained *P. caprae*, an ungulate malaria parasite known to infect domestic goats.

Besides that, earlier studies reported avian haemosporida in *Culex*, *Anopheles*, *Aedes*, and *Psorophora* (Fryxell et al. 2014; Carlson et al. 2015; Noden et al. 2021), while only one study identified cervid haemosporida in *An. punctipennis* (Martinsen et al. 2016).

12.3.4 Fungi

Four fungal species were identified from the gut of *Ae. aegypti* and *Ae. albopictus* in culture-dependent assay, like *Meyerozyma guilliermondii*, *Rhodotorula glutinis*, *Sporobolomyces roseus*, and *Candida parapsilosis* (Bozic et al. 2017). *Candida parapsilosis* and *Meyerozyma guilliermondii* became opportunistic human digestive tract pathogens. These fungi were isolated from bloodstream of immunocompromised infected patients in many countries of the Asia-Pacific region (Singh and Parija 2012; Tan et al. 2016). *Candida parapsilosis* was also reported from nonhuman sources, like domestic animals (Santin et al. 2013) and *Culex pipiens* (Steyn et al. 2016).

12.3.5 Nematode

12.3.5.1 Lymphatic Filariasis

Another significant tropical disease is lymphatic filariasis, also known as elephantiasis, which is mainly caused by the nematode *Wuchereria bancrofti* and in turn transmitted by the bites of *Culex*, *Anopheles*, and *Aedes* spp. (McGreevy et al. 1982). The most common symptoms of these illnesses are pain and swelling of the genitals, whereas lymphangitis, lymphadenitis, and elephantiasis are less frequent (Wijers and McMahon 1976). The WHO launched the "Global Programme to Eliminate Lymphatic Filariasis" to eradicate this plague, targeting its complete eradication by 2020. Besides that, the most studied mosquito-borne filarioid nematodes are *Brugia* spp. and *Setaria* spp., which cause human morbidity in different areas of the world along with *Dirofilaria immitis* causing heartworm in canines (Mullen and Durden 2018). *Wuchereria bancrofti*, *Brugia timori*, and *B. malayi*, considered as human parasites, are animal reservoirs of epidemiologic importance, and some primates are the primary reservoir hosts of zoonotic *B. pahangi*. In Southeast Asia, the principal vectors of *B. malayi* are *Mansonia uniformis*, *Ma. indiana*, *Ma. annulifera*, and *Ma. onneae* (Mulyaningsih et al. 2019; Siriyasatien et al. 2023). *Brugia pahangi* has also been connected to *Armigeres* spp. and *Mansonia* spp. *Aedes* mosquitoes have been related to the parasites *Setaria*, *Brugia*, and *Wuchereria* (Siriyasatien et al. 2023). *Setaria* spp. is mainly found in bovids and cervids and can be transmitted by different mosquitoes (Cancrini et al. 1995, 1997; Übleis et al. 2018). They also observed five species of *Setaria* parasite (*S. cervi*, *S. digitata*, *S. equina*, *S. tundra*, and *S. labiatopapillosa*) related to *Anopheles* spp., *Culex* spp., *Armigeres* spp., and *Aedes* spp. *Setaria digitata* was associated with *Anopheles* spp., *Culex* spp., and *Aedes* spp., whereas *S. labiatopapillosa* was linked with *Anopheles* spp., *Culex* spp., *Armigeres* spp., and *Aedes* spp.

Dirofilaria immitis was detected from *Aedes albopictus* and *Culex pipiens* complex (Noden et al. 2021). *Culex tarsalis* and *Ae. albopictus* were recorded in transmitting this canine heartworm (Noden et al. 2015).

12.4 Prevention and Control

These mosquito vector control programs greatly depend on using chemicals, such as organophosphate and pyrethroid insecticides. Most of the insecticides are nonbiodegradable and have some undesirable effects on living beings. With time, vectors become resistant to these insecticides. The major strategies for controlling these vectors are as follows:

1. Eradication of breeding sites by eliminating all uncovered water-holding containers can be the most effective method in reducing the population by preventing them from being laid by mosquitoes.
2. Release of some larvivorous fishes or other biological organisms as predators or parasites of mosquito larvae.
3. Biodegradable insecticide spraying may be useful for adult mosquitoes.
4. *Wolbachia* infection is becoming a popular alternative strategy for controlling the pathogen transmissions by reducing the competence of mosquito populations.

12.5 Conclusion

Mosquito-borne diseases are a noteworthy threat to human and animal health. Mosquitoes serve as vectors of many of the deadly diseases. They infect approximately 500 million people, leading to approximately one million deaths globally per year besides affecting animals. Recently, many mosquito-borne diseases have been in control through vector control programs. Still, despite these strategies, some are uncontrolled and dangerous, causing a significant loss to animals and humans. There is an urgent need to control these vector populations as a better way to reduce the disease burden.

Points to Remember

1. Vector-borne diseases account for more than 17% of all infectious diseases.
2. *Anopheles*, *Culex*, and *Aedes* spp. are mainly implicated as the primary vectors of malaria, lymphatic filariasis, Japanese encephalitis, West Nile fever, dengue, yellow fever, Zika, and Rift Valley fever, respectively.
3. Besides that, they also transmit several other veterinary important viruses, bacteria, fungi, protozoa and nematodes to animals.
4. Their broad host ranges include cattle, sheep, horses, nonhuman primates, pigs, dogs, cats, birds, and even reptiles increasing infection rates and chances of zoonotic transmission.
5. Do not allow stagnation of water; discard unused containers to eradicate their breeding habitats.

References

Abdellahoum Z, Maurin M, Bitam I (2020) Tularemia as a mosquito-borne disease. Microorganisms 9(1):1–14

Adugna T, Yewhelew D, Getu E (2021) Bloodmeal sources and feeding behavior of anopheline mosquitoes in Bure district, northwestern Ethiopia. Parasit Vectors 14:1–12

Aitken TH, Tesh RB, Beaty BJ, Rosen L (1979) Transovarial transmission of yellow fever virus by mosquitoes (*Aedes aegypti*). Am J Trop Med Hyg 28(1):119–121

Ali A, Wahid B, Rafique S, Idrees M (2017) Advances in research on Zika virus. Asian Pac J Trop Med 10(4):321–331

Althouse BM, Hanley KA, Diallo M, Sall AA, Ba Y, Faye O, Diallo D, Watts DM, Weaver SC, Cummings DA (2015) Impact of climate and mosquito vector abundance on sylvatic arbovirus circulation dynamics in Senegal. Am J Trop Med Hyg 92(1):88–97

Amarela CF (2017) Monitoramento dos casos e óbitos de febre amarela no Brasil. INFORME-NÂ 43:2017

Armstrong PM, Andreadis TG (2010) Eastern equine encephalitis virus in mosquitoes and their role as bridge vectors. Emerg Infect Dis 16(12):1869–1874

Auerswald H, Maquart PO, Chevalier V, Boyer S (2021) Mosquito vector competence for Japanese encephalitis virus. Viruses 13(6):1154

Beasley DW, McAuley AJ, Bente DA (2015) Yellow fever virus: genetic and phenotypic diversity and implications for detection, prevention and therapy. Antivir Res 115:48–70

Boreham P, Chandler J, Highton R (1975) Studies on the feeding patterns of mosquitoes of the genera *Ficalbia*, *Mimomyia* and *Uranotaenia* in the Kisumu area of Kenya. Bull Entomol Res 65:69–74

Bozic J, Capone A, Pediconi D, Mensah P, Cappelli A, Valzano M, Mancini MV, Scuppa P, Martin E, Epis S, Rossi P (2017) Mosquitoes can harbour yeasts of clinical significance and contribute to their environmental dissemination. Environ Microbiol Rep 9(5):642–648

Brady OJ, Gething PW, Bhatt S, Messina JP, Brownstein JS, Hoen AG, Moyes CL, Farlow AW, Scott TW, Hay SI (2012) Refining the global spatial limits of dengue virus transmission by evidence-based consensus. PLoS Negl Trop Dis 6:e1760

Brault AC, Foy BD, Myles KM, Kelly CLH, Higgs S, Weaver SC, Olson KE, Miller BR, Powers AM (2004) Infection patterns of o'nyong nyong virus in the malaria-transmitting mosquito, *Anopheles gambiae*. Insect Mol Biol 13(6):625–635

Brown WC, Akey DL, Konwerski JR, Tarrasch JT, Skiniotis G, Kuhn RJ, Smith JL (2016) Extended surface for membrane association in Zika virus NS1 structure. Nat Struct Mol Biol 23(9):865–867

Bruce-Chwatt LJ, Garrett-Jones C, Weitz B (1966) Ten years' study (1955-64) of host selection by anopheline mosquitos. Bull World Health Organ 35(3):405

Brustolin M, Talavera S, Nuñez A, Santamaría C, Rivas R, Pujol N, Valle M, Verdún M, Brun A, Pages N, Busquets N (2017) Rift Valley fever virus and European mosquitoes: vector competence of *Culex pipiens* and *Stegomyia albopicta* (= *Aedes albopictus*). Med Vet Entomol 31(4):365–372

Butler D (2016) Fears rise over yellow fever's next move: scientists warn vaccine stocks would be overwhelmed in the event of large urban outbreaks. Nature 532(7598):155–157

Cancrini G, Pietrobelli M, Frangipane di Regalbono AF, Tampieri MP, Della Torre A (1995) Development of *Dirofilaria* and *Setaria* nematodes in *Aedes albopictus*. Parassitologia 37(2–3):141–145

Cancrini G, Pietrobelli M, Di Regalbono AF, Tampieri MP (1997) Mosquitoes as vectors of *Setaria labiatopapillosa*. Int J Parasitol 27(9):1061–1064

Carlson JS, Walther E, TroutFryxell R, Staley S, Tell LA, Sehgal RN, Barker CM, Cornel AJ (2015) Identifying avian malaria vectors: sampling methods influence outcomes. Parasit Vectors 8(1):1–16

Chevillon C, Briant L, Renaud F, Devaux C (2008) The chikungunya threat: an ecological and evolutionary perspective. Trends Microbiol 16(2):80–88

da Silva Neves NA, da Silva Ferreira R, Morais DO, Pavon JAR, de Pinho JB, Slhessarenko RD (2022) Chikungunya, Zika, Mayaro, and equine encephalitis virus detection in adult Culicinae from South Central Mato Grosso, Brazil, during the rainy season of 2018. Braz J Microbiol 53:1–8

de Mendonça SF, Rocha MN, Ferreira FV, Leite THF, Amadou SC, Sucupira PH, Marques JT, Ferreira AG, Moreira LA (2021) Evaluation of *Aedes aegypti*, *Aedes albopictus*, and *Culex quinquefasciatus* mosquitoes competence to Oropouche virus infection. Viruses 13(5):755

Diallo M, Thonnon J, Traore-Lamizana M, Fontenille D (1999) Vectors of chikungunya virus in Senegal: current data and transmission cycles. Am J Trop Med Hyg 60(2):281–286

Díaz S, Camargo C, Avila FW (2021) Characterization of the reproductive tract bacterial microbiota of virgin, mated, and blood-fed *Aedes aegypti* and *Aedes albopictus* females. Parasit Vectors 14(1):1–12

Dieme C, Rotureau B, Mitri C (2017) Microbial pre-exposure and vectorial competence of *Anopheles* mosquitoes. Front Cell Infect Microbiol 7:508

Erlanger TE, Weiss S, Keiser J, Utzinger J, Wiedenmayer K (2009) Past, present, and future of Japanese encephalitis. Emerg Infect Dis 15(1):1

Escobar D, Ascencio K, Ortiz A, Palma A, Fontecha G (2020) Distribution and phylogenetic diversity of *Anopheles* species in malaria endemic areas of Honduras in an elimination setting. Parasit Vectors 13(1):1–12

Fryxell RTT, Lewis TT, Peace H, Hendricks BB, Paulsen D (2014) Identification of avian malaria (*Plasmodium* sp.) and canine heartworm (*Dirofilaria immitis*) in the mosquitoes of Tennessee. J Parasitol 100(4):455–462

Gao H, Cui C, Wang L, Jacobs-Lorena M, Wang S (2020) Mosquito microbiota and implications for disease control. Trends Parasitol 36(2):98–111

George J, Blanford S, Thomas MB, Baker TC (2014) Malaria mosquitoes host-locate and feed upon caterpillars. PLoS One 9(11):e108894

Gibrail MM, Fiaccadori FS, Souza M, Almeida TNV, Chiang JO, Martins LC, Ferreira MS, Cardoso DDDDP (2016) Detection of antibodies to Oropouche virus in non-human primates in Goiânia City, Goiás. Rev Soc Bras Med Trop 49:357–360

Gould J, Valdez JW (2024) A little on the nose: a mosquito targets the nostrils of tree frogs for a blood meal. Ethology 130(2):e13424

Gould E, Pettersson J, Higgs S, Charrel R, De Lamballerie X (2017) Emerging arboviruses: why today? One Health 4:1–13

Gratz NG (2004) Critical review of the vector status of *Aedes albopictus*. Med Vet Entomol 18(3):215–227

Gunathilaka N, Denipitiya T, Hapugoda M, Abeyewickreme W, Wickremasinghe R (2016) Determination of the foraging behaviour and blood meal source of malaria vector mosquitoes in Trincomalee District of Sri Lanka using a multiplex real time polymerase chain reaction assay. Malar J 15:1–10

Gwee SXW, St John AL, Gray GC, Pang J (2021) Animals as potential reservoirs for dengue transmission: a systematic review. One Health 12:100216

Hanson RP (1957) An epizootic of equine encephalomyelitis that occurred in Massachusetts in 1831. Am J Trop Med Hyg 6(5):858–862

Heinen LBDS, Zuchi N, Serra OP, Cardoso BF, Gondim BHF, Santos MAMD, Souto FJD, Paula DAJD, Dutra V, Dezengrini-Slhessarenko R (2015) Saint Louis encephalitis virus in Mato Grosso, Central-Western Brazil. Rev Inst Med Trop Sao Paulo 57:215–220

Hennessey MJ, Fischer M, Panella AJ, Kosoy OI, Laven JJ, Lanciotti RS, Staples JE (2016) Zika virus disease in travelers returning to the United States, 2010–2014. Am J Trop Med Hyg 95(1):212

Huang W, Wang S, Jacobs-Lorena M (2020) Use of microbiota to fight mosquito-borne disease. Front Genet 11:196

Hubálek Z, Halouzka J (1999) West Nile fever—a reemerging mosquito-borne viral disease in Europe. Emerg Infect Dis 5:643

Jansen CC, Beebe NW (2010) The dengue vector *Aedes aegypti*: what comes next. Microbes Infect 12(4):272–279

Jelinek T, Dobler G, Hölscher M, Löscher T, Nothdurft HD (1997) Prevalence of infection with dengue virus among international travelers. Arch Intern Med 157(20):2367–2370

Kaavya K, Tharakan J, Joshi CO, Aneesh EM (2022) Role of vertically transmitted viral and bacterial endosymbionts of *Aedes* mosquitoes. Does paratransgenesis influence vector-borne disease control? Symbiosis 86(2):139–153

Kading RC, Borland EM, Cranfield M, Powers AM (2013) Prevalence of antibodies to alphaviruses and flaviviruses in free-ranging game animals and nonhuman primates in the greater Congo basin. J Wildl Dis 49(3):587–599

Karamchandani PV (1946) Dengue group of fevers in India. Lancet 247(6386):92–93

Komar N, Spielman A (1994) Emergence of eastern encephalitis in Massachusetts. Ann N Y Acad Sci 740:157–168

Kraemer MU, Sinka ME, Duda KA, Mylne A, Shearer FM, Brady OJ, Messina JP, Barker CM, Moore CG, Carvalho RG, Coelho GE (2015) The global compendium of *Aedes aegypti* and *Ae. albopictus* occurrence. Sci Data 2(1):1–8

Ladreyt H, Durand B, Dussart P, Chevalier V (2019) How central is the domestic pig in the epidemiological cycle of Japanese encephalitis virus? A review of scientific evidence and implications for disease control. Viruses 11(10):949

Lindh JM, Terenius O, Faye I (2005) 16S rRNA gene-based identification of midgut bacteria from field-caught *Anopheles gambiae* sensu lato and *A. funestus* mosquitoes reveals new species related to known insect symbionts. Appl Environ Microbiol 71(11):7217–7223

Mancini MV, Damiani C, Accoti A, Tallarita M, Nunzi E, Cappelli A, Bozic J, Catanzani R, Rossi P, Valzano M, Serrao A (2018) Estimating bacteria diversity in different organs of nine species of mosquito by next generation sequencing. BMC Microbiol 18(1):1–10

Manguin S, Bangs MJ, Pothikasikorn J, Chareonviriyaphap T (2010) Review on global cotransmission of human *Plasmodium* species and *Wuchereria bancrofti* by *Anopheles* mosquitoes. Infect Genet Evol 10(2):159–177

Mansfield KL, Hernández-Triana LM, Banyard AC, Fooks AR, Johnson N (2017) Japanese encephalitis virus infection, diagnosis and control in domestic animals. Vet Microbiol 201:85–92

Marra PP, Griffing S, Caffrey C, Kilpatrick MA, McLean R, Brand C, Saito EMI, Dupuis AP, Kramer L, Novak R (2004) West Nile virus and wildlife. Bioscience 54(5):393–402

Martinsen ES, McInerney N, Brightman H, Ferebee K, Walsh T, McShea WJ, Forrester TD, Ware L, Joyner PH, Perkins SL, Latch EK (2016) Hidden in plain sight: cryptic and endemic malaria parasites in North American white-tailed deer (*Odocoileus virginianus*). Sci Adv 2(2):e1501486

McCrae AWR, Henderson BE, Kirya BG, Sempala SDK (1971) Chikungunya virus in the Entebbe area of Uganda: isolations and epidemiology. Trans R Soc Trop Med Hyg 65:152–168

McGreevy PB, Kolstrup N, Tao J, McGreevy MM, Marshall TDC (1982) Ingestion and development of *Wuchereria bancrofti* in *Culex quinquefasciatus*, *Anopheles gambiae* and *Aedes aegypti* after feeding on humans with varying densities of microfilariae in Tanzania. Trans R Soc Trop Med Hyg 76(3):288–296

McGregor BL, Connelly CR, Kenney JL (2021) Infection, dissemination, and transmission potential of North American *Culex quinquefasciatus*, *Culex tarsalis*, and *Culicoides sonorensis* for Oropouche virus. Viruses 13(2):226

McIntosh BM (1970) Antibody against chikungunya virus in wild primates in Southern Africa. S Afr J Med Sci 35(3):65–74

Mendoza-Roldan JA, Mendoza-Roldan MA, Otranto D (2021) Reptile vector-borne diseases of zoonotic concern. Int J Parasitol Parasites Wildl 15:132–142

Monath TP (2007) Dengue and yellow fever-challenges for the development and use of vaccines. N Engl J Med 357(22):2222–2225

Moutailler S, Krida G, Schaffner F, Vazeille M, Failloux AB (2008) Potential vectors of Rift Valley fever virus in the Mediterranean region. Vector Borne Zoonotic Dis 8(6):749–754

Mullen G, Durden L (2018) Medical and veterinary entomology, 3rd edn. Academic Press, New York

Mulyaningsih B, Umniyati SR, Hadisusanto S, Edyansyah E (2019) Study on vector mosquito of zoonotic *Brugia malayi* in Musi Rawas, South Sumatera, Indonesia. Vet World 12(11):1729

Murphy FA, Fauquet CM, Bishop DH, Ghabrial SA, Jarvis AW, Martelli GP, Mayo MA, Summers MD (eds) (2012) Virus taxonomy: classification and nomenclature of viruses, vol 10. Springer Science & Business Media, Berlin

Noden BH, Coburn L, Wright R, Bradley K (2015) An updated checklist of the mosquitoes of Oklahoma including new state records and West Nile virus vectors, 2003–06. J Am Mosq Control Assoc 31(4):336–345

Noden BH, Bradt DL, Sanders JD (2021) Mosquito-borne parasites in the Great Plains: searching for vectors of nematodes and avian malaria parasites. Acta Trop 213:105735

Oliveira AR, Cohnstaedt LW, Cernicchiaro N (2018) Japanese encephalitis virus: placing disease vectors in the epidemiologic triad. Ann Entomol Soc Am 111(6):295–303

Oliver J, Lukacik G, Kokas J, Campbell SR, Kramer LD, Sherwood JA, Howard JJ (2018) Twenty years of surveillance for eastern equine encephalitis virus in mosquitoes in New York State from 1993 to 2012. Parasit Vectors 11(1):1–12

Olson JG, Ksiazek TG, Gubler DJ, Lubis SI, Simanjuntak G, Lee VH, Nalim S, Juslis K, See R (1983) A survey for arboviral antibodies in sera of humans and animals in Lombok, Republic of Indonesia. Ann Trop Med Parasitol 77(2):131–137

Paupy C, Delatte H, Bagny L, Corbel V, Fontenille D (2009) *Aedes albopictus*, an arbovirus vector: from the darkness to the light. Microbes Infect 11(14–15):1177–1185

Pavela R (2015) Essential oils for the development of eco-friendly mosquito larvicides: a review. Ind Crop Prod 76:174–187

Pinheiro FdP (1982) Situação Das Arboviroses Na Região Amazônica. Fundação Serviços Saúde Pública 25:27–48

Pinheiro FP, Travassos DR (1981) Oropouche virus. I. A review of clinical, epidemiological, and ecological findings. Am J Trop Med Hyg 30:149–160

Pinheiro FP, Travassos da Rosa AP, Vasconcelos PF (2004) Oropouche fever. In: Feigin RD (ed) Textbook of pediatric infectious diseases. Saunders, Philadelphia, pp 2418–2423

Platonov AE, Rossi G, Karan LS, Mironov KO, Busani L, Rezza G (2012) Does the Japanese encephalitis virus (JEV) represent a threat for human health in Europe? Detection of JEV RNA sequences in birds collected in Italy. Eur Secur 17(32):20241

Pohlit AM, Lopes NP, Gama RA, Tadei WP, de Andrade Neto VF (2011) Patent literature on mosquito repellent inventions which contain plant essential oils—a review. Planta Med 77(6):598–617
Possas C (2016) Zika: what we do and do not know based on the experiences of Brazil. Epidemiol Health 38:e2016023
Powers AM, Logue CH (2007) Changing patterns of chikungunya virus: re-emergence of a zoonotic arbovirus. J Gen Virol 88(9):2363–2377
Pruetz JD, Socha A, Kante D (2010) New range record for the lesser spot-nosed guenon (*Cercopithecus petaurista*) in southeastern Senegal. Afr Primates 7:64–66
Rappole JH, Hubálek Z (2003) Migratory birds and West Nile virus. J Appl Microbiol 94:47–58
Ravanini P, Huhtamo E, Ilaria V, Crobu MG, Nicosia AM, Servino L, Rivasi F, Allegrini S, Miglio U, Magri A, Minisini R (2012) Japanese encephalitis virus RNA detected in *Culex pipiens* mosquitoes in Italy. Eur Secur 17(28):20221
Reis LAM, Pampolha ABO, Nascimento BLSD, Dias DD, Araújo PADS, Silva FSD, Silva LHDSE, Reis HCF, Silva EVPD, Nunes Neto JP (2023) Genus *Culex* Linnaeus, 1758 (Diptera: Culicidae) as an important potential arbovirus vector in Brazil: an integrative review. Life 13(11):2179
Reisen WK (2003) Epidemiology of St. Louis encephalitis virus. Adv Virus Res 61:139–184
Rogers K (2023) *Aedes*. Encyclopedia Britannica. https://www.britannica.com/animal/Aedes
Ryan SJ, Carlson CJ, Mordecai EA, Johnson LR (2019) Global expansion and redistribution of *Aedes*-borne virus transmission risk with climate change. PLoS Negl Trop Dis 13(3):e0007213
Sah R, Siddiq A, Al-Ahdal T, Maulud SQ, Mohanty A, Padhi BK, El-Shall NA, Chandran D, Emran TB, Hussein NR, Dhama K (2023) The emerging scenario for the eastern equine encephalitis virus and mitigation strategies to counteract this deadly mosquito-borne zoonotic virus, the cause of the most severe arboviral encephalitis in humans—an update. Front Trop Dis 3:1077962
Sanfelice V (2022) Mosquito-borne disease and newborn health. Health Econ 31(1):73–93
Santin R, Mattei AS, Waller SB, Madrid IM, Cleff MB, Xavier MO, Nobre MDO, Nascente PDS, Mello JRBD, Meireles MCA (2013) Clinical and mycological analysis of dog's oral cavity. Braz J Microbiol 44:139–144
Singh R, Parija SC (2012) *Candida parapsilosis*: an emerging fungal pathogen. Indian J Med Res 136(4):671–673
Siriyasatien P, Intayot P, Sawaswong V, Preativatanyou K, Wacharapluesadee S, Boonserm R, Sor-Suwan S, Ayuyoe P, Cantos-Barreda A, Phumee A (2023) Description of potential vectors of zoonotic filarial nematodes, *Brugia pahangi*, *Setaria digitata*, and *Setaria labiatopapillosa* in Thai mosquitoes. Heliyon 9(2):1–9
Song BH, Yun SI, Woolley M, Lee YM (2017) Zika virus: history, epidemiology, transmission, and clinical presentation. J Neuroimmunol 308:50–64
Stanaway JD, Shepard DS, Undurraga EA, Halasa YA, Coffeng LE, Brady OJ, Hay SI, Bedi N, Bensenor IM, Castañeda-Orjuela CA, Chuang TW (2016) The global burden of dengue: an analysis from the global burden of disease study 2013. Lancet Infect Dis 16(6):712–723
Staples JE, Gershman M, Fischer M, Centers for Disease Control and Prevention (CDC) (2010) Yellow fever vaccine: recommendations of the Advisory Committee on Immunization Practices (ACIP). MMWR Recomm Rep 59(RR-7):1–27
Steyn A, Roets F, Botha A (2016) Yeasts associated with *Culex pipiens* and *Culex theileri* mosquito larvae and the effect of selected yeast strains on the ontogeny of *Culex pipiens*. Microb Ecol 71:747–760
Subbarao SK, Nanda N, Rahi M, Raghavendra K (2019) Biology and bionomics of malaria vectors in India: existing information and what more needs to be known for strategizing elimination of malaria. Malar J 18:1–11
Tan TY, Hsu LY, Alejandria MM, Chaiwarith R, Chinniah T, Chayakulkeeree M, Choudhury S, Chen YH, Shin JH, Kiratisin P, Mendoza M (2016) Antifungal susceptibility of invasive *Candida* bloodstream isolates from the Asia-Pacific region. Sabouraudia 54(5):471–477

Thongyuan S, Kittayapong P (2017) First evidence of dengue infection in domestic dogs living in different ecological settings in Thailand. PLoS One 12(8):e0180013

Tyagi BK, Munirathinam A, Venkatesh A (2015) A catalogue of Indian mosquitoes. Int J Mosq Res 2(2):50–97

Übleis SS, Cuk C, Nawratil M, Butter J, Schoener E, Obwaller AG, Zechmeister T, Duscher GG, Rubel F, Lebl K, Zittra C (2018) Xenomonitoring of mosquitoes (Diptera: Culicidae) for the presence of filarioid helminths in Eastern Austria. Can J Infect Dis Med Microbiol 2018:1–6

Van Beurden E (1980) Mosquitoes (*Mimomyia elegans* (Taylor)) feeding on the introduced cane toad *Bufo marinus* (Linnaeus): implications for control of a toad pest. Aust J Zool 20:501–504

Veronesi E, Paslaru A, Silaghi C, Tobler K, Glavinic U, Torgerson P, Mathis A (2018) Experimental evaluation of infection, dissemination, and transmission rates for two West Nile virus strains in European *Aedes japonicus* under a fluctuating temperature regime. Parasitol Res 117:1925–1932

Villegas LE, Campolina TB, Barnabe NR, Orfano AS, Chaves BA, Norris DE, Pimenta PF, Secundino NFC (2018) Zika virus infection modulates the bacterial diversity associated with *Aedes aegypti* as revealed by metagenomic analysis. PLoS One 13(1):e0190352

Vorou R (2016) Zika virus, vectors, reservoirs, amplifying hosts, and their potential to spread worldwide: what we know and what we should investigate urgently. Int J Infect Dis 48:85–90

Vu DM, Jungkind D, LaBeaud AD (2017) Chikungunya virus. Clin Lab Med 37(2):371–382

Waddell LA, Greig JD (2016) Scoping review of the Zika virus literature. PLoS One 11(5):e0156376

Wijers DJB, McMahon JE (1976) Early signs and symptoms of bancroftian filariasis in males at the east African coast. East Afr Med J 53(2):57–63

Wimalasiri-Yapa BR, Stassen L, Huang X, Hafner LM, Hu W, Devine GJ, Yakob L, Jansen CC, Faddy HM, Viennet E, Frentiu FD (2019) Chikungunya virus in Asia–Pacific: a systematic review. Emerg Microbes Infect 8(1):70–79

World Malaria Report (2023) World Health Organization. https://www.who.int/teams/global-malaria-programme/reports/worldmalaria-report-2023

Yanmanee S, Seethamchai S, Kuamsab N, Karaphan S, Suwonkerd W, Jongwutiwes S, Putaporntip C (2023) Natural vectors of *Plasmodium knowlesi* and other primate, avian and ungulate malaria parasites in Narathiwat Province, Southern Thailand. Sci Rep 13(1):8875

Zeller H (2012) Is Japanese encephalitis emerging in Europe? Eur Secur 17(32):20242

Management of Mosquitoes: Past, Present, and Future 13

Goutam Chandra and Sunanda Burman

Abstract

Over 40% of the world's population is at risk of mosquito-borne diseases, which are a growing global health challenge. The only effective method for preventing certain mosquito-borne diseases is to reduce the vector population. Developing pesticide resistance has made it impossible to control mosquito vectors in a sustainable way, which has prompted the development of new techniques. Diverse and cutting-edge mosquito control technologies are being created to combat this expanding issue. Conceptually, these include an astounding range of strategies, from use of synthetic and plant-based products that may cause a mosquito population to collapse or limit its capacity to transmit disease by developing genetically modified mosquitoes and creating sterile mosquitoes to using green synthesized nanoparticles and also attractive toxic sugar bait (ATSB) which attracts sugar-seeking mosquitoes, which is a new and promising method of controlling mosquitoes. Despite using a variety of techniques to keep these vector populations below the threshold level, including chemical, biological, and genetic strategies, we are still unable to stop the spread of illness. In order to effectively manage mosquitoes, an interdisciplinary approach is therefore necessary, and this theme issue will include essays highlighting the variety of studies on mosquito management.

Keywords

Mosquito-borne diseases · Vector mosquitoes · Mosquito management · Effective strategy

G. Chandra (✉) · S. Burman
Department of Zoology, The University of Burdwan, Burdwan, West Bengal, India

© The Author(s), under exclusive license to Springer Nature Singapore Pte Ltd. 2024

Omkar (ed.), *Mosquitoes*, https://doi.org/10.1007/978-981-97-4163-2_13

Learning Objectives

1. To learn various measures of mosquito management.

13.1 Introduction

The most frequent infections among the world's poorest individuals are neglected tropical diseases. Arthropods are extremely threatening carriers of viruses and parasites that could cause pandemics or epidemics among the growing global population of people and animals. More than 40% of the world's population is at risk from mosquito-borne diseases, which constitute a severe global health concern (Deng et al. 2023). According to Caraballo and King (2014), mosquito-borne diseases afflict more than 700 million people worldwide and claim 1 million lives each year. Innovative approaches to combat mosquito-borne disease are urgently required due to the restrictions on eradication progress imposed by insecticide and drug resistance, along with the absence of vaccinations. In recent years, it has become clear that using mosquito microbiota to reduce the spread of infections carried by mosquitoes has significant potential (Huang et al. 2020). A major hazard to millions of people worldwide is posed by mosquitoes (Diptera: Culicidae), which spread diseases like malaria, dengue, yellow fever, filariasis, Japanese encephalitis, and the Zika virus. According to a World Health Organization (WHO 2023a) estimate, malaria caused 247 million illnesses and 619,000 fatalities in 2021, and existing malaria control efforts have made no appreciable headway. In addition, Culicidae spread important infections and parasites that horses and dogs are particularly vulnerable to, including dog heartworm, West Nile virus, and eastern equine encephalitis. Dengue incidence and reported epidemic areas have both increased significantly on a global scale (WHO 2023b). In addition to recent outbreaks of novel arboviruses, the emergence of resistance in a number of Culicidae species, and the fast global spread of highly invasive mosquitoes, mosquito control programs are currently confronting significant and critical challenges. Sadly, there is now no known cure for the majority of arboviruses spread by mosquitoes, particularly dengue. Additionally, there are considerable obstacles that still prevent the successful control of other diseases transmitted by mosquitoes, such as malaria. It is absolutely essential to develop new strategies and vector control tools that target aquatic life stages and adults. *Anopheles*, *Aedes*, and *Culex*, the three major mosquito genera, spread the pathogens that cause a wide range of serious illnesses to both humans and animals. Control of mosquito-borne disease vectors or cessation of human-vector interaction is the only means of preventing or minimizing these diseases. The occurrence of pesticide resistance, which has grown significantly in recent years, is these products' biggest flaw. A perplexing task for public health organizations around the world is the effective and long-term management of these arthropod vectors focusing in particular on small-scale communities in developing nations who have limited access to adequate diagnoses, prevention, and treatment for infectious diseases (Benelli 2019).

Controlling vector populations is frequently the most effective method of preventing the spread of these diseases, if not the only one. On the other hand, despite the fact that mass drug administration (MDA) has mostly stopped the major spread of diseases like lymphatic filariasis and malaria, the disease vectors are still a nuisance to people. Because of all those factors, mosquito management is still a crucial part of many nations' public health systems. Chemical, biological, and genetic techniques are among the best at controlling mosquitoes. Insecticide-treated bed nets, adult repellents, biological control agents against mosquito young instars (primarily fish, amphibians, and copepods), sterile insect technique (SIT), "boosted SIT," symbiont-based techniques, and transgenic mosquitoes are the foundations of today's control strategies. The use of eave tubes, appealing toxic sugar baits (ATSBs), and biocontrol chemicals has been extensively studied in relation to mosquitoes. A further potential strategy for vector control is the use of genetically modified (GM) mosquitoes. At the moment, none of these individual approaches are completely effective. There is an urgent need for new environmentally friendly mosquito vector management techniques. In comparison with chemical and physical methods, plant-mediated nanoparticle manufacturing is cheaper and simpler and doesn't involve the use of very toxic substances, high pressure, energy, or temperature. A rising variety of plant-based chemicals have been developed in recent years for the efficient and quick extracellular synthesis of metal nanoparticles that are effective against mosquitoes at very low levels (i.e., 1–30 ppm) (Benelli et al. 2017).

In the present study, I tried to provide a brief and up-to-date assessment of the current status of mosquito management research.

13.2 Global Impact of Major Mosquito-Borne Diseases

Increasing levels of globalization, global warming, and climate change and also human migration have caused invasive species to spread ecologically. These invasive species, which include arthropods, spread fatal illnesses that can spark pandemics or epidemics. Mosquitoes (Diptera: Culicidae), which serve as carriers of numerous dangerous infections, are crucial in that regard (Onen et al. 2023).

Malaria, dengue, chikungunya, yellow fever, Zika, West Nile, lymphatic filariasis, etc. are some of the most prevalent diseases that are spread through mosquito bites throughout the world. The majority of these diseases have neither a vaccination nor a treatment available (Qureshi 2017).

Dengue, yellow fever, chikungunya, and Zika have produced serious illness epidemics in many metropolitan areas. Mosquitoes alone are responsible for the death of over a million people each year due to the spread of malaria (WHO 2019). The rate of infection transmission has sharply increased in recent years, and an increasing number of scientists are worried that global warming could enable the spread of mosquito-borne diseases to explode globally. Parasite infected anopheline mosquitoes spread malaria. More than 400,000 people die from it each year, while dengue, yellow fever, chikungunya, and Zika have produced catastrophic illness epidemic in numerous urban areas (Jones et al. 2021).

Nematodes that affect the lymphatic system, muscles, connective tissue, or blood in their vertebrate hosts are the cause of filariasis. *Wuchereria bancrofti*, *Brugia malayi*, and *Brugia timori* are the principal species. They are mostly spread by the *Culex quinquefasciatus* mosquito. Although it is most common in Africa and India, filariasis can also be found in China, Japan, Sri Lanka, and a number of Pacific island nations (Cano et al. 2014). In response to growing global concerns about filariasis, the WHO launched the "Global Programme to Eliminate Lymphatic Filariasis" in 2000 (Ottesen et al. 2008).

Dengue, an arbovirus that subtly entered the Western Hemisphere over several decades, grew more active in the 1990s and infected more people, leading to a significant international public health concern. *Aedes* mosquitoes, specifically *Aedes aegypti* and to a lesser extent *Ae. albopictus*, are the primary carriers of dengue.

The viral infection caused by *Aedes* mosquitoes that is spread most frequently is dengue. An estimated 96 million symptomatic cases and 40,000 fatalities from dengue are reported annually, affecting more than 3.9 billion people in over 129 countries (WHO 2020).

Yellow fever is an acute viral hemorrhagic disease that is prevalent in tropical regions of Central and South America, Africa, and the Caribbean (WHO 2016a). *Aedes* mosquitoes, specifically *Ae. aegypti*, afflicted with the disease are the main carriers. As of right now, there is no specific antiviral medication for the treatment of yellow fever.

The most recent of four significant arbovirus introductions into the Western Hemisphere over the past 20 years, the current outbreaks of Zika virus infections in South America, Central America, and the Caribbean (Fauci and Morens 2016), constitutes a new threat to human health. Zika is a mosquito-borne illness linked to neurological problems and has lately been the source of numerous significant outbreaks encompassing about 89 countries (WHO 2016b).

According to Attar (2016), Benelli and Mehlhorn (2016), and WHO (2016a), the emergence of the Zika virus came after that of dengue, the 1999 West Nile virus, and the 2013 chikungunya virus.

Mosquitoes spread the encephalitis viruses that cause Japanese encephalitis, La Crosse encephalitis, Murray Valley encephalitis, eastern equine encephalitis, western equine encephalitis, St. Louis encephalitis, West Nile, Mayaro fever, etc. (Go et al. 2014; Gould et al. 2017; Qureshi and Saeed 2019). They also spread Rift Valley fever virus, Kunjin virus, Ross River virus, Barmah Forest virus, and Jamestown Canyon virus which are among the infections that have been reported (Hills 1996; Pepin et al. 2010; Sarwar 2016).

13.3 The Scenario of Mosquito-Borne Diseases in the Past and Their Management

13.3.1 Mosquito Control

13.3.1.1 Early Twentieth Century

Source reduction, use of petroleum oils (kerosene), and the encouragement of guppies and other fish were the main strategies used in the first quarter of the twentieth century to manage mosquitoes (Howard 1892, 1902). Evidence supporting the manner of action of larviciding oil mixes was presented by Freeborn and Atsatt (1918). Additionally, Freeborn promoted biological control agents in 1922.

13.3.1.2 Second Quarter of the Twentieth Century

The vector control technology of the twentieth century continued to emphasize source reduction, oil application, and botanical preparation in the second part of the century. The main line of defense was presented as source reduction and naturalistic and biological control techniques. The more enduring and long-lasting tactics were thought to be supplemented by oiling and the use of chemical adulticides and larvicides. In order to improve public comfort and well-being, Herms and Gray (1944) suggested reducing or eliminating pestiferous species. They also mentioned the issue of pest mosquitoes following the cessation of disease transmission.

13.3.1.3 Third Quarter of the Twentieth Century

Following the introduction of DDT at the end of the second quarter of the twentieth century, an extraordinary rush to discover, create, and test new forms of synthetic chemicals occurred in the first half of the third quarter. A large number of newly constructed or altered buildings were made accessible for testing against various harmful pests. Insecticides of the first, second, and third generations were developed throughout the 1950s and 1960s.

There was also a surge in research on genetic regulation of vectors throughout the 1960s and 1970s. Many genetic procedures were explored, including translocations, cytoplasmic incompatibility, meiotic drive, and sterile hybrids (Whitten and Foster 1975), but none of these strategies were practical. Similarly, much research was conducted on chemosterilants, but none of these compounds became operational in vector control programs.

13.3.1.4 Fourth Quarter of the Twentieth Century

Scientists focused on developing selective biorational insecticides and integrated pest control strategies in the third quarter of this century, which went into effect in the fourth quarter, while the environmental movement was at its height in the 1960s and 1970s. The IGRs diflubenzuron, fenoxycarb, and methoprene were manufactured and approved at this time for various applications as a result of these efforts.

The invention of spore-forming and toxin-producing bacteria for vector control and their isolation represent another significant advancement in the field of vector control. *Bacillus thuringiensis* subsp. *israelensis* (*B.t.i.*), an entomopathogenic bacterium, was registered in 1980 for the purpose of controlling mosquitoes and blackflies. In 1992, *B. sphaericus*, another bacterial agent, was registered.

13.4 The Presently Used Mosquito Management Strategies

Vector-based interventions are the main strategies for lowering the public health burden of the majority of diseases spread by mosquitoes. Prior to the invention of pesticides, these treatments relied on environmental management and were centered on eliminating mosquito breeding grounds and enhancing housing with screens to stop mosquitoes from entering through windows and doors. These vectors can be effectively controlled by using insecticides, larvicidal chemicals, bed nets, as well as drugs for chemoprevention and illness treatment. Depending on the circumstance, the type and severity of the mosquito problem, the environment, and the resources available, different mosquito management tactics will be used. The WHO expects innovative vector control products that are not part of a recognized intervention class to demonstrate epidemiological proof of their effects on public health. Through a phase III field study that includes epidemiological results, the effectiveness of a vector control tool used in the best possible circumstances can be evaluated. In order to evaluate the performance of the vector control tool when it is supplied and utilized in "real-world" circumstances, the WHO will make suggestions for pilot deployment based on the findings of phase III trials (WHO 2018). It is noted that the average monthly production rate of *Ae. aegypti* in car tires (0.737 m), bus/truck tires (1.041 m) filled with rainwater, and bus/truck tires lodged with pondwater is reported to be deficient. A single rainwater-filled bus or truck tire could transmit an average of 3957 *Ae. aegypti*, which is much more than the average number generated in a rainwater-filled car tire and a pondwater-laden bus or truck tire over a wet month. The difference between the tires of a bus or truck and those on a vehicle might be due to their higher water retaining capacity. On the other hand, it appeared that the various physiochemical differences between pondwater and rainfall determined the variation in mosquito breeding potential in a bus or truck tire. Based on this investigation, it is hypothesized that a large-scale generation of *Ae. aegypti* may be caused by an open tire disposal site. It is advised to store used tires in a shaded area, cover them with a polythene sheet, or create openings on the underside of each tire to allow rainwater to collect and drain (Malla et al. 2019).

13.4.1 Personal Protection

Personal protection, which includes the use of clothing, a mat, a repellent, a mosquito coil, or other devices to guard against mosquito bites (Srivastava et al. 2014).

13.4.2 Environmental Manipulations, Habitat Modification, and Source Reduction

Environmental manipulations include the installation of a dependable piped water supply in communities, including household connections, or other long-lasting physical changes that limit the habitats used by vector larvae and also by source reduction by getting rid of containers, tires, and other things that can collect water and harbor mosquito breeding grounds, to stop the breeding by changing the environment such that mosquitoes cannot breed there (Al-Akel and Suliman 2011). One well-known instance is subsoil drainage, in which drains are buried to obstruct mosquito access. Additionally, environments should be changed, such as by regarding drainage ditches, so that water drains swiftly and doesn't collect as standing water. Environmental management aims to alter the environment by eliminating or recycling nonessential containers that serve as larval dwellings in order to prevent or reduce vector propagation and human contact with the vector-pathogen. To cut down on mosquito breeding grounds, experts remove as many oviposition sites as they can. Getting rid of mosquito breeding grounds can reduce the need for insecticides (Christian et al. 2021; Forsyth et al. 2020).

13.4.3 Mosquito Surveillance

To determine what kinds and numbers of mosquitoes are present in an area and whether they are carrying pathogens, experts keep an eye on mosquitoes, pathogens, and host animals like birds. Experts in mosquito control use surveillance to acquire these data. These procedures aid experts in deciding whether, when, and where control measures are required to manage mosquito populations before illnesses start to spread. If experts find that local mosquitoes are transmitting viruses, they begin carrying out further measures listed in their mosquito control programs (Hribar et al. 2022).

13.4.4 Use of Mosquito Traps

One of the best ways to lower the likelihood of contracting a mosquito-borne illness is to avoid mosquito bites. Bed nets with pesticide and indoor residual sprays primarily target mosquitoes that feed inside throughout the night. Mosquito traps are a safe, chemical-free technique of controlling mosquitoes and are operated by electricity or gas (Chandra 2000; Poulin et al. 2017).

13.4.5 Mechanical Methods

Yellow light is also proven to repel mosquitoes less than white light.

Devices like mosquito magnets and electric mosquito zappers are used in the mechanical method (Kalita et al. 2013).

13.4.5.1 Use of Electric Zapper

An electric zapper kills insects when they come into touch with its fatal dosage of electrical charge by first luring them in with ultraviolet light.

13.4.5.2 Mosquito Magnet

Through the release of carbon dioxide, heat, and wetness, the mosquito magnet imitates animals. These are mixed with an attractant called octenol, which is a naturally occurring plant pheromone, and as the mosquito comes too close to the magnet, it is dragged in and ultimately dies of dehydration. Mosquito magnet has the advantage of capturing mosquitoes (Kalita et al. 2013; Jeyaprakasam et al. 2021; Maia and Moore 2011).

13.4.6 Reproductive Manipulations Through Understanding Mating Behavior

The management of mosquito populations is increasingly turning to mosquito control tactics that involve modifying or taking advantage of mating behaviors (Diabate and Tripet 2015). These methods often involve chemically sterilized mosquitoes being raised in laboratories. According to evidence, males can have several matings while females only do it once. In order to regulate mosquito populations, modifying or taking advantage of mating habits is quickly emerging as a key strategy. In order to better understand sexual selection in mosquitoes and eventually ensure the effectiveness of novel control tactics that target mosquito reproduction, combining cutting-edge technology with sexual selection theory is a potent approach. In swarms, male mating success varies greatly, and research suggests that this variation is likely caused by scrambling competition between males as well as female preference. Implementing this new knowledge will enhance reproductive control tool stability over the long term (Cator et al. 2021).

Mosquito behavior is crucial to vector control programs, and in order to improve control measures, further understanding of the chemical ecology of mate hunting, swarming landmarks, and mate selection at swarming sites is needed (Benelli et al. 2016).

13.4.7 Insect Growth Regulators

It has been claimed that insect growth regulators such as Isostearyl alcohol, petroleum distillates, spinosad, methoprene, pyriproxyfen, and triflumuron can *control An. stephensi, Cx. quinquefasciatus, Ae. Albopictus, Ae. Aegypti,* etc. (Lau et al. 2015; Nayar and Ali 2003; Kamal and Khater 2010).

13.4.8 Application of Sterile Insect Technique (SIT) and Boosted SIT

It is an environmentally friendly, species-specific approach of controlling insects that releases vast quantities of sterile insects. If sterile males are discharged in sufficient numbers and for long enough periods of time, they can reduce or eradicate the pest population locally by mating with native females and reducing their capacity for reproduction (Alphey et al. 2010).

To breed with wild female mosquitoes, sterile, irradiated male mosquitoes are frequently released into the wild. Male mosquitoes are not harmful. Eggs produced as a result do not hatch. An area's targeted mosquito population gradually declines over time. The population of *Culex pipiens fatigans* was wiped out in various

Ceylonese islands through the discharge of sterile males, despite the low success rate of this procedure in the field application. The introductions of cytoplasmic incompatibility, hybrid sterility, lethal factors, or altered sex ratios are all being tested as ways to reduce the mosquito population.

Additionally, SIT has recently been paired with auto-dissemination, a technology that has recently shown to be particularly effective for controlling *Aedes* species but cannot be applied on a broad scale. A novel control strategy known as "boosted SIT" has been developed as a result, and it may be able to completely eradicate mosquitoes and other vector-borne diseases in a given region (Bouyer and Lefrançois 2014).

13.4.9 Use of Insecticides and Repellents

13.4.9.1 Use of Insecticides

Chemicals known as insecticides are used to kill mosquitoes.

13.4.9.1.1 Use of Insecticides of Synthetic Origin

Adult mosquitoes are killed using chemical insecticides, such as via fogging. To kill mosquito adults, chemical insecticides can be sprayed indoors or outdoors directly onto mosquito resting areas. Four different types of insecticides can chemically control adults. The first is the application of residual pesticides like DDT, dieldrin, endrin, and others that kill mosquitoes by penetrating their cuticle or mouthparts. The other three mosquitocides are stomach poisons like Paris green; contact poisons like pyrethrum, rotenone, and nicotine; fumigants like HCN; and methyl bromide. Organochlorine, carbonates, pyrethroids, and organophosphates are mainly chemical insecticides that are used as chemical mosquitocides. Mosquito eggs can be destroyed by organophosphorus substances. Malathion, abate, DDT, dieldrin, fenthion, fenitrothion, lindane, dursban, and other emulsifiable or oil-based chemicals can be used to treat larval water. Mosquito larvae can be eliminated with a 2% dusting of Paris green. Using mineral oils like crude oil, kerosene, light diesel oil, etc. on the surface of breeding water will also suffocate mosquito larvae and kill them. Environmental risks are significant when chemical substances are used in control operations. Most commercial mosquito management repellents are made with ingredients like *N,N*-diethyl-meta-toluamide (DEET), allethrin, *N,N*-diethyl mandelic acid amide, and dimethyl phthalate (Chandra 2000).

13.4.9.1.2 Use of Insecticides of Botanical Origin

Many of the insecticides currently in use have a botanical origin. Some examples of botanical pesticides used for mosquito control include chrysanthemum, azadirachtin, nicotine, d-limonene, camphor, anabasine, turpentine, etc. (Ghosh et al. 2012; Ojewumi et al. 2021).

13.4.9.2 Mosquito Repellent

Mosquito repellent refers to any substance that is applied to the skin, clothing, or other surfaces to prevent mosquitoes from landing on it and biting.

13.4.9.2.1 Use of Chemical Repellent

Thus, insect repellents function by disguising human odor or by employing a fragrance that insects ordinarily avoid. Most commercial mosquito management repellents are made with ingredients like *N,N*-diethyl-meta-toluamide (DEET), allethrin, *N,N*-diethyl mandelic acid amide, and dimethyl phthalate. In comparison with DEET, the most powerful and long-lasting repellent is IR3535 (3-[*N*-butyl-*N*-acetyl]-aminopropionic acid) (Ojewumi et al. 2021; Poolprasert et al. 2019; Asadollahi et al. 2019).

13.4.9.2.2 Use of Plant-Based Repellent

The importance of searching of plants as sources of novel metabolites for mosquitocidal properties has been highlighted by the discovery of the plant-derived drug artemisinin for the treatment of malaria and the subsequent granting of the Nobel Prize in 2015 (Benelli et al. 2016). Since ancient times, people have used plant-based repellents as a form of personal defense against various mosquito species. The development of novel natural products as a substitute to chemical-based repellents benefits greatly from knowing about the traditionally used repellent plants. Commercial repellents made from plant-based chemicals have recently grown in popularity among customers since they are frequently thought of as "safe" in contrast to well-known synthetic repellents; however, this perception is occasionally mistaken. The normal WHO Pesticide Evaluation Scheme procedures for assessing repellents have not yet been sufficiently applied in enough trials. Additional, standardized research is required to more accurately assess repellent substances and create new goods that are both highly repellent and safe for consumers. Another inexpensive and frequently successful way to repel mosquitoes is to use mosquito coils constructed of dried plants and flammable materials like sawdust due to their smoke toxic effects (Patel et al. 2012). Many plant species are reported to have smoke toxic effects (Singha et al. 2011; Ghosh et al. 2016a, b; Vivekanandhan et al. 2018).

Since ancient times, people have used plant-based repellents as a kind of personal protection against various mosquito species (Ojewumi et al. 2021). The development of novel natural products as an alternative to chemical repellents benefits greatly from knowledge of conventional repellent plants. The most volatile chemical components are found in essential oils, which are utilized as personal defense against bloodsucking mosquitoes. In the future, plant oils may be a good replacement for synthetic repellents because they are generally safe, affordable, and widely accessible. Numerous researches have documented evidence of the ability of plant extracts or essential oils to repel mosquito vectors globally. For generations, people have burned plants as crude fumigants to ward off pesky insects. Later, they have applied oil formulations made from plants to the skin, a practice first documented by Roman and Indian researchers (Fradin and Day 2002; Johnson 1998).

13.4.9.2.2.1 Smoke Toxicity of Plant Extracts

The smoke produced by the plant-based insecticide-infused mosquito repellent sticks demonstrated successful results, demonstrating that natural insecticides are

Table 13.1 List of some of reported plants having smoke toxic effects

Plant showing smoke toxicity	Effective on which mosquito species	References
Azadirachta indica, *Eucalyptus camaldulensis*, and *Ocimum forskolin* leaves	*Anopheles arabiensis* and *Aedes aegypti*	Wendimu and Tekalign (2021)
Acanthospermum hispidum leaf	*Anopheles stephensi*, *Aedes aegypti*, and *Culex quinquefasciatus*	Vivekanandhan et al. (2018)
Punica granatum and *Borago officinalis* leaf	*Culex quinquefasciatus*	Murugan et al. (2023)
Cipadessa baccifera (Roth) plant extracts	*Anopheles stephensi*, *Aedes aegypti*, and *Culex quinquefasciatus*	Ramkumar et al. (2015)
Seeds of *Carica papaya*	*Culex quinquefasciatus*, *Anopheles stephensi*	Rawani et al. (2012)
Leaf extracts of *Swietenia mahagoni*	*Anopheles stephensi*	Adhikari and Chandra (2014)

always as effective as synthetic repellents. Table 13.1 includes some of the information demonstrating the mosquito toxicity of smoke from plant extracts.

13.4.9.2.2.2 Plant Extracts and Essential Oils as Mosquito Repellents

Insect-repelling abilities have been discovered in a variety of plant extracts and essential oils, and during the past few years, more research has emerged proving that plant-based insect repellents are at least as effective as synthetic ones. Some of the research data on plant-based repellents are shown in Table 13.2.

13.4.10 Green Synthesized Nanoparticles for Mosquito Control

A simple and affordable solution for vector control has emerged using green nanoparticle manufacturing. Recently, a rising number of plant species have been suggested for the effective and quick synthesis of metal nanoparticles outside of cells, with good biotoxicity against mosquitoes. Numerous studies have shown that metal nanoparticles made from plant extracts are effective against vector mosquitoes as a larvicide, ovideterrent, and ovicide (Madhiyazhagan et al. 2017; Benelli 2016; AlQahtani et al. 2017; Haldar et al. 2013).

13.4.11 Biological Control Using Natural Predators

Initially, it was intended to use larvicidal methods of action to decrease the overall vector population by using entomopathogenic bacteria as control agents. At various stages of the mosquito life cycle, naturally occurring microbial infections can take place and produce various physiological effects. Numerous microorganisms, such as fungi, bacteria, nematodes, and protozoa, have been found to be successful in killing mosquitoes, particularly in their larval stage (Huang et al. 2017).

Table 13.2 List of plant-based mosquito repellents

Plant extract as repellent	**Affected mosquito species**	**References**
Leaf extract of *Cymbopogon winterianus*, *Cinnamomum zeylanicum* bark extract, and *Thymus vulgaris* leaf extract	*Anopheles gambiae*	Deletre et al. (2013)
Leaf extract of *Swietenia mahagoni*	*Culex quinquefasciatus*	Adhikari et al. (2012)
Seed extract of *Cuscuta chinensis*	*Culex quinquefasciatus*	Chakraborty et al. (2022)
Fruit extract of *Drypetes roxburghii*	*Culex quinquefasciatus* and *Anopheles stephensi*	Halder and Chandra (2015)
Mesua ferra L. leaf extract	*Culex quinquefasciatus*	Singha et al. (2011)
Plant-derived essential oils as repellent	**Affected mosquito species**	**References**
Citronella oil obtained from lemongrass	*Anopheles culicifacies*	Amer and Mehlhorn (2006)
Garlic (*Allium sativum*), clove (*Syzygium aromaticum*), lemongrass (*Cybopogon citratus*), citronella grass (*Cymbopogon nardus*), peppermint (*Mentha piperita*), eucalyptus (*Eucalyptus globulus*), orange (*Citrus sinensis*), and sweet basil (*Ocimum basilicum*) and their combinations against	*Aedes aegypti* and *Anopheles dirus*	Sritabutra et al. (2011)
Essential oil composition of *Ocimum basilicum* L. and *Ocimum minimum*	*Anopheles minimus*	Özcan and Chalchat (2002)
Essential oils extracted from *Mentha longifolia*, *Zanthoxylum armatum*, *Erigeron bonariensis*, and *Erigeron canadensis*	*Aedes aegypti*	Abbas et al. (2022)
Securidaca longepedunculata, *Lantana camara*, *Ricinus comminus*	*Anopheles gambiae*	Maurice et al. (2022)
Essential oils of *Cinnamomum zeylanicum*, *Citrus grandis*, *Citrus medica*, and *Citrus sinensis*	*Anopheles gambiae*	Akono et al. (2016)
Oil from narrow-leaved eucalyptus, lemon-scented eucalyptus, broad-leaved eucalyptus, and *Eucalyptus globulus*	*Anopheles stephensi*	Asadollahi et al. (2019)
Essential oils (EOs) extracted from *Cymbopogon citratus*, *Cymbopogon winterianus*, *Eucalyptus citriodora*, and *Eucalyptus camaldulensis*	*Aedes aegypti*	Manh et al. (2020)
Asteraceae oil of *Ajania tenuifolia*, Rutaceae oil from the leaves of *Citrus sinensis*	*Culex pipiens quinquefasciatus*	Yang et al. (2005)
Rosmarinus officinalis	*Anopheles stephensi*, *Aedes aegypti*	Pratiwi (2021)
Essential oil of *Syzygium aromaticum*	*Anopheles stephensi*	Osanloo et al. (2018)
Perovskia atriplicifolia, *Citrus reticulata* (fruit peels), *C. reticulata* (leaves), *Mentha longifolia*, and *Dysphania ambrosioides*	*Aedes aegypti*	Haris et al. (2023)
Essential oils of *Cananga odorata* (ylang ylang), *Citrus sinensis* (orange), *Cymbopogon citratus* (lemongrass), *Cymbopogon nardus* (citronella grass), *Eucalyptus citriodora* (eucalyptus), *Ocimum basilicum* (sweet basil), and *Syzygium aromaticum* (clove)	*Aedes aegypti* (Linn.), *Culex quinquefasciatus* (Say), and *Anopheles dirus*	Phasomkusolsil and Soonwera (2011)

Biological control techniques have already contributed significantly to reducing the mosquito population, primarily by controlling the vector. The introduction of mosquito predators has frequently been essential to the success of biological mosquito control. The fall of mosquito populations in the wild may be caused by predatory animal species. Data from numerous biological control initiatives demonstrate how introducing particular species can drastically lower mosquito populations, which in turn results in a reduction in disease (Hegazy et al. 2022). In order to lessen the existing reliance on insecticide-based mosquito control, biocontrol solutions strive to be sustainable and target a variety of different mosquito species.

13.4.11.1 Use of Larvivorous Fish

Of the least expensive vector management strategies that effectively suppress the mosquito population over a lengthy period of time is the release of larvae eating fish.

Utilizing fish predators is the oldest way of mosquito control. The mosquito larvae eating fish enables us to control mosquito populations. The main species utilized for controlling mosquito larvae are *Gambusia affinis* and others from the family Cyprinodontidae. When compared to chemical control, using larvae eating fish is thought to be more successful (Chandra et al. 2008; Naseem et al. 2016).

In India, *Aplocheilus*, *Aphanius dispar*, *Colisa*, and *Chanda nama* are indigenous larval fishes that are utilized as malaria preventative measures. *Gambusia*, *Oreochromis*, *Carassius*, *Xenentodon*, and *Poecilia* (Fig. 13.1) are exotic larval fishes that are utilized as malaria control agents (Kumar and Hwang 2006).

13.4.11.2 Control Through Microorganisms

A number of bacteria are investigated and discovered to be efficient biological control agents, such as *Bacillus thuringiensis morrisoni*, *Asticcacaulis excentricus*, *Clostridium bifermentans malaysia*, *Bacillus thuringiensis niedellin*, *Bacillus*

Fig. 13.1 The mosquito larva eating fish *Poecilia* sp. (*guppy fish*) which has been recognized as an important predator of mosquito larvae

thuringiensis canadensis, *Bacillus thuringiensis malaysiensis*, and *Synechococcus* (Poopathi et al. 2014). By taking into account the demands of contemporary environmental protection, the control of mosquitoes with microbial control agents such as *Bacillus thuringiensis israelensis* (Bti) and *Bacillus sphaericus* (Bsp) offers man a way to serve all of his interests, protect himself from mosquitoes, and protect nature reserves (Becker 2000). Additionally, microbial control agents like *Wolbachia* and *Asaia* that prevent pathogen development or transmission from mosquito to human are being researched; however, unanticipated interactions with mosquito gut bacteria can limit their efficacy (Maia and Moore 2011). *Bacillus* strains are practical, readily usable, locally produced, and inexpensive and have the capacity for rapid dissemination. Bs and Bt had a greater impact on *Aedes*, *Cx. quinquefasciatus*, and *An. Arabiensis* (Dacey and Chain 2020).

13.4.11.3 Use of Entomopathogenic Fungi

The significance of entomopathogenic fungus in the management of the malarial vector *Anopheles* sp. is crucial. These usages involve fungus from the genera *Metarhizium*, *Beauveria*, *Culicinomyces*, *Coelomomyces*, *Entomophthora*, and *Lagenidium*. Fungi alter the fitness conditions and behavior of the vector, which has an adverse influence on the transmission of malaria. Metarhizium has been found to inhibit the transmission of the malarial parasite from the vector by inducing the development of antimalarial peptides in mosquitoes (Naseem et al. 2016).

13.4.11.4 Nematodes

Many nematode species, including *Romanomermis iyengari* and *R. culicivorax*, have the potential to be biocides for mosquito larvae. They are said to successfully parasitize anophelines in rainwater pools and paddy fields (Chandra 2000). The efficiency of the nematode *R. iyengari* in controlling *Anopheles gambiae* was assessed in a study by Abagli et al. (2019).

13.4.11.5 Protozoa

Using gregarines and microsporidians to control larvae is an effective method. *Thelohania*, *Stempellia*, *Nosema*, and other genera of microsporidians are significant and beneficial (Chandra 2000). Chronic and fatal infection of mosquito species *Culex pseudovishnui* and *Cx. tritaeniorhynchus* was reported by Das (2003).

13.4.11.6 Plants

Azolla pinnata, a water fern, inhibits mosquito oviposition and egg hatchability in order to minimize mosquito populations. Less mosquito larvae have been discovered in the rice fields, pools, ponds, and wells, among other places (Chandra 2000; Ansari and Sharma 1991).

In Table 13.3, a number of naturally occurring mosquito-eating predator species are listed.

Table 13.3 Mosquito predators as tools for biological control of mosquitoes

Predator organisms/predatory species		Target prey mosquito species	References
Larvivorous fish	*Gambusia affinis*	For *Gambusia affinis* nonspecific because of nutrition and predatory tendencies	Huang et al. (2017), Yadav et al. (1992)
	Carassius auratus		Chatterjee et al. (1997)
	Poecilia reticulata	*Aedes albopictus*, *Aedes aegypti*, and *Culex quinquefasciatus*, *Anopheles stephensi*	Saleeza et al. (2014), Deacon et al. (2019)
	Aplocheilus dayi *Aplocheilus parvus* *Aplocheilus lineolatus* *Aplocheilus blockii* *Aplocheilus panchax*	Anopheline and culicine larvae	Ranathunge et al. (2021), Fernando et al. (2018) Chandra et al. (2008), Manna et al. (2011) Yadav et al. (1992), Aditya et al. (2012)
	Pethia conchonius and *Trichogaster chuna*	*Culex quinquefasciatus*	Chandra et al. (2015)
	Aphanius dispar	*Anopheles stephensi*, *Aedes aegypti*, and *Culex quinquefasciatus*	Al-Akel and Suliman (2011), Haq and Yadav (2011)
	Oryzias melastigma, *Oreochromis mossambicus*, and *Esomus danricus*	Anopheline and culicine larvae	Yadav et al. (1992)
	Danio rerio	*Aedes* mosquito larvae	Yadav et al. (1992), Aditya et al. (2012), Jafari et al. (2019)
	Ambassis (=Chanda) Nama, *Parambassis (=Chanda) ranga*, *Colisa fasciatus*	*Culex quinquefasciatus*	Aditya et al. (2012)
	Puntius tetrazona and *Hyphessobrycon rosaceus*	*Culex vishnui* subgroup larvae	Barik et al. (2018)

(continued)

Table 13.3 (continued)

Predator organisms/predatory species		Target prey mosquito species	References
	Air breathing fish *Anabas testudineus*, *Clarias batrachus*, and *Heteropneustes fossilis*	*Culex quinquefasciatus*	Bhattacharjee et al. (2009)
	Aquarium fish *Betta splendens*, *Pseudotropheus tropheops*, *Osphronemus gorami*, and *Pterophyllum scalare*	*Anopheles stephensi*	Ghosh et al. (2004)
	Cyprinus carpio, *Ctenopharyngodon idella*, *Oreochromis niloticus*, and *Clarias gariepinus*	*Anopheles stephensi*	Ghosh et al. (2005)
	Notopterus notopterus	*Culex vishnui*	Adhikari et al. (2021)
	Hypophthalmichthys molitrix	*Culex vishnui* group	Rawani et al. (2022)
Tadpoles		Eggs of *Aedes aegypti*	Bowatte et al. (2013)
Copepods Mainly *Mesocyclops* and *Macrocyclops* species, *Mesocyclops scrassus*, *Cyclops varicans*, and *C. languides*		*Culex* and *anopheles* larvae *Aedes aegypti*, *Aedes albopictus*	Marten and Reid (2007), Ranathunge et al. (2019)
Dragonfly nymph		Larvae of *Anopheles stephensi*, *Culex quinquefasciatus*, and *Aedes aegypti*	Singh et al. (2003), Ellis and Ladeau (2013) Chatterjee et al. (2007)
Aquatic hemipterans *Laccotrephes griseus* *Anisops sardea*		*Culex quinquefasciatus* *Anopheles stephensi*, *Culex quinquefasciatus*, *Culex vishnui*	Sareein et al. (2019), Ghosh and Chandra (2011) Mondal et al. (2014; 2017)
Elephant mosquito, *Toxorhynchites* (Fig. 13.2) Larvae of many species of *Toxorhynchites* like *Toxorhynchites splendens*, *Toxorhynchites amboinensis*, *Toxorhynchites rutilus*, *Toxorhynchites brevipalpis*, *Toxorhynchites moctezuma*		*Ae. albopictus* *Aedes aegypti*, *Aedes albopictus*, and *Culex quinquefasciatus* *Aedes aegypti* *Aedes aegypti* *Aedes aegypti*	Malla et al. (2023) Digma et al. (2019) Alomar and Alto (2022) Wood et al. (2022) Tikasingh and Eustace (1992)

Fig. 13.2 Mosquito predator *Toxorhynchites splendens* larva and adult

13.4.12 Mosquito Control with Genetic Engineering

Unfortunately, the disease burden caused by vector mosquitoes has not been completely eliminated by existing control strategies centered on insecticides and environmental management. We're looking for genetically based, scalable, and transportable solutions that reduce the likelihood of spreading these diseases. To solve these issues, methods like releasing genetically altered mosquitoes into the environment or genome engineering-based mosquito control strategies like gene drives have been developed. Even before the methods for permanently altering mosquitoes were developed, the biological management of diseases transmitted by mosquitoes by genetically modified mosquitoes was advocated. A recent increase in the development of genetic control technologies has involved modified mosquitoes intended to either reduce the population or modify the population. The sterile insect technique (SIT), the incompatible insect technique (IIT), and several transgene-based technologies, such as gene drives, are examples of suppression tactics (WHO and UNICEF 2010).

An effective and long-lasting method for controlling mosquito-borne diseases is to use transgenic insects created through genetic engineering. This tactic depends on the widespread release of transgenic mosquitoes into the wild, where it is anticipated that the transgene will either last permanently or briefly in the population of mosquitoes in the environment. Pathogen-resistant ("refractory") mosquitoes are created with the intention of being released into wild populations where they can disperse their heritable changes to stop the transfer of pathogens. Using effector genes that reduce offspring production in a particular mosquito population, genetically sterile insects are released to reduce mosquito populations. The most popular tactic based on this theory is the introduction of insects that have a dominant deadly gene (RIDL) (Wang et al. 2021).

The ability to drive endonuclease genes that lower female mosquito fertility is one of the most prospective uses for CRISPR-Cas9-based gene drives, which are methods for lowering mosquito populations' capability to transmit malaria (North et al. 2020).

There are reasons to be hopeful that this approach might succeed, putting aside worries about the introduction of genetically engineered mosquitoes. Genetic engineering technologies enable quick and exciting advancements, but they also call for an early and open conversation to involve regulatory bodies, the scientific community, and the general public. The ultimate goal of genetic engineering for mosquito control is to ensure that any potential unintended ecological and environmental repercussions are minimized while still giving future generations the undeniable advantages of a world free of vector-borne illnesses.

Additionally, novel vector control technologies based on RNAi are currently being used to manage mosquitoes (Müller et al. 2023; Gabrieli et al. 2014).

13.4.13 Attractive Targeted Sugar Bait Method

A new mosquito control technique called attractive targeted sugar bait (ATSB) technology involves inviting mosquitoes to feed on a sugar suspension that contains a toxin (Njoroge et al. 2023).

13.4.14 Integrated Mosquito Management

Numerous methods, each with benefits and drawbacks, have been used to prevent the transmission of diseases by mosquitoes. However, strategies like integrated vector management, which takes a variety of goals into account, seem to provide the finest feasible solutions for control that are simultaneously quick and long-lasting as well as the most affordable. In order to minimize mosquito populations while preserving a quality environment, it uses a comprehensive strategy for controlling mosquito populations that both reduces annoyance and safeguards the general public's health (Fouet and Kamdem 2019; Bouyer et al. 2020; Rupp 2001; Zhao and Xue 2022).

13.5 Emerging Challenges and Approaches for Mosquito Control in the Present

As carriers of deadly infections and parasites, mosquitoes (Diptera: Culicidae) pose a serious threat to millions of people worldwide (Nebbak et al. 2022). Many different mosquito species spread diseases, some of which are fatal. *Aedes aegypti* and *Aedes albopictus* are capable of transmitting the dengue, chikungunya, and Zika viruses. While numerous *Culex* mosquitoes are known to carry the Japanese encephalitis virus, anopheline mosquitoes transmit both human and monkey malaria. Due to their bites, other *Culex* mosquitoes are the primary source of nuisance. Vector control is essential in this situation to reduce the spread of various diseases and annoyances, in addition to providing treatment and healthcare. The use of insecticides is the major technique to combat these deadly diseases, but mosquitoes are evolving and spreading pesticide resistance, impeding control efforts. To manage mosquito populations effectively, sustainably, and safely, new mosquito control methods are being developed and used (Onen et al. 2023). The fundamental objective of these strategies, which have been developed over the past few decades, is to lower mosquito populations in order to combat epidemic scenarios.

Currently, a number of genetically based strategies are being utilized to tackle mosquitoes in addition to conventional management methods. In order to be included in vector control programmes for mosquito population control, a number of novel approaches have recently been evaluated for their cost-effectiveness and efficacy. These include GM mosquitoes, paratransgenesis, ATSB, gene drives, and the conventional radiation-based sterile insect method (SIT), which releases irradiated sterile males. The *Wolbachia*-based incompatible insect method (IIT), which relies on the release of *Wolbachia*-infected males, and the antibiotic-based release of insects bearing a dominant lethal gene (RIDL) are two alternatives of mosquito control (Benelli 2019; Javed et al. 2022; Martín-Park et al. 2022).

Numerous plant extracts, largely based on laboratory assays, have been reported to exhibit mosquitocidal or repellent activity against mosquito vectors; however, there are restrictions on their efficacy and usability in the field. It is necessary to resolve problems with their formulation and commercialization, nonstandard evaluation of their bioactivities, and their persistence over long periods of time in order to develop effective and long-lasting methods for their usage. Without a doubt, more bioactive chemicals exist that need to be studied, and future research should focus on finding plant-based products with the ultimate goal of employing them as a reliable remedy to control mosquito populations and mosquito-borne diseases.

Roadblocks still persist despite the fact that some control techniques are already in place and have been found to be, in varied degrees, successful in reducing the mosquito's ability to spread disease and act as a nuisance. The epidemics and the spread of diseases carried by mosquitoes persist for a variety of complex reasons, such as insecticide resistance, improper design of control programs, ineffective coverage, a lack of and inadequately trained human resources, as well as a lack of funding and infrastructure (Dahmana and Mediannikov 2020). The genomic era has provided the chance to reconsider time-tested tactics and create cutting-edge, intriguing new ways. However, none of these individual strategies are fully successful at the time. The vector control programs for the control of mosquitoes can use an integrated vector management (IVM) method. The application of IVM techniques assists vector control programs in locating and utilizing more local evidence, integrating interventions when necessary, and collaborating with the health sector, other sectors, households, and communities.

13.6 Future Direction in Sustainable Mosquito Control

In addition to recent outbreaks of novel arboviruses, the emergence of resistance in a number of Culicidae species, and the swift global spread of highly invasive mosquitoes, mosquito control programs are currently confronting significant and critical challenges. Insecticide-treated bed nets, adult repellents, biological control agents against mosquito young instars (primarily fish, amphibians, and copepods), sterile insect technique (SIT), "boosted SIT," symbiont-based techniques, and transgenic mosquitoes are the mainstays of today's control strategies. A variety of new diseases spread by mosquitoes have emerged in recent years as a result of climate change, a sharp rise in population, environmental degradation, and heightened pesticide resistance. However, the usage of pesticides may have a detrimental effect on other life forms, which may subsequently cause an ecosystem to become unbalanced. Therefore, it is crucial to concentrate on developing cutting-edge, practical, safe, and affordable techniques that have no adverse effects on groups that are not the target of the strategy. To combat this growing problem, new and innovative mosquito control solutions are now being developed. Programs for biological control have shown encouraging results, although certain extremely successful techniques, like genetic engineering, are still not fully taken into account as a means of control. Enhancing current tactics, such as the discharge of dominantly lethal insects, the sterile insect technique, or transgenesis, may offer crucial options for averting outbreaks, reducing the risk of at-risk populations, and combating resistance. From a conceptual standpoint, these tactics span a wide range, from intrusive transgenic cassettes that might drastically reduce mosquito numbers or a population's ability to spread disease.

New tactics based on alluring poisonous sugar baits and novel agents like *Wolbachia* and *Asaia* are currently being thoroughly researched for prospective use as chemical substitutes. Meanwhile, promising strategies, like those covered in this article, have previously demonstrated their efficacy but are rarely implemented, necessitating further consideration and attention in vector control strategies. Other strategies, such lawmaking and health education, also play a significant and supplemental role in mosquito control.

13.7 Conclusion

As the traditional ways of controlling mosquito populations, control tactics based on chemical, biological, and physical aspects have all been put in place to stop the spread of diseases transmitted by mosquitoes. Scientists are now able to investigate unconventional mosquito management strategies based on genetic modification, nanotechnology, etc., thanks to innovation and cutting-edge research. The best strategy to control the mosquito population is to use a variety of management techniques in cooperation.

While highlighting the breadth of study on the topic, this article on mosquito management will demonstrate the necessity of an interdisciplinary approach. This chapter will highlight a number of mosquito control methods that are in various stages of development and will include both opinion pieces and research articles with information from the lab and the field on control strategy development. As a majority of the current control measures are being developed and then tested in the field for their actual efficacy, I believe that secure academic collaboration between many academic domains will have a significant impact on the future of research on mosquitoes.

Controlling mosquito populations necessitates cooperation between the disciplines of physics, phytochemistry, vector biology, physiology, behavior, and ecology.

Points to Remember

1. As effective medications and vaccinations are currently unavailable, vector management is the only means of preventing and controlling illnesses spread by mosquitoes.
2. The increasing spread of illnesses and their widespread prevalence have made mosquito control an international concern.
3. The globe is witnessing a constant growth in these diseases despite the use of many conventional tactics, such as the employment of biological agents, environmental management, chemical-based interventions, human-vector interaction regulation, etc.
4. As a result, these control measures must be implemented as effectively as possible.
5. At the moment, chemical pesticides are the mainstay of mosquito control.
6. Nonetheless, human health, nontargets, and the environment have all suffered as a result of the frequent and careless use of these toxicants.
7. Therefore, researchers are always investigating the other novel approaches.
8. The measures now in place can be strengthened, and the mosquito population can be reduced with the aid of the actions.
9. However, implementing any new measure necessitates tracking its effectiveness, which is a logistical barrier in and of itself.
10. It is advised to apply the integrated mosquito control plan and conduct routine mosquito monitoring in the regions where mosquitoes are most common tactics.

References

Abagli AZ, Alavo TB, Perez-Pacheco R, Platzer EG (2019) Efficacy of the mermithid nematode, *Romanomermis iyengari*, for the biocontrol of Anopheles gambiae, the major malaria vector in sub-Saharan Africa. Parasit Vectors 12(1):1–8

Abbas MG, Haris A, Binyameen M, Nazir A, Mozūratis R, Azeem M (2022) Chemical composition, larvicidal and repellent activities of wild plant essential oils against *Aedes aegypti*. Biology 12(1):8

Adhikari U, Chandra G (2014) Larvicidal, smoke toxicity, repellency and adult emergence inhibition effects of leaf extracts of *Swietenia mahagoni* Linnaeus against *Anopheles stephensi* Liston (Diptera: Culicidae). Asian Pac J Trop Dis 4:S279–S283

Adhikari U, Singha S, Chandra G (2012) In vitro repellent and larvicidal efficacy of *Swietenia mahagoni* against the larval forms of *Culex quinquefasciatus* Say. Asian Pac J Trop Biomed 2(1):S260–S264

Adhikari U, Ghosh A, Mallick S, Chandra G (2021) Predation potentiality of *Notopterus notopterus* on JE vector, *Culex vishnui*. Int J Mosq Res 8:1–5

Aditya G, Pal S, Saha N, Saha GK (2012) Efficacy of indigenous larvivorous fishes against *Culex quinquefasciatus* in the presence of alternative prey: implications for biological control. J Vector Borne Dis 49(4):217

Akono PN, Mbida JA, Dongmo PM, Tonga C, Tchamga LA, Mounbain F, Magne GT, Mache PN, Kekeunou S (2016) Chemical composition and insecticidal activity of essential oils of *Cinnamomum zeylanicum*, *Citrus grandis*, *Citrus medica* and *Citrus sinensis* leaves from Cameroon on *Anopheles gambiae* Giles, 1902. J Entomol Zool Stud 4(6):17–23

Al-Akel AS, Suliman EM (2011) Biological control agent for mosquito larvae: review on the killifish, *Aphanius dispar dispar* (Rüppel, 1829). Afr J Biotechnol 10(44):8683–8688

Alomar AA, Alto BW (2022) Evaluation of pyriproxyfen effects on *Aedes aegypti* and predatory mosquito *Toxorhynchites rutilus* (Diptera: Culicidae). J Med Entomol 59(2):585–590

Alphey L, Benedict M, Bellini R, Clark GG, Dame DA, Service MW, Dobson SL (2010) Sterile-insect methods for control of mosquito-borne diseases: an analysis. Vector Borne Zoonotic Dis 10(3):295–311

AlQahtani FS, AlShebly MM, Govindarajan M, Senthilmurugan S, Vijayan P, Benelli G (2017) Green and facile biosynthesis of silver nanocomposites using the aqueous extract of *Rubus ellipticus* leaves: toxicity and oviposition deterrent activity against Zika virus, malaria and filariasis mosquito vectors. J Asia Pac Entomol 20(1):157–164

Amer A, Mehlhorn H (2006) Repellency effect of forty-one essential oils against *Aedes*, *Anopheles*, and *Culex* mosquitoes. Parasitol Res 99:478–490

Ansari MA, Sharma VP (1991) Role of *Azolla* in controlling mosquito breeding in Ghaziabad district villages (UP). Indian J Malariol 28(1):51–54

Asadollahi A, Khoobdel M, Zahraei-Ramazani A, Azarmi S, Mosawi SH (2019) Effectiveness of plant-based repellents against different *Anopheles* species: a systematic review. Malar J 18(1):1–20

Attar N (2016) ZIKA virus circulates in new regions. Nat Rev Microbiol 14(2):62

Barik M, Bhattacharjee I, Ghosh A, Chandra G (2018) Larvivorous potentiality of *Puntius tetrazona* and *Hyphessobrycon rosaceus* against *Culex vishnui* subgroup in laboratory and field based bioassay. BMC Res Notes 11:1–5

Becker N (2000) Bacterial control of vector-mosquitoes and black flies. In: Entomopathogenic bacteria: from laboratory to field application. Springer, Berlin, pp 383–398

Benelli G (2016) Plant-mediated synthesis of nanoparticles: a newer and safer tool against mosquito-borne diseases? Asian Pac J Trop Biomed 6(4):353–354

Benelli G (2019) Managing mosquitoes and ticks in a rapidly changing world—facts and trends. Saudi J Biol Sci 26(5):921–929

Benelli G, Mehlhorn H (2016) Declining malaria, rising of dengue and Zika virus: insights for mosquito vector control. Parasitol Res 115:1747–1754

Benelli G, Jeffries CL, Walker T (2016) Biological control of mosquito vectors: past, present, and future. Insects 7(4):52

Benelli G, Caselli A, Canale A (2017) Nanoparticles for mosquito control: challenges and constraints. J King Saud Univ Sci 29(4):424–435

Bhattacharjee I, Aditya G, Chandra G (2009) Laboratory and field assessment of the potential of larvivorous, air-breathing fishes as predators of culicine mosquitoes. Biol Control 49(2):126–133

Bouyer J, Lefrançois T (2014) Boosting the sterile insect technique to control mosquitoes. Trends Parasitol 30(6):271–273

Bouyer J, Yamada H, Pereira R, Bourtzis K, Vreysen MJ (2020) Phased conditional approach for mosquito management using sterile insect technique. Trends Parasitol 36(4):325–336

Bowatte G, Perera P, Senevirathne G, Meegaskumbura S, Meegaskumbura M (2013) Tadpoles as dengue mosquito (*Aedes aegypti*) egg predators. Biol Control 67(3):469–474

Cano J, Rebollo MP, Golding N, Pullan RL, Crellen T, Soler A, Hope LA, Lindsay SW, Hay SI, Bockarie MJ, Brooker SJ (2014) The global distribution and transmission limits of lymphatic filariasis: past and present. Parasit Vectors 7:1–9

Caraballo H, King K (2014) Emergency department management of mosquito-borne illness: malaria, dengue, and West Nile virus. Emerg Med Pract 16(5):1–23; quiz 23–4. PMID: 25207355

Cator LJ, Wyer CA, Harrington LC (2021) Mosquito sexual selection and reproductive control programs. Trends Parasitol 37(4):330–339

Chakraborty S, Burman S, Chandra G (2022) Repellent and oviposition-deterrent efficacies of *Cuscuta chinensis* on filarial vector *Culex quinquefasciatus*. Int J Mosq Res 9(3):48–52
Chandra G (2000) Mosquito. Sribhumi Publishing Co., Calcutta, pp 1–102
Halder KM, Chandra G (2015) Adulticidal bioassay and repellency potentiality of fruits of *Drypetes roxburghii* on *Culex quinquefasciatus* and *Anopheles stephensi*. J Mosq Res 5(12)
Chandra G, Bhattacharjee I, Chatterjee SN, Ghosh A (2008) Mosquito control by larvivorous fish. Indian J Med Res 127(1):13–27
Chandra G, Mandal B, Bhattacharjee I, Kundu JK (2015) Biocontrol efficacy of two indigenous fishes *Pethia conchonius* and *Trichogaster chuna* against larvae of the mosquito, *Culex quinquefasciatus* Say, 1823. Asian Acad Res J Multidiscip 2(5):38–52
Chatterjee SN, Das S, Chandra G (1997) Gold fish (*Carrasius auratus*) as a strong larval predator of mosquito. Trans Zool Soc India 1:112–114
Chatterjee SN, Ghosh A, Chandra G (2007) Eco-friendly control of mosquito larvae by *Brachytron pratense* nymph. J Environ Health 69(8):44–49
Christian UA, Kayode OI, Chinenye UC, Sule UB (2021) Environmental manipulation: a potential tool for mosquito vector control. In: The wonders of diptera-characteristics, diversity, and significance for the world's ecosystems. IntechOpen, London
Dacey DP, Chain FJ (2020) The challenges of microbial control of mosquito-borne diseases due to the gut microbiome. Front Genet 11:504354
Dahmana H, Mediannikov O (2020) Mosquito-borne diseases emergence/resurgence and how to effectively control it biologically. Pathogens 9(4):310
Das BP (2003) *Chilodonella uncinata*—a protozoa pathogenic to mosquito larvae. Curr Sci 85:483–489
Deacon AE, Ghosh SK, Bhat A, Magurran AE (2019) Predatory behaviour of female guppies (*Poecilia reticulata*) in a mosquito control context: the importance of social and habitat factors. Aquat Invasions 14(3):478–489
Deletre E, Martin T, Campagne P, Bourguet D, Cadin A, Menut C, Bonafos R, Chandre F (2013) Repellent, irritant and toxic effects of 20 plant extracts on adults of the malaria vector *Anopheles gambiae* mosquito. PLoS One 8(12):e82103
Deng SQ, Khater EIM, Tambo E, Wang DQ (2023) Editorial: emerging mosquito-borne diseases and novel biocontrol strategies. Front Cell Infect Microbiol 13:1143165. https://doi.org/10.3389/fcimb.2023.1143165. PMID: 36844405; PMCID: PMC9950743
Diabate A, Tripet F (2015) Targeting male mosquito mating behaviour for malaria control. Parasit Vectors 8:1–3
Digma JR, Sumalde AC, Salibay CC (2019) Laboratory evaluation of predation of *Toxorhynchites amboinensis* (Diptera: Culicidae) on three mosquito vectors of arboviruses in The Philippines. Biol Control 137:104009
Ellis MR, Ladeau S (2013) Influence of dragonfly larvae on mosquito development and survival, vol 12545(8). Cary Institute of Ecosystem Studies, Millbrook
Fauci AS, Morens DM (2016) Zika virus in the Americas—yet another arbovirus threat. N Engl J Med 374(7):601–604
Fernando GK, Jayakody S, Wijenayake WM, Galappaththy GN, Yatawara M, Harishchandra J (2018) A comparison of the larvivorous habits of exotic *Poecilia reticulata* and native *Aplocheilus parvus*. BMC Ecol 18(1):1–2
Forsyth JE, Mutuku FM, Kibe L, Mwashee L, Bongo J, Egemba C, Ardoin NM, LaBeaud AD (2020) Source reduction with a purpose: mosquito ecology and community perspectives offer insights for improving household mosquito management in coastal Kenya. PLoS Negl Trop Dis 14(5):e0008239
Fouet C, Kamdem C (2019) Integrated mosquito management: is precision control a luxury or necessity? Trends Parasitol 35(1):85–95
Fradin MS, Day JF (2002) Comparative efficacy of insect repellents against mosquito bites. N Engl J Med 347(1):13–18
Freeborn SB, Atsatt RF (1918) The effects of petroleum oils on mosquito larvae. J Econ Entomol 11:299–307

Gabrieli P, Smidler A, Catteruccia F (2014) Engineering the control of mosquito-borne infectious diseases. Genome Biol 15(11):1–9

Ghosh A, Chandra G (2011) Functional responses of *Laccotrephes griseus* (Hemiptera: Nepidae) against *Culex quinquefasciatus* (Diptera: Culicidae) in laboratory bioassay. J Vector Borne Dis 48(2):72

Ghosh A, Bhattacharjee I, Ganguly M, Mondal S, Chandra G (2004) Efficacy of some common aquarium fishes as biocontrol agent of preadult mosquitoes. Indonesian Bull Health Res 32(4):65308

Ghosh A, Mandal S, Bhattacharjee I, Chandra G (2005) Biological control of vector mosquitoes by some common exotic fish predators. Turk J Biol 29(3):167–171

Ghosh A, Chowdhury N, Chandra G (2012) Plant extracts as potential mosquito larvicides. Indian J Med Res 135(5):581

Ghosh A, Chandra G, Bandyopadhyay S, Mondal RP (2016a) Effect of habitat modifications on predation potential of *Anisops sardea* (Hemiptera: Notonectidae) against larvae of *Culex vishnui*, vector of Japanease encephalitis. J Mosq Res 6

Ghosh P, Mondal RP, Haldar KM, Chandra G (2016b) Larvicidal, pupicidal and smoke toxic activity of Alangium salvifolium leaf extracts against *Culex vishnui* group mosquitoes. J Mosq Res 6

Go YY, Balasuriya UB, Lee CK (2014) Zoonotic encephalitides caused by arboviruses: transmission and epidemiology of alphaviruses and flaviviruses. Clin Exp Vaccine Res 3(1):58–77

Gould E, Pettersson J, Higgs S, Charrel R, De Lamballerie X (2017) Emerging arboviruses: why today? One Health 4:1–3

Haldar KM, Haldar B, Chandra G (2013) Fabrication, characterization and mosquito larvicidal bioassay of silver nanoparticles synthesized from aqueous fruit extract of Putranjiva, *Drypetes roxburghii* (Wall.). Parasitol Res 112:1451–1459

Haq S, Yadav RS (2011) Geographical distribution and evaluation of mosquito larvivorous potential of *Aphanius dispar* (Rüppell), a native fish of Gujarat, India. J Vector Borne Dis 48(4):236

Haris A, Azeem M, Abbas MG, Mumtaz M, Mozūratis R, Binyameen M (2023) Prolonged repellent activity of plant essential oils against dengue vector, *Aedes aegypti*. Molecules 28(3):1351

Hegazy MI, Hegazy AM, Saad AM, Salem HM, El-Tahan AM, El-Saadony MT, Soliman SM, Taha AE, Alshehri MA, Ahmed AE, Swelum AA (2022) Some biologically active microorganisms have the potential to suppress mosquito larvae (*Culex pipiens*, Diptera: Culicidae). Saudi J Biol Sci 29(4):1998–2006

Herms WB, Gray HF (1944) Mosquito control, practical methods for abatement of disease vectors and pests, 2nd edn. Oxford University Press, New York

Hills S (1996) Ross River virus and Barmah Forest virus infection. Commonly asked questions. Aust Fam Physician 25(12):1822–1824

Howard LO (1892) An experiment against mosquitoes. Insect Life 5(1):12–14

Howard LO (1902) Mosquitoes. How they live, how they carry disease, how they are classified, how they may by destroyed. McClure Phillips and Co., New York

Hribar LJ, Boehmler MB, Murray HL, Pruszynski CA, Leal AL (2022) Mosquito surveillance and insecticide resistance monitoring conducted by the Florida keys mosquito Control District, Monroe County, Florida, USA. Insects 13(10):927

Huang YJ, Higgs S, Vanlandingham DL (2017) Biological control strategies for mosquito vectors of arboviruses. Insects. 8(1):21

Huang W, Wang S, Jacobs-Lorena M (2020) Use of microbiota to fight mosquito-borne disease. Front Genet 11:196

Jafari A, Enayati A, Jafari F, Haghi FM, Hosseini-Vasoukolaei N, Sadeghnezhad R, Azarnoosh M, Fazeli-Dinan M (2019) A narrative review of the control of mosquitoes by Larvivorous fish in Iran and the world. Iran J Health Sci 7(2):49–60

Javed Z, Mansha S, Saleem U, Mangat A, Rasool B, Imran M et al (2022) Innovative methods of mosquito management. In: Advances in diptera—insight, challenges and management tools. IntechOpen, London. https://doi.org/10.5772/intechopen.107364

Jeyaprakasam NK, Pramasivan S, Liew JW, Van Low L, Wan-Sulaiman WY, Ngui R, Jelip J, Vythilingam I (2021) Evaluation of mosquito magnet and other collection tools for *Anopheles* mosquito vectors of simian malaria. Parasit Vectors 14:1–3

Johnson T (1998) CRC ethnobotany desk reference. CRC Press, Boca Raton, p 60

Jones RT, Ant TH, Cameron MM, Logan JG (2021) Novel control strategies for mosquito-borne diseases. Philos Trans R Soc Lond B Biol Sci 376(1818):20190802. https://doi.org/10.1098/rstb.2019.0802. Epub 2020 Dec 28. PMID: 33357056; PMCID: PMC7776938

Kalita B, Bora S, Sharma AK (2013) Plant essential oils as mosquito repellent—a review. Int J Res Dev Pharm Life Sci 3(1):741–747

Kamal HA, Khater EI (2010) The biological effects of the insect growth regulators; pyriproxyfen and diflubenzuron on the mosquito *Aedes aegypti*. J Egypt Soc Parasitol 40(3):565–574

Kumar R, Hwang JS (2006) Larvicidal efficiency of aquatic predators: a perspective for mosquito biocontrol. Zool Stud Taipei 45(4):447–460

Lau KW, Chen CD, Lee HL, Norma-Rashid Y, Sofian-Azirun M (2015) Evaluation of insect growth regulators against field-collected *Aedes aegypti* and *Aedes albopictus* (Diptera: Culicidae) from Malaysia. J Med Entomol 52(2):199–206. https://doi.org/10.1093/jme/tju019

Madhiyazhagan P, Murugan K, Kumar AN, Nataraj T, Subramaniam J, Chandramohan B, Panneerselvam C, Dinesh D, Suresh U, Nicoletti M, Alsalhi MS (2017) One pot synthesis of silver nanocrystals using the seaweed Gracilaria edulis: biophysical characterization and potential against the filariasis vector *Culex quinquefasciatus* and the midge *Chironomus circumdatus*. J Appl Phycol 29:649–659

Maia MF, Moore SJ (2011) Plant-based insect repellents: a review of their efficacy, development and testing. Malar J 10(3):1–14

Malla RK, Mandal KK, Dutta M, Chandra G (2019) An estimation of monthly propagation of denguevector *Aedes aegypti* in rainwater filled tires. Int J Pest Manag 66:239. https://doi.org/10.1080/09670874.2019.1616130

Malla RK, Mandal KK, Burman S, Das S, Ghosh A, Chandra G (2023) Numerical analysis of predatory potentiality of *Toxorhynchites splendens* against larval Aedes albopictus in laboratory and semi-field conditions. Sci Rep 13(1):7403

Manh HD, Hue DT, Hieu NT, Tuyen DT, Tuyet OT (2020) The mosquito larvicidal activity of essential oils from *Cymbopogon* and *eucalyptus* species in Vietnam. Insects 11(2):128

Manna B, Aditya G, Banerjee S (2011) Habitat heterogeneity and prey selection of *Aplocheilus panchax*: an indigenous larvivorous fish. J Vector Borne Dis 48(3):144

Marten GG, Reid JW (2007) Cyclopoid copepods. J Am Mosq Control Assoc 23(sp2):65–92

Martín-Park A, Che-Mendoza A, Contreras-Perera Y, Pérez-Carrillo S, Puerta-Guardo H, Villegas-Chim J, Guillermo-May G, Medina-Barreiro A, Delfín-González H, Méndez-Vales R, Vázquez-Narvaez S (2022) Pilot trial using mass field-releases of sterile males produced with the incompatible and sterile insect techniques as part of integrated *Aedes aegypti* control in Mexico. PLoS Negl Trop Dis 16(4):e0010324

Maurice NW, Pierre S, Nukenine NE (2022) Mosquitocidal properties of three essential oil plants tested against *Anopheles gambiae* Sensu Stricto, Giles 1902 (Diptera: Culicidae, Anophelinae), main malarial vector. J Trop Dis Public Health 10(2):1000311

Mondal RP, Ghosh A, Bandyopadhyay S, Chandra G (2014) Functional response analysis of *Anisops sardea* (Hemiptera: Notonectidae) against *Culex quinquefasciatus* in laboratory condition. Indian J Med Res 140(4):551

Mondal RP, Chandra G, Bandyopadhyay S, Ghosh A (2017) Effect of temperature and search area on the functional response of *Anisops sardea* (Hemiptera: Notonectidae) against *Anopheles stephensi* in laboratory bioassay. Acta Trop 166:262–267

Müller R, Bálint M, Hardes K, Hollert H, Klimpel S, Knorr E, Kochmann J, Lee KZ, Mehring M, Pauls SU, Smets G (2023) RNA interference to combat the Asian tiger mosquito in Europe: a pathway from design of an innovative vector control tool to its application. Biotechnol Adv 8:108167

Murugan K, Rajaganesh R, Hwang JS, Wang L, Vasanthakumaran M, Dahms HU, Panneerselvam C, Mohanta YK, Muthupandian S, Janarthanam RB, Chen F (2023) Smoke toxicity effect of bio-fabricated mosquito coil for the sustainable management of mosquito vectors. J Nat Pesticide Res 21:100048

Naseem S, Malik MF, Munir T (2016) Mosquito management: a review. J Entomol Zool Stud 4(5):73–79

Nayar JK, Ali A (2003) A review of monomolecular surface films as larvicides and pupicides of mosquitoes. J Vector Ecol 28(2):190–199

Nebbak A, Almeras L, Parola P, Bitam I (2022) Mosquito vectors (Diptera: Culicidae) and mosquito-borne diseases in North Africa. Insects 13(10):962

Njoroge TM, Hamid-Adiamoh M, Duman-Scheel M (2023) Maximizing the potential of attractive targeted sugar baits (ATSBs) for integrated vector management. Insects 14(7):585

North AR, Burt A, Godfray HC (2020) Modelling the suppression of a malaria vector using a CRISPR-Cas9 gene drive to reduce female fertility. BMC Biol 18(1):1–4

Ojewumi ME, Obanla OR, Atauba DM (2021) A review on the efficacy of *Ocimum gratissimum*, *Mentha spicata*, and *Moringa oleifera* leaf extracts in repelling mosquito. Beni-Suef Univ J Basic Appl Sci 10:1–2

Onen H, Luzala MM, Kigozi S, Sikumbili RM, Muanga CJ, Zola EN, Wendji SN, Buya AB, Balciunaitiene A, Viškelis J, Kaddumukasa MA (2023) Mosquito-borne diseases and their control strategies: an overview focused on green synthesized plant-based metallic nanoparticles. Insects 14(3):221

Osanloo M, Sedaghat MM, Esmaeili F, Amani A (2018) Larvicidal activity of essential oil of *Syzygium aromaticum* (Clove) in comparison with its major constituent, eugenol, against *Anopheles stephensi*. J Arthropod Borne Dis 12(4):361

Ottesen EA, Hooper PJ, Bradley M, Biswas G (2008) The global programme to eliminate lymphatic filariasis: health impact after 8 years. PLoS Negl Trop Dis 2(10):e317

Özcan M, Chalchat JC (2002) Essential oil composition of *Ocimum basilicum* L. Czech J Food Sci 20(6):223–228

Patel EK, Gupta A, Oswal RJ (2012) A review on: mosquito repellent methods. Int J Pharm Chem Biol Sci 2(3):310–317

Pepin M, Bouloy M, Bird BH, Kemp A, Paweska J (2010) Rift Valley fever virus (Bunyaviridae: Phlebovirus): an update on pathogenesis, molecular epidemiology, vectors, diagnostics and prevention. Vet Res 41(6):61

Phasomkusolsil S, Soonwera M (2011) Efficacy of herbal essential oils as insecticide against *Aedes aegypti* (Linn.), *Culex quinquefasciatus* (Say) and *Anopheles dirus* (Peyton and Harrison). Southeast Asian J Trop Med Public Health 42(5):1083

Poolprasert P, Tanruean K, Napiroon T, Phusing S, Torres JR, Villanueva PM (2019) Larvicidal effects of *Paederia pilifera* Hook. f. Leaft and *Cuscuta reflexa* Roxb. Stem extracts against the dengue vector mosquito *Aedes aegypti* Linn. J Appl Sci 18(1):31–38

Poopathi S, Thirugnanasambantham K, Mani C, Ragul K, Sundarapandian SM (2014) Isolation of mosquitocidal bacteria (*Bacillus thuringiensis*, *B. sphaericus* and *B. cereus*) from the excreta of arid birds. Indian J Exp Biol 52:739–747

Poulin B, Lefebvre G, Muranyi-Kovacs C, Hilaire S (2017) Mosquito traps: an innovative, environmentally friendly technique to control mosquitoes. Int J Environ Res Public Health 14(3):313

Pratiwi MA (2021) The repellent activity test of rosemary leaf (Rosmarinus officinalis L) essential oil gel preparations influence on *Aedes aegypti* mosquito. J Phys Conf Ser 1788(1):012016

Qureshi AI (2017) Zika virus disease: from origin to outbreak. Academic Press, New York

Qureshi AI, Saeed O (eds) (2019) Dengue virus disease: from origin to outbreak. Academic Press, New York

Ramkumar G, Karthi S, Muthusamy R, Natarajan D, Shivakumar MS (2015) Adulticidal and smoke toxicity of *Cipadessa baccifera* (Roth) plant extracts against *Anopheles stephensi*, *Aedes aegypti*, and *Culex quinquefasciatus*. Parasitol Res 114:167–173

Ranathunge T, Abeyewickreme W, Iqbal MC, Hapugoda M (2019) Use of cyclopoid copepods for control of *anopheles* (Diptera: Culicidae) mosquito larvae to prevent re-emergence of malaria in Sri Lanka. J Vector Borne Dis 56:200–206

Ranathunge T, Kusumawathie PH, Abeyewickreme W, Udayanga L, Fernando T, Hapugoda M (2021) Biocontrol potential of six locally available fish species as predators of *Aedes aegypti* in Sri Lanka. Biol Control 160:104638
Rawani A, Ghosh A, Laskar S, Chandra G (2012) Aliphatic amide from seeds of *Carica papaya* as mosquito larvicide, pupicide, adulticide, repellent and smoke toxicant. J Mosq Res 2(2)
Rawani A, Ghosh A, Chandra G (2022) Functional response of fingerlings of *Hypophthalmichthys molitrix* (Silver carp) as mosquito larval predator against *Culex vishnui* group. Proc Zool Soc 75(4):413–417
Rupp HR (2001) Integrated mosquito management: no new thing. Emerg Infect Dis 7(4):761
Saleeza SN, Norma-Rashid Y, Sofian-Azirun M (2014) Guppies as predators of common mosquito larvae in Malaysia. Southeast Asian J Trop Med Public Health 45(2):299
Sareein N, Phalaraksh C, Rahong P, Techakijvej C, Seok S, Bae YJ (2019) Relationships between predatory aquatic insects and mosquito larvae in residential areas in northern Thailand. J Vector Ecol 44(2):223–232
Sarwar M (2016) Mosquito-borne viral infections and diseases among persons and interfering with the vector activities. Int J Vaccines Vaccination 3(2):00063
Singh RK, Dhiman RC, Singh SP (2003) Laboratory studies on the predatory potential of dragon-fly nymphs on mosquito larvae. J Commun Dis 35(2):96–101
Singha S, Adhikari U, Chandra G (2011) Smoke repellency and mosquito larvicidal potentiality of *Mesua ferra* L. leaf extract against filarial vector *Culex quinquefasciatus* Say. Asian Pac J Trop Biomed 1(1):S119–S123
Sritabutra D, Soonwera M, Waltanachanobon S, Poungjai S (2011) Evaluation of herbal essential oil as repellents against *Aedes aegypti* (L.) and *Anopheles dirus* Peyton & Harrion. Asian Pac J Trop Biomed 1(1):S124–S128
Srivastava PK, Sharma RS, Sharma SN, Singh S, Das Gupta RK, Baruah K (2014) Integrated vector management: policy and implementation under National Vector Borne Disease Control Programme. Indian J Commun Dis 46(2):46–50
Tikasingh ES, Eustace A (1992) Suppression of Aedes aegypti by predatory *Toxorhynchites moct-ezuma* in an Island habitat. Med Vet Entomol 6(3):272–280
Vivekanandhan P, Senthil-Nathan S, Shivakumar MS (2018) Larvicidal, pupicidal and adult smoke toxic effects of *Acanthospermum hispidum* (DC) leaf crude extracts against mosquito vectors. Physiol Mol Plant Pathol 101:156–162
Wang GH, Gamez S, Raban RR, Marshall JM, Alphey L, Li M, Rasgon JL, Akbari OS (2021) Combating mosquito-borne diseases using genetic control technologies. Nat Commun 12(1):4388
Wendimu A, Tekalign W (2021) Field efficacy of ethnomedicinal plant smoke repellency against *Anopheles arabiensis* and *Aedes aegypti*. Heliyon 7(7):e07373
Whitten MJ, Foster GG (1975) Genetical methods of pest control. Annu Rev Entomol 20:461467
WHO (2016) Zika virus. Fact sheet no 1 updated January 2016. Accessed 7 Aug 2023
Wood MJ, Alkhaibari AM, Butt TM (2022) Stress-mediated responses of Aedes aegypti (Diptera: Culicidae) larvae when exposed to *Metarhizium brunneum* (Hypocreales: Clavicipitaceae) and *Toxorhynchites brevipalpis* (Diptera: Culicidae). J Med Entomol 59(5):1732–1740
World Health Organization (2016a) Zika virus. Fact sheet no 1 updated January 2016. Accessed 30 Jul 2023
World Health Organization (2016b) Neglected tropical diseases. Geneva. <http://www.who.int/neglected_diseases/diseases/en/>. Accessed 30 Jul 2023
World Health Organization (2018) The evaluation process for vector control products. WHO, Geneva
World Health Organization (2019) World malaria report 2019. World Health Organization, Geneva
World Health Organization (2020) Vector borne diseases. https://www.who.int/news-room/fact-sheets/detail/vector-borne-diseases. Accessed 7 Aug 2023
World Health Organization (2023a) Dengue and severe dengue. https://www.who.int/news-room/fact-sheets/detail/dengue-and-severe-dengue. Accessed 7 Aug 2023

World Health Organization (2023b) World malaria report 2022. https://www.who.int/publications/i/item/9789240064898. Accessed 29 Jul 2023

World Health Organization and UNICEF (2010) Progress and prospects for the use of genetically modified mosquitoes to inhibit disease transmission

Yadav RS, Padhan K, Sharma VP (1992) Fishes of district Sundargarh, Orissa, with special reference to their potential in mosquito control. Indian J Malariol 29(4):225–233

Yang P, Ma Y, Zheng S (2005) Adulticidal activity of five essential oils against *Culex pipiens quinquefasciatus*. J Pestic Sci 30(2):84–89

Zhao TY, Xue RD (2022) Integrated mosquito management in rice field in China. Wetl Ecol Manag 30(5):963–973

Eco-Friendly Management of Mosquitoes

14

Devinder Kaur Kocher, Sarita Kumar, and Neera Kapoor

Abstract

Management of mosquitoes has always relied on the use of chemical insecticides, which either kill or prevent the adults from biting or eradicate the larvae at breeding sites. DEET (N,N-diethyl-meta-toluamide), picaridin, permethrin, allethrin, malathion, and temephos are some of the synthetic insecticides used frequently for the mosquito control. However, frequent application of these chemicals has resulted in insecticide resistance among the mosquito species, adverse effects on environmental quality, biological magnification of toxic substances through the food chain, and lethal effects on nontarget organisms including human health. Though chemical insecticides generally use compounds at low concentrations, studies have shown that extensive and recurrent use of these compounds cause residual effects affecting the nontarget organisms, humans, and the environment. Due to their potential virulent effects, high operational cost, and environmental pollution, there is an urgent need of developing and implementation of alternate eco-safe biorational approaches to control mosquitoes, which can manage resistance and prevent resurgence of mosquito-borne diseases. Thus, the current scenario encourages the use of alternative mosquito control approaches, which are effective and economical, and have negligible impact on the environment. It is believed that mosquito control can be achieved

D. K. Kocher (✉)
Department of Zoology, Punjab Agricultural University, Ludhiana, Punjab, India
e-mail: dkocher2002@pau.edu

S. Kumar
Department of Zoology, Acharya Narendra Dev College, University of Delhi, New Delhi, India
e-mail: saritakumar@andc.du.ac.in

N. Kapoor
School of Sciences, IGNOU, New Delhi, India
e-mail: neerakapoor@ignou.ac.in

© The Author(s), under exclusive license to Springer Nature Singapore Pte Ltd. 2024
Omkar (ed.), *Mosquitoes*, https://doi.org/10.1007/978-981-97-4163-2_14

faster using these vector control measures, which can complement chemical-based interventions. This chapter aims to highlight ecofriendly mosquito control measures, which have been broadly categorized into three types: (1) phytochemicals, (2) biological control agents, and (3) genetic approaches. These methods will give a better understanding of developing an integrated approach for effective ecofriendly mosquito control that will reduce the dependency on insecticide-based control measures.

Keywords

Biocontrol agents · Essential oils · Genetic manipulations · Mosquito control · Plant-based products · Transgenics

Learning Objectives

This chapter has been written with the following important objectives:

1. To highlight the problems associated with chemical insecticides being commonly used for the control of mosquitoes.
2. To recognize the need for ecofriendly mosquito control approaches, alternative to chemical insecticides.
3. To discuss the principles of various eco-safe mosquito control strategies and, understand their merits and demerits.
4. To comprehend these approaches by using suitable examples.

14.1 Introduction

Mosquitoes are well-known cosmopolitan disease vectors, which transmit diseases, like malaria, dengue, chikungunya, Japanese encephalitis, filaria, West Nile fever, and yellow fever, and are responsible for millions of deaths each year globally. Since time immemorial, several control interventions have been used for control of mosquitoes. Presently, most of the control strategies are based on the use of chemical insecticides because of the lack of effective vaccination and medication against mosquito-borne diseases. However, the continuous and extensive use of synthetic chemicals is associated with a number of drawbacks, including harmful environmental impacts, damaging effects on nontarget species, and the development of insecticide resistance in vectors. Therefore, practicing IMM (integrated mosquito management) is the need of the hour. The integrated control of mosquitoes is the combined use of biotic, physical, and chemical controls in addition to the use of natural enemies and other population-regulating variables. People now prefer the new green mosquito control, which mainly includes ecological control methods. Several newer and safer tools have been developed to control mosquitoes. This chapter deals with the use of phytochemicals, biological and genetic control methods for the management of mosquitoes.

Phytochemicals are widely explored and employed to suppress mosquito population. These are secondary metabolites extracted from plants and utilized against mosquitoes at a concentration which does not alter the physiology of nontarget organisms. Several groups of phytochemicals, such as alkaloids, essential oils (EOs), phenolics, steroids, and terpenoids, from different plants have been reported for their mosquitocidal activities. Biological control comprises the use of beneficial insects, such as predators and parasitoids, or pathogens such as fungi and viruses, to control mosquitoes. The goal of biological control is to bring their population down, below the economic threshold, instead of its eradication. The biological agents, being present naturally in the environment, are considered safe and effective to use. Genetic control is an advanced approach, which involves sterile insect technique, release of insect carrying a dominant lethal, *Wolbachia* bacteria-based control, population replacement strategy, population suppression strategy, gene drive, etc.

Many intervention tactics have been applied globally due to the diversity of mosquito species, their biology, and the diseases they transmit. It is crucial to identify the right interventions for a given vector in order to control it. This chapter discusses all ecofriendly strategies with their merits and demerits and suitable examples.

A brief overview of these approaches is given below:

14.2 Phytochemicals in Mosquito Control

Phytochemicals arise from the Greek word *phyton*, which means "plant"; thus, these are chemicals which are derived from the plants or produced by the plants through metabolism. Phytochemicals are extracted either from the whole plant or from definite plant parts, such as flowers, leaves, stems, bark, fruits, seeds, and roots (Shaalan et al. 2005). Generally, these chemicals are involved in various biological activities of the host plant, like growth, hormonal action, stimulation of enzymes, antibacterial effect, and disease preventive mechanisms (Forni et al. 2019). Thousands of phytochemicals are well known; out of these, many have been used in preparing traditional medicines, some being used as poisons and a few being utilized for protection against mosquitoes (Sadiq et al. 2017). The plant world comprises a rich undiscovered pool of phytochemicals that may be widely utilized in place of synthetic insecticides in mosquito control program.

14.2.1 History of Phytochemicals

Since the prehistoric days, plants whether herbs, shrubs, or trees in parts or in whole were being commonly used by our ancestors for the treatment of many diseases and disorders. Prior to the discovery of synthetic pesticides, plants and their products were the only pest-managing agents available to the farmers across the world. To prevent mosquito bites and create a mosquito-free environment, burning of herbs and the bark of trees has been practiced from the ancient times (Kihampa 2011). Since 1920s, plants have been used to control mosquitoes (Raghavendra et al. 2011), but the development of synthetic insecticides, like DDT, in 1939 put an end to the use of phytochemicals in mosquito control programs. The decision to focus again on phytochemicals that are readily biodegradable and have no negative effects on nontarget organisms as well as to the environment was appreciated after experiencing numerous issues as a result of the excessive and careless application of synthetic insecticides. Since then, efforts have been made to identify the structure of new bioactive chemicals derived from the plants as well as to start their commercial production (Shajahan et al. 2022).

Historically, the commercial development of botanical insecticides is credited to a lady of Ragusa, Dalmatia, who noticed dead insects on a discarded bouquet of pyrethrin flowers. She then started milling pyrethrum into powder leading to the birth of pyrethrin industry (Hartzell and Wilcoxon 1941). Since then, pyrethrins from chrysanthemum flowers and many synthetic derivatives have been commercialized as effective pesticides. One of the earliest reports showing the use of plant extracts against mosquito larvae is credited to Campbell et al. (1933) who found that plant alkaloids, like nicotine, anabasine, methyl anabasine and lupinine, extracted from the Russian weed, *Anabasis aphylla*, killed larvae of different species of *Culex* mosquito.

14.2.2 Biochemical Nature of Phytochemicals

Phytochemicals are basically secondary metabolites that are evolved to defend the plants against constant selection pressure of herbivore predators and other environmental conditions. These can be majorly categorized into five groups: (1) alkaloids, (2) phenolic acids, (3) terpenoids, (4) carbohydrates, and (5) lipids (Bilal and Hassan 2012). Plant secondary metabolites are vital as anti-nutritional components of food and animal feed, which are helpful in protecting the plant from insect pests, pathogens, and diseases along with showing their involvement in plant-plant communication and also serving as attractants for pollinators (Rattan 2010). Alkaloids, steroids, terpenoids, EOs, and phenolics are only a few of the classes of phytochemicals from various plants that have been reported for their insecticidal properties in the past (Isman 1997), and specifically secondary metabolites, such as ocimenone, rotenone, capillin, quassin, and thymol, showed good ovicidal, larvicidal, and pupicidal activities against mosquitoes (Domínguez-Martínez et al. 2003; Bobadilla et al. 2005; Shaalan et al. 2005). Such natural insecticidal properties of plants are of great importance and thus play a pivotal role in vector control strategies. Therefore, researchers have started to work on the various aspects of botanicals, so as to utilize these secondary metabolites in mosquito control program.

14.2.3 Mode of Action of Phytochemicals

Generally, the active toxic ingredients of plant extracts are secondary metabolites that evolved to protect them from herbivores. When these toxic substances are encountered by the mosquitoes, a relatively unambiguous response is triggered that has a nonspecific influence on a wide range of molecular targets, such as proteins (enzymes, receptors, signaling molecules, ion channels, and structural proteins), nucleic acids, and biomembranes, besides added cellular components (Rattan 2010). Plant metabolites basically result in abnormal functioning of the nervous system (such as in neurotransmitter synthesis, storage, release, binding and reuptake; receptor activation and function; and on enzymes involved in signal transduction pathway), thus, in turn, affecting the insect physiology in many different ways and at various receptor sites. Several affected vital physiological functions include inhibition of acetylcholinesterase (AChE), GABA-gated chloride channel, sodium and potassium ion exchange, and cellular respiration. Such disruption also includes the blockage of calcium channels, nerve cell membrane action, octopamine receptors, hormonal balance disruption, mitotic poisoning, interference in the molecular events of morphogenesis, and alteration in the behavior and memory of cholinergic system (Rattan 2010). Out of these, the most important activity is the inhibition of AChE activity, as it is a key enzyme responsible for terminating the nerve impulse transmission through synaptic pathway. It is well known that the alteration in functioning of AChE is one of the main reasons for resistance mechanisms, especially playing a significant role in the development of organophosphorus and carbamate resistance among insect pests (Senthil-Nathan et al. 2008a).

Larvicidal compounds present in phytochemicals act by absorption through the cuticle, via the respiratory tract, and/or they enter by ingestion via the gastrointestinal tract of larva, after which they reach the site of action causing systemic effects by diffusion into different tissues (Cantrell et al. 2010). Many EOs extracted from plants even may deter insect feeding, while some others control the growth of insects by acting as insect growth regulators through analogs or antagonistic effects to endogenous hormones (Franzios et al. 1997), while some results in delayed development (Kaur and Kocher 2021). The rapid effect of some natural larvicides is indicative of their neurotoxic mode of action appearing in the form of symptoms, like excitation, convulsions, hyperactivity, and hyperextension of the legs and abdomen followed by paralysis (fast knockdown) and death (Enan 2001). These effects are similar to those produced by conventional insecticides, like as that of temephos (an organophosphate) which is commonly used in vector-borne disease control programs (WHO 2009).

Generally, gut region is the target of numerous insecticidal complexes, and its integrity is dynamic for digestion and conferring resistance against toxins (Stenfors et al. 2008). The midgut of the mosquito larvae is the main part, which is involved in digestion, ion transport, absorption, and osmoregulation processes and is an important site for synthesis of digestive enzymes (Elumalai et al. 2016). Hence, plant-derived molecules primarily target the gut epithelium layer, decrease the digestive enzyme activity, and exert a serious impact on the gastric caeca and the

Malpighian tubules (Senthil-Nathan et al. 2008b). This has also been confirmed by histological studies of the mosquito larvae that displayed midgut cell damage, after being exposed to various botanical compounds (Yu et al. 2015; Riat and Kocher 2017). Further, treatment with phytochemicals has also been associated with altered protein (Fallatah 2014) and biochemical profiles in mosquitoes (Senthilkumar et al. 2013). Another important aspect is mosquito-olfactory sensory system, which is also affected by the plant-based compounds. Actually, phytochemicals bind to the odorant chemoreceptors, subsequently hindering the flight orientations of the mosquitoes (Bohbot et al. 2010). Plant-based semiochemicals can be exploited to lure the mosquitoes to an insecticide trap, thereby forming an integral part of an integrated vector control program (Kamala-Jayanthi et al. 2015).

Detoxifying and metabolic enzymes act as important biomarkers, and their expression has a significant role in evaluating the impact of toxic chemicals on physiology of insects. Esterase and phosphatase are mainly the two major classes of enzymes playing a key role in several physiological events of the mosquitoes. Generally, detoxifying enzymes are involved in digestion, reproduction, juvenile hormone metabolism, neuronal conduction, molting, and more importantly detoxification of toxic chemicals. Phosphatases are basically involved in tissue development, cellular differentiation, carbohydrate metabolisms, and synthesis of ATP (Koodalingam et al. 2014). The CYP450s, glutathione-S-transferase, sodium oxide dismutase, and esterase are recognized as the foremost enzymes accountable for the metabolic resistance of the insects (Hemingway et al. 2004). Any change in the activity of these enzymes due to toxicity of phytochemicals results in modifying the normal functioning; thus, it can be utilized for management of mosquitoes.

14.2.4 Potential of Phytochemicals as Bioinsecticides

So far, >2000 indigenous plant species are known to produce chemical factors and metabolites showing marvelous biological activities in terms of good ovicidal, larvicidal, pupicidal, and adulticidal potential along with their repellent properties against mosquitoes (Singh and Bansal 2003; Shaalan et al. 2005). Some phytochemicals act as general toxicants to all life stages of mosquitoes, whereas others interfere with growth and reproduction or act on the olfactory receptors, eliciting responses of attractancy or repellency (Sukumar et al. 1991). Phytochemical compounds extracted from the plants belonging to families such as Asteraceae, Lamiaceae, Solanaceae, Miliaceae, and Rutaceae have been reported to exhibit larvicidal, adulticidal, or repellent activities against different mosquito species (Raveen et al. 2017). Phytochemicals can be extracted from either whole plants or specific parts of the plant, depending on the activity of the derivatives. Some plants accumulate bioactive chemicals differentially in the various parts of the plant, such as leaves, roots, stem, fruits, fruit peel, seeds, rhizome, flowers, bark, etc., which have been reported to be used as a source of botanicals in mosquito control with varying efficacy. Researchers have investigated that the effectiveness of chemicals derived from specific plant parts often varies with the mosquito species (Tehri and Singh

2015). Sukumar et al. (1991) have discussed the existence of variations in the extent to which phytochemical compounds are effective on target mosquito species relative to plant parts from which these were extracted, responses in species and their developmental stages against the specified extract, solvent used for extraction, geographical origin of the plant, photosensitivity of some of the compounds in the extract, and effect on growth and reproduction of mosquitoes. The activation of plant secondary substances by light, and their subsequent photosensitizing effects on insects, especially mosquito larvae, is an important factor contributing to the enhancement of toxicity.

Mosquito larval control is the critical component and backbone of most IMM programs throughout the world. The larval stages are the main target for insecticides because mosquitoes breed in water, so it is easy to handle them in a confined place in comparison with that of flying adults. The larvicides are primarily designed to influence mosquito larvae and decrease the hazards to nontarget organisms. In recent years, EOs extracted from plants are under much considerable attention as potential mosquito larvicides because they constitute a rich source of bioactive compounds, such as flavonoids, alkaloids, and monoterpenes, and these compounds are biodegradable into nontoxic products, have little or no harmful effects on nontarget organisms and environment, and thus are potentially suitable for use in IMM programs (Żukowska and Durczyńska 2024). In fact, many researchers have reported the effectiveness of many EOs (like basil, neem, cinnamon, camphor, citronella, lemon, eucalyptus, and pine) against mosquito larvae. Several indigenous plants in India, subtropical parts of Asia, and many parts of the world have been studied for their larvicidal potential against different stages of larvae of various mosquitoes (Ghosh et al. 2012). In the case of larvicidal activity, efficacy and potency of phytochemicals seem to be dependent on relative absorption of these compounds on the larval surface (Enan 2001). However, it could be attributed to different species of mosquitoes or variation in composition of EOs used. In the past few years, larvicidal activity of plant-based EOs from different plants including neem (Nour et al. 2012), eucalyptus (Cavalca et al. 2010), citronella (Soonwera and Phasomkusolsil 2016), citrus (Simon and Akeju 2017), cinnamon (Jantan et al. 2005), and many more has been demonstrated (Pavela 2015). Larvicidal potential of 361 EOs extracted from 269 different plant species tested against *Ae. aegypti* concluded that out of these, >60% EOs with larvicidal property were found to be mainly derived from plant species belonging to Myrtaceae, Lamiaceae, and Rutaceae (Dias and Moraes 2014). Some reports are available regarding the synergistic effect of botanicals indicating the higher larvicidal efficacy of phytochemical-mixed formulations than individual phytochemical, resulting in development of less resistance (Cotchakaew and Soonwera 2020).

Several laboratory studies have focused on determination of biopotency, efficacy, residual activity, irritant or repellent properties, diagnostic concentration, and possible cross-resistance of candidate larvicides or adulticides. In laboratory bioassays, mosquito larvae are exposed to various concentrations of larvicides, and a mortality rate based on lethal concentration (LC) of the larvicide for 50% and 90% mortality (LC_{50} and LC_{90}) or for 50% and 90% inhibition of adult emergence (IE_{50} and IE_{90}), has been recorded. Candidate larvicides and adulticides are also tested against multiresistant mosquito strains and a susceptible reference strain to assess the cross-resistance and, if detected, biochemical, immunological, and molecular methods are used to determine the mechanism of resistance as recommended by the World Health Organization (WHO 1998). In 2005, WHO published the guidelines for "Laboratory and Field Testing of Mosquito Larvicides," aiming to standardize the procedures for testing the mechanisms of action of larvicides. This document defines the sequential phases for evaluating larvicidal compounds and their preparation, and illustrates the potency of a chemical against a particular species of mosquito larvae that must be compared with other insecticides. As per the guidelines (WHO 2005), the selected plants or their extracts/products/EOs, as the case may be, are required to meet two essential conditions: (a) they had to cause sufficient mortality in standard larvicidal tests to achieve $LC_{50} \leq 100$ ppm, and (b) their chemical composition had to be known. It is important that inherent activity of candidate bioinsecticides should be assessed before they can be effectively used against mosquito populations. WHO has established methods to screen the efficacy and field application acceptability of new compounds as potential mosquito larvicides and adulticides (WHO 2005; 2013). Once candidate compounds are selected from laboratory tests, they are subjected to small-scale field testing in natural breeding sites (such as drains, sewage water tanks, ponds, rice plots, etc.) or under simulated field conditions (artificial containers filled with water, experimental huts, etc.). Those larvicides and adulticides that show promise in small-scale field trials should be validated in larger-scale field trials against mosquito populations in natural breeding habitats using optimum field dosages. At this stage, the storage, handling, and application of the insecticide formulation should be considered for proper functioning of application and dispersal of the bioinsecticides in natural ecosystems (Meryem et al. 2022).

14.2.5 Repellent Action of Phytochemicals

The repellency of plant materials has been exploited for thousands of years by man, most simply by hanging bruised plants in houses, a practice that is still in wide use throughout the developing countries. Plants have also been used for centuries in the form of crude fumigants, where plants were burned to drive away nuisance mosquitoes and later as oil formulations applied to the skin or clothes. Plant-based repellents are still extensively used in the traditional ways throughout rural communities in the tropics, because for many of the poorest communities, these are the only available means of protection from mosquito bites (Moore et al. 2006). "Natural" smelling repellents are preferred because plants are perceived as a safe and trusted means of mosquito bite prevention. The field of repellent development from plants is extremely fertile due to wealth of insecticidal compounds found in plants as defense against insects.

Monoterpenes, cineole, eugenol, limonene, terpinolene, citronellol, citronellal, camphor, and thymol are incessant fixings in various EOs that have mosquito repellent properties (Nerio et al. 2010). Insecticide-treated nets (ITNs) are one of the major intervention methods to control mosquitoes along with the incorporation of plant-based insect repellents in fabrics which seems to be a prompt and alternative way to provide safer protection against mosquito bites. The repellent phytochemicals connect with the female mosquito's olfactory receptors, hindering the impression of smell, thus keeping the mosquitoes from perceiving the host (Tripathi et al. 2009). It can also disrupt correspondence in mating conduct by impeding the capacity of antennal sensilla, and fruitless mating could eventually bring down the population of insect pests. Mosquito repellents derived from basil, castor oil, cedar, clove, fennel, citronella, eucalyptus, neem, rosemary, and celery extracts have already been tested and found to be effective (Shukla et al. 2018; Dhillon et al. 2020). The Environmental Protection Agency (EPA) has registered citronella, lemon, and eucalyptus oils as mosquito repelling agents being harmless for the skin and having low toxicity and comparable efficacy (Katz et al. 2008).

14.2.6 Advantages and Limitations of Phytochemicals

Due to frequent and unsystematic use of chemical insecticides, mosquitoes are becoming resistant; also this has led to serious human health and environment related issues. The natural plant products are considered as the best candidates to overcome these negative trends of synthetic insecticides. In the recent past, use of natural insecticides from plant origin for mosquito control is considered as an environmental friendly approach (Ghosh and Chandra 2006). Phytochemicals offer not only effective mosquito control agents but also are biorational alternatives to organic synthetic pesticides. These chemicals are from natural sources, and with a high degree of biodegradation, this makes them environmentally safe control agents. With an ever-increasing public interest and awareness in the environment, in both developed and developing countries, positive public perception of natural pesticides is an added incentive for their development and use. Additionally, plant-derived pesticides contain mixture of chemical compounds that operate concomitantly through multiple mode of actions working on both behavioral and physiological processes, in contrast to conventional insecticides, which are based on a single active ingredient. Thus, the chance of resistance development by the insects to such compounds is extremely low (Regnault-Roger et al. 2012). Phytochemicals have short residual half-life, which could be advantageous when synergistically used together with other biological control agents (Isman et al. 2011). It is encouraging that these features of phytochemicals make them alternative natural solutions for the development of suitable products to interfere with the mosquito-host interaction and reduce disease transmission.

Bioinsecticides are though considered safe because they are target-oriented, reduce environmental pollution risk, and eliminate resistance, but despite of all these beneficial attributes, they have certain drawbacks. Among the phytochemicals, EOs are extensively studied, and their larvicidal and repellent activities against mosquitoes make them favorable natural chemicals. Although the potential of EOs as larvicidal agents against many species of mosquitoes is well documented, however no EO-based larvicide is commercially available till date. This is because of the major challenges being faced by the researchers for the successful application of such products mainly relating to their solubility and sustainability in the aquatic environment. A variety of EOs have significant repellent effects also; however, these effects tend to dissipate quickly due to their high volatility. EOs generally act in the vapor phase being active only for a short period of time (Islam et al. 2017). There are also other potential limitations to the efficacy of bioinsecticides, such as environmental conditions, mosquito fitness, mosquito resistance, parts of the plants used, solvents used in extraction steps, insecticide dose, and exposure time (Namias et al. 2021). These effects should be considered for successful assessment of novel bioinsecticides in mosquito control. Finally, it is important to monitor effects of phytochemicals on nontarget organisms, such as aquatic plankton, invertebrates or small vertebrates. Although plant-derived products are considered environmentally safe, some exhibit toxicity against small crustaceans, which are an important component of aquatic fauna (Conti et al. 2014); even slight sedative effects on fish have also been observed (Silva et al. 2015). However, as discovered in recent years, toxicity of plant products against these nontarget organisms depends on the applied concentration and time of exposure. Natural products are not always safer, and some might cause adverse reactions, such as skin irritation, mainly due to the presence of constituents present in the plant EOs, limiting their extensive usage (Benelli and Pavela 2018). Moreover, sometimes, the available toxicity and safety information is limited and contradictory.

14.2.7 Future Perspectives of Phytochemicals in Mosquito Control

Throughout the world, phytochemicals are comparatively safe to nontarget organisms, economical, and easily available. At present, botanical insecticides make <1% of the world's pesticide market (Sola et al. 2014). Isolation of secondary metabolites from plants for their evaluation against mosquitoes is an important tool for the management of mosquito-borne diseases. The positive results of initial studies on mosquitocidal potential of botanicals encourage further interest to investigate these bioactive compounds. Identifying botanical insecticides that are effective, appropriate, adaptive to overcome ecological hazards, biodegradable and have a broad spectrum of larvicidal properties as a new defense in the arsenal of insecticides, can help to formulate an appropriate alternative product in order to fight against vector-borne diseases. Problems associated with their formulation and commercialization, non-standardization in evaluation of their bioactivities, and their persistence for longer durations should be resolved for the development of effective and sustainable methods for their usage. There is no doubt that there is need for exploring new bioactive compounds and future research should focus on searching for plant-based products with the ultimate goal of deploying them as a reliable remedy to control mosquito populations and mosquito-borne diseases.

Recent rapid progress in biotechnology has led to a massive increase in the use of multiomics tools including genomics, proteomics, and metabolomics data. However, these tools have not been utilized to find novel and more effective bioinsecticides. The newly advanced techniques are required for isolation and characterization of phytocompounds through NMR spectroscopy, FT-IR spectroscopy, UV scan, LC-MS, GC-MS, HPLC, XRD, SEM analysis, and chromatography techniques. This will enhance the interest to search for the effective new phytocompounds from plants, as a source of novel ovicidal, larvicidal, pupicidal, and adulticidal agents, which may be helpful in managing the mosquito population in the coming future. At present, the development of natural product-based insecticides with more effective and long-lasting protection is required, ensuring the development of natural repellents, larvicides, and adulticides with improved efficacy, long-lasting protection, and enhanced safety. In particular, current formulation technology, nanotechnology, and recent techniques, like microencapsulation, may accelerate the emergence of novel and effective plant-based insecticides with long-lasting mosquitocidal effects in future.

14.3 Biological Control of Mosquitoes

One of the cornerstones of preventing mosquito-borne illnesses and safeguarding the population from mosquitoes is the use of biological control techniques. Integrated vector management (IVM) strategies use biological control as one of the most important elements. It is described as the process in which vector population is lowered down by natural enemies, usually with human intervention. As a potential method of lowering mosquito vector population, biological control is growing in popularity these days. The efficacy, environmental impact, and safety to nontarget organisms of several mosquito-killing biocontrol techniques have been evaluated. The potential of fish, invertebrate predators, bacteria, nematodes, viruses, fungi, and protozoa as vector control agents has been investigated (Davidson and Becker 1996).

14.3.1 Biological Control Agents (BCAs) Used for Mosquitoes

It has been discovered that biological control methods play a significant role in reducing mosquito populations. Recently, there has been an increase in the use of biocontrol agents for vector control. This chapter will address in-depth the function of biological agents in controlling mosquito vectors.

14.3.1.1 Predators

Mosquito population can be significantly decreased by natural enemies that prey on mosquito larvae and pupae in aquatic habitats (Griffin and Knight 2012). Numerous aquatic creatures, such as fish, amphibians, copepods, odonate young instars, water bugs, and even larvae of different mosquito species, prey on mosquito young instars.

14.3.1.1.1 Platyhelminthes (Flatworms)

A few species of Turbellaria have the ability to consume mosquito larvae as food (Dambach 2020). Although they are present in phytotelmata, it is unknown if turbellarians can affect mosquito abundance (Kitching 2000). Certain Turbellaria species, which are not found in phytotelmata, such as *Giardia anceps* (Tricladida: Dugesiidae), *Mesostoma ehrenbergii*, and *Bothromesostoma evelinae* (Rhabdocoela: Typhloplanidae), thrive in artificial micro-water bodies and consume mosquitoes. It is likely possible to utilize them for mosquito control (Tranchida et al. 2009).

14.3.1.1.2 Arachnida

Aquatic and semiaquatic spiders, including the global *Dolomedes* (Pisauridae), *Pirata*, *Pardosa* (Lycosidae), and the European *Argyroneta* (Cybaeidae), consume mosquito larvae (Perevozkin et al. 2004). However, it is unusual for micro-aquatic environments to have these predators present (Kitching 2000). *Paracyrba wanlessi* is the only spider species known to feed on mosquito larvae in phytotelmata or bamboo stem chambers (Żabka and Kovac 1996). Water mites of the genus *Arrhenurus* (Arrhenuridae) are found in phytotelmata (Kitching 2000) and are capable of feeding on mosquito larvae in their first instar stage (Rajendran and Prasad 1994).

14.3.1.1.3 Crustaceans

Predatory Copepoda are widespread crustaceans found in freshwaters. They are thought to be the most efficient means of reducing the population of mosquitoes (Marten et al. 1989). Copepod species that have been identified as active mosquito young instar predators include *Cyclops vernalis* (Anbu et al. 2016), *Megacyclops formosanus*, *M. aspericornis*, *M. edax*, *M. guangxiensis*, *M. longisetus* (Manrique-Saide et al. 1998), and *M. thermocyclopoides* (Schaper 1999; Kumar et al. 2012; Murugan et al. 2015a). Copepod predators have the potential to be economical, mass-reared biocontrol agents, since they are labor-intensive and inexpensive to use against mosquitoes in urban and semi-urban environments (Chitra et al. 2013). Being smaller than mosquito larvae, copepods have to employ certain strategies. According to Schaper and Hernández-Chavarría (2006), copepods attack larvae from the side or behind, consequently damaging the abdomen or syphon. Because of the disruption of the respiratory system, any kind of trauma to the siphon is the most precarious to *Ae. aegypti* (Awasthi et al. 2012). Copepods typically cling on to their victim until it dies and then they consume it.

14.3.1.1.4 Insects

This group of hexapods comprises predators, which prey upon mosquitoes and their larvae. Different types of insects belonging to various orders are very helpful in controlling the mosquito population (Vinogradov et al. 2022). Importance of dragonfly, beetles, and bugs is discussed in detail in the following sections.

- **Damselflies and dragonflies (Odonata).** People have appreciated the benefits of dragonflies (Fig. 14.1) and damselflies (Fig. 14.2) (Order: Odonata) as predators of mosquitoes. Both adults and nymphs devour mosquitoes (Shaalan and Canyon

Fig. 14.1 Dragonflies - predator of mosquitoes https://pixabay.com/photos/insects-wings-dragonfly-macro-591710/, accessed on March 6, 2024

Fig. 14.2 Damselfly - a biocontrol agent of mosquitoes. https://pixabay.com/photos/dragonfly-insect-macro-wings-4293710/, accessed on March 6, 2024

2009; Jacob et al. 2017). Odonata larvae have a long developmental history, and also they are limited to permanent bodies of water, which makes them ideal biocontrol agents; nevertheless, the species needs to be resistant to drying out in transitory water bodies.

- **Hemipteran predators**. The majority of water bugs eat mosquitoes. Typically, *Sigara striata*, *Corixa punctata*, and *Cymatia coleoptrata* are found in mosquito breeding sites. However, they are not very significant mosquito predators due to their predominantly omnivorous eating habits. The main families of hymenopteran predators that feed on mosquitoes are Nepidae, Hydrometridae, Notonectidae, Pleidae, Gerridae, and Naucoridae (Murdoch et al. 1984).
- **Coleopteran predators:** Predatory beetles feed on mosquito larvae at the larval and/or imago stage (Becker et al. 2010). Being numerous and ferocious, water

beetles are considered as excellent aquatic predators. The ability of these predators to thrive and proliferate in large numbers throughout a variety of mosquito breeding habitats enhances their capabilities. Dytiscids (water beetles) are the most significant predators of mosquitoes. Moreover, more than 100 *Ae. vexans* fourth instar larvae can be swallowed daily by their tiny larvae. The three main examples of predatory beetles are *Guignotus pusillus*, *Colymbetes fuscus*, and *Hydrophilus caraboides*.

- **Dipteran predators:** Although some mosquito larvae are obligate or facultative predators that prey on other mosquito larvae (Shaalan and Canyon 2009), however, their effectiveness as a control measure for potentially harmful mosquito species has not received much attention. The only exceptions are those belonging to the genus *Toxorhynchites*, which eat older larvae and pupae more frequently, boosting the effectiveness of the control. Representatives of the genus *Lutzia* are thought to be promising mosquito control agents (Singh et al. 2014; Moirangthem and Singh 2018). Larvae of the Chaoboridae family play a significant role as consumers of mosquitoes in micro-aquatic environments. Unlike larger predators, they consume all stages of mosquito larvae, not only the last one (Cuthbert et al. 2019).

14.3.1.1.5 Predatory Fish

Insectivorous fish, present in vast bodies of water, are thought to be one of the primary regulators of the mosquito populations and are commonly employed in practice (Griffin and Knight 2012). However, fish do not typically inhabit phytotelmata and other micro-aquatic environments (Kitching 2000). The two most well-known predators of mosquito larvae are *Poecilia reticulata* and *Gambusia affinis*. In almost 60 countries, they have been introduced with the intention of controlling mosquitoes (Chandra et al. 2008). *G. affinis* (Fig. 14.3) is the most widely employed mosquito-controlling organism. In 1937, Hackett published a study on the effectiveness of *G. affinis* in reducing malaria in Europe. *G. affinis* played a

Fig. 14.3 *Gambusia affinis* - a widely used predator - of mosquito larvae. https://www.istockphoto.com/photo/guppy-gm544116840-97759791, accessed on March 6, 2024

significant role in Turkey and Iran for reducing the malaria cases (Tabibzadeh et al. 1970). Fish are also occasionally introduced since they are able to eat aquatic plants. They can lower mosquito populations as predators, but they can also lower mosquito populations by consuming aquatic vegetation, which in turn lowers the habitats available for mosquito larvae to thrive in. Other fish species used to manage mosquito populations include the common carp, *Cyprinus carpio*; the grass carp, *Ctenopharyngodon idella*, *Aplocheilus panchax*, and *Cynolebias bellotti*; and three subtropical cichlid species, *Tilapia zillii*, *Oreochromis mossambicus*, and *Oreochromis hornor* (Takken 1991; Asimeng and Mutiga 1992).

14.3.1.1.6 Amphibians

Several researches have stated that anurans, especially frogs and toads, can be used to reduce mosquito populations (Bowatte et al. 2013; Murugan et al. 2015b). Tadpoles aggressively hunt for *Ae. aegypti* eggs. It has been demonstrated that *Aedes* prefers to lay its eggs in tadpole water and that the eggs are preyed upon by tadpoles belonging to the *Polypedates cruciger*, *Bufo*, *Ramanella*, *Ephlyctis*, and *Hoplobatrachus* genera (Bowatte et al. 2013). However, other studies have also shown that larvae of *Cx. quinquefasciatus* can be eaten by three common Thai anuran *species* (*B. melanostictus*, *Kaloula pulchra*, and *Hylarana raniceps*) (Weterings 2015).

14.3.1.2 Parasites

- **Nematodes.** Insect parasitic nematodes can be classified as either obligatory or facultative parasitic nematodes (Pirali-Kheirabadi 2012). Despite the phylum Nematoda having 5 orders and 14 families of obligatory parasites, only the Mermithidae has been found in wild mosquito populations (Platzer 1981). These nematodes are interesting because of their potential as biocontrol agents. In addition to the Mermithidae family, there are eight other significant nematode families, which attack, sterilize, and kill insects or alter host growth. These families are Allantonematidae, Diplogasteridae, Neotylenchidae, Rhabditidae, Heterorhabditidae, Sphaerulariidae, Steinernematidae, and Tetradonematidae (Petersen 1985). The larger and more widely used nematode is the mermithid for controlling mosquito larvae (Figs. 14.4 and 14.5). Mermithids are especially desirable since they are minimally or nonhazardous to the environment. Because

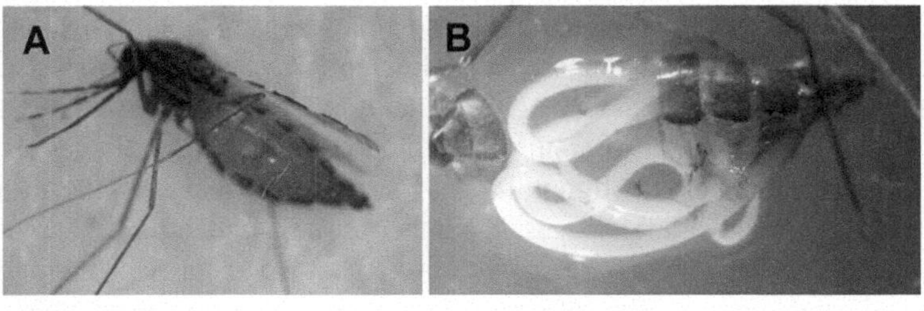

Fig. 14.4 Picture of a mermithid parasitized *Anopheles rufipes* (**a**) and the dissected mosquito abdomen with mermithid parasite (**b**) (Source: Kobylinski et al. 2012)

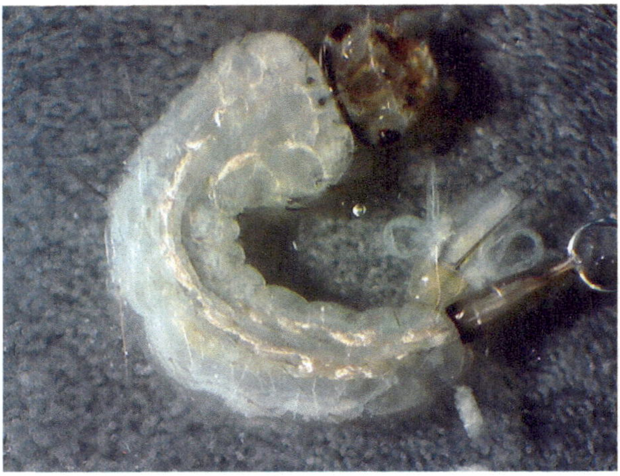

Fig. 14.5 Dead larvae with post-parasitic nematode emergence (Source: Abagli et al. 2019)

of their long lifespan, they do not really threaten to drive out other helpful species through competitive displacement. Several species of mermithids have been discovered as mosquito larvae controllers (Petersen 1985). The species infect its host by invasion by spiracles or anus, ingestion by the host insect, or infiltration through the cuticle (Pirali-Kheirabadi 2012).

14.3.1.3 Pathogens

14.3.1.3.1 Entomopathogenic Fungus

Fungal infections that impact arthropods are frequent and crucial to preserving the equilibrium of arthropod populations in tropical forests (Evans et al. 2018). Similar to almost all other insect orders, dipterans are prone to fungal diseases. The first entomopathogenic fungus to be used in the management of *An. gambiae* larvae was *Coelomomyces* (Muspratt 1963). This is the largest genus of fungi that parasitizes insects. They are obligatory parasites of chironomids and mosquitoes, which means they need their hosts to finish their life cycle. In addition to *Coelomomyces*, several studies have shown that *Lagenidium* and *Culicinomyces* can also influence mosquito vectors. This entomopathogenic fungus alters the physiology of mosquito larvae and reduces their ability to feed on blood (Darbro et al. 2012).

In California, *Lagenidium giganteum* was employed as biocontrol agent to manage the vectors of Western equine encephalitis and West Nile virus (Kerwin and Washino 1988). In North America, *L. giganteum* in combination with a copepod was found to be successful in controlling *Anopheles* and *Culex* (Federici 1995). There are a large number of different fungal species that infect and kill mosquitoes throughout their larval or adult stages. A number of fungal infections that have been found to target and manipulate *Ae. aegypti* in African woods provided a long-term, economical, and ecologically safe remedy for the flavivirus pandemics in the Americas (Evans et al. 2018). The phylum Chytridiomycota contains a number of entomopathogenic fungal species, but only two genera—*Coelomomyces* and *Coelomycidium*—have been investigated for their potential to biologically control mosquitoes and blackflies (Tanada and Kaaya 1993). In terms of controlling pests, entomophthoralean fungi that have been studied the most are *Conidiobolus*, *Entomophthora*, *Erynia*, and *Neozygites*.

14.3.1.3.2 Bacteria

Numerous attempts have been made worldwide to locate suitable mosquitocidal bacteria in the natural environment since they are an environmentally friendly alternative to chemical insecticides for controlling mosquitoes (Poopathi et al. 2014). When ecological concerns, like safety for humans and other nontarget organisms, pesticide residue reductions in aquatic ecology, improved activity of other natural enemies, and improved biodiversity in aquatic environments, are taken into account, two mosquitocidal bacteria, *Bacillus thuringiensis* (Bti) and *B. sphaericus* (Bs), have been used as broad spectrum biolarvicides under a variety of conditions with little to no ecological harm. Rather than being actual biological agents that recycle and sustain themselves in the environment, both *Bacillus* species function as microbial pesticides (Mullen and Durden 2009).

Bacillus sp.

A new phase in the battle against mosquitoes began with the identification of the Gram-positive, endospore-forming soil bacterium, *B. thuringiensis* spp. *israelensis* (Bti), in Israel's Negev desert (Goldberg and Margalit 1977). The creation of the potent *B. sphaericus* strain has been crucial in the management of mosquitoes (Poopathi et al. 2013). *B. thuringiensis* is a Gram-positive spore-forming bacterium that generates a wide variety of biocidal toxins. Cry and Cyt proteins are delta endotoxins, which have a wide range of activity against worms, insects, and even human cancer cells. Similar crystalline toxins are produced by *B. sphaericus*, and the two bacteria are often used in conjunction (Poopathi and Tyagi 2006). The most popular microbial insecticide in the world is made from *B. thuringiensis* (Bti) preparation (Ramírez-Lepe and Ramírez-Suero 2012). It is a highly effective pathogen because it can be mass-produced with ease and it is toxicologically safe for wildlife and humans (Ingabire et al. 2017). When Bti is consumed, the bacterial spores' crystal proteins release an endotoxin, which creates pores/nonselective channels in the apical membrane of larval midgut. Since nonselective channels are permeable to neutral solutes, cations, and anions, water enters the channels and leads to swelling and cell lysis, which kills the organism (Knowles and Ellar 1987). The species, *B. sphaericus*, works better in biologically contaminated water, and it works particularly well against *Culex* species (Mullen and Durden 2009). Other bacteria that are employed to control mosquito vectors include *B. alvei*, *B. brevis*, *B. amyloliquefaciens*, *B. circulans*, *B. subtilis*, *Brevibacill leterosporous*, *Pseudomonas fluorescens*, *P. aeruginosa*, and *Clostridium bifermentans* (Lalithambika and Vani 2016).

Pseudomonas frederiksbergensis

A rod-shaped, Gram-negative bacterium called *P. frederiksbergensis* was initially discovered in Frederiksberg, Denmark. In Saudi Arabia and Riyadh, third instar larvae of the most prevalent lymphatic filariasis vector, *Cx. pipiens,* were tested with an extract of this bacteria. The extract was found to have a potent larvicidal impact that was mediated by the shrinkage of the intestinal mucosa and the destruction of the midgut epithelium. According to Ahmed et al. (2014), it had been tested for biocidal activity against aedine and anopheline, with encouraging results.

Wolbachia Bacteria

A particular kind of bacteria that inhabits insect cells is becoming more and more important in halting the development to diseases carried by mosquitoes, such as dengue fever and the Zika virus. *Wolbachia* is a genus of Gram-negative bacteria. *Wolbachia* bacteria would be discussed in detail in the section of genetic control.

14.3.1.4 Advantages and Disadvantages of Biocontrol Methods

14.3.1.4.1 Advantages

As a vector management technique, biocontrol has several advantages over chemical insecticides.

- One of the biggest benefits of biocontrol is that it is an environmentally friendly method that does not release poisons into the environment (Kok and Kok 1999).
- Another important benefit is the selectivity of the target species, which reduces the likelihood of harm to nontarget species.
- Since BCAs reproduce and disperse on their own, they can grow in number and spread, which is another benefit of using biocontrol methods.
- The vector is unlikely to develop resistance, or it takes a long time to do so, according to biological treatment (Kwenti 2017).
- Since self-propagation of biocontrol agents will keep the target vector at a manageable level for a considerable amount of time, biocontrol is also a cost-effective solution (Kok and Kok 1999).
- Additionally, a small number of biocontrol agents have the capacity to develop to extremely high densities and continuously control pests over a large region. The cost of deploying BCAs typically makes biocontrol less expensive than chemical control (Hamed et al. 2022).

14.3.1.4.2 Disadvantages

Though, there are several benefits of biocontrol strategies, but this method has some drawbacks also, which are enumerated below:

- Establishing and maintaining biocontrol is more difficult than insecticidal control.
- BCAs are more susceptible to environmental influences than chemical control.

- Furthermore, because natural enemies are frequently impacted by environmental conditions, biocontrol may be unpredictable (Emden 2004).
- Another significant problem is that conventional insecticides are incompatible with them. This is because applying insecticides in a particular environmental niche changes its biological makeup, upsetting the delicate equilibrium between BCA and their targets (Emden 2004).
- Additionally, biocontrol takes long time to function (Mullen and Durden 2009). Moreover, one of the objectives of biocontrol is not the regulation of pests. In most situations, the objective is to maintain the insect population below the economic injury level (EIL) (Hamed et al. 2022).

14.4 Genetic Control of Mosquitoes

Because of hazardous effects of chemical-based interventions, there has been an expansion in the development of genetic control technologies. A few of the genetic approaches are described under with their pros and cons.

14.4.1 Sterile Insect Technique (SIT)

Sterile insect technique was first proposed by E. F. Knipling in 1955 for the management of insect pests (Knipling 1955). The technique comprises four steps: mass production of insects in the laboratory, segregating adult males, inducing sterility in males by irradiating, and finally release of the sterile males in the target area (Fig. 14.6). Once released, the sterile males compete with the wild males for mating with wild females. Since the released males are sterile, the mating does not produce

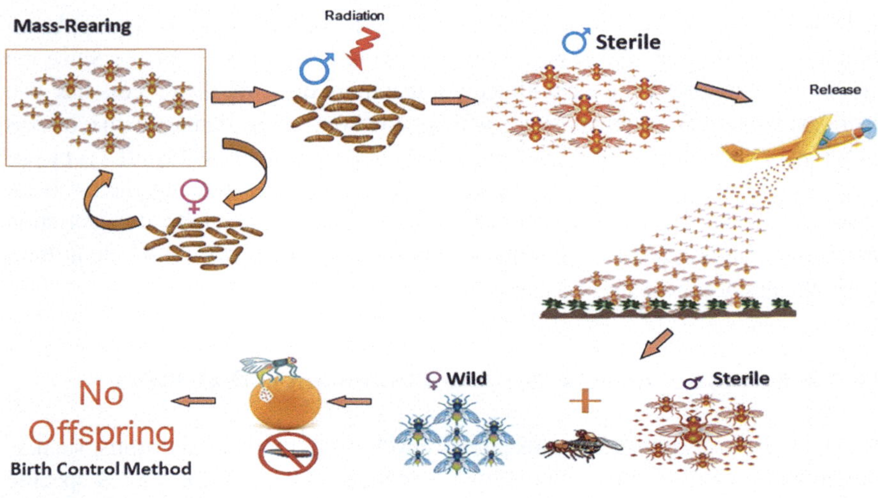

Fig. 14.6 Principle of sterile insect technique (Source: Vreysen et al. 2021) (CC-BY-4.0)

any offspring, and gradually, the population of insects declines in the area. In fact, the approach becomes more efficacious with the reduction in population. Consequently, continued release of sterile males reduces the insect population progressively.

The approach was initially used to eradicate *Cochliomyia hominivorax* (screw-worm flies) and *Ceratitis capitata* (fruit flies) which was found successful. Later, it was also attempted for the management of mosquito population. The earlier attempts gave mixed results but a few cases proved to be fruitful. SIT approach has also been trialed to manage *Ae. albopictus* at four different sites in Italy. The study showed the sterility rates of 18–68% in the eggs collected from the SIT areas as compared to those collected from the untreated areas (Bellini et al. 2013).

The SIT technique is species-specific and ecologically benign. Since it involves the release of only male mosquitoes, it is acceptable among community and health workers due to no impact on the biting rates. Nevertheless, it cannot be used as a sole approach and instead is used as a part of the integrated vector management program. It requires an understanding of the population size, spatial patterns of abundance, and high-quality laboratory-reared sterile males for the successful results (Dyck et al. 2021). The major limitation of the approach is that it is not sustainable in nature. It requires continuous release of sterile mosquitoes as once release of SIT male mosquitoes is stopped, the targeted species of mosquito eventually returns to normal. Further, it can be successful in an area only when the sterile males are more abundant than the wild males. As continuous inflow of the wild male mosquitoes and gravid females increases their population, the approach can prove to be successful only in isolated areas.

Another major constraint is that it requires continuous rearing and production of sufficient number of sterile males for their continuous release in the fields. It is a highly skillful and time-consuming process (Benedict and Robinson 2003). The heavy expenditure involved in maintenance of rearing and irradiation facility limits its use. Further, sex separation in adults has always been problematic. Earlier, male and female pupae used to be sorted mechanically according to the size dimorphism, which principally segregated male-only population (Ansari et al. 1977). Since the irradiation at adult stage is operationally much more difficult, the process has also been experimented at pupal stage which, however, often damaged the insects (Andreasen and Curtis 2005). In addition, radiation overdose possibly affects physiology, longevity, and other parameters of the mosquitoes making their release unsuccessful. Other issues associated with the technique include transportation problems, slow action, non-persistence, and planning for long-term implementation.

14.4.2 Release of Insects Carrying a Dominant Lethal (RIDL)

The approach uses transgenic mosquitoes developed with particular traits, such as conditional genetic sterility or lethality. These mosquito strains carry a sex-specific lethal gene that causes lethality in their offspring causing population reduction. The release of insects carrying a dominant lethal (RIDL) approach has been attempted in both male and female mosquitoes engineered genetically with specific genes. One the most common examples include the OX513A *Ae. aegypti* strains, which were genetically modified mosquitoes by introduction of LA 513 transposons (Aldridge 2008). Males are modified in such a way that their larvae are unable to reach maturity unless they are provided tetracycline-containing diet. The mating between male OX513A *Ae. aegypti* and wild females results in transmission of this trait in their offspring. Since tetracycline is not present in the wild, the mosquito larvae are unable to survive and die before reaching to the adulthood. In Brazil, the sustained release of OX513A *Ae. aegypti* males declined its population by 95% in comparison with the control site (Carvalho et al. 2015). Such approach can overcome the issues associated with irradiation of males required in SIT.

14.4.3 Development of Flightless Females

In another novel approach, females of *Ae. aegypti* have been made flightless by genetically engineering with the objective that females will not be able to mate, bite the humans, and search the oviposition sites (Macias et al. 2017). The female-specific lethal gene was inserted into adult females, which caused destruction of wing muscles resulting in flightless females. A flight muscle promoter gene, Actin-4, has been identified in *Ae. aegypti* and *Ae. albopictus* to drive a dominant lethal gene in the indirect flight muscles, resulting in the development of flightless phenotypes in females (Fu et al. 2010; Labbé et al. 2012). O'Leary and Adelman (2020) found that knocking Actin-4 and Myosin-fem genes using CRISPR-Cas9 resulted in 100% flightless females.

14.4.4 Development of Refractory Mosquitoes

Development of mosquitoes, refractory to diseases, is another well-thought approach as it prevents the disease transmission by them. It does not allow mosquitoes to either acquire the pathogen or their replication in their body. The approach involves upregulation/downregulation of particular antibodies and proteins or manipulation of cell signaling. It is believed that replacing existing vector populations with the refractory mosquitoes, unable to transmit specific pathogens, will reduce the disease prevalence and morbidity.

Marshall and Taylor (2009) developed *An. stephensi* mosquitoes, refractory to *Plasmodium berghei*. As sporozoites of *Plasmodium* pass through the gut wall of the mosquitoes into haemocoel and finally enter their salivary glands, they identified target protein receptors in these sites required by *P. berghei*. The specific gene constructs were introduced into the mosquitoes, which prevented the release of sporozoites from the gut and consequently their entry in the salivary glands. In *Ae. aegypti*, a trypsin inhibitor gene has been identified, which could play role in making them refractory (Barón et al. 2010).

With the advancement in technology, several techniques have been developed, among which RNAi seems to be a plausible tool for suppressing the replication of arboviruses transmitted by *Aedes* mosquitoes (Alphey et al. 2013). Transgene-based expression of a hairpin RNA in either the midgut (Franz et al. 2006) or salivary glands (Mathur et al. 2010) of *Aedes* blocked the dengue virus (DENV) transmission strongly. However, the reduced fitness cost of the mosquitoes due to refractory gene has diverted the attention of scientists toward other strategies.

14.4.5 Distortion of Sex Ratio

Researchers discovered a male-determining factor (M-factor) on the Y chromosome representing highly repetitive M locus (male-determining locus). These factors include *gYG2/Yob* in *An. gambiae* (Hall et al. 2016), *Nix* (Fig. 14.7) in *Ae. aegypti* (Hall et al. 2015), and *Guy1* in *An. stephensi* (Criscione et al. 2016). They reported that a dominant M locus establishes the male sex (M/m) in *Ae. aegypti*. It was proposed that this M-factor can be tapped to transform female offspring into harmless nonbiting males and can be used in mosquito control strategies (Hall et al. 2015). Further, ectopic expression of the gene can develop masculinized females, while knocking the *Nix* gene in male develops feminine males.

The role of M locus in determination of male sex was also demonstrated by Aryan et al. (2020). They also reported that the harmless males transformed from females lack *myo-sex* gene present on M locus. Since this gene is required for flight, the transgenic males were unable to fly, reducing the mating incidences. They

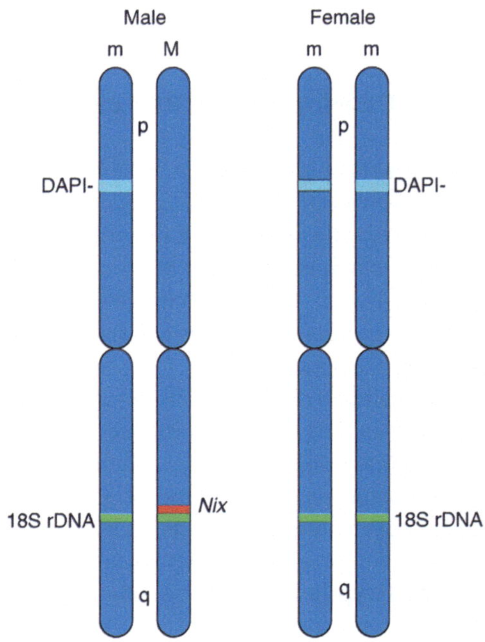

Fig. 14.7: Schematic representation of the *nix* gene on the male chromosome of *Aedes albopictus* (Source: Gomulski et al. 2018) (CC-BY-3.0)

recommended that this strategy can be used as a complement strategy of SIT and can be a significant intervention.

Another approach exploited the specificity of the homing endonuclease I-PpoI of *An. gambiae*. It selectively cleaved the ribosomal gene sequences present on the X chromosome, which inhibited its inheritance producing >95% male offspring (Galizi et al. 2014). Slicing of the X chromosome prevents its transmission to the next generation.

14.4.6 *Wolbachia*-Based Control

Wolbachia is an intracellular bacterium found in more than 50% of insect species and with ability of vertical transmission (Hilgenboecker et al. 2008). Several studies have been carried out to use the bacterium in mosquito management programs, especially. It needs to be introduced in the mosquito through transinfections as naturally mosquitoes do not possess *Wolbachia*. The studies have revealed that *Wolbachia*-based control approaches can disrupt replication and transmission of arboviruses (Hoffmann et al. 2015). The strategy also suppresses the mosquito population as the *Wolbachia* transinfection decreases the physiological and reproductive fitness of mosquito. Thus, it diminishes the population by causing significant mortality and also reducing lifespan to half (McMeniman et al. 2009; Axford et al. 2016).

The first *Wolbachia* strain (*wPip-Wolbachia pipientis*) was discovered in *Cx. pipiens* (Hertig and Wolbach 1924) followed by identification of other strains, *wAlbA* and *wAlbB*, in *Ae. albopictus* (Zhou et al. 1998). In northern Australia, a *wAlbB2-F4* strain was generated, which was found incompatible with the wild strain as well as *Wolbachia*-infected *Ae. aegypti* (Beebe et al. 2021). The release of three million males in 600 houses, ~50 males/house, 3 times a week for 20 weeks, declined 80% of the population relative to the control. *Wolbachia*-based mosquito control strategy acts in two ways, population replacement or population suppression (Yen and Failloux 2020).

- **Population replacement strategy**. This approach is used to modify mosquito populations for pathogen resistance. The *Wolbachia*-infected female mosquitoes are released in the field. These mosquitoes, after mating with males (*Wolbachia*-infected or not), produce viable offspring, allowing a wide spread of *Wolbachia* in the field population that harbors less competent individuals, even as the total number of mosquitoes remains unchanged (Fig. 14.8). It has been shown that *Wolbachia* may block pathogens in several ways. These include competition for fatty acids, regulation of vector miRNAs, and upregulated natural immunity pathways (Geoghegan et al. 2017). In addition, it can directly interact with viral RNA, which restricts pathogen infection, propagation, and transmission.
- **Population suppression strategy**. It aims to release *Wolbachia*-infected male mosquitoes that, after mating with wild females, do not produce viable offspring, thus reducing the total number of mosquitoes (Fig. 14.9). Because of competitive

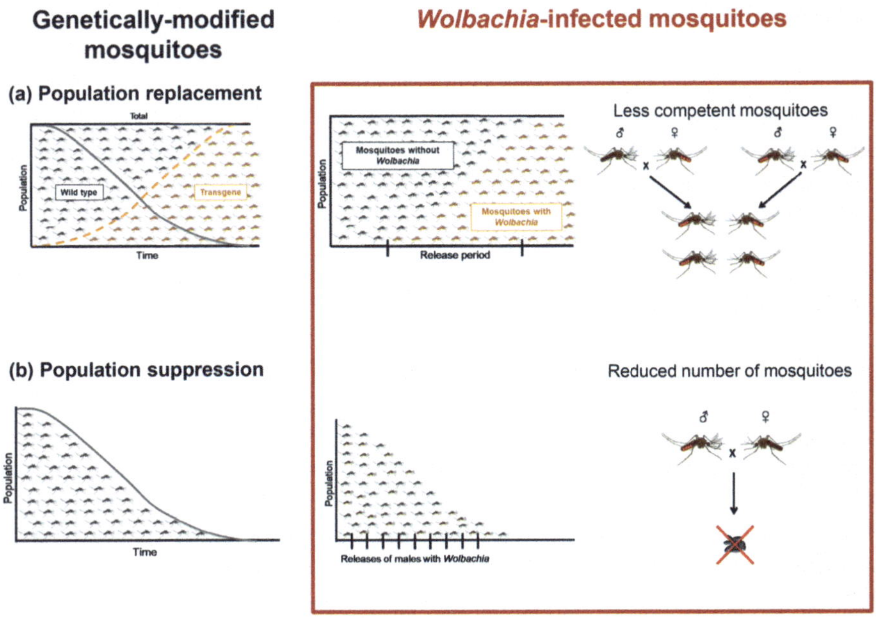

Fig. 14.8 *Wolbachia*-based mosquito control strategy (Source: Yen and Failloux 2020) (CC-BY-4.0)

advantage of *Wolbachia*-infected variants over wild types, the mating between *Wolbachia*-infected males and wild females produces sterile eggs (Flor et al. 2007). This has been attributed to the unidirectional cytoplasmic incompatibility (CI). On the other hand, mating between both infected sexes but with different *Wolbachia* strains also produce nonviable offspring (bidirectional incompatibility) (Zhong and Li 2014). Hence, multiple releases of infected males can suppress mosquito populations and potentially interrupt disease transmission. However, accidental releases of *Wolbachia*-infected females can compromise the strategy as these are fertile irrespective of the fertility status of males and, thus, there are chances of vertical transmission of the infection. A study has shown that in Singapore, *Wolbachia* strain established in the fields due to accidental releases of just three infected females. Thus, *Wolbachia*-based control can be a three-pronged approach:

(a) Production of sterile offspring from mating between infected males and uninfected females due to CI.
(b) Reduced fitness cost of *Wolbachia*-infected strain with decreased lifespan.
(c) Blockage of virus transmission by *Wolbachia*-infected strain.

The success of approach also depends upon the traits of environment in which mosquitoes are being released. In Rio de Janeiro, Brazil, the *Wolbachia*-infected mosquitoes failed to develop due to heavy use of insecticides by local authorities (Garcia et al. 2019). Further, the technique required continuous monitoring of

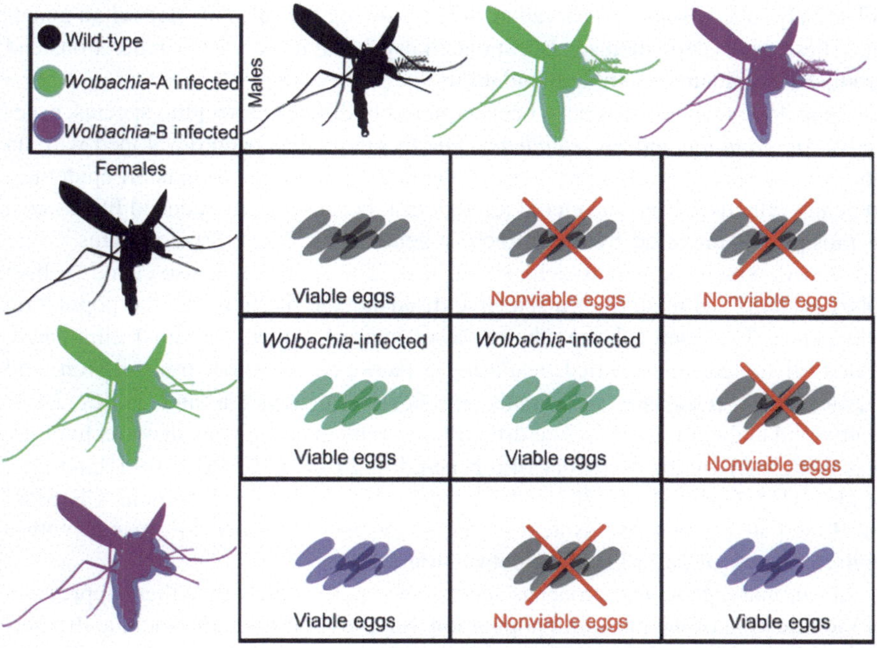

Fig. 14.9 *Wolbachia* induced cytoplasmic incompatibility (CI) (Source: Wang et al. 2021) (CC-BY-4.0)

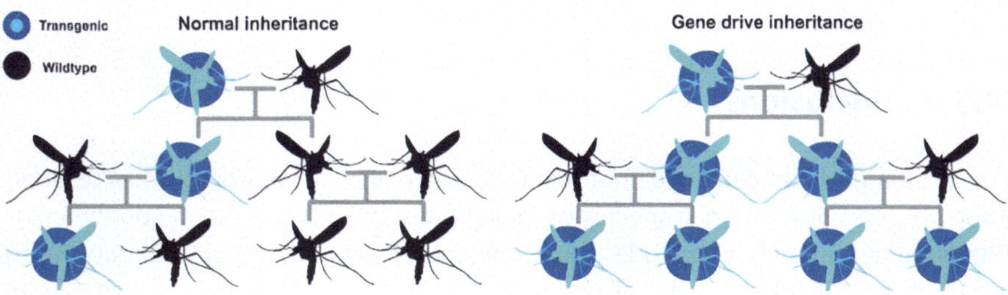

Fig. 14.10 Inheritance pattern in gene drive approach: (**a**) normal inheritance; (**b**) gene drive inheritance. Source: Wang et al. (2021) (CC-BY-4.0)

released mosquitoes, as they may endure in the environment for several years. The monitoring also helps to know if there is need of rerelease of infected mosquitoes with same or different *Wolbachia* strain (Ritchie et al. 2018).

14.4.7 Gene Drive

Gene drive technologies help in the spread of a genetic modification through a population at higher inheritance rates than the normal (Fig. 14.10). The mosquitoes selectively inherit some genes, rather than following the rules of Mendelian genetics. These genes copy themselves exponentially through successive generations and gradually dominate the whole population.

Gene drive technologies have been explored in different mosquito species, especially *An. gambiae* and *An. stephensi*. The technique has been developed with an objective to spread a refractory gene in the target population to make it epidemiologically effective. It is considered a tough task because of the reduced fitness cost in mosquitoes imparted by the refractory gene. Thus, "gene drive systems" were developed, which can help to achieve this task. The approach is also called "selfish DNA systems" as in spite of a selective disadvantage, it augments the population prevalence. The approach taps the genes of interest, which are either eliminated, added, disrupted, or modified, resulting in pathogen infection, transmission, and elevated host resistance to a pathogen (Carballar-Lejarazú and James 2017; Hammond and Galizi 2017). The difficulty in implementing gene drive technology was made easy by the discovery of a gene-editing tool, CRISPR-Cas9 (Hammond et al. 2021). Gantz et al. (2015) developed an efficacious autonomous CRISPR-Cas9-mediated gene drive system in *An. stephensi*. It resulted in progeny with a high frequency of germ-line gene conversion.

In summary, genetic approaches achieve mosquito control by either suppression or modification of the mosquito population. Suppression strategies include the sterile insect technique (SIT), *Wolbachia*-based strategy (based on cytoplasmic incompatibility), and various transgenic technologies including gene drives. The population modification strategies include generation of pathogen-resistant (refractory) mosquitoes to prevent pathogen transmission.

14.5 Conclusions

Excessive, indiscriminate, and continuous use of synthetic insecticides has resulted in the progress of resistance species of mosquitoes and adverse effects on the environment and human health. Today, environmental safety and ecofriendly approach in nature are considered to be of paramount importance. This has necessitated an urgent need for developing and implementation of different alternate approaches to control mosquitoes. Several phytochemicals/secondary metabolites present in plants are capable of inducing acute and chronic harmful effects on mosquitoes. Out of these, alkaloids, steroids, terpenoids, essential oils, and phenolics are a few classes of phytochemicals responsible for their insecticidal properties. Approximately 2000 plant species have been reported to produce secondary metabolites with natural larvicidal, adulticidal, and repellent properties, thus showing an important role associated with pest control programs as bioinsecticides. In the last few decades, increased attention has been paid worldwide to biological means for the control of mosquitoes, which include the utilization of natural predators, pathogens, and parasites. Many organisms are known to prey on mosquito larvae and thus act as effective biocontrol agents, like crustaceans, insects, spiders, flatworms, mosquito fish (mainly *Gambusia* and guppy), dragonflies, frogs, birds, and bats. Parasites, like mermithids and entomopathogenic fungus, parasitize on mosquitoes and affect their physiology. *Bacillus* and *Wolbachia* have proven to be the most effective pathogens in controlling the mosquitoes. Recently, another method that has come up in the trend is genetic control approach in which mosquito control is done by genetic manipulations either by population suppression or by population replacement or refractory insect strategy. The most famous technique is SIT, which involves mass rearing of sterile male mosquitoes followed by their release in the fields to reduce the reproduction capacity of the target population. Transgenic flightless female mosquitoes, refractory mosquitoes, and *Wolbachia*-based mosquitoes have also been developed to reduce disease transmission. No control measure is 100% accurate, so to achieve effective control success, an integrated mosquito management (IMM) program has to be planned and properly followed.

Points to Remember

1. Extensive use of chemical insecticides for the control of mosquitoes has raised many human health and environmental issues along with creating problem of insecticidal resistance among mosquitoes. So, there is need for development and implementation of some alternative approaches.
2. Ecofriendly and efficient mosquito control methods can be broadly divided into three types: (a) control by phytochemicals, (b) biological control, and (c) genetic control.
3. Phytochemicals are basically secondary metabolites accumulated differentially in the various parts of the plant, such as leaves, roots, stem, fruits, fruit peel, seeds, rhizome, flowers, bark, etc., and out of these phytochemicals, alkaloids, steroids, terpenoids, EOs, and phenolics are a few classes of phytochemicals responsible for their insecticidal properties.
4. Phytotoxic compounds trigger influence on a wide range of molecular targets such as enzymes, receptors, signaling molecules, ion channels, structural proteins, nucleic acids, and biomembranes, besides added cellular components, thus in turn affecting the insect physiology in many different ways and at various receptor sites.
5. Mechanism of action of these phytochemicals is basically by affecting several vital physiological functions in insects including inhibition of AChE, GABA-gated chloride channel, sodium and potassium ion exchange, as well as cellular respiration.
6. So far, >2000 indigenous plant species generally belonging to families Asteraceae, Lamiaceae, Solanaceae, Meliaceae, and Rutaceae have been known to produce chemical factors and metabolites showing marvelous biological activities in terms of good ovicidal, larvicidal, pupicidal, and adulticidal potential along with their repellent properties against different mosquito species.
7. As per the guidelines of WHO, the selected plants or their extracts/products/EOs, as the case may be, are required to fulfill two essential conditions: (a) these had to cause sufficient mortality in standard larvicidal tests to achieve $LC_{50} \leq 100$ ppm, and (b) their chemical composition had to be known.

8. As phytochemicals are from natural sources and are highly biodegradable, this makes them environmentally safe control agents in comparison with the synthetic chemical insecticides, which are creating serious human health and environment related issues. Also, the chances of resistance development against these plant-derived insecticides among mosquitoes are extremely low, because these contain mixtures of chemical compounds that operate cooperatively through multiple mode of actions in contrast to conventional insecticides, which are based on a single active ingredient.
9. The major drawback of using plant-based insecticides is their instability due to the high volatility and thermal decomposition. It is, therefore, the need of the hour to focus on improving their stability to make them as effective as synthetic insecticides.
10. Biological control can be defined as the control of one type of living organism with the application of another. Biocontrol agents are basically predators, parasites, and pathogens.
11. Predators are mainly free living organisms that prey upon mosquitoes (eggs, larval or adult stage) and help to manage the population of mosquitoes. Among invertebrates, crustaceans, insects, spiders and flatworms are main mosquito predators, and among vertebrates, fish like *Gambusia* and *Poecilia,* and some amphibians are predators.
12. Parasites are the organisms, which may complete either a part or the whole life in their hosts and affect the host's physiology. Mermithids are found to be the most effective parasites of mosquitoes (both larvae and adults).
13. Entomopathogenic fungus alters the physiology of mosquito larvae and reduces their ability to feed on blood. *Coelomomyces* is the largest genus of fungi that parasitizes insects.
14. Bacteria are comparatively more pathogenic to mosquitoes than any other pathogen. The bacteria affect the digestive tracts of mosquitoes, cause cell lysis, and ultimately lead to the death. *Bacillus* and *Wolbachia* have proven to be the most effective in controlling the mosquitoes.
15. Biological control is an ecofriendly, target specific, and cost-effective method; in spite of having many benefits, the main constraint of using this method is that it needs longer time to acquire a control over the pest/vector population. Therefore, it becomes difficult to manage the pest/vector on large scale in short period of time.
16. Genetic control approaches against mosquitoes are strategized to reduce the number of individual vector mosquitoes (population suppression) or to reduce the ability of individual mosquitoes within the population to transmit the pathogens (population replacement or refractory insect strategy).
17. The sterile insect technique (SIT) is a species-specific and ecofriendly method of mosquito control. It involves mass rearing of sterile male mosquitoes and releasing them in the fields to reduce the reproduction capacity of the target population.
18. Transgenic flightless female mosquitoes have been developed encoding a repressible, late-acting, sex-specific lethal gene.

19. The refractory mosquitoes have been developed to reduce disease transmission by blocking the acquisition of pathogens, their replication, expression of specific antibodies/peptides, or manipulating cell signaling.
20. The mosquitoes with female-specific dominant lethal transgenes and disordered X chromosome can be used to develop majorly males for release in the fields.
21. The male-determining M-factor has been tapped in the mosquitoes to transform female adults into harmless nonbiting males with lower fitness.
22. *Wolbachia*-based mosquito control strategy is used for population replacement or population suppression.
23. Gene drive technologies help in spread of a genetic modification through a population at higher inheritance rates than the normal.

References

Abagli AZ, Alavo TB, Perez-Pacheco R, Platzer EG (2019) Efficacy of the mermithid nematode, *Romanomermis iyengari*, for the biocontrol of *Anopheles gambiae*, the major malaria vector in sub-Saharan Africa. Parasit Vectors 12(1):1–8

Ahmed AM, Abdel-Megeed AA, Alqahtani HM (2014) A novel mosquitocidal bacterium as a biocontrol agent in Saudi Arabia: II-A promising larvicide against *Culex pipiens* mosquito. Pak J Zool 46(3):773–782

Aldridge S (2008) Genetically modified mosquitoes. Nat Biotechnol 26:828–830

Alphey L, McKemey A, Nimmo D, Oviedo MN, Lacroix R, Matzen K, Beech C (2013) Genetic control of *Aedes* mosquitoes. Pathog Glob Health 107(4):170–179

Anbu P, Murugan K, Madhiyazhagan P, Dinesh D, Subramaniam J, Panneerselvam C, Suresh U, Alarfaj AA, Munusamy MA, Higuchi A, Hwang JS, Kumar S, Nicoletti M, Benelli G (2016) Green-synthesised nanoparticles from *Melia azedarach* seeds and the cyclopoid crustacean *Cyclops vernalis*: an eco-friendly route to control the malaria vector *Anopheles stephensi*? Nat Prod Res 30(18):2077–2084

Andreasen MH, Curtis CF (2005) Optimal life stage for radiation sterilization of *Anopheles* for sterile insect releases. Med Vet Entomol 19:238–244

Ansari MA, Singh KR, Brooks GD, Malhotra PR, Vaidyanathan V (1977) The development of procedures and techniques for mass rearing of *Aedes aegypti*. Indian J Med Res 65(Suppl):91–99

Aryan A, Anderson MAE, Biedler JK, Qi Y, Overcash JM, Naumenko AN, Sharakhova MV, Mao C, Adelman ZN, Tu Z (2020) *Nix* alone is sufficient to convert female *Aedes aegypti* into fertile males and *myo-sex* is needed for male flight. Proc Natl Acad Sci USA 117(30):17702–17709

Asimeng EJ, Mutiga MJ (1992) Field evaluation of *Tilapia zilli* (Gervais) as a biological control agent for mosquito control. Biol Control 2:317–320

Awasthi AK, Wu CH, Tsai KH, King CC, Hwang JS (2012) How does the ambush predatory copepod *Megacyclops formosanus* (Harada, 1931) capture mosquito larvae of *Aedes aegypti*? Zool Stud 51(7):927–936

Axford JK, Callahan AG, Hoffmann AA, Yeap HL, Ross PA (2016) Fitness of wAlbB *Wolbachia* infection in *Aedes aegypti*: parameter estimates in an outcrossed background and potential for population invasion. Am J Trop Med Hyg 94:507–516

Barón OL, Ursic-Bedoya RJ, Lowenberger CA, Ocampo CB (2010) Differential geneexpression from midguts of refractory and susceptible lines of the mosquito, *Aedes aegypti*, infected with Dengue-2 virus. J Insect Sci 10:41

Becker N, Petric D, Zgomba M, Boase C, Madon M, Dahl C, Kaiser A (2010) Mosquitoes and their control. Springer Science & Business Media, Heidelberg, p 577

Beebe NW, Pagendam D, Trewin BJ, Bomer A, Bradford M, Ford A, Liddington C, Bondarenco A, De Barro PJ, Gilchrist J, Paton C, Staunton KM, Johnson B, Maynard AJ, Devine GJ, Hugo LE, Rasic G, Cook H, Massaro P, Snoad N, Crawford JE, White BJ, Xi Z, Ritchie SA (2021) Releasi*ng* incompatible males drives strong suppression across populations of wild and *Wolbachia*-carrying *Aedes aegypti* in Australia. Proc Natl Acad Sci USA 118(41):e2106828118

Bellini R, Medici A, Puggioli A, Balestrino F, Carrieri M (2013) Pilot field trials with *Aedes albopictus* irradiated sterile males in Italian urban areas. J Med Entomol 50(2):317–325

Benedict MQ, Robinson AS (2003) The first releases of transgenic mosquitoes: an argument for the sterile insect technique. Trends Parasitol 19(8):349–355

Benelli G, Pavela R (2018) Beyond mosquitoes-essential oil toxicity and repellency against bloodsucking insects. Ind Crop Prod 117:382–392

Bilal H, Hassan SA (2012) Plants secondary metabolites for mosquito control. Asian Pac J Trop Dis 2:168

Bobadilla M, Zavala F, Sisniegas M, Zavaleta G, Mostacero J, Taramona L (2005) Evaluación larvicida de suspension esacuosas de *Annona muricata* Linnaeus «guanábana» sobre *Aedes aegypti* Linnaeus (Diptera, Culicidae*)*. Rev Peru Biol 12:145–152

Bohbot JD, Lu T, Zwiebel LJ (2010) Molecular regulation of olfaction in mosquitoes. In: Takken W, Knols BGJ (eds) Olfaction in vector-host interactions. Wageningen Academic Publishers, Wageningen, pp 17–38

Bowatte G, Perera P, Senevirathne G, Meegaskumbura S, Meegaskumbura M (2013) Tadpoles as dengue mosquito (*Aedes aegypti*) egg predators. Biol Control 67(3):469–474

Campbell FL, Sullivan WW, Smith LN (1933) The relative toxicity of nicotine, anabasine, methyl anabasine and lupinine for culicine mosquito, larvae. J Econ Entomol 26:500–509

Cantrell CL, Pridgeon JW, Fronczek FR, Becnel JJ (2010) Structure–activity relationship studies on derivatives of Eudesmanolides from *Inula helenium* as toxicants against *Aedes aegypti* larvae and adults. Chem Biodivers 7:1681–1697

Carballar-Lejarazú R, James AA (2017) Population modification of *Anopheline* species to control malaria transmission. Pathog Glob Health 111:424–435

Carvalho DO, McKemey AR, Garziera L, Lacroix R, Donnelly CA, Alphey L, Malavasi A, Capurro ML (2015) Suppression of a field population of *Aedes aegypti* in Brazil by sustained release of transgenic male mosquitoes. PLoS Negl Trop Dis 9(7):e0003864

Cavalca PAM, Lolis MIG, Reis B, Bonato CM (2010) Homeopathic and larvicide effect of *Eucalyptus cinereal* essential oil against *Aedes aegypti*. Braz Arch Biol Technol 53:835–843

Chandra G, Bhattacharjee I, Chatterjee SN, Ghosh A (2008) Mosquito control by larvivorous fish. Indian J Med Res 127(1):13–27

Chitra T, Murugan K, Kumar AN, Madhiyazhagan P, Nataraj T, Indumathi D, Hwang JS (2013) Laboratory and field efficacy of *Pedalium murex* and predatory copepod, *Mesocyclops longisetus* on rural malaria vector, *Anopheles culicifacies*. Asian Pac J Trop Dis 3(2):111–118

Conti B, Flamini G, Cioni PL, Ceccarini L, Leonardi MM, Benelli G (2014) Mosquitocidal essential oils: are they safe against non-target aquatic organisms? Parasitol Res 113:251–259

Cotchakaew N, Soonwera M (2020) Toxicity of several botanical essential oils and their combinations against females of *Aedes albopictus* (Skuse) and *Anopheles minimus* (Theobald): oviposition deterrent, ovicidal and adulticidal efficacies. Asian Pac J Trop Biomed 9:29–39

Criscione F, Qi Y, Tu Z (2016) GUY1 confers complete female lethality and is a strong candidate for a male-determining factor in *Anopheles stephensi*. elife 5:e19281

Cuthbert RN, Ortiz-Perea N, Dick JT, Callaghan A (2019) Elusive enemies: consumptive and ovipositional effects on mosquitoes by predatory midge larvae are enhanced in dyed environments. Biol Control 132:116–121

Dambach P (2020) The use of aquatic predators for larval control of mosquito disease vectors: opportunities and limitations. Biol Control 150:104357

Darbro JM, Johnson PH, Thomas MB, Ritchie SA, Kay BH, Ryan PA (2012) Effects of *Beauveria bassiana* on survival, blood-feeding success, and fecundity of *Aedes aegypti* in laboratory and semi-field conditions. Am J Trop Med Hyg 86(4):656

Davidson EW, Becker N (1996) Microbial control of vectors. In: Beaty BJ, Marquardt WC (eds) The biology of disease vectors. University Press of Colorado, USA, pp 549–563

Dhillon G, Bains S, Kocher DK (2020) Microencapsulated citronella (*Cymbopogon nardus*) essential oil as mosquito repellent finish for cotton. Agric Res J 57(2):275–279

Dias CN, Moraes DFC (2014) Essential oils and their compounds as *Aedes aegypti* L. (Diptera: Culicidae) larvicides: review. Parasitol Res 113:565–592

Domínguez-Martínez VG, Martínez-Vázquez M, Collar Gómez E, AtzinGarcía J, Chimal-popoca FL (2003) Pupicidal activity of annonacin for *Aedes aegypti* (L.) (Diptera: Culicidae). Folia Entomol Mex 42:349–358

Dyck VA, Hendrichs JP, Robinson AS (eds) (2021) The sterile insect technique: principles and practice in area-wide integrated pest management. Second edition. CRC Press, Boca Raton

Elumalai D, Kaleena PK, Ashok K, Suresh A, Hemavathi M (2016) Green synthesis of silver nanoparticle using *Achyranthes aspera* and its larvicidal activity against three major mosquito vectors. Eng Agric Environ Food 9:1–8

Emden VHF (2004) Pest and vector control. Cambridge University Press

Enan EE (2001) Insecticidal activity of essential oils: octopaminergic sites of action. Comp Biochem Physiol Part C 130:325–337

Evans HC, Elliot SL, Barreto RW (2018) Entomopathogenic fungi and their potential for the management of *Aedes aegypti* (Diptera: Culicidae) in the Americas. Mem Inst Oswaldo Cruz 113:206–214

Fallatah HI (2014) Noninvasive biomarkers of liver fibrosis: an overview. Adv Hepatol, Article ID 357287

Federici BA (1995) The future of microbial insecticides as vector control agents. J Am Mosq Control Assoc (Mosquito News) 11(2):260–268

Flor M, Hammerstein P, Telschow A (2007) *Wolbachia*-induced unidirectional cytoplasmic incompatibility and the stability of infection polymorphism in parapatric host populations. J Evol Biol 20:696–706

Forni C, Facchiano F, Bartoli M, Pieretti S, Facchiano A, D'Arcangelo D, Norelli S, Valle G, Nisini R, Beninati S (2019) Beneficial role of phytochemicals on oxidative stress and age-related diseases. Biomed Res Int 8748253

Franz AW, Sanchez-Vargas I, Adelman ZN, Blair CD, Beaty BJ, James AA, Olson KE (2006) Engineering RNA interference-based resistance to dengue virus type 2 in genetically modified *Aedes aegypti*. Proc Natl Acad Sci USA 103:4198–4203

Franzios G, Mirotsou M, Hatziapostolou E, Kral J, Scouras ZG, Mavragani-Tsipidou P (1997) Insecticidal and genotoxic activities of mint essential oils. J Agric Food Chem 45:2690–2694

Fu G, Lees RS, Nimmo D, Aw D, Jin L, Gray P, Berendonk TU, White-Cooper H, Scaife S, Phuc HK, Marinotti O, Jasinskiene N, James AA, Alphey L (2010) Female-specific flightless phenotype for mosquito control. Proc Natl Acad Sci USA 107(10):4550–4554

Galizi R, Doyle L, Menichelli M, Bernardini F, Deredec A, Burt A, Stoddard BL, Windbichler N, Crisanti A (2014) A synthetic sex ratio distortion system for the control of the human malaria mosquito. Nat Commun 5:3977

Gantz VM, Jasinskiene N, Tatarenkova O, Fazekas A, Macias VM, Bier E, James AA (2015) Highly efficient Cas9-mediated genedrive for population modification of the malaria vector mosquito *Anopheles stephensi*. Proc Natl Acad Sci USA 112(49):E6736–E6743

Garcia GDA, Sylvestre G, Aguiar R, Da Costa GB, Martins AJ, Lima JBP, Petersen MT, Lourenço-de-Oliveira R, Shadbolt MF, Rašić G, Hoffmann AA, Villela DAM, Dias FBS, Dong Y, O'Neill SL, Moreira LA, Maciel-de-Freitas R (2019) Matching the genetics of released and local *Aedes aegypti* populations is critical to assure *Wolbachia* invasion. PLoS Negl Trop Dis 13:e0007023

Geoghegan V, Stainton K, Rainey SM, Ant TH, Dowle AA, Larson T, Hester S, Charles PD, Thomas B, Sinkins SP (2017) Perturbed cholesterol and vesicular trafficking associated with dengue blocking in *Wolbachia*-infected *Aedes aegypti* cells. Nat Commun 8(1):526

Ghosh A, Chandra G (2006) Biocontrol efficacy of *Cestrum diurnum* L. (Solanaceae: Solanales) against the larval forms of *Anopheles stephensi*. Nat Prod Res 20(04):371–379

Ghosh A, Chowdhury N, Chandra G (2012) Plant extracts as potential mosquito larvicides. Indian J Med Res 135(5):581–598

Goldberg LJ, Margalit J (1977) A bacterial spore demonstrating rapid larvicidal activity against *Anopheles sergentii, Uranotaenia unguiculata, Culex univitattus, Aedes aegypti* and *Culex pipiens* complex. Mosq News 37:355–358

Gomulski LM, Mariconti M, Di Cosimo A, Scolari F, Manni M, Savini G, Malacrida AR, Gasperi G (2018) The *Nix* locus on the male-specific homologue of chromosome 1 in *Aedes albopictus* is a strong candidate for a male-determining factor. Parasit Vectors 11(Suppl 2):647

Griffin LF, Knight JM (2012) A review of the role of fish as biological control agents of disease vector mosquitoes in mangrove forests: reducing human health risks while reducing environmental risk. Wetl Ecol Manag 20:243–252

Hackett LW (1937) Malaria in Europe. An ecological study. Oxford Univ. Press, p 336

Hall AB, Basu S, Jiang X, Qi Y, Timoshevskiy VA, Biedler JK, Sharakhova MV, Elahi R, Anderson MAE, Chen X-G, Sharakhov IV, Adelman ZN, Tu ZJ (2015) A male-determining factor in the mosquito *Aedes aegypti*. Science 348:1268–1270

Hall AB, Papathanos P-A, Sharma A, Cheng C, Akbari CC, Assour L, Bergman NH, Cagnetti A, Crisanti A, Dottorini T, Florentini E, Galizi R, Hnath J, Jiang X, Koren S, Nolan T, Radune D, Sharakhova MV, Steele A, Timoshevsky VA, Windbichler N, Zhang S, Hahn MW, Phillippy AM, Emrich SJ, Sharakov IV, Tu ZJ, Besansky NJ (2016) Radical remodeling of the Y chromosome in a recent radiation of malaria mosquitoes. Proc Natl Acad Sci USA 113:E2114–E2123

Hamed AM, El-Sherbini MS, Abdeltawab MS (2022) Eco-friendly mosquito-control strategies: advantages and disadvantages. Egypt Acad J Biol Sci, E Med Entomol Parasitol 14(1):17–31

Hammond AM, Galizi R (2017) Gene drives to fight malaria: current state and future directions. Pathog Glob Health 111:412–423

Hammond A, Pollegioni P, Persampieri T, North A, Minuz R, Trusso A, Bucci A, Kyrou K, Morianou I, Simoni A, Nolan T, Müller R, Crisanti A (2021) Gene-drive suppression of mosquito populations in large cages as a bridge between lab and field. Nat Commun 12:4589

Hartzell A, Wilcoxon F (1941) A survey of plant products for insecticidal properties. Contributed by Boyce Thompson Institute, 12: 127–141

Hemingway J, Hawkes N, Mccarroll L, Ranson H (2004) The molecular basis of insecticide resistance in mosquitoes. Insect Biochem Mol Biol Educ 34:653–655

Hertig M, Wolbach S (1924) Studies on rickettsia-like microorganisms in insects. J Med Res 44:329–374

Hilgenboecker K, Hammerstein P, Schlattmann P, Telschow A, Werren JH (2008) How many species are infected with *Wolbachia*?—a statistical analysis of current data. FEMS Microbiol Lett 281:215–220

Hoffmann AA, Ross PA, Rašić G (2015) *Wolbachia* strains for disease control: ecological and evolutionary considerati*ons*. Evol Appl 8:751–768

Ingabire CM, Hakizimana E, Rulisa A, Kateera F, Van Den Borne B, Muvunyi CM, Mutesa L, Van Vugt M, Koenraadt CJ, Takken W (2017) Community based biocontrol of malaria mosquitoes using *Bacillus thuringiensis* var. *israelensis* (Bti) in Rwanda: community awareness, acceptance and participation. Malar J 16:399

Islam J, Zaman K, Duarah S, Raju PS, Chattopadhyay P (2017) Mosquito repellents: an insight into the chronological perspectives and novel discoveries. Acta Trop 167:216–230

Isman MB (1997) Neem and other botanical insecticides: barriers to commercialization. Phytoparasitica 25:339–344

Isman MB, Miresmailli S, Machial C (2011) Commercial opportunities for pesticides based on plant essential oils in agriculture, industry and consumer products. Phytochem Rev 10:197–204

Jacob S, Thomas AP, Manju EK (2017) Bio control efficiency of odonata nymphs on *Aedes aegypti* larvae. IOSR J Environ Sci Toxicol Food Technol 11:1

Jantan IB, Yalvema MF, Ahmad NW, Jamal JA (2005) Insecticidal activities of the leaf oils of eight *Cinnamomum* species against *Aedes aegypti* and *Aedes albopictus*. Pharm Biol 43:526–532

Kamala-Jayanthi PD, Aurade RM, Kempraj V, Chakravarthy AK, Verghese A (2015) Glimpses of semiochemical research applications in Indian horticulture: present status and future perspectives. In: Chakravarthy A (ed) New horizons in insect science: towards sustainable pest management. Springer, New Delhi, pp 239–257

Katz TM, Miller JH, Hebert AA (2008) Insect repellents: historical perspectives and new developments. J Am Acad Dermatol 58:865–871

Kaur N, Kocher DK (2021) Effect of nanoemulsified *Eucalyptus globulus* oil on development, emergence and survival of *Aedes aegypti* L. Indian J Entomol 84(4):804–808

Kerwin JL, Washino RK (1988) Field evaluation of *Lagenidium giganteum* (Oömycetes: Lagenidiales) and description of a natural epizoötic involving a new isolate of the fungus. J Med Entomol 25(6):452–460

Kihampa C (2011) Tanzanian botanical derivatives in the control of malaria vectors: opportunities and challenges. J Appl Sci Environ Manag 15(1):155–160

Kitching RL (2000) Food webs and container habitats: the natural history and ecology of phytotelmata. Cambridge University Press, p 431

Knipling E (1955) Possibilities of insect control or eradication through use of sexually sterile males. J Econ Entomol 48:459–462

Knowles BH, Ellar DJ (1987) Colloid-osmotic lysis is a general feature of the mechanism of action of *Bacillus thuringiensis* δ-endotoxins with different insect specificity. Biochim Biophys Acta (BBA) General Subjects 924(3):509–518

Kobylinski KC, Sylla M, Black W, Foy BD (2012) Mermithid nematodes found in adult Anopheles from southeastern Senegal. Parasit Vectors 5:1–7

Kok LT, Kok V (1999) Biological control for the public. Cabi Publishing

Koodalingam A, Deepalakshmi R, Ammu M, Rajalakshmi A (2014) Effects of neemazal on marker enzymes and hemocyte phagocytic activity of larvae and pupae of the vector mosquito *Aedes aegypti*. J Asia Pac Entomol 17:175–181

Kumar PM, Murugan K, Kovendan K, Panneerselvam C, Kumar KP, Amerasan D, Subramaniam J, Kalimuthu K, Nataraj T (2012) Mosquitocidal activity of *Solanum xanthocarpum* fruit extract and copepod *Mesocyclops thermocyclopoides* for the control of dengue vector *Aedes aegypti*. Parasitol Res 111:609–618

Kwenti TE (2017) Biological control of parasites. In: Khater H et al (eds) Natural remedies in the fight against parasites. IntechOpen, United Kingdom, pp 24–58

Labbé GM, Scaife S, Morgan SA, Curtis ZH, Alphey L (2012) Female-specific flightless (fsRIDL) phenotype for control of *Aedes albopictus*. PLoS Negl Trop Dis 6:e1724

Lalithambika B, Vani C (2016) *Pseudomonas aeruginosa* KUN2, extracellular toxins-A potential source for the control of dengue vector. J Vector Borne Dis 53(2):105–111

Macias VM, Ohm JR, Rasgon JL (2017) Gene drive for mosquito control: where did it come from and where are we headed? Int J Environ Res Public Health 14(9):1006

Manrique-Saide P, Ibanez-Bernal S, Delfin-Gonzalez H, Tabla PV (1998) *Mesocyclops longisetus* effects on survivorship of *Aedes aegypti* immature stages in car tyres. Med Vet Entomol 12:386–390

Marshall JM, Taylor CE (2009) Malaria control with transgenic mosquitoes. PLoS Med 6(2):e20

Marten GG, Astaiza R, Suarez MF, Monje C, Reid JW (1989) Natural control of larval *Anopheles albimanus* (Diptera: Culicidae) by the Predator *Mesocyclops* (Copepoda: Cyclopoida). J Med Entomol 26:624–627

Mathur G, Sanchez-Vargas I, Alvarez D, Olson KE, Marinotti O, James AA (2010) Transgene-mediated suppression of dengue viruses in the salivary glands of the yellow fever mosquito, *Aedes aegypti*. Insect Mol Biol 19:753–763

McMeniman CJ, Lane RV, Cass BN, Fong AWC, Sidhu M, Wang YF, O'Neill SL (2009) Stable introduction of a life-shortening *Wolbachia* infection into the mosquito *Aedes aegypti*. Science 323:141–144

Meryem S, Demirak S, Canpolat E (2022) Plant-based bioinsecticides for mosquito control: impact on insecticide resistance and disease transmission. Insects 13(2):162

Moirangthem BD, Singh SN (2018) Comparative studies of three potent bioagent against mosquito larvae. Int J Mosq Res 5:10

Moore SJ, Lenglet A, Hill N (2006) Plant-based insect repellents. In: Debboun M et al (eds) Insect repellents: principles methods, and use. CRC Press, Boca Raton, FL, pp 276–305

Mullen GR, Durden LA (2009) Medical and veterinary entomology. Academic Press, Elsevier

Murdoch WW, Bence JR, Chesson JA (1984) Effects of the general predator, *Notonecta* (Hemiptera) upon a freshwater community. J Anim Ecol 53:791–808

Murugan K, Benelli G, Ayyappan S, Dinesh D, Panneerselvam C, Nicoletti M, Hwang JS, Kumar PM, Subramaniam J, Suresh U (2015a) Toxicity of seaweed-synthesized silver nanoparticles against the filariasis vector *Culex quinquefasciatus* and its impact on predation efficiency of the cyclopoid crustacean *Mesocyclops longisetus*. Parasitol Res 114:2243–2253

Murugan K, Priyanka V, Dinesh D, Madhiyazhagan P, Panneerselvam C, Subramaniam J, Suresh U, Chandramohan B, Roni M, Nicoletti M, Alarfaj AA, Higuchi A, Munusamy MA, Khater HF, Messing RH, Benelli G (2015b) Predation by Asian bullfrog tadpoles, *Hoplobatrachus tigerinus*, against the dengue vector, *Aedes aegypti*, in an aquatic environment treated with mosquitocidal nanoparticles. Parasitol Res 114:3601–3610

Muspratt J (1963) Destruction of the larvae of *Anopheles gambiae* Giles by a *Coelomomyces* fungus. Bull World Health Organ 29(1):81

Namias A, Jobe NB, Paaijmans KP, Huijben S (2021) The need for practical insecticide-resistance guidelines to effectively inform mosquito-borne disease control programs. elife 10:e65655

Nerio LS, Olivero-Verbel J, Stashenko E (2010) Repellent activity of essential oils: a review. Bioresour Technol 101(1):372–378

Nour AH, Sandanasamy J, Nour AH (2012) Larvicidal activity of extracts from different parts of neem (*Azadirachta indica*) against *Aedes aegypti* mosquitoes' larvae. Sci Res Essays 7:2810–2815

O'Leary S, Adelman ZN (2020) CRISPR/Cas9 knockout of female-biased genes *AeAct-4* or *myo-fem* in *Ae. aegypti* results in a flightless phenotype in female, but not male mosquitoes. PLoS Negl Trop Dis 14(12):e0008971

Pavela R (2015) Essential oils for the development of eco-friendly mosquito larvicides: a review. Ind Crop Prod 76:174–187

Perevozkin VP, Lukyantsev SV, Gordeev MI (2004) Comparative analysis of foraging behavior in aquatic and semiaquatic spiders of the genera *Argyroneta, Dolomedes, Pirata*, and *Pardosa*. Russ J Ecol 35:103

Petersen JJ (1985) Nematodes as biological control agents. In: Baker JR, Muller R (eds) Advances in parasitology. Academic Press, London, pp 307–344

Pirali-Kheirabadi K (2012) Biocontrol of parasites. In: Shah MM (ed) Parasitology. InTechOpen, United Kingdom, pp 52–66

Platzer E (1981) Biocontrol of mosquitoes with mermithids. J Nematol 13:257

Poopathi S, Tyagi BK (2006) The challenge of mosquito control strategies: from primordial to molecular approaches. Biotechnol Mol Biol Rev 1(2):51–65

Poopathi S, Mani C, Thirugnanasambantham K, Praba VL, Ahangar NA, Balagangadharan K (2013) Identification and characterization of a novel marine *Bacillus cereus* for mosquito control. Parasitol Res 113(1):323–332

Poopathi S, Thirugnanasambantham K, Mani C, Ragul K, Sundarapandian S (2014) Isolation of mosquitocidal bacteria (*Bacillus thuringiensis, B. sphaericus* and *B. cereus*) from excreta of arid birds. Indian J Exp Biol 52(7):739–747

Raghavendra K, Barik TK, Reddy BP, Sharma P, Dash AP (2011) Malaria vector control: from past to future. Parasitol Res 108(4):757–779

Rajendran R, Prasad RS (1994) A laboratory study on the life cycle and feeding behaviour of *Arrenurus madaraszi* (Acari: Arrenuridae) parasitizing *Anopheles* mosquitoes. Ann Trop Med Parasitol 88(2):169–174

Ramírez-Lepe M, Ramírez-Suero M (2012) Biocontrol of mosquito larvae by a *Bacillus thuringiensis* subsp. *israelensis*. In: Perveen F (ed) Insecticides Pest Engineering. IntechOpen, United Kingdom, pp 239–264

Rattan RS (2010) Mechanism of action of insecticidal secondary metabolites of plant origin. Crop Prot 29:913–920

Raveen R, Ahmed F, Pandeeswari M, Reegan D, Tennyson S, Arivoli S, Jayakumar M (2017) Laboratory evaluation of a few plant extracts for their ovicidal, larvicidal and pupicidal activity against medically important human dengue, chikungunya and zika virus vector, *Aedes aegypti* Linnaeus 1762 (Diptera: Culicidae). Int J Mosq Res 4(4):17–28

Regnault-Roger C, Vincent C, Arnason JT (2012) Essential oils in insect control: low-risk products in a high-stakes world. Annu Rev Entomol 57:405–424

Riat A, Kocher DK (2017) Study of histoarchitectural changes in *Anopheles stephensi* larvae following exposure to *Eucalyptus globulus* and *Aloe vera* oils. Turkish J Zool 41(5):763–773

Ritchie SA, Van Den Hurk AF, Smout MJ, Staunton KM, Hoffmann AA (2018) Mission accomplished? We need a guide to the 'post release' world of *Wolbachia* for *Aedes*-borne disease control. Trends Parasitol 34:217–226

Sadiq MB, Tharaphan P, Chotivanich K, Tarning J, Anal AK (2017) *In vi*tro antioxidant and antimalarial activities of leaves, pods and bark extracts of *Acacia nilotica* (L.) Del. BMC Complement Alternat Med 17(1):372

Schaper S (1999) Evaluation of Costa Rican copepods (Crustacea: Eudecapoda) for larval *Aedes aegypti* control with special reference to *Mesocyclops thermocyclopoides*. J Am Mosq Control Assoc 15:510–519

Schaper S, Hernández-Chavarría F (2006) Scanning electron microscopy of damage caused by *Mesocyclops thermocyclopoides* (Copepoda: Cyclopoidea) on larvae of the dengue fever vector *Aedes aegypti* (Diptera:Culicidae). Rev Biol Trop 54:847

Senthilkumar A, Jayaraman M, Venkatesalu V (2013) Chemical constituents and larvicidal potential of *Feronia limonia* leaf essential oil against *Anopheles stephensi, Aedes aegypti* and *Culex quinquefasciatus*. Parasitol Res 112:1337–1342

Senthil-Nathan S, Choi MY, Seo HY, Paik CH, Kalaivani K, Kim JD (2008a) Effect of azadirachtin on acetylcholinesterase (AChE) activity and histology of the brown planthopper *Nilaparvata lugens* (Stål). Ecotoxicol Environ Saf 70:244–250

Senthil-Nathan S, Hisham A, Jayakumar G (2008b) Larvicidal and growth inhibition of the malaria vector *Anopheles stephensi* by triterpenes from *Dysoxylum* spp. (Meliaceae). Fitoterapia 76:106–111

Shaalan EA, Canyon DV (2009) Aquatic insect predators and mosquito control. Trop Biomed 26:223–261

Shaalan EA-S, Canyon D, Younes MWF, Abdel-Wahab H, Mansour A-H (2005) A review of botanical phytochemicals with mosquitocidal potential. Environ Int 31(8):1149–1166

Shajahan A, Zahir-Hussian MI, Kumar SR, Ajintha K (2022) A review on plant phytochemicals potential for mosquito control. Int J Mosq Res 9(6):47–54

Shukla D, Wijayapala S, Vankar PS (2018) Effective mosquito repellent from plant based formulation. Int J Mosq Res 5:19–24

Silva LL, Garlet QI, Koakoski G, Oliveira TA, Barcellos LJG, Baldisseroto B, Pereira AMS, Heinzmann BM (2015) Effects of anesthesia with the essential oil of *Ocimum gratissimum* L. in parameters of fish stress. Rev Bras Plant Medicinal 17:215–223

Simon OIA, Akeju AV (2017) Laboratory evaluation of extract from peels and seeds of some *citrus* species against *Anopheles* mosquitoes (Diptera: Culicidae). Int J Mosq Res 4:48–54

Singh KV, Bansal SK (2003) Larvicidal properties of a perennial herb *Solanum xanthocarpum* against vectors of malaria and dengue/DHF. Curr Sci 84:749–751

Singh H, Marwa LR, Mishra A, Singh KV (2014) Predatory habits of *Lutzia* (*Metalutzia*) *fuscana* (Wiedmann) (Diptera: Culicidae) in the arid environments of Jodhpur, western Rajasthan, India. Arthropods 3:70

Sola P, Mvumi BM, Ogendo JO, Mponda O, Kamanula JF, Nyirenda SP, Belmain SR, Stevenson PC (2014) Botanical pesticide production, trade and regulatory mechanisms in sub-Saharan Africa: making a case for plant-based pesticidal products. Food Secur 6:369–384

Soonwera M, Phasomkusolsil S (2016) Effect of *Cymbopogon citrates* (lemongrass) and *Syzygium aromaticum* (clove) oils on the morphology and mortality of *Aedes aegypti* and *Anopheles dirus* larvae. Parasitol Res 115:1691–1703

Stenfors ALP, Fagerlund A, Granum PE (2008) From soil to gut: *Bacillus cereus* and its food poisoning toxins. FEMS Microbiol Rev 32:579–606

Sukumar K, Michael JP, Boobar LR (1991) Botanical derivatives in mosquito control: a review. J Am Mosq Control Assoc 7(2):210–237

Tabibzadeh I, Behbehani C, Nakhai R (1970) Use of *Gambusia* fish in the malaria eradication programme of Iran. Bull World Health Organ 43:623–626

Takken W (1991) The role of olfaction in host-seeking of mosquitoes: a review. Insect Sci Appl 12:287–295

Tanada Y, Kaaya HK (1993) In: Vega FE, Kaya HK (eds) Insect pathology. Academic Press, San Diego

Tehri K, Singh N (2015) The role of botanicals as green pesticides in integrated mosquito management—a review. Int J Mosq Res 2(1):18–23

Tranchida MC, Macia A, Brusa F (2009) Predation potential of three flatwormspecies (Platyhelminthes: Turbellaria) on mosquitoes (Diptera: Culicidae). Biol Control 49:270

Tripathi AK, Sha W, Shulaev V, Stins MF, Sullivan DJ (2009) *Plasmodium falciparum*–infected erythrocytes induce NF-κB regulated inflammatory pathways in human cerebral endothelium. Blood 114(19):4243–4252

Vinogradov DD, Sinev AY, Tiunov AV (2022) Predators as control agents of mosquito larvae in micro-reservoirs. Inland Water Biol 15(1):39–53

Vreysen MJB, Abd-Alla AMM, Bourtzis K, Bouyer J, Caceres C, de Beer C, Carvalho D, Maiga H, Mamai W, Nikolouli K, Yamada H, Pereira R (2021) The insect pest control laboratory of the joint FAO/IAEA programme: ten years (2010–2020) of research and development, achievements and challenges in support of the sterile insect technique. Insects 12:346

Wang GH, Gamez S, Raban RR, Marshall JM, Alphey L, Li M, Rasgon JL, Akbari OS (2021) Combating mosquito-borne diseases using genetic control technologies. Nat Commun 12(1):4388

Weterings R (2015) Tadpoles of three common anuran species from Thailand do not prey on mosquito larvae. J Vector Ecol 40:230–232

WHO (1998) Techniques to detect insecticide resistance mechanisms (field and laboratory manual). World Health Organization, Geneva

WHO (2005) Guidelines for laboratory and field testing of mosquito larvicides. World Health Organization, Geneva

WHO (2009) Temephos in drinking-water: use for vector control in drinking-water sources and containers. World Health Organization, Geneva

WHO (2013) Guidelines for laboratory and field-testing of long-lasting insecticidal nets. World Health Organization, Geneva

Yen P-S, Failloux AB (2020) A review: *Wolbachia*-based population replacement for mosquito control shares common points with genetically modified control approaches. Pathogens 9(5):404

Yu CH, Dang YK, Zhou ZP, Wu C, Zhao FZ, Sachs MS, Liu Y (2015) Codon usage influences the local rate of translation elongation to regulate co translational protein folding. Mol Cell 59:744–754

Żabka M, Kovac D (1996) *Paracyrba wanlessi*-a new genus and species of Spartaeinae from Peninsular Malaysia, with notes on its biology (Arachnida: Araneae: Salticidae). Senckenberg Biol 76:153

Zhong Y, Li Z-X (2014) Bidirectional cytoplasmic incompatibility induced by cross-order transfection of *Wolbachia*: implications for control of the host population. Microb Ecol 68:463–471

Zhou W, Rousset F, O'Neill SL (1998) Phylogeny and PCR-based classification of *Wolbachia* strains using *wsp* gene sequences. Proc R Soc B: Biol Sci 265:509–515

Żukowska G, Durczyńska Z (2024) Properties and applications of essential oils: a review. J Ecol Eng 25(2):333–340

Insecticidal Resistance and Counter Mechanism in Mosquitoes: A Coevolutionary Trade-off

15

Ida Anandi Mukhiya and Amlan Das

Abstract

Mosquitoes are a global health risk, yet eradication efforts have been rendered futile due to the exponential rise of insecticide resistance arising in mosquitoes. Toxic chemical resistance can manifest in various ways in mosquitoes, including behavioural, physiological, metabolic and genetic mechanisms. They have developed precise strategies to combat the insecticidal action in maintaining their population, and, in doing so, they have acquired tolerance to diverse toxic chemicals through different ways including enhanced metabolic detoxification genes, such as P450, esterase, GST, overexpression, and alterations of cuticular proteins and structural modifications in voltage-gated sodium channel, GABA receptors, and AChE genes. However, the depth of knowledge and research on gene regulation and mechanisms and how quickly they can develop resistance to insecticides is remarkably low. These genetic and metabolic alterations provide mosquitoes with a greater level of toxicant resistance.

Using insecticides against mosquitoes entails trade-off costs, such as a reduction in body size, an elongation of the larval development period and a decrease in the female adult longevity. As natural selection affects mutations in coding gene sequences, comprehensive genome sequencing in various mosquito strains will open new possibilities in designing revolutionary pesticides to kill mosquitoes selectively.

Keywords

Mosquito · Insecticides · Behaviour-physiochemical-genetic resistance

---

I. A. Mukhiya · A. Das (✉)
Entomology Laboratory, Department of Zoology, University of Calcutta, Kolkata, West Bengal, India

© The Author(s), under exclusive license to Springer Nature Singapore Pte Ltd. 2024

Omkar (ed.), *Mosquitoes*, https://doi.org/10.1007/978-981-97-4163-2_15

Learning Objectives

1. Delivering functional mechanism and counteraction on insecticidal resistance in mosquitoes.
2. Analyzing the trade-off cost towards insecticidal resistance.
3. Discussion on target-specific application of green pesticides against innate and acquired resistance.

15.1 Introduction

Mosquitoes (*Culicidae*: Diptera) are a worldwide menace to human health due to their notorious spreading of fatal diseases. Since the emergence of human civilization, various cultural, biological, and chemical weapons have often been designed to eliminate its population and dispersal. Although organic and inorganic insecticides have been a success in controlling mosquito populations, their efficacy is temporary and limited due to recurring adaptations in mosquitoes, making the eradication efforts futile.

There has been a constant back-and-forth between mosquito adaptations and efforts to control vectors since the initial inquiry on insecticidal resistance of mosquitoes against chlorinated hydrocarbons (Gjullin and Peters 1952). Mosquitoes are known to be the primary carriers of various disease-causing microorganisms, including the highly lethal *Plasmodium falciparum* (unicellular Plasmodiidae protozoa) and *Flavivirus* (a single-stranded RNA virus), which are responsible for causing malaria and dengue, respectively. Additionally, mosquitoes are also responsible for transmitting other fatal diseases, such as chikungunya (a ss RNA *Togaviridae* virus), yellow fever (a ss RNA *Flaviviridae* virus), *Zika* (a ss RNA *Flaviviridae* virus), and Japanese encephalitis (a ss RNA Flaviviridae virus). Each of these mosquito-borne diseases causes millions of deaths worldwide annually (WHO 2022). Efforts to combat fatalities have led to the introduction of intravenous and extra-venous drugs, including vaccinations and medications, which are still being researched and improved (Murrell et al. 2011).

Until now, different combinations and doses of organochlorines, organophosphates, carbamates, pyrethroids, pyrroles, and phenyl pyrazoles have been used as effective insecticides. The most widely used and efficient green insecticides are the Pyrethroids and their derivatives [active ingredients: trans-chrysanthemic and trans-pyrethroid acid; source: *Chrysanthemum* (Staba and Zito 1985) and *Tanacetum* plants (Lybrand et al. 2020) from the plant family Asteraceae] because of their sustainability and biodegradability. On the other hand, organophosphates (parathion, chlorpyrifos, malathion) are more effective synthetic insecticides due to their acute neurotoxic nature (Tsai and Lein 2021). Acetylcholinesterase (AChE) in the autonomic ganglia of the sympathetic and parasympathetic nervous systems is inhibited by organophosphate pesticides. It cleaves acetylcholine into choline and acetic acid for normal body functioning. Organophosphates bind to serine residues of acetylcholinesterase and inactivate the enzyme; as a result, huge amount of acetylcholine molecules are accumulated in the postsynaptic receptors stimulating the effector organ for longer period, which ultimately leads the insect to death (Table 15.1).

Table 15.1 The mechanism of actions underlying the effects of insecticides on mosquitoes and the mechanisms that counteract the effects of those insecticides applied to them

Insecticide	Chemistry	Modus operandi	Counter mechanism	Species witnessed	References
DDT (Dichloro diphenyl trichloroethane)	$C_{14}H_9Cl_5$	Targets voltage-sensitive Na-channels (VSSC) → affects nervous system by increasing depolarization time → keeps VSSC open for longer period → leading to insect's death	Cuticular thickening by overexpression of CP proteins (CPLCG3; CPR124; CPR127; CPR129)	*An. gambiae*	Bharati and Saha (2021), Yahouedo (2017)
Pyrethroids (Fen valerate)	$C_{22}H_{19}Cl_2NO_3$	do	Overexpression of P450 → Upregulation of CpLac2 gene	*An. funestus* *Cx. pipiens pallens*	Soderlund (2010), Amenya et al. (2008), Chengyuan Pan and Mo (2008)
Organophosphates (Malathion; Temephos; Chlorpyrifos)	$C_{10}H_{19}O_6PS_2$	Inhibits acetylcholinesterase (AChE) → increased accumulation of Ach at synaptic ends → release neurotoxins → leading to insect's death	A296G substitution in GABA receptor → G119S mutation in AChE → F455W substitution in ace 1	*An. gambiae* *Cx. tritaeniorhynchus*	Milesi et al. (2022), Du et al. (2005), Karunaratne et al. (2013), Nabeshima et al. (2004)
Carbamates (Carbaryl)	$C_{12}H_{11}NO_2$	do	do	do	
Phenyl pyrazoles (Fipronil)	$C_{12}H_4Cl_2F_6N_4OS$	Block glutamate → activate chloride channels → cell dysfunction → leading to insect's death	Point mutation in GABA receptor	*An. stephensi*	Cole et al. (1993)
Organochlorines or Cyclodiene (Dieldrin; Aldrin)	$C_{12}H_8Cl_6$	Functional disruption of Na and Ca channels → transporter disruption → interfere γ-aminobutyric acid (GABA) neurotransmission by blocking specific GABA receptors → leading to insect's death	A296S substitution in GABA receptor → Single BP substitution of Ala to Ser → Change of polarity through addition of OH group → steric hinderance to cyclodine's binding site	*An. funestus* *Ae. aegypti* *Cx. pipiens pallens*	Karunaratne et al. (2013), Wondji et al. (2011), Thompson et al. (1993)

Current evidence on the resurgence of mosquito-borne diseases caused by insecticide-resistance development is alarming and threatens human welfare (WHO 2022). Numerous intricate studies and investigations have demonstrated rapid adaptability and efficient mosquito-resistance capabilities through physiological (Xu et al. 2022), behavioural (Gatton et al. 2013), metabolic (Vulule et al. 1999), and genetic changes (Knipple et al. 1991). It is likely to be realistic that the future direction of vector control research will be on comprehending an extensive grasp of knowledge and mechanism of an integrated science of all cohorts. As a result, it is crucial to comprehend how mosquitoes gain insecticide resistance through counter mechanisms, genetic regulation, coevolution, and the trade-offs associated with developing resistance.

15.2 Insecticidal Resistance in Mosquitoes

15.2.1 Behavioural Resistance

Behavioural resistance in mosquitoes is the adaptive capacity of their behaviours when exposed to insecticides or other control methods, thereby decreasing their susceptibility to these measures. This type of resistance can pose significant challenges for controlling their population. The following studies on behavioural resistance in mosquitoes explain how the insect overcomes the adverse effects of insecticides (Fig. 15.1).

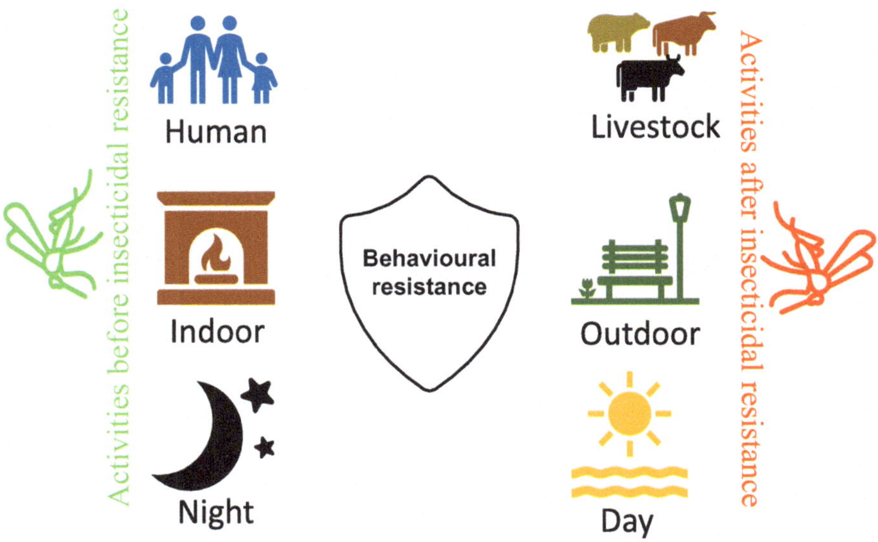

Fig. 15.1 The three modes of behavioural resistance in mosquitoes on pesticide application—trophic avoidance (shifting of biting preference), spatial avoidance (shifting of biting place), and temporal avoidance (shifting of biting time)

(a) **Avoidance conduct:** Mosquitoes may exhibit avoidance behaviour by actively fleeing places that have been treated with pesticides or where control measures are being implemented. This behaviour may involve modifications in feeding, mating, or resting activities to minimize the likelihood of being exposed to insecticides. 'Spatial avoidance' refers to their ability to detect and avoid certain areas or objects based on sensory cues, and as a result, mosquitoes alter their feeding preference outdoors rather than indoors (Gatton et al. 2013), while "temporal avoidance", as described by Gatton et al. (2013), is the behaviour of shifting biting time from night to early morning or evening to avoid long-lasting insecticidal nets (LLINs) or indoor residual sprayings (IRS). For example, the peak biting hours of *An. funestus* was found to be delayed till daybreak (Moiroux et al. 2012). However, 'trophic avoidance' bypasses host near insecticides (particularly humans) and chooses alternate food sources, mostly livestock (Ndenga et al. 2016).

(b) **Enhanced mobility:** Increased pesticide pressure can lead to enhanced mobility in mosquitoes causing them to scatter across wider regions to evade control measures. They may also alter their behaviour at various times of the day in response to insecticidal sprayings. The increased movement may provide difficulties in precisely targeting mosquito populations with pesticides (Pates and Curtis 2005).

(c) **Reproductive shift:** Mosquitoes possess the capacity to modify their reproductive behaviour in reaction to varying levels of exposure to pesticides. This may entail alterations in mating behaviour, such as heightened courtship or mating activity, as a means of compensating for the mortality caused by insecticides (Pates and Curtis 2005).

(d) **Habitat alteration:** Mosquitoes can modify their sheltering behaviour by seeking refuge in secure locations or adopting strategies to reduce their vulnerability to insecticides. These strategies may include changing the locations or habitats where they lay eggs or adjusting their flying altitudes and latitudes (Pates and Curtis 2005) (Table 15.2).

Table 15.2 Behavioural resistance in mosquitos. Temporal resistance is the shift in biting time; spatial resistance is the shift in biting location; and trophic resistance is the shift in biting host

Species showed behavioural resistance				
Temporal	Spatial	Trophic	Studied location	References
An. farauti			Solomon Islands	Thomsen et al. (2017)
			Papua New Guinea	Taylor (1975)
An. funestus			Kenya	Russell et al. (2011)
			Tanzania	Bayoh et al. (2014)
An. arabiensis			Kenya	Russell et al. (2011)
				Bayoh et al. (2014)
	An. gambiae		Bioko Island	Meyers et al. (2016)

(continued)

Table 15.2 (continued)

Species showed behavioural resistance				
Temporal	Spatial	Trophic	Studied location	References
	An. funestus		Tanzania	Russell et al. (2011)
		Cx. quinquefasciatus	Kenya	Bøgh et al. (1998)
		An. funestus	Kenya	Bøgh et al. (1998)
		An. gambiae	Kenya	Bøgh et al. (1998)
				Ndenga et al. (2016)
		An. fluviatilis	India	Waite et al. (2017)
		An. culicifacies	India	Waite et al. (2017)

15.3 Metabolic Resistance

Metabolic resistance in mosquitoes is the mechanism by which mosquitoes gain pesticide resistance through increased metabolic detoxification. This type of resistance entails the increased efficacy of enzymes engaged in the metabolism and detoxification of insecticides. Mosquitoes usually generate metabolic enzymes at a basal level, but insecticide exposure can trigger their overexpression, resulting in enhanced detoxification and decreased susceptibility to insecticides (Strode et al. 2008; Marinotti et al. 2013; Yang and Liu 2011; Yan et al. 2012). However, these metabolic activities are thought to be mediated by three gene families—cytochrome P450s, esterases, and glutathione S-transferases. The research indicates that overexpression of these genes causes increased insecticide resistance. In the context of transcriptional upregulation, insects rely on cytochrome P450s and GSTs to augment protein synthesis and enhance enzyme functions. It ultimately results in elevated metabolic detoxification of insecticides, enabling the insects to develop resistance against harmful substances (Vulule et al. 1999). These assertions have been recorded in *An. albimanus* (Brogdon et al. 1999), *An. funestus* (Casimiro et al. 2014), *An. gambiae* (Vulule et al. 1999), *Ae. aegypti* (Marcombe et al. 2012), *An. stephensi* (Enayati et al. 2003), and *An. annularis* (Gunasekaran et al. 2011) (Fig. 15.2a).

Several enzyme groups are linked to metabolic resistance in mosquitoes. The enzymes specifically bind to specific insecticides as substrates and thereafter through catalytic actions degrade the toxins.

(a) Enzymes from the cytochrome P450 monooxygenases family can oxidize various substrates, including harmful insecticides. Their catalytic effects enhance the solubility of pesticides in water and facilitate their excretion. Cytochrome P450 enzymes play a crucial role in cellular metabolism and the detoxification of xenobiotics (Feyereisen 2005). Overexpression of P450 genes has been reported in deltamethrin-resistant *An. minimus* (Rongnoparut et al. 2005), pyrethroid-resistant *An. funestus* (Amenya et al. 2008), permethrin-resistant *Cx.*

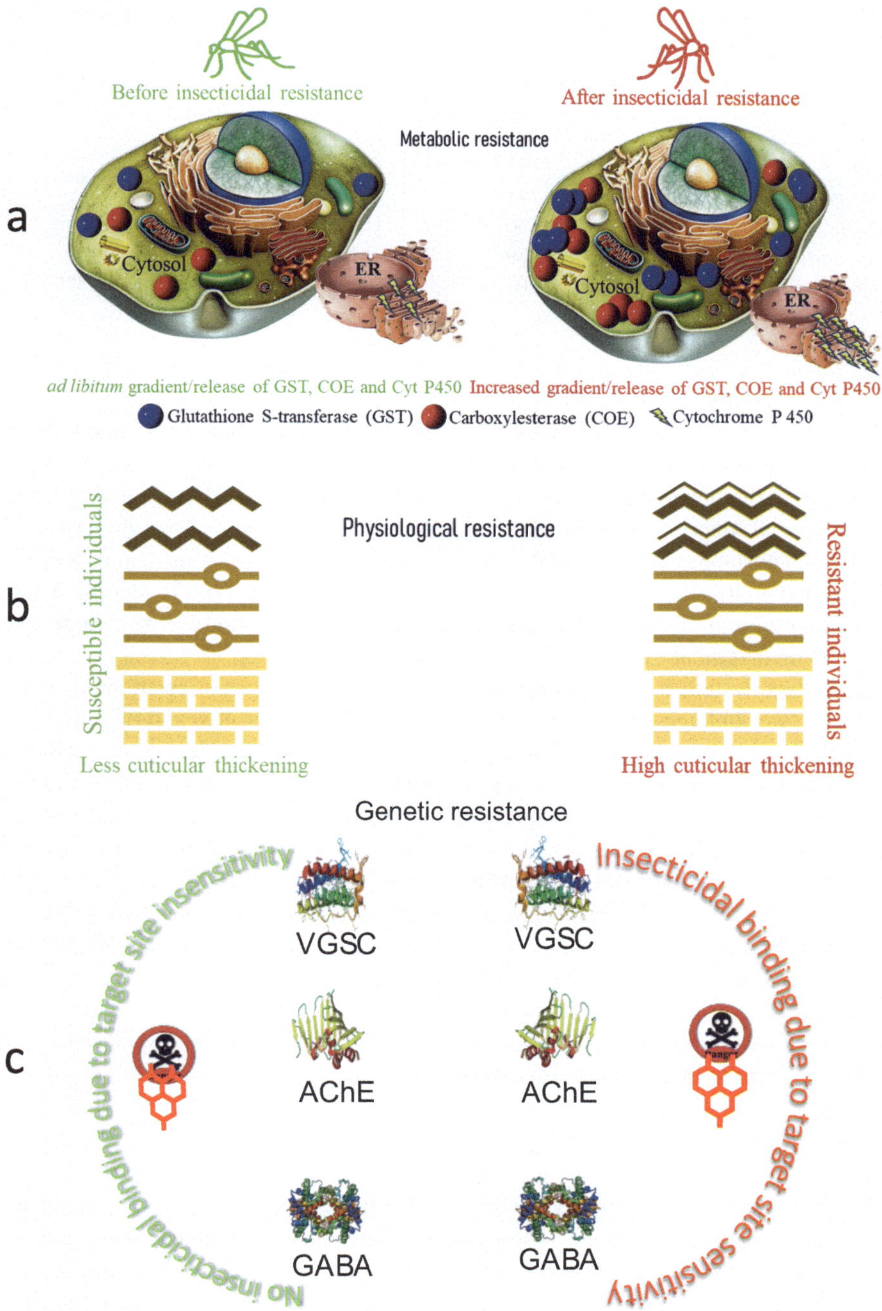

Fig. 15.2 (**a**) Metabolic resistance due to excessive synthesis of detoxifying enzymes including Cytochrome P450 in mitochondria, Glutathione-S-Transferases (GST), and Carboxylesterase (CoE) in cytosol, (**b**) Physiological resistance due to cuticular thickening resulting primary protection of insecticides, (**c**) Genetical resistance due to specific target-site mutations at Voltage Gated Sodium Channel (VGSC), Acetylcholinesterases (AChE), and GABA receptors, resulting insecticidal inefficiency to target site binding

Table 15.3 Metabolic resistance in mosquitos and the responsible overexpressed genes. The numbers in each column represent the total number of gene sequences involved

Gene expression observed in species	Overexpressed genes and gene sequence counts identified for metabolic resistance			References
	Esterase	P450	GST	
An. gambiae	51	111	31	Ranson et al. (2002)
Ae. aegypti	49	160	26	Strode et al. (2008)
An. darlingi	20	89	30	Marinotti et al. (2013)
Cx. quinquefasciatus	71	204	35	Yang and Liu (2011), Yan et al. (2012)

quinquefasciatus (Gong et al., 2013), and deltamethrin-resistant *Cx. pipiens pallens* (Gong et al. 2005).

(b) Glutathione S-transferases (GSTs) are crucial in binding-reduced glutathione to specific electrophilic substrates, such as insecticides, through chemical conjugation, thereby avoiding oxidative damage within the cell. Therefore, they have a crucial function in the breakdown, removal of toxins, and elimination of several substances that originate from within the body or come from outside sources (Ranson and Hemingway 2005). Ortelli et al. (2003) identified the upregulation of the GSTE2 gene in DDT-resistant *An. Gambiae* that supported the involvement of GST in insecticide resistance.

(c) Carboxylesterase (COE) enzymes facilitate the hydrolysis of ester, amide, and thioester linkages from insecticides, specifically pyrethroids, and organophosphates that progressively disintegrate the insecticides into less-toxic metabolites, such as alcohols and acids, which leads to the detoxification of insecticides through metabolism. A comprehensive investigation has been carried out by Pasteur and Raymond (1996) to examine the overproduction of detoxifying proteins resulting from the amplification of esterase genes. Further, *Cx. quinquefasciatus* has also been shown to overproduce esterase genes by amplification (Mouchès et al. 1986), and Pyrethroids have been found to be metabolized by carboxy/cholinesterase (Grigoraki et al. 2016) (Table 15.3).

15.4 Physiological Resistance

15.4.1 Cuticular Thickening

Mosquito cuticles are primarily composed of chitin polysaccharides, cuticular proteins, hydrocarbons, phenolic biopolymers, sclerotin, and melanin. Thus, cuticle modifications have led to increased survival strategies of insects against insecticides by reducing insecticides' ability to penetrate the insect's body and extending the period needed for insects to develop metabolic and genetic resistance. The upregulation of cuticular proteins (CPs), laccase (an enzyme that catalyses oxidation reaction), and ABC (ATP-binding cassette) transporters, which generally function as efflux pumps in eukaryotes that facilitate the outflow of cuticular components towards the cuticle, are the major contributors to cuticular thickening. The cuticle protein (CP) encoded by *CPR63* was found to play a role in deltamethrin resistance in *Cx. pipiens pallens* (Xu et al. 2022). An increased expression of the CpLac-2 gene in *Cx. pipiens pallens* made it resistant to Fenvalerate, a synthetic pyrethroid insecticide (Chengyuan Pan and Mo 2008). Pignatelli et al. (2018) found legs of mosquitoes having higher levels of cuticular hydrocarbons (CHCs) due to the higher expression of ABC transporters (Fig. 15.2b).

15.4.2 Increased Chemosensory Protein Expression

When mosquitoes are exposed to neurotoxic insecticides, the chemicals interfere with neurotransmission, leading to paralysis, convulsions, and ultimately death. Sensory appendage protein (SAP2) is a soluble, chemosensory protein (CSP) found exclusively in the sensory appendages of many arthropods, primarily facilitating the transportation of hydrophobic compounds through the sensillum lymphatic system. Insecticide resistance has been reported to be positively correlated with increased expression of SAP2 in mosquito legs, whereas silencing SAP2 protein resulted in full vulnerability to pyrethroid. These SAP2 proteins tightly adhere to pyrethroids when the insecticide comes in contact with mosquitoes. It was studied by Ingham et al. 2020 in *An. gambiae*.

15.5 Genetic Resistance

Mosquitoes have evolved methods to evade the impact of insecticides at the genetic level as well. Those insecticides that target specific regions on the insect body have become a redundant tactic since mosquitoes have surpassed such selection pressure enforced by insecticides by altering the genetic component of those target areas. It is also referred to as target site mutation or target site insensitivity and is developed by means of point mutations in the genes responsible for encoding the protein composition of the insect body targeted by insecticides. Such point mutations deciphered in mosquitoes affect the voltage-gated sodium channels (VGSCs) or voltage-sensitive sodium channels (VSSCs), acetylcholinesterase (AChE), and GABA receptors (Ffrench-Constant et al. 1993) (Fig. 15.2c).

15.5.1 Voltage-Gated Sodium Channels (VGSC)

Insecticides, such as DDT and pyrethroids, can bind to sodium channels in the nervous system, thereby causing VGS channels to stay open for a longer period, resulting depolarization of the neural path, which ultimately leads to the death of the insect (Davies et al. 2007; Dong 2007; Liu et al. 2005). However, resistance to those insecticides (DDT and pyrethroids), known as knockdown resistance (KDR), may occur due to decreased sensitivity to the sodium channels at the target sites (Knipple et al.

1991; Li et al. 2012; Xu et al. 2012). While studying genes responsible for sensitising the voltage-gated sodium channel (VGSC), it was identified that V1016G, I1011M, V410L, and F1534 genes are responsible for altering the Na-channel influx causing an overall imbalance of neural transmission in *Ae. aegypti* (Srisawat et al. 2010; Du et al. 2005; Haddi et al. 2017). The V1016G mutation and other KDRs reduce the susceptibility to permethrin and deltamethrin by altering the amino acid composition in the VGSC (Plernsub et al. 2013; Hirata et al. 2014).

So far, eleven synonymous mutations have been detected in mosquitoes, with four occurring in different species or strains (Du et al. 2013). In such cases, mutations for amino acid substitutions are as follows: leucine to phenylalanine, or cysteine, or serine, or tryptophan at position 1014 (L1014F/C/S/W); isoleucine to methionine or valine at position 1011 (I1011M/V); valine to glycine at position 1016 (V1016G); and phenylalanine to cysteine at position 1534 (F1534C) (Du et al. 2013). Additional six more mutations have also been identified and they are S989P (Ser to Pro), I1011M/V (Ile to Met/Val), L1014F/S (Leu to Phe/Ser), V1016G/I (Val to Gly/Ile), F1534C (Phe to Cys), and D1763Y (Asp to Tyr) (Du et al. 2013). In an investigation of genetic variations in sodium channels, specifically comparing resistant and susceptible strains, six synonymous (Leu852, Gly891, Ala1241, Asp1245, Pro1249, and Gly1733) and three non-synonymous mutations (Ala109Ser, Leu982Phe, and Trp1573Arg) were identified from two different field strains of *Cx. quinquefasciatus* and their permethrin-resistant offspring. The evolutionary impact is assessed by examining how these mutations affect the fitness of mosquitoes in the presence of insecticides, as well as the spread of resistance alleles in natural populations over time. In summery, the findings likely explore how synonymous and nonsynonymous genetic variations contribute to the adaptation of mosquitoes to insecticides, offering valuable insights into the evolutionary dynamics of resistance and informing strategies for sustainable mosquito control (Xu et al. 2012; Li et al. 2012).

15.5.2 Acetylcholinesterase (AChE)

AChE is crucial for transmitting nerve impulses as it hydrolyses the neurotransmitter acetylcholine. However, inhibition of AChE causes an accumulation of acetylcholine, which induces continuous neural excitement and ends in insect paralysis (Table 15.1). Insecticides, such as organophosphates and carbamates, exert their detrimental effects by inhibiting the activity of AChE. Point mutations in AChE have been documented to decrease susceptibility to carbamates and organophosphates (OPs) (Silberman and Taylor 2023). The ace-1 and ace-2 genes are responsible for encoding the AChE1 and AChE2 proteins, respectively (Cha and Lee 2015). The active site of AChE1 undergoes two structural modifications that are known to inhibit the binding of the insecticide to the target site. Amino acid substitution in the ace-1 gene (G119S) has been identified in *An. albimanus* (Alout et al. 2008), *Cx. vishnui* (Rinkevich et al. 2013), *Cx. pipiens* (Weill et al. 2002), and *Cx. quinquefasciatus* (Weill et al. 2003). As a result, mosquitoes carrying these muted forms of AChE experience reduced susceptibility to insecticides, enabling their survival and reproduction, thereby transmitting the resistant genes to their progeny.

15.5.3 GABA Receptors

Gamma-aminobutyric acid (GABA) acts as an inhibitory neurotransmitter by blocking the opening of the pentameric transmembrane chloride channel (Cole et al. 1993, 1995). It contributes to the decrease in the level of neuronal excitability across the entire nervous system by inhibiting nerve transmission. The GABA receptor serves as a specific location for cyclodiene insecticides to act against. The structure comprises five subunits, each containing an extracellular Cys loop and four transmembrane domains (M1, M2, M3, and M4). The M2 domain significantly contributes to the development of ion channels. The point mutation resulting in the substitution of Alanine 296 with serine or glycine (A296S/G) causes dieldrin (a cyclodiene) resistance in mosquitoes (Ffrench-Constant et al. 1993).

The genetic control of insecticide resistance in mosquitoes is observed to have been through *trans* and *cis*-regulation of genes. For example, *GST-2* gene upregulation in *Ae. Aegypti* was found to be *trans* regulated (Grant and Hammock 1992), and Cyp9m10 in *Cx. quinquefasciatus* is *cis*-regulated by CuRE1transposable element (Wilding et al. 2012). Such regulations can correlate with gene overexpression that confers resistance (Wilding et al. 2012). The regulatory pathway of the P450 genes in *Cx. quinquefasciatus* was studied employing knowledge of GPCR gene overexpression (Li et al. 2015). Though the intense involvement of the GPCR intercellular pathway in developing resistance to permethrin is known (Liu et al. 2007; Reid et al. 2012), more investigations on GPCR regulatory pathways are to be unfolded to identify the controlling genes associated with insecticide resistance.

15.6 Trade-off of Insecticidal Resistance

A mutation is neither naturally advantageous nor harmful; instead, it is dependent upon the genetic background, the ecological circumstances, and other alleles that are considered for evaluation (Lenormand et al. 2018). Resistance mutations are proposed to have pleiotropic effects on different traits. When such variation occurs, the idea of 'resistance cost' becomes important, because these traits were previously advantageous due to natural selection. A positive change in one feature that is linked to a negative change in another is referred to as a trade-off, where the costs are defined in relation to fitness (Stearns 1989). The costs of such trade-offs have been addressed in mosquitoes as they develop pesticide resistance (Fig. 15.3).

Mosquitoes that develop resistance to insecticides, like DDT and pyrethroids, due to insensitivity at the target site are referred to as knockdown resistance (KDR), which is commonly found in the voltage-gated sodium channel (VGSC). Two well-known KDRs in *Ae. aegypti*, Val1016Ile and Phe1534Cys, were investigated to assess the fitness cost of these mutations on life history traits in mosquitoes (Brito et al. 2013). The resistant strain, CIT-32, was crossed with a susceptible strain, Rockfeller, for eight generations to generate Rock-Kdr, a susceptible strain with

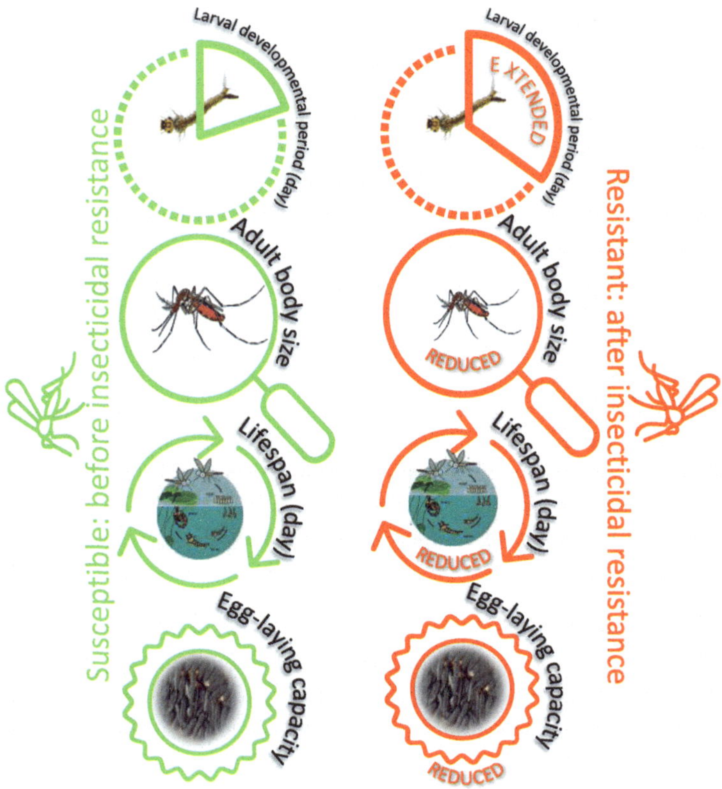

Fig. 15.3 The costs of resistance to pesticides in mosquitoes include prolonged larval developing time, decreased adult body size, shortened longevity and reduced fecundity

Kdr mutations. Consequently, a comparison was made between Rock-Kdr and Rockfeller to analyze the trade-off costs. Adult *Ae. aegypti* showed increased movement and a drop in egg-laying females, resulting in lower overall egg production. In addition, the larval development period turned longer compared to the susceptible strain Rockfeller (Brito et al. 2013). Further, three strains of *Aedes aegypti*, PMD, PMD-R, and UPK-R, along with a hybrid were examined for fitness traits including larval development, fecundity, blood-feeding capability, storage egg viability, adult body size, and longevity (Saingamsook et al. 2019). Despite lacking any KDR-related alleles, PMD was resistant to DDT. In contrast, PMD-R possessed the homozygous 1534C Kdr allele and was resistant to both permethrin and DDT. In addition to being resistant to DDT and permethrin, UPK-R also carries the Kdr alleles 989P and 1016G, which allow them to withstand deltamethrin. The outcomes showed a decline in the body size, fecundity, longevity of females, and hatchability of eggs. While there was no discernible impact on fitness in terms of blood feeding capability, the eggs' viability did decrease after storage (Saingamsook et al. 2019). While conducting other experiments with *Ae. aegypti* focusing on V1016G and S989P Kdr mutations, it was found that exposure to pyrethroid resulted in longer larval development, smaller adult body size, and reduced longevity in female adults, reflecting the fitness costs (Rigby et al. 2020).

15.7 Conclusion

The formulation of pesticides has progressively become more intricate over time, leading to substantial selective pressure on mosquitoes in their natural habitat. Nevertheless, mosquitoes have developed a co-evolutionary relationship with insecticides and successfully adapted to the selective pressure of natural selection, allowing them to survive in a natural environment. Insecticide technology has moved away from traditional chemical pesticides in favour of new technologies, such as nano pesticides. Many formulations, such as emulsions (nanoemulsions), nanocapsules (with polymers), and nanoparticles (metals, metal oxides, and nanoclays) have been proposed in developing effective insecticides for mosquitoes (Rai et al. 2014). The declining effectiveness of chemically synthesized insecticides and their potential harm to the environment, nontarget organisms, and humans due to residual pollution highlights the importance of nano insecticides and their use in mosquito control (Lang et al. 2021). As insecticides are always changing their mode of action, to understand how mosquitoes resist insecticides, it is important to delve into how their biophysiochemical resistance mechanisms have evolved over time (hence, their survival strategies). Therefore, developing efficient insecticides to prevent resistance mechanisms could help enhance mosquito management. The impact of natural selection on changes in coding gene sequences can be explored through full genome sequencing of different mosquito strains. This will provide new opportunities for developing innovative insecticides that can specifically target and eliminate mosquitoes.

Points to Remember

1. Vector-borne diseases, especially transmitted through mosquitos, are persistent because of their prevalence and eradication limitations. Controlling mosquito populations is becoming increasingly difficult because of the emergence of insecticidal resistance and subsequent adaptations.
2. The cytotoxic and neurotoxic effects on mosquitos are restricted over time. Insecticidal resistance in mosquitoes is favoured by behavioural changes (such as shifting of biting time, location, and host), metabolic amendments (such as metabolite detoxification and enzyme kinetics), physiological adjustments (such as increased sensory proteins and thicker cuticular tissue), and genetic resistance (such as target-site insensitivity in ACHEs, VGSC, and GABA).
3. The trade-off between life history traits and insecticide resistance favours co-evolution.

Acknowledgement The authors are thankful to the Head, Department of Zoology, University of Calcutta for providing necessary laboratory facilities and support.

References

Research Papers and Review Articles

Alout H, Djogbenou L, Berticat C, Chandre F, Weill M (2008) Comparison of *Anopheles gambiae* and *Culex pipiens*acetycholinesterase 1 biochemical properties. Comp Biochem Physiol B(150):271–277

Amenya DA, Naguran R, Lo TC, Ranson H, Spillings BL, Wood OR, Brooke BD, Coetzee M, Koekemoer LL (2008) Over expression of a cytochrome P450 (CYP6P9) in a major African malaria vector, *Anopheles funestus*, resistant to pyrethroids. Insect Mol Biol 17(1):19–25

Bayoh MN, Walker ED, Kosgei J, Ombok M, Olang GB, Githeko AK, Killeen GF, Otieno P, Desai M, Lobo NF (2014) Persistently high estimates of late night, indoor exposure to malaria vectors despite high coverage of insecticide treated nets. Parasit Vectors 7:1–13. https://doi.org/10.1186/1756-3305-7-380

Bharati M, Saha D (2021) Insecticide resistance status and biochemical mechanisms involved in *Aedes* mosquitoes: a scoping review. Asian Pac J Trop Med 14(2):52–63

Bøgh C, Pedersen EM, Mukoko DA, Ouma JH (1998) Permethrin impregnated bednet effects on resting and feeding behaviour of lymphatic filariasis vector mosquitoes in Kenya. Med Vet Entomol 12:52–59

Brito LP, Linss JG, Lima-Camara TN, Belinato TA, Peixoto AA, Lima JB, Valle D, Martins AJ (2013) Assessing the effects of *Aedes aegypti*kdr mutations on pyrethroid resistance and its fitness cost. PLoS One 8(4):e60878

Brogdon WG, McAllister JC, Corwin AM, Cordon-Rosales C (1999) Independent selection of multiple mechanisms for pyrethroid resistance in Guatemalan *Anopheles albimanus* (Diptera: Culicidae). J Econ Entomol 92(2):298–302

Casimiro S, Coleman M, Mohloai P, Hemingway J, Sharp B (2014) Insecticide resistance in *Anopheles funestus* (Diptera:Culicidae) from Mozambique. J Med Entomol 43(2):267–275

Cha DJ, Lee SH (2015) Evolutionary origin and status of two insect acetylcholinesterases and their structural conservation and differentiation. Evol Dev 17(1):109–119

Chengyuan Pan YZ, Mo J (2008) The clone of laccase gene and its potential function in cuticular penetration resistance in *Culex pipienspallens* to fenvalerate. Pestic Biochem Physiol 93:105–111

Cole LM, Nicholson RT, Casida JE (1993) Action of phenyl pyrazole insecticides at the GABA-gated chloride channel. Pestic Biochem Physiol 46:47–54

Cole LM, Roush TT, Casida JE (1995) Drosophila GABA-gated chloride channel: modified [3H] EBOB binding site associated with Ala → Ser or Gly mutants of Rdl subunit. Life Sci 56(10):757–765

Davies TGE, Field LM, Usherwood PNR, Williamson MS (2007) A comparative study of voltage-gated sodium channels in the Insecta: implications for pyrethroid resistance in *Anopheline* and other *Neopteran* species. Insect Mol Biol 16(3):361–375

Dong K (2007) Insect sodium channels and insecticide resistance. Invertebr Neurosci 7:17–30

Du W, Awolola TS, Howell P, Koekemoer LL, Brooke BD, Benedict MQ, Coetzee M, Zheng L (2005) Independent mutations in the Rdl locus confer dieldrin resistance to *Anopheles gambiae* and *Anopheles arabiensis*. Insect Mol Biol 14(2):179–183

Du Y, Nomura Y, Satar G, Hu Z, Nauen R, He SY, Zhorov BS, Dong K (2013) Molecular evidence for dual pyrethroid-receptor sites on a mosquito sodium channel. Proc Natl Acad Sci 110(29):11785–11790

Enayati AA, Vatandoost H, Ladonni H, Townson H, Hemingway J (2003) Molecular evidence for a kdr-like pyrethroid resistance mechanism in the malaria vector mosquito *Anopheles stephensi*. Med Vet Entomol 17(2):138–144

Feyereisen R (2005) Insect cytochrome P450. Comprehensive molecular insect science, pp. 1–77

Ffrench-Constant RH, Rocheleau TA, Steichen JC, Chalmers AE (1993) A point mutation in a Drosophila GABA receptor confers insecticide resistance. Nature 363(6428):449–451

Gatton LM, Chitnis N, Churcher T, Donnelly JM, Ghani CAH, Godfray JCH, Gould F, Hastings I, Marshall J, Ranson H, Rowland M, Shaman J, Lindsay WS (2013) The importance of mosquito behavioural adaptations to malaria control in Africa. Evolution 67(4):1218–1230

Gjullin CM, Peters RF (1952) Recent studies of mosquito resistance to insecticides in California. Mosq News 12:1–7

Gong MQ, Gu Y, Hu XB, Sun Y, Ma L, Li XL, Sun LX, Sun J, Qian J, Zhu CL (2005) Cloning and overexpression of CYP6F1, a cytochrome P450 gene, from deltamethrin-resistant *Culex pipienspallens*. Acta Biochim Biophys Sin 37(5):317–326

Gong Y, Li T, Zhang L, Gao X, Liu N (2013) Permethrin induction of multiple cytochrome P450 genes in insecticide resistant mosquitoes, *Culex quinquefasciatus*. International journal of biological sciences 9(9):863

Grant DF, Hammock BD (1992) Genetic and molecular evidence for a trans-acting regulatory locus controlling glutathione S-tranferase-2 expression in *Aedes aegypti*. Mol Gen Genet, Springer 234:169–176

Grigoraki L, Balabanidou V, Meristoudis C, Miridakis A, Ranson H, Swevers L, Vontas J (2016) Functional and immunohistochemical characterization of CCEae3a, a carboxylesterase associated with temephos resistance in the major arbovirus vectors Aedes aegypti and *Aedes albopictus*. Insect Biochem Mol Biol 74:61–67

Gunasekaran K, Muthukumaravel S, Sahu SS, Vijayakumar T, Jambulingam P (2011) Glutathione S transferase activity in Indian vectors of malaria: a defense mechanism against DDT. J Med Entomol 48(3):561–569

Haddi K, Tomé HV, Du Y, Valbon WR, Nomura Y, Martins GF, Dong K, Oliveira EE (2017) Detection of a new pyrethroid resistance mutation (V410L) in the sodium channel of *Aedes aegypti*: a potential challenge for mosquito control. Sci Rep 7(1):46549

Hirata K, Komagata O, Itokawa K, Yamamoto A, Tomita T, Kasai S (2014) A single crossing-over event in voltage-sensitive Na+ channel genes may cause critical failure of dengue mosquito control by insecticides. PLoS Negl Trop Dis 8(8):3085

Ingham VA, Anthousi A, Douris V, Harding NJ, Lycett G, Morris M, Vontas J, Ranson H (2020) A sensory appendage protein protects malaria vectors from pyrethroids. Nature 577(7790):376–380

Karunaratne SH, Weeraratne TC, Perera MD, Surendran SN (2013) Insecticide resistance and, efficacy of space spraying and larviciding in the control of dengue vectors *Aedes aegypti* and *Aedes albopictus* in Sri Lanka. Pestic Biochem Physiol 107(1):98–105

Knipple DC, Doyle KE, Marsella-Herrick PA, Soderlund DM (1991) Tight genetic linkage between the kdr insecticide resistance traiand a voltage-sensitive sodium channel gene in the house fly. Proc Natl Acad Sci USA 91:2483–2487

Lang C, Mission EG, Fuaad AA, Shaalan M (2021) Nanoparticle tools to improve and advance precision practices in the agrifoods sector towards sustainability—a review. J Clean Prod 293:126063

Lenormand T, Harmand N, Gallet R (2018) Cost of resistance: an unreasonably expensive concept. bioRxiv, 276675 https://doi.org/10.1101/276675

Li T, Zhang L, Reid WR, Xu Q, Dong K, Liu N (2012) Multiple mutations and mutation combinations in the sodium channel of permethrin resistant mosquitoes, *Culex quinquefasciatus*. Sci Rep 2(1):781

Li T, Cao C, Yang T, Zhang L, He L, Xi Z, Bian G, Liu N (2015) A G-protein-coupled receptor regulation pathway in cytochrome P450-mediated permethrin-resistance in mosquitoes, *Culex quinquefasciatus*. Sci Rep 5(1):17772

Liu H, Xu Q, Zhang L, Liu N (2005) Chlorpyrifos resistance in mosquito *Culex quinquefasciatus*. J Med Entomol 42(5):815–820

Liu N, Liu H, Zhu F, Zhang L (2007) Differential expression of genes in pyrethroid resistant and susceptible mosquitoes, *Culex quinquefasciatus*. Gene 394:61–68

Lybrand DB, Xu H, Last RL, Pichersky E (2020) How plants synthesize Pyrethrins: safe and biodegradable insecticides. Trends Plant Sci 25(12):1240–1251

Marcombe S, Mathieu RB, Pocquet N, Riaz MA, Poupardin R (2012) Insecticide resistance in the dengue vector *Aedes aegypti* from Martinique: distribution, mechanisms and relations with environmental factors. PLoS One 7:e30989

Marinotti O, Cerqueita GC, de Almeida LGP, Ferro MI, da Silva Loreto EL (2013) The genome of *Anopheles darlingi*, the main neotropical malaria vector. Nucleic Acids Res 41:7387–7400

Meyers JI, Pathikonda S, Popkin-Hall ZR, Medeiros MC, Fuseini G, Matias A, Garcia G, Overgaard HJ, Kulkarni V, Reddy VP (2016) Increasing outdoor host-seeking in *Anopheles gambiae* over 6 years of vector control on Bioko Island. Malar J 15:1–13. https://doi.org/10.1186/s12936-016-1286-6

Milesi P, Claret J, Unal S, Weill MP, Labbé P (2022) Evolutionary trade-offs associated with copy number variations in resistance alleles in *Culex pipiens* mosquitoes. Parasit Vectors 15484. https://doi.org/10.1186/s13071-022-05599-8

Moiroux N, Gomez MB, Pennetier C, Elanga E, Djenontin A, Chandre F, Djegbe I, Guis H, Corbel V (2012) Changes in *Anopheles funestus* biting behavior following universal coverage of long-lasting insecticidal nets in Benin. J Infect Dis 206:1622–1629. https://doi.org/10.1093/infdis/jis565

Mouchès C, Pasteur N, Bergé JB, Hyrien O, Raymond M, de Saint R, Vincent B, De Silvestri M, Georghiou GP (1986) Amplification of an esterase gene is responsible for insecticide resistance in a California *Culex* mosquito. Science 233(4765):778–780

Murrell S, Wu SC, Butler M (2011) Review of dengue virus and the development of a vaccine. Biotechnol Adv 29:239–247

Nabeshima T, Mori A, Kozaki T, Iwata Y, Hidoh O (2004) An amino acid substitution attributable to insecticide-insensitivity of acetylcholinesterase in a Japanese encephalitis vector mosquito, *Culex tritaeniorhynchus*. Biochem Biophys Res Commun 313:7

Ndenga BA, Mulaya NL, Musaki SK, Shiroko JN, Dongus S, Fillinger U (2016) Malaria vectors and their blood-meal sources in an area of high bed net ownership in the western Kenya highlands. Malar J 15:1–10. https://doi.org/10.1186/s12936-016-1115-y

Ortelli F, Rossiter LC, Vontas J, Ranson H, Hemingway J (2003) Heterologous expression of four glutathione transferase genes genetically linked to a major insecticide-resistance locus from the malaria vector *Anopheles gambiae*. Biochem J 373(3):957–963

Pasteur N, Raymond M (1996) Insecticide resistance genes in mosquitoes: their mutations, migration, and selection in field populations. J Hered 87(6):444–449

Pates H, Curtis C (2005) Mosquito behavior and vector control. Annu Rev Entomol 50:53–70

Pignatelli P, Ingham VA, Balabanidou V, Vontas J, Lycett G, Ranson H (2018) The *Anopheles gambiae* ATPbinding cassette transporter family: phylogenetic analysis and tissue localization provide clues on function and role in insecticide resistance. Insect Mol Biol 27(1):110–122

Plernsub S, Stenhouse SA, Tippawangkosol P, Lumjuan N, Yanola J, Choochote W, Somboon P (2013) Relative developmental and reproductive fitness associated with F1534C homozygous knockdown resistant gene in *Aedes aegypti* from Thailand. Trop Biomed 30(4):621–630

Rai SK, Alistair BAB, Philip TR, Roman A, Sabine B, Qasim C, Geert C, Teresa FF, Jay G, Melanie K, Iseult L, James R, Chris S, David S, Karen T, Van den Brink PJ (2014) J Agric Food Chem 62(19):4227–4240

Ranson H, Hemingway J (2005) Mosquito glutathione transferases. Methods Enzymol 401:226–241. https://doi.org/10.1016/S0076-6879(05)01014-1

Ranson H, Claudianos C, Ortelli F, Abgrall C, Hemingway J (2002) Evolution of supergene families associated with insecticide resistance. Science 298:179–181

Reid WR, Zhang L, Liu N (2012) The whole genome transcriptome analysis of the mosquito *Culex quinquefasciatus* following permethrin selection. PLoS One 7:e47163

Rigby LM, Rašić G, Peatey CL, Hugo LE, Beebe NW, Devine GJ (2020) Identifying the fitness costs of a pyrethroid-resistant genotype in the major arboviral vector *Aedes aegypti*. Parasit Vectors 13(1):1–12

Rinkevich FD, Du Y, Dong K (2013) Diversity and convergence of sodium channel mutations involved in resistance to pyrethroids. Pestic Biochem Physiol 106:93–100

Rongnoparut P, Ugsang DM, Baimai V, Honda K, Sithiprasasna R (2005) Use of a remote sensing-based geographic information system in the characterizing spatial patterns for *Anopheles minimus* A and C breeding habitats in western Thailand. Southeast Asian J Trop Med Public Health 36(5):1145

Russell TL, Govella NJ, Azizi S, Drakeley CJ, Kachur SP, Killeen GF (2011) Increased proportions of outdoor feeding among residual malaria vector populations following increased use of insecticide-treated nets in rural Tanzania. Malar J 10:80

Saingamsook J, Yanola J, Lumjuan N, Walton C, Somboon P (2019) Investigation of relative development and reproductivity fitness cost in three insecticide-resistant strains of *Aedes aegypti* from Thailand. Insects 10:265

Silberman J, Taylor A (2023) Carbamate toxicity. In: StatPearls. StatPearls Publishing, Treasure Island, FL

Soderlund DM (2010) Toxicology and mode of action of pyrethroid insecticides. In: Hayes' handbook of pesticide toxicology. Academic Press, pp 1665–1686

Srisawat R, Komalamisra N, Eshita Y, Zheng M, Ono K, Itoh TQ, Matsumoto A, Petmitr S, Rongsriyam Y (2010) Point mutations in domain II of the voltage-gated sodium channel gene in deltamethrin-resistant *Aedes aegypti* (Diptera: Culicidae). Appl Entomol Zool 45(2):275–282

Staba EJ, Zito SW (1985) The production of pyrethrins by chrysanthemum cinerariaefolium (Trev) Bocc. In: Primary and secondary metabolism of plant cell cultures Berlin. Springer Berlin Heidelberg, Heidelberg, pp 209–214

Stearns SC (1989) Trade-offs in life-history evolution. Funct Ecol JSTOR 3(3):259–268

Strode C, Wondji CS, David J-P, Hawkes NJ, Lumjuan N (2008) Genomic analysis of detoxification genes in the mosquito *Aedes aegypti*. Insect Biochem Mol Biol 38:113–123

Taylor B (1975) Changes in the feeding behaviour of a malaria vector, *Anopheles farauti* Lav., following use of DDT as a residual spray in houses in the British Solomon Islands protectorate. Trans R Entomol Soc Lond 127:277–292

Thompson M, Shotkoski F, Ffrench-Constant R (1993) Cloning and sequencing of the cyclodiene insecticide resistance gene from the yellow fever mosquito *Aedes aegypti*. Conservation of the gene and resistance associated mutation with Drosophila. FEBS Lett 325(3):187–190. https://doi.org/10.1016/0014-5793(93)81070-g. PMID: 8391473

Thomsen KE, Koimbu G, Pulford J, Jamea-Maiasa S, Ura Y, Keven BJ, Siba MP, Mueller I, Hetzel WM, Reimer JL (2017) Mosquito behavior change after distribution of Bednets results in decreased protection against malaria exposure. J Infect Dis 215:790–797. https://doi.org/10.1093/infdis/jiw615

Tsai YH, Lein PJ (2021) Mechanisms of organophosphate neurotoxicity. Curr Opin Toxicol 26:49–60

Vulule JM, Beach RF, Atieli FK, McAllister JC, Brogdon WG (1999) Elevated oxidase and esterase levels associated with permethrin tolerance in *Anopheles gambiae* from Kenyan villages using permethrin impregnated nets. Med Vet Entomol 13:239–244

Waite JL, Swain S, Lynch PA, Sharma SK, Haque MA, Montgomery J, Thomas MB (2017) Increasing the potential for malaria elimination by targeting zoophilic vectors. Sci Rep 7:1–10. https://doi.org/10.1038/srep40551

Weill M, Fort P, Berthomieu A, Dubois MP, Pasteur N (2002) A novel acetylcholinesterase gene in mosquitoes codes for the insecticide target and is non-homologous to the *ace* gene in *Drosophila*. Proc R Soc Lond B 269:2007–2016

Weill M, Lutfalla G, Mogensen K, Chandre F, Berthomieu A (2003) Insecticide resistance in mosquito vectors. Nature 423:136–137

WHO (2022) Report on insecticide resistance in Aedes mosquitoes (*Ae. aegypti, Ae. albopictus, Ae. vittatus*) in WHO South-East Asia Region countries SEA-CD-334 October

Wilding CS, Smith I, Lynd A, Yawson AE, Weetman D (2012) A cis-regulatory sequence driving metabolic insecticide resistance in mosquitoes: functional characterisation and signatures of selection. Insect Biochem Mol Biol 42:699–707

Wondji CS, Dabire RK, Tukur Z, Irving H, Djouaka R (2011) Identification and distribution of a GABA receptor mutation conferring dieldrin resistance in the malaria vector *Anopheles funestus* in Africa. Insect Biochem Mol Biol 41(7):484–491

Xu Q, Zhang L, Li T, Zhang L, He L, Dong K, Liu N (2012) Evolutionary adaptation of the amino acid and codon usage of the mosquito sodium channel following insecticide selection in the field mosquitoes. https://doi.org/10.1371/journal.pone.0047609

Xu Y, Xu J, Zhou Y, Li X, Meng Y, Ma L, Zhou D, Shen B, Sun Y, Zhu C (2022) CPR63 promotes pyrethroid resistance by increasing cuticle thickness in *Culex pipienspallens*. Parasit Vectors 15(1):1–7

Yahouedo GA (2017) Contributions of cuticle permeability and enzyme detoxification to pyrethroid resistance in the major malaria vector *Anopheles gambiae*. Sci Rep 7(1):11019

Yan L, Yang P, Jiang F, Cui N, Ma E (2012) Transcriptomic and phylogenetic analysis of *Culex pipiensquinquefasciatus* for three detoxification gene families. BMC Genomics 13(609):558

Yang T, Liu N (2011) Genome analysis of cytochrome P450s and their expression profiles in insecticide resistant mosquitoes, *Culex quinquefasciatus*. PLoS One 6:e29418

Mosquito Research in India

16

B. K. Tyagi

Abstract

Mosquitoes have been associated with humans since times immemorial. Mosquitoes have been figured out in the scriptures of yore, and the Indian literature is not only the oldest reference to mosquitoes, but it also has the most complete treatment on the epidemiology of malaria, with many mosquito species identified with tertian and quartan types of malarial fevers. The modern science of mosquito research began with Manson (J Linn Soc (Zool) 14:304–311, 1878) discovery of mosquito (*Culex pipiens*) and filariasis worm (*Wuchereria bancrofti*) connection. This was furthered in cases of malaria (Ross. Br Med J 2:1786–1788, 1897; Grassi. Acc. Lincei, Rend Sc. Mat Fis Nat. 8:21–8, 1899) and other mosquito-borne infections. One of the most promising modern discoveries has been the RIDL gene transfer technology in the male mosquito of *Aedes aegypti* through genetic manipulation and control of vector populations in many countries such as Brazil and the USA. Inter alia, many inventions, innovations, and discoveries in Culicidae have been made by a large number of scientists which have impacted a great deal of the overall understanding and control of mosquitoes in India.

Keywords

Mosquitoes · Research · Control · India

B. K. Tyagi (✉)
Department of Biosciences, University Institute of Biotechnology, Chandigarh University, Gharuan, Mohali, Punjab, India

© The Author(s), under exclusive license to Springer Nature Singapore Pte Ltd. 2024

Omkar (ed.), *Mosquitoes*, https://doi.org/10.1007/978-981-97-4163-2_16

Learning Objectives

1. Mosquitoes are serious pest and vectors of diseases.
2. Many deadly infections like malaria and dengue are transmitted by mosquitoes.
3. Control of mosquito is largely possible through integrated vector control (IVC) strategy.
4. New control approaches such as genetic manipulation and use of *Wolbachia* must be intensified.

16.1 Introduction

Mosquitoes concerned humans for thousands of years, and their reference in connection with human sufferings and maladies could be traced in the literature of yore of several countries such as Bharat (India), China, Spain, Greece, etc. (Rao 1984; Tyagi 2020). After the advent of binomial system nomenclature by Linnaeus (1758), in his magnum opus *Systema Naturae*, tenth edition, mosquitoes began to be named scientifically; albeit Culicidae, a dipteran family dedicated to mosquitoes was a melange of several different kinds of dipterans: the true mosquitoes (Culicinae) and the mosquito-like other insects Chaoboridae and Dixidae (Christophers 1933; Barraud 1934). Since the beginning of the twentieth century, the classification of mosquitoes (Culicidae) has undergone several changes, and the family Culicidae is defined to include only two subfamilies Anophelinae and Culicinae, the latter including also the once separately erected subfamily Toxorhynchitinae (Theobald 1901a, b, 1903, 1907, 1910; Theobald 1908; Wilkerson et al., 2021; Tyagi et al. 2024). Built, the classificatory system of Culicidae is still in a dynamic state.

16.2 The Earliest Naming of Mosquitoes

Linnaeus (1758, 1762) included many mosquito species in the tenth edition of his classic *Systema Naturae*. Two of these species, namely, *Culex pipiens* and *Aedes aegypti*, were the only Indian species supposed to be then occurring in the country (Rao 1984; Controversies arising out of naming of Aedes aegypti has been analytically discussed by Clements and Harbach 2018). The first Indian species described from a sample collected within the country was *Anopheles stephensi.*

16.3 Species Complexes

Since mid-1950s, several mosquito species, especially *Anopheles*, have been found as a complex of many indistinguishable taxa such as *Anopheles culicifacies* (in India), *An. gambiae* (in Africa), and *An. maculipennis* (in Europe). In addition, *Culex pipiens* (global) and *Aedes albopictus* (global) are also characterized as species complexes. Such sibling species complexes of certain vector mosquito species carry tremendous biomedical significance as there are some, and not all, species of the complex that are capable to transmit the human disease pathogens such as malaria parasites. Lately, while *An. gambiae* complex from Africa had been taxonomically described as good species (*An. arabiensis*, *An. bwambae*, *An. melas*, *An. merus*, *An. quadriannulatus*, *An. gambiae s.s.*, *An. Coluzzii*, and *An. amharicus*), the same is still awaited in India for the *An. culicifacies* species complex and many others in the country (Table 16.1).

Table 16.1 Sibling species complex of *Anopheles* mosquitoes in India

Species	Sibling species complex	Designated species
Anopheles culicifacies	5	A, B, C, D, and E
Anopheles fluviatilis	3	S, T, and U
Anopheles subpictus	4	A, B, C, and D
Anopheles annularis	2	A and B
Anopheles dirus	5 (in India, only 3 species occur)	*An. dirus* s.s. (species A), *An. cracens* (species B), *An. scanloni* (species C), *An. baimaii* (species D), *An. elegans* (species E), *An. nemophilous* (species F), *An. takasagoensis*, and a cryptic species tentatively designated as *An.* aff. *Takasagoensis*
An. sundaicus	4	*An. sundaicus* s.s., *An. epiroticus* (formerly species A), *An. sundaicus* species D, and *An. sundaicus* species E.

Table 16.2 Indian mosquito species arranged under subfamilies, tribes, genera, and subgenera (*Source*: Adapted from Tyagi et al., 2015)

Subfamily Tribe	No. of genera	No. of subgenera	No. of species
Anophelinae			
Anophelini	1	2	64
Culicinae			
Toxorhynchitini	1	1	9
Aedeomyiini	1	1	2
Aedini	34	18	11
Culicini	2	8	90
Culisetini	1	2	3
Ficalbiini	2	3	7
Hodgesiini	1	0	2
Mansoniini	2	2	7
Orthopodomyiini	1	0	5
Sabethini	3	4	16
Uranotaeniini	1	2	34
	50	**43**	**420**

As far as Indian wealth of mosquito fauna is concerned, there are currently 420 (>11% of the world's total taxa) identifiable species, although some of them have sibling species complexes which are not incorporated (Table 16.2).

16.4 Definitions of Mosquitoes

Mosquitoes are the true flies, i.e., with one pair of functional wings, and classified under the family Culicidae of the order Diptera (class Insecta; phylum Arthropoda) (Clements 1992, 1999, 2012). Paleontological studies suggested that earliest mosquitoes had originated a little before the Mesozoic era, in the late Carboniferous Period. The forefathers of modern day mosquitoes were apparently staunch vegetarian, and in the course of evolution, they seemed to have evolved the habit of hematophagy (blood-feeding). However, somewhere along the diversified evolutionary pathways, they lost this character of blood-feeding on animals and resorted to survive on plant juices instead. The reversal to phytophagy (living on plant products like various kinds of exudations and dews) was also ephemeraland; a proportion of mosquito species regained the old attribute of blood-feeding (Fig. 16.1; Tyagi et al. 2024).

It is interesting to know that all mosquitoes (>3600 species) are not in the habit of blood-feeding animals. Interestingly, only 75% of the world's mosquitoes (2700 species) feed on blood of a variety of animals, both vertebrates and invertebrates, and there are, surprisingly enough, even less than 10% species that are actually

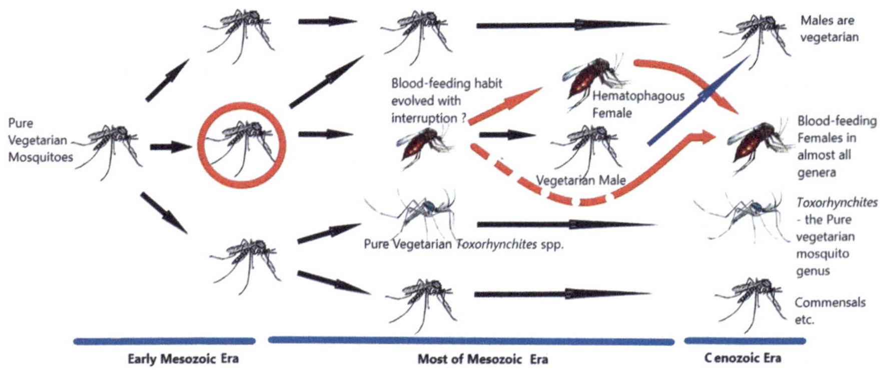

Fig. 16.1 Diagrammatic representation of a possible hypothesis depicting pathways in different geological ages the habit of hematophagy in some species (*Source*: Tyagi et al. 2024)

Table 16.3 Some of the deadliest and debilitating mosquito-borne diseases to man (*Source*: Tyagi 2020)

Arthropod	Vector	Virus	Disease	Geographical distribution
Mosquito	*Aedes aegypti*, *Ae. albopictus*	Dengue virus 1–4	Dengue	Tropical region
	Aedes aegypti, *Ae. albopictus*	Yellow fever virus	Yellow fever	Africa and South America
	Culex tritaeniorhynchus	Japanese encephalitis virus	Japanese encephalitis	Asia, Pacific
	Culex annulirostris	Murray Valley encephalitis virus	Murray Valley encephalitis	Australia and Papua New Guinea
	Culex spp.	Rocio virus	Rocio	America and Brazil
	Culex pipiens and *Cx. quinquefasciatus*	St. Louis encephalitis virus	St. Louis encephalitis	North and South America
	Culex spp.	West Nile virus	West Nile	Africa, Asia, Europe, and America
	Culex spp.	Rift Valley virus	Rift Valley fever	Africa, Middle East
	Aedes triseriatus	La Crosse virus	La Crosse encephalitis	North America
	Aedes triseriatus	California encephalitis virus	California encephalitis	North America, Europe, Asia
	Aedes aegypti, *Aedes albopictus*	Chikungunya virus	Chikungunya	Africa and Asia region
	Culex annulirostris, *Ae. vigilax*, *Ae. camptorhynchus*	Ross River virus	Ross River virus	Australia, South Pacific
	Culicidae *Haemagogus* spp.	Mayaro virus	Mayaro	South America
	Anopheles funestus, *Anopheles gambiae*	O'nyong-nyong virus	O'nyong-nyong fever	Africa
	Culex spp.	Sindbis virus	Sindbis	Africa, Egypt, Israel, Philippines, and Australia
	Aedes notoscriptus	Barmah Forest virus	Barmah Forest virus	Australia
	Culiseta melanura and *Cs. morsitans*	Eastern equine encephalitis virus	Eastern equine encephalitis	North, Central, and South America and the Caribbean
	Culex tarssalis	Western equine encephalitis virus	Western equine encephalitis	Americas
	Aedes taeniorhynchus	Venezuelan equine encephalitis virus	Venezuelan equine encephalitis	Americas
	Anopheles spp.	*Plasmodium* spp.	Malaria	Tropical and subtropical region
	Anopheles and *Culex* spp.	*Wuchereria bancrofti*, *Brugia malayi*, and *Brugia timori*	Lymphatic filariasis	Southeast Asia and Africa

involved in transmitting disease pathogens belonging to malaria, filariasis, yellow fever, dengue, chikungunya, Zika, Japanese encephalitis, etc. (Table 16.3).

These mosquito-borne diseases levy a huge physical, financial, and psychological burden on humans as well as countries. They affect not only the health but also reduce economy and emaciate intelligentsia.

16.5 Mosquito Diversity in Time and Space

Mosquitoes have a viable strength of over 3700 species, with an estimate of worldwide population of more than 100–200 trillion. They are found virtually in all ecosystems except the small Atlantic country of Iceland (an island country in Northern Europe, straddling the Eurasian and North American plates between the Greenland Sea and the North Atlantic Ocean, northwest of the British Isles, with a total area of 103,125 km^2) where not a single mosquito of any type has ever been reported (Harbach 2018; Lippi et al. 2023). Some mosquitoes like living near people (e.g., *Anopheles stephensi*, *Aedes aegypti*), while others prefer (e.g., *Aedes albopictus*) marshes (e.g., *Aedes sollicitans*, *Ae. cantator*, and *Ae. taeniorhynchus*) or tall grasses. They live in all environments ranging from rural and urban areas, ravines, caves, and mountains to deserts (Tyagi 2023b). They occur at altitudes above 2000, e.g., *Aedes* spp. mosquitoes in Colombia have been recently reported to survive a life cycle indoors at elevations as great as 2200 m (Suárez and Nelson 2024) and may survive subterraneously deep inside coalmines with very little oxygen and light (Rao 1984). There are also cavernicolous (troglobionts or cave-dwelling) mosquitoes (Bodharamik et al. 2021).

16.6 Mosquito Anatomy and Physiology

Mosquitoes are unisexual animals, that is, male and female reproductive systems are present in different individuals (Tyagi 2003, 2004, 2008). They are bilaterally symmetrical (Tyagi 2023a). The adult mosquito's body, covered with an exoskeleton, is organized in three distinct body regions: head, thorax, and abdomen. The head is ovoid in shape, with two large compound eyes and five appendages such as two antennae, two maxillary palpi, and the proboscis. The head is the sensory center for the mosquito, where the compound eyes and antennae are located, as well as the proboscis used for feeding. The thorax has three segments: the prothorax, mesothorax, and metathorax. Each segment has a pair of jointed legs, while the mesothorax has a pair of functional wings and the metathorax, a pair of knobbed structures called halteres. The abdomen has ten segments, of which the three terminal segments are specialized for reproduction and excretion. Female mosquitoes are strikingly identifiable from the males by shape and the form of their antennae and mouthparts, both attached on the head (Fig. 16.2). The head of an adult mosquito, e.g. *Aedes togoi*, is equipped with three different types of appendages: antennae, maxillary palps, and proboscis. Both antennae and maxillary palps are involved in

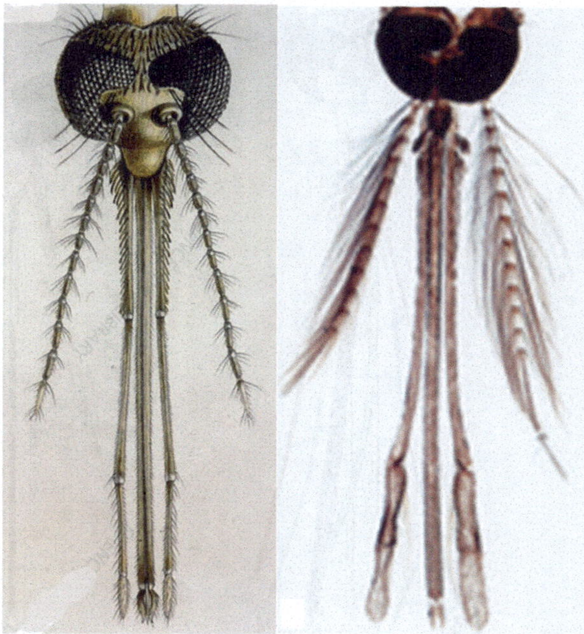

(a) (b)

Fig. 16.2 Head and mouthparts of (**a**) female and (**b**) male *Anopheles* mosquito; the latter as seen in a mounted slide, while the former handmade drawing

host detection. There is a pair of antennae on the front of head. They are bushy in males and sparsely haired in females. Only female mosquitoes blood-feed and are involved in disease transmission (Tyagi 2020).

While females possess a proboscis that is much more needlelike in appearance, allowing it to pierce the skin of animals to access blood, in males, the proboscis is wider and branched and has a feathery appearance which aids their hearing capabilities (Fig. 16.2). The mouthpart (proboscis) of a mosquito consists of six piercing stylets that include the labrum, paired mandibles, hypopharynx, and paired maxillae. The labrum, approximately 30 μm in diameter, functions as food canal and is the largest and stiffest stylet among the fascicles. An opaque cuticle wrapping the six stylets prevents the stylets to be observed clearly. After the labrum was removed with the help of microsurgical treatment using ultrafine forceps to strip off the opaque cuticle, the salivary duct and food canal were viewable. The cross-sectional image of the proboscis shows the labrum, food canal, salivary canal, and laciniae.

Hematophagous female mosquitoes bear highly specialized piercing and sucking type of mouthparts suitable for feeding on blood of their host, duly supported by a pair of the salivary glands (Fig. 16.3). The piercing-cum-sucking tools, known collectively as the fascicle, contain one pair of teeth-bearing maxillary stylets, one pair of mandibular stylets, a labrum, and a hypopharynx with its salivary canal. All these structures are packed in the gutterlike labium and constitute a proboscis. The labrum forms a food canal and provides rigidity for the fascicle, whereas the maxillae anchor themselves to the host skin by their teeth. Whether the mandibles play an active role during piercing as in other bloodsucking nematoceran flies (Downes

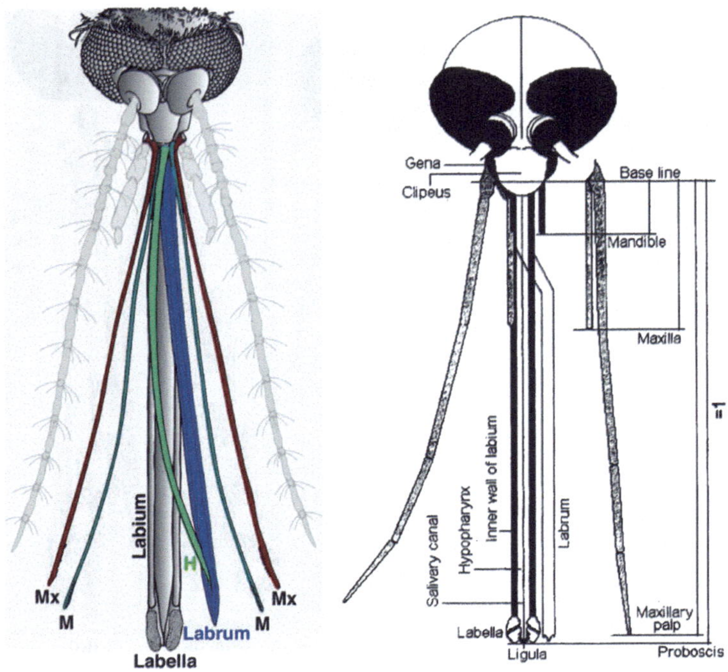

(a) (b)

Fig. 16.3 (**a**) Illustration of head appendages from the southern house mosquito. In this computer reconstruction, the two main olfactory appendages, the antennae and the maxillary palps, were de-emphasized in light gray. For clarity, the proboscis is depicted with the stylet removed from the labium, and the six fascicles forming the stylet are separated and displayed in color. During blood-feeding, the labium (in gray) bends and remains outside with the two lobes on the tip (labella) making direct contact with the host's skin. All six fascicles of the stylet are inserted in the host's skin, i.e., a pair of maxillae (Mx, cayenne), a pair of mandibles (M, fern), the hypopharynx (H, clover), and the labrum (blueberry). (**b**) Head and mouthparts of a male *Aedes* mosquito

1978) or simply serve as a closure of the distal opening of the labral food canal (Clements 1992) is not clear. The hypopharynx, which partially makes the ventral closure of the labral food canal, enters the host skin together with other parts and releases saliva from a gutter on its surface (the salivary canal) during feeding.

As compared to that of the female mosquito, male mosquitoes do not take blood. Their food sources are mainly floral and extrafloral nectars, honeydew, or even plant tissue. Accordingly, their mouthparts are not developed for piercing. In fact, mouthparts of adult males have been quite meagerly examined still with plenty of erroneous information. Most of the information available comes from aedine mosquitoes, mainly *Aedes aegypti* and *Ae. albopictus*. However, advanced microscopy has resolved many issues associated with the mouthparts of male mosquitoes since the accurate measurement of lengths of maxillae, mandibles, maxillary palpi, and proboscises became scrupulously possible; their lengths were presented as ratios to proboscis lengths. However, their range in length (maxillae 0.13–0.50, mandibles

0.04–0.17 length of the proboscis) may be regarded as a subgeneric attribute (Wahid et al., 2002). Their detailed structures were further possible by scanning electron microscopy. It is irony that in contrast to previous reports, mandibles were found in the male *Ae. aegypti*. Variations in both maxillary and mandibular lengths were significant in *Ae. aegypti* and *Ae. albopictus*. High variation of these structures among and within species indicates that the average length of these structures in only one species may not be a reliable representative of a subgenus and those of one strain may not be reliable for a species. Maxillae and mandibles are distinctly shorter than the proboscis. Their delicate structures and the large coefficient of variation suggest that they exist only as vestigial structures. However, a positive correlation exists between lengths of maxillae and those of mandibles, but mandibles are usually shorter than maxillae. The hypopharynx is discernible from the labium wall by its texture and border, and this suggests that it was a free stylet in the past.

16.7 Mosquito Research on Disease Connection

16.7.1 Discovery of Mosquito-Filariasis Connection

Mosquitoes around the world had been conceived to cast evils on man for centuries, and by 1870–1880s, some concrete evidence also began to emerge manifesting mosquito's role in transmitting certain diseases. The first important work on the role of insects was done by Dr. Patrick Manson in China during the 1870s when he showed that mosquitoes might be vectors of the disease colloquially known as elephantiasis from the gross swellings, particularly of the legs and scrotum, that it often produces (Harrison 1978). The hypothesis which he forwarded was absolutely irrefutable: he found out the young microfilariae, which unlike their procreator moved freely in the bloodstream and were always enclosed in a delicate membrane, could neither feed nor move of their own volition and therefore must complete their life cycles somewhere else—the mosquito? In early 1877, Manson succeeded in demonstrating that mosquito, *Culex pipiens*, was responsible to effect man to man transfer of the highly incapacitating infection in humans—the lymphatic filariasis (Manson 1878). This discovery was by all standards a great victory in furthering human knowledge about mosquito's placement in the man-mosquito-disease triangle.

16.7.2 Discovery of Mosquito-Malaria Connection

Taking a cue from Manson's "filariasis-mosquito connection" theory, Ross (1897) began studying malaria in India without even having seen a parasite until much later time of Laveran's discovery of *Plasmodium falciparum/malariae*, in 1880, due mostly to his initial contretemps to this new finding. In addition to this, he was also unable to identify mosquitoes which compounded his problems manifold. However, he candidly called his focused mosquitoes as "dappled wings," possibly *Anopheles*

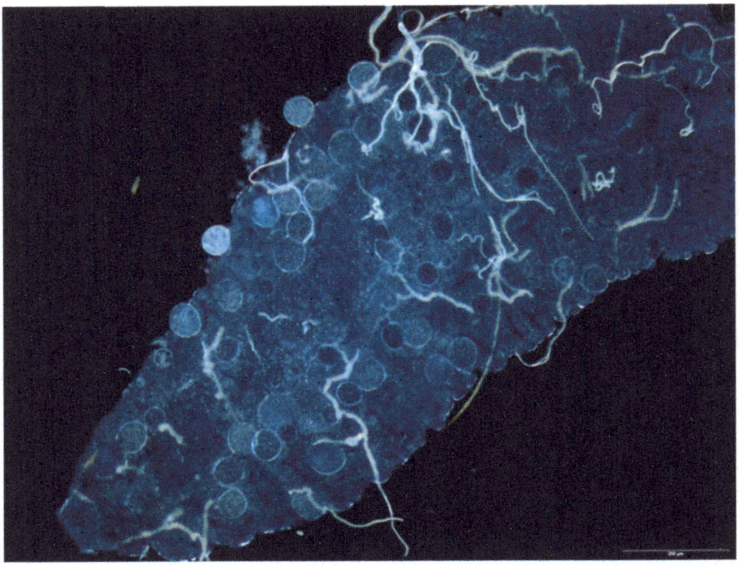

Fig. 16.4 A microscope view of the midgut of an *A. stephensi* mosquito infected with malaria parasites (the round circles)

stephensi. He worked relentlessly about finding the mosquito's role in malaria transmission. Though unaware of mosquito identification and dissection techniques, he burned midnight oil dissecting all the mosquitoes thought of carrying the pathogen (Fig. 16.4). He finally succeeded to establish an inseparable link between mosquito (*Anopheles*) and the malaria parasite. For his work (Ross 1897, 1898a, b), he was awarded the Nobel Prize in 1902.

16.7.3 Discovery of Mosquito-Yellow Fever Connection

Yellow fever had been the scourge of the Caribbean and the coastal and Southern United States for centuries. In the eighteenth and nineteenth centuries, devastating epidemics of yellow fever had claimed thousands of lives. The government of the USA formed the Yellow Fever Commission in Cuba consisting of Walter Reed, James Carroll, A. Agramonte, and Jesse Lazear. In the first few years, unsuccessful attempts were made to prove a relationship between yellow fever and the bacteria *Bacillus icteroides*. It failed, and the Commission members focused their efforts on determining how yellow fever was spread. They had found mounting evidence against the commonly held fomite theory of transmission. Among the several other theories, Dr. Jesse Lazear was most intrigued by the possibility of a mosquito vector. The mosquito theory of transmission of yellow fever had been around for many years.

Possibly the first to discuss it was Dr. Josiah Nott of Mobile, Ala., in 1848. It had been proposed again by Dr. Carlos Finlay of Havana in 1881. But try as he might, Finlay had been unable to prove his theory in almost 100 experiments on human subjects. Finlay's mosquito theory remained in abeyance until attention was again drawn to it by the demonstration in recent years of the part played by these insects, the mosquitoes, in the causation of other tropical diseases. Lazear, who had previous experience with mosquitoes during his malaria research, was most interested in the theory. Finlay was convinced that only the female *Culex fasciatus* mosquito (now called *Aedes aegypti*) transmitted the disease. Lazear carefully hatched the mosquitoes and allowed himself, along with another Army doctor, A.S. Pinto, to be bitten by mosquitoes that had fed on yellow fever patients on the fifth day of illness which had been 12 days earlier. They remained well. Seven other volunteers were bitten; they also remained well. The researchers were beginning to doubt there was anything to this mosquito theory. On Aug. 27, 1900, Lt. James Carroll and Lazear were discussing their lack of results when Carroll suggested that they try one on the mosquitoes on him. Carroll's mosquito had bitten a yellow fever patient on the second day of illness, 12 days before. Two days after the bite, Carroll was not feeling well, and on the third day, he was clearly ill with fever. A blood test for malaria was negative, and as Carroll became sicker, it was obvious he had yellow fever. Lazear and Aristides Argamonte were very concerned about Carroll's health as he was somewhat frail and well past 40 years of age. However, they were excited that the mosquito bite could have caused his yellow fever. This would be impossible to prove conclusively, as Carroll had been in contact with other yellow fever patients at the hospital in Havana and in the autopsy room. He could have gotten it in any of these places; no one could know for sure. On Aug. 31, the day Carroll was diagnosed with yellow fever, Lazear was in the post hospital at Columbia Barracks. Lazear and Argamonte were much relieved several days later when it became apparent that Carroll was going to survive. There had been no cases of yellow fever on the post or in the hospital. Pvt. William H. Dean of the Seventh Cavalry was about to be discharged from the hospital for a minor condition when Lazear asked him if he would be willing to submit to an experiment involving his mosquitoes. Dean has newly arrived in Cuba and had never been off post. He agreed and was experimentally bitten that same day. Being careful not to revel in someone's illness, they must have been elated when Dean became ill and was diagnosed with yellow fever. Dean was the first experimental case of yellow fever transmitted by the bite of a mosquito that was irrefutable as to its origin.

This discovery was, however, subjected to much harsh treatment and was always in the center of stormy debates. Even celebrated men of science became locked in fierce disagreement. These differences in opinion gave rise to wildly different approaches to prevention and treatment. Around 1793, in Washington, D.C., Dr. Albert King hypothesized that mosquitoes were the real culprits, but he was ridiculed and repudiated openly for his ideas to protect his nation from such infectious mosquito bites. Another hundred years would pass before scientists realized that King was on the right track. In 1881, a Cuban physician, Dr. Carlos Finlay, made a discovery that would revolutionize disease control and ultimately make possible the construction of the Panama Canal. He identified the *Aedes aegypti* mosquito as the carrier of yellow fever. Finlay's breakthrough had glitches, however. He correctly had surmised that the mosquito was spreading the disease by biting people previously infected with the disease and then transmitting the disease agent to other people through its bite. But when he attempted to prove his theory by having infected mosquitoes inoculate healthy volunteers with the disease, he failed. What Finlay didn't know was that the virus needed to incubate for approximately 12 days inside the insect in order to develop. He was attempting to inoculate his volunteers with the virus using mosquitoes that had only been infected a few hours. Finlay formulated a detailed plan for combating both yellow fever and malaria. Initially, Reed did not put much stock in Finlay's mosquito theory; efforts focused initially on ridding the Cuban capital of the filth and contamination presumed to cause the disease. But when the yellow fever death toll rose despite a massive clean-up effort, Reed decided to test Finlay's theory. Reed constructed two small buildings through which air could circulate freely to first show that the disease was not airborne. In one building, he had volunteers sleep for 20 nights in the foul bedsheets and garments of yellow fever patients. In the other building, volunteers slept next to infected mosquitoes but were separated by screens. In this experiment, Reed did not use recently infected mosquitoes like Finlay had, but rather insects that had been infected for some time. Then one volunteer was exposed to infected mosquitoes for three successive nights. This volunteer was the only one to develop yellow fever. Proof at last was conclusive: *Aedes aegypti* was the culprit. Reed had yielded the knowledge that would free Havana from yellow fever, although even now certain geniuses of the time, like Gorgas, remained unconvinced of Finlay's theory but resorted to eliminate the mosquito and its breeding places, besides quarantining the patients to prevent them from the remaining infective mosquitoes. The results were spectacular as within less than a year, yellow fever had been virtually eradicated. Final conclusion certainly is that it was Dr. Carlos Finlay of Havana who, in 1881, propounded the theory that mosquitoes were the carriers of yellow fever.

16.7.4 Discovery of Mosquito-Dengue Fever Connection

Incidentally, dengue is among the main mosquito-borne infections to be causing great concern to both the people who are at risk of infection with its deadly forms, the dengue hemorrhagic fever (DHF) and dengue shook syndrome (DSS), and the health planners all over the world, but particularly in the tropics, due to disease's tendency to spread to non-endemic areas in all the world's five continents. Owing to the growing concern associated with its causing enormous morbidity and mortality, and the subsequent sequelae to health retardation at various levels, the Tropical Disease Research of the World Health Organization has in recent past incorporated dengue within its fold of priority research.

As it does even today, the disease was often mistaken for either yellow fever or other similar illnesses in Asia, Africa, and America. Dengue came to be recognized as a separate entity only during the first part of the nineteenth century when a severe epidemic occurred in the Caribbean country. Finally, the mosquito *Aedes aegypti* was firmly confirmed to be playing a vector's role in transmitting dengue fever virus in 1906. Subsequently, other species were also identified to carry the transmission, but much less seriously, and the most notable such a species is the Asian tiger mosquito, *Ae. albopictus*. The latter species, from whom dengue virus was detected in India as late as 1988, has in recent years gained tremendous significance due to its implication in disease transmission in Kerala State, hitherto largely a terra incognito for epidemic dengue (Tyagi et al. 2006). In Malaysia, a simian zoonosis has been identified which has rendered an altogether new dimension to the dengue epidemiology.

16.7.5 Discovery of Mosquito-Encephalitis Connection

Almost each of the mosquito-encephalitis theories, all of which have been floored during the twentieth century, is as full of scintillating and stimulating scientific zeal, charm, and obsession as were characterized any of their predecessors with respect to other mosquito-borne infections. Yet, nearly all these certainly went on to be accepted in the concerned research community without an unreasonable confabulation or consternation. The Japanese encephalitis mosquito theory dates back to 1930, while that of the infamous St. Louis encephalitis is credited to 1942.

16.7.6 Discovery of Mosquito-Zika Fever Connection

Zika, a flavivirus transmitted mainly by mosquitos of the genus *Aedes*, was discovered in 1947 in Uganda. From the 1960s to 1980s, human infections were found across Africa and Asia, typically accompanied by mild illness. The first large outbreak of disease caused by Zika infection was reported from the island of Yap (Federated States of Micronesia) in 2007, as the virus moved from Southeast Asia across the Pacific. During an outbreak in French Polynesia in 2013–2014, Guillain-Barré syndrome was linked to Zika infection, and cases of microcephaly in newborn children were also retrospectively linked to this outbreak. The World Health Organization (WHO) received the first reports of locally transmitted infection from Brazil in May 2015. In July 2015, health ministry officials from Brazil reported an association between Zika virus infection and Guillain-Barré syndrome in adults. In October 2015, the WHO received reports from Brazil of microcephaly in babies whose mothers had been exposed to Zika during pregnancy. At this time, there was no proof of a causal link between Zika infection and these neurological complications. In February 2016, as infection moved rapidly through the range occupied by *Aedes* mosquitos in the Americas, the WHO declared that Zika infection associated with microcephaly and other neurological disorders constituted a public health emergency of international concern (PHEIC). By the start of February 2016, local transmission of Zika infection had been reported from more than 20 countries and territories in the Americas, and an outbreak numbering thousands of cases was underway in Cabo Verde in western Africa. Beyond the range of its mosquito vectors, Zika virus infections are expected to be carried worldwide by people as they travel and be transmitted by travelers to sexual partners who have not been to places where the virus is endemic.

In 2015, Brazil saw a dramatic increase in the number of babies born with microcephaly, a condition in which a baby's head is much smaller than normal due to abnormal brain development while in the womb, often resulting in developmental delays later in life. Normally microcephaly is a relatively rare occurrence usually caused by exposure of a pregnant woman to drugs, alcohol, or environmental toxins, but nearly 3000 cases were reported in Brazil, about 20 times the number of cases reported in the prior year 2014. Initial epidemiological investigations implicated an infection during pregnancy with the Zika virus as the cause. Prior to the microcephaly epidemic, Zika virus was a relatively unknown pathogen.

Transmitted through the bite of an infected female mosquito *Aedes aegypti*, Zika virus normally causes few or no symptoms in children and adults, although in rare instances it can result in an autoimmune disease, Guillain-Barre syndrome, in which the body's immune system attacks part of the nervous system leading to muscle weakness and paralysis. The past several decades have seen an alarming increase in numbers of *Aedes aegypti*, which also carries the viruses for yellow fever, dengue, and chikungunya, affecting millions of people worldwide. This enormous increase in numbers of mosquitoes most likely reflects the explosion of human-created trash that litters the world. Needing only small amounts of water that puddle in abandon tires, nonbiodegradable containers, and plastic bags to lay their eggs, the mosquito has evolved to live in close quarters with humans, especially in poorer regions of the world where trash removal, mosquito netting, air-conditioning, and insect control are limited or nonexistent. Zika virus has been observed by virus hunters making slow progress from the Zika Forest in Uganda to Indonesia to Micronesia and the Polynesian Islands mirroring the movement of populations across the Pacific. Until 2013, Zika virus seemed mostly innocuous and not worth worrying about. Then in 2015 it was clear that Zika virus was becoming a very serious concern. The virus is believed to have arrived in the blood of infected travelers attending the 2014 World Soccer Cup. Because most Zika infections show few or no symptoms, these travelers may not even have realized they were infected.

The combination of many individuals carrying the virus in their blood coming in contact with and being bitten by the mosquito vector present in great numbers resulted in an unprecedented rise in Zika infections and the terrible consequences for unborn babies. It has not been determined whether Zika has a nonhuman animal reservoir.

The rapid spread of Zika throughout Central and South America may also be the result of a change in the virus itself. Mutations may have altered the virus enabling it to infect humans more easily or increasing the number of Zika virus particles in human blood. Either of these changes could increase the rate of Zika spread by increasing the chances of a mosquito picking up the virus from an infected individual and transmitting it to another person.

16.8 Major Mosquito Control Research

Vector mosquito control is an integral part of the disease management. Mosquito control can be divided into two areas of responsibility: individual and public. Mosquito control remains the only means available to combat some medically important, vector-borne pathogens, such as West Nile, dengue, and chikungunya viruses, because no approved vaccine, therapeutant, or prophylaxis exists. Pesticides continue to hold the main plank against vector-borne diseases especially during epidemics. Unfortunately, many mosquito vector species have lately developed resistance against most of conventional insecticides, and the likelihood of availability of newer insecticides is highly precarious, cost challenging, tedious, and a long exercise. They may be applied to control larvae (larvicides) or adults (adulticides). Applications of adulticides or larvicides are made after the presence of mosquitoes has been demonstrated by surveillance procedures. Application is made by prescribed standards. All insecticides must have the name and amount of active ingredient (AI) appearing on the label; examples are DEET and pyrethroids. No pesticide is 100% safe and care must be exercised in the use of any pesticide. Material Safety Data Sheet (MSDS) contains basic information about a product intended to help you work safely with the material.

Chemical insecticides are used most commonly in mosquito control, with formulations that include larvicides and adulticides (e.g., space sprays, residual indoor applications, and insecticide-treated bed nets). With each of these approaches, however, their efficacy can be reduced by an inability to achieve adequate coverage that is needed to effectively reduce pathogen transmission. Larvicides are applied to aquatic habitats of developing immature mosquitoes ("breeding sites") and are demonstrated to reduce mosquito-borne disease transmission. At times larvicidal treatments fail in certain aquatic habitats which can be small, sheltered, and difficult to locate and treat ("cryptic breeding sites"). Because financial and human resources are limited, it can be difficult to achieve a coverage level sufficient to reduce disease. Furthermore, some larvicide formulations are expensive, and the large quantities required for broadcasting strategies can be cost prohibitive. Unequivocally, the greatest arsenal fight of the vector and pest mosquitoes in recent past has been the discovery of DDT using which many countries overcame the mosquito menace. However, the euphoria was only short-lived since many mosquito vectors developed resistance against the pesticide. Subsequently, insecticide impregnated bed nets using pyrethroids of variety became popular, but even these are not immune to cause resistance development in mosquitoes (Jambulingam et al. 1989). The discovery of *Bacillus thuringiensis* var. *israelensis* was found very significant (Becker et al. 2010), although resistance developed by some species against it raised serious doubts on its longtime usage in vector control. Autodissemination strategies have attracted attention as a method to reduce mosquito populations and benefit from mosquito behavior, in which a female mosquito visits multiple breeding sites. As practiced currently, "dissemination stations" are attractive to adult females, which enter the station, become contaminated with a juvenile hormone analog (JHA), exit, and then contaminate breeding sites with levels of JHA that are lethal to immature mosquitoes. The autodissemination method is particularly attractive for those species that breed within small, cryptic sites, which serve as refugia from existing insecticidal measures.

The males deliver JHA to breeding sites, either directly or indirectly, i.e., via the cross contamination of females, which subsequently transfer JHA to breeding sites. The examined autocidal method can be used preemptively, e.g., in areas with low densities of indigenous mosquitoes and in advance of the natural population increase. Unlike autodissemination approaches that rely upon the indigenous population, an approach based on artificially reared males can made more intensive, because the number of males released is limited only by the logistics of male rearing and release and methods for mass production of mosquitoes are developed. The autodissemination approach has been shown effective at treating cryptic refugia that remain unaffected by existing mosquito control methods. This approach relies on adult mosquito behavior to spread larvicide to breeding sites at levels that are lethal to immature mosquitoes. Prior studies demonstrate that "dissemination stations," deployed in mosquito infested areas, can contaminate adult mosquitoes, which subsequently deliver the larvicide to breeding sites. In some situations, however, preventative measures are needed, e.g., to mitigate seasonal population increases. Here we examine a novel approach that combines elements of autocidal and autodissemination strategies by releasing artificially reared, male mosquitoes that are contaminated with an insecticide. For the above practical constraints in controlling vectors of diseases, autodissemination has attracted attention due to its potential to address important gaps with existing mosquito control methods. Autodissemination is a method of pesticide "self- delivery," which is premised upon the use of insects as the delivery agent. Insects carrying small amounts of insecticide can deliver an active ingredient to cryptic refugia, rather than human applicators, and this method can require less pesticide relative to broadcasting. For this reason, autodissemination approaches have become an important pesticidal method for termites, beetles, and moths.

16.8.1 Strategies Based on Genetic Manipulation

Ample biomedical significance tools and products have been derived in context with the male mosquitoes. Actually, they are most important in designing strategies to control their deadly counterpart, the female mosquitoes. There are a number of ways in which the males can be manipulated to execute the hammering on the hematophagous and pathogen-transmitting fair sex, for example, male mosquitoes can work as conveyer for insecticide (Mains et al. 2015) or a lethal gene (Alphey 2002), on one hand, and provide a battery of molecules of highly significant pharmacological utility like anesthesia, on the other. Genetic modification (GM) of vector mosquitoes is one of such technologies, which may be mainly used either to suppress or replace the wild populations of a vector so as to decrease vector populations or reduce the vector's ability to transmit. Among various genetic manipulations or GM technologies becoming available slowly to control vector mosquitoes are as follows:

1. Transgenesis, using Release of Insect population having Dominant Lethal (RIDL) gene system, is pioneering in the sense that this technology has been applied in many countries including Brazil and the USA. A more advanced and fool-proof technology evolving in mosquito domain is gene drive-based genome editing, e.g., CRISPER/Cas9, to suppress or replace the vector population.
2. Paratransgenesis (e.g., deploying endosymbiont *Wolbachia*-induced cytoplasmic incompatibility for replacing natural vector population).

In particular, the advent of CRISPR technology has excited the potential to engineer new game-changing technologies and innovative systems that can be used to control wild populations of mosquitoes. Two developments of particular interest are a self-limiting system termed precision-guided sterile insect technique (pgSIT) and a homing-based gene drive (HGD). The unique features of these systems can make them valuable tools to control vector mosquitoes in the future. All these biotechnological advancements in vector control are designed to fit well in the multi-methodical integrated vector management (IVM) strategy.

These applications include the release of reared mosquitoes (GM) in the environment to introduce modified genetic traits in wild population. GM encompasses multiple approaches, which are broadly categorized into two types. The first category includes sterilized insect technique (SIT) for population suppression, and the second category is the gene drive systems for population replacement or manipulation. The sterilized insect technique (SIT) includes the dominant lethal gene systems (RIDL), *Wolbachia*-mediated cytoplasmic incompatibility (CI), and classical radiation-induced male sterility, while population manipulative technologies include Medea-based gene drive, under-dominance gene drive, homing endonuclease, *Wolbachia*-mediated heritable biocontrol or genetically modified midgut bacteria, and transposable element like Piggy Bac. These techniques have been put into field application in many countries (Cayman Islands, Florida in the USA, Brazil, Mexico, Malaysia, etc.). However, in India, most of them are still in laboratory development and/or contained stage. Two private institutions, namely, the International Institute of Biotechnology and Toxicology (IIBAT) in Tamil Nadu and Gangabishan Bhikulal Investment and Trading Limited (GBIT)'s Hybrid Seeds Co. (MAHYCO), Dawalwadi, in Maharashtra State, began laboratory based investigations on the RIDL *Aedes aegypti* during mid-2000. While IIBAT discontinued after a few of years of investigation, MAHYCO progressed satisfactorily from laboratory to containment studies with splendid results (Patil et al., 2022). Further, these investigations require different region-based field testing under independent monitoring in order to prove the result to gain the consent of society to implement. These techniques require large volume of mosquitoes to be released into the environment at different period of interval to either suppress or replace mosquito population.

16.8.2 *Wolbachia*-Based Control of Mosquito Vectors

The *Wolbachia*-based research (Mains et al. 2016) for controlling *Aedes aegypti* and dengue has been recently taken up at the ICMR-Vector Control Research Centre, Puducherry, in collaboration with Monash University, Australia, whereby naturally occurring *Wolbachia* bacterium will be introduced into dengue virus-carrying mosquitoes to "inhibit" viral transmission. The efficacy of the disease control method will be tested at the VCRC before undertaking any field trial (cf. The Economics Times, Feb. 07, 2017).

16.9 Summary of Innovative Research by Indian Scientists

16.9.1 Dr. Ronald Ross's Discovery of Oocysts in the Midgut of a Mosquito

The most famous discovery in mosquito science ever has crowned Dr. Ronald Ross for his demonstration of the inextricable link between mosquito and malaria, paving way for a rather easier and direct way of eradication of the disease through vector control (Ross 1897, 1898a, b).

16.9.2 Mosquito Pupae Separator

Dr. V.P. Sharma developed a new technique for sex separation of mosquitoes at pupal stage for which he was awarded the Young Scientist Award.

16.9.3 Mosquito Sampler

Sampling is among the most important research activities warranted to determine a lot of various bioecological characteristics about the mosquitoes, especially ones acting as vector species for communicable disease pathogens. In mosquito science, adult sampling, represented by per man-hour density, has for long been in use, and in India it is still practiced using a mouth-operated aspirator, though outdated, unethical, and not practiced in other countries. This aspirator device, comprising a plastic or glass tube connected to one end with a rubber tube, and separated with latter by a thin mosquito plastic or metallic netting at the joint (so that mosquitoes sucked are held there at the end of the tube), was in the earlier days recommended by even the World Health Organization for all countries combating malaria scourge. Even though this device has in course undergone enormous modification and, in some developed countries, like the USA and UK, has also been successfully replaced by certain motorized sampling devices, however, in India the same mouth-operated device had more or less been exclusively employed in the exercise (except in some institutes where, too, the motorized samplers were obtained by purchase from abroad!)—totally unaware of the fact that the aspirator was a health hazard, besides of course not fool-proof to basic entomological vagaries generally encountered during adult sampling in field. This mouth-operated aspirator, though simple in appearance, could be operated up to nearly 2 feet above the erected hand (meaning a sampling person could collect adult mosquitoes everywhere but the roof, which in most urban setup is about 9–10 ft. above the ground), but with the sampler held aloft, he could barely access 8 ft., considering average human height of 5′6″. Consequently, only a part of the adults actually present indoors could be sampled during a specified period of time (meaning insufficient and misleading sampling), and, above all, the risk of contracting throat infection by the sampling human (the inhaled mosquito carries a lot of dust fouled with fungi, bacteria, and/or similar unwanted things). The entire sampling procedure involving mouth-operated aspirator seemed a farce in the name of scientific investigation.

B.K. Tyagi (1993) invented a "Mechanical Mosquito Sampler" (or "Tyagi Sampler") that was highly effective in sampling all kinds of resting mosquitoes, economical, and transportable. He also obtained patent for the same (India patent # 191635) in 2005.

16.9.4 Transfer of Technologies Based on Mosquitoes

B.K. Tyagi had executed a Transfer of Technology (ToT), namely, "TheJE detection in Wild-caught mosquitoes using TOXO-IFA System for JEV surveillance." The true promoter and exponent for this methodology, accomplished in two stages, is V. Thenmozhi (cf. Thenmozhi et al. 2016), even though A. Gajanana could be credited to have invented it (Gajanana et al. 1997). Under this technique, wild caught mosquitoes are separated species-wise, and while unfed mosquitoes were pooled (1–50) on the same day of collection, engorged female mosquitoes were held for 48 h for digestion of blood meals before pooling. Mosquito (only females) abundance was calculated as density (number collected per man-hour). Mosquitoes were pooled (minimum of five specimens per pool) and stored at −80 °C until processed for virus detection and isolation. Two systems were used:

1. Antigen capture ELISA. Monoclonal antibody 6B4A-10 (reactive against all viruses in JE/WN/SLE/MVE complex) was used as capture antibody and monoclonal antibody peroxidase conjugate SLE MAB 6B6C-1 (reactive against all flaviviruses) as detector antibodies (supplied by T.F. Tsai, Centres for Disease Control and Prevention, Fort Collins Co., USA). A mosquito pool was considered ELISA positive if its optical density value was mean + four standard deviation of the six normal pools.
2. Insect bioassay (Toxo-IFA). *Toxorhynchites splendens* mosquito larvae were inoculated with ELISA positive pools intracerebrally and incubated for 7–10 days

at 29 °C and then tested by the indirect immunofluorescence assay (IFA) on head squeeze preparations (Toxo-IFA). Smears were tested with JEV-specific monoclonal antibody, MAB 112 (supplied by Dr. Kimura Kuroda, Tokyo Metropolitan Institute of Neurosciences, Japan), and detected by fluorescein isothiocyanate (FITC) conjugated anti-mouse immunoglobulin (Dakoppats, Denmark).

Virus infection rate in mosquitoes was expressed as minimum infection rate (MIR, i.e., number of positive pools/total number of mosquitoes tested × 1000).

16.9.5 N,N-Diethylphenylacetamide (DEPA)

Mosquito-borne diseases, including malaria, filariasis, dengue, and encephalitis viruses, continue to cause morbidity and mortality in humans, domestic animals, and wildlife. Arthropod "eradication" measures envisaged five decades ago failed to make any significant impact. A change to the philosophy of insect "control" or "management" has continued to evolve. The use of insect repellents seems to be one of the best available means for personal protection against annoyance and diseases associated with hematophagous organisms, especially because repellents protect subjects without having a major adverse impact on the environment. Amides, imides, esters, ethers, and polyfunctional compounds are known to be good repellents. Selected petroleum oil fractions and essential oils repel *Ae. aegypti* (L.). Dimethyl phthalate (DMP) was the first synthetic chemical that had considerable repellent efficacy and has been used widely as a mosquito repellent. N,N-diethyl m-toluamide (DEET) emerged in 1954 as a superior arthropod repellent. Cost considerations arising from the unavailability of m-toluic acid for the production of DEET in India prompted investigation into an array of alternative compounds bearing high repellence, safety, and cost-effectiveness. Consequently, this resulted in the development of a new insect repellent, N,N-diethylphenylacetamide (DEPA) (Kalyanasundaram 1982). N,N-diethylphenylacetamide, popularly called as DEPA, has repellent activity against hematophagous insects including mosquitoes, blackflies, horse flies, muscoid flies, rat fleas, and ticks, as well as land leeches and cockroaches. The efficacy of DEPA is comparable to that of N,N-diethyl-3-methylbenzamide (DEET) in both laboratory and field tests (Kandasamy 2021). Toxicological studies indicate that DEPA is safe for human use. Different formulations as well as fabrics impregnated with DEPA are effective for personal protection from insect bites. Dr. Vijay Veer, former Director, DRDO, during his deputation in the DRDO's Defence Research Development Establishment (DRDE) at Gwalior, sweated hard to swing the use of this great repellent in favor of Indian armed personnel serving in difficult pest- and vector-infested front areas, saving the nation in the process crores of rupees otherwise going out in the purchase of DEET.

Fig. 16.5 *AlboTyag*, a simple and manually operated trap developed to prevent breeding by *Aedes albopictus* in the rubber-collecting cups tethered with the main plank to harvest rubber (*Source*: Tyagi 2020)

16.9.6 "MosTyag": An *Aedes albopictus* Breeding Preventer for Rubber-Harvesting Cups

Tyagi et al. (2006) found out that in the Kerala State, where dengue had erupted in a severe epidemic during 2003, the vector mosquito *Aedes albopictus* bred predominantly in the latex collecting cups in the rubber plantation fields grown over nearly 1000,000 ha in the mountainous and sylvatic terrains. The mosquito biology in the sylvan environment of the rubber plantation posed several challenges in controlling it. Tyagi (2020) invented a trap called *AlboTyag* which maneuvered to completely check the breeding of the vector in the rubber-collecting cups even during the more often than not intermittent rains (Fig. 16.5).

16.10 Conclusion

Mosquitoes are reasons for not only their irritating and painful bites but also their role in the transmission of several life-threatening and debilitating diseases like malaria, dengue, chikungunya, Zika, West Nile, filariasis, and Japanese encephalitis. All these vices are associated with the female—the femme fatale! They, that is female mosquitoes only, are branded objects of deep abhorrence and dislike for all they offer to mankind—the annoyance, the pain, the nuisance, and the deadly infections, which have been eventually responsible in wasting precious national resources, e.g., human health, economy, and intelligentsia. A mosquito is small in size but is giant of a killer. In fact, no animal on Earth has affected human lives so directly and profusely as is brought about by the mosquito. Mosquitoes have, therefore, rightfully earned the sobriquet "deadliest predators on the planet."

Mosquitoes belong to the family Culicidae (i.e., culicids, bearing an elongate proboscis, comprised three subfamilies, i.e., Anophelinae, Toxorhynchitinae, and Culicinae, embracing all "true" mosquitoes) in class Insecta and are an enormous source of a great deal of scientific information on life. The dominance of mosquitoes among the world's hematophagous animals is a fundamental scientific insight which means that in chorogeographical and bioecological attributes of biodiversity beyond our comprehension, these creatures not only surpass to function as a self-sustaining biological system but permeate our lives inextricably. They are, in fact, an integral and complex part of the terrestrial, aerial, and freshwater ecosystems with which the future of humans is inseparably interlinked. In the whole animal kingdom, fewer can match the popularity and importance that the mosquito carries to the human life! It is indeed no small enemy of humans to whom they carry viruses, bacteria, and parasites, which they, along with their other arthropod allies, transmit through bites, infecting some 700 million people and killing more than one million every year.

Male mosquitoes are far away from spready any of the above mischiefs. In contrast to their fair sex, they are not only innocuous in purest sense of expression but also beneficial to both nature and humans. Male mosquitoes are often good but less known pollinators. They are currently exploited through transgenic strain development for the control of mosquito populations where females are the culprit. Male mosquitoes have also contributed significantly in developing local anesthesia from the proteins in their saliva. Truly, male mosquitoes, by their side of biology, have a lot of fables to tell for a better tomorrow—in context with sound sleep!

Mosquitoes are among the most studied animals in the world. Many discoveries have won the discoverers many laurels including the Nobel Prize (Dr. Ronald Ross), Darwin Award (Dr. Batista Grassi), Dr. M.O.T. Iyengar Award (Dr. B.K. Tyagi), etc. While some mosquito-borne diseases, for example, malaria and lymphatic filariasis, have been controlled a great deal, nevertheless they still pose many challenges to managers and planners in the country. Malaria in India is targeted for elimination by 2030.

Points to Remember

1. Mosquitoes may be small organisms but they are giant killers.
2. Protection from mosquito bites should be practiced, using repellents, full sleeve shirts, etc.
3. Prevent water storage in and around houses and other living conditions.

References

Alphey L (2002) Re-engineering the sterile insect technique. Insect Biochem Mol Biol 32:1243–1247

Barraud PJ (1934) The fauna of British India, including Ceylon and Burma. Diptera: family Culicidae; Tribes Megarhinini and Culicini, vol 5. Taylor & Francis, London, p 463

Becker N, Petric D, Zgomba M, Boase C, Madon M, Dahl C, Kaiser A (2010) Mosquitoes and their control. Springer, Berlin, Heidelberg, p 577. https://doi.org/10.1007/978-3-540-92874-4

Bodharamik T, Sungvornyothin S, Juntarajumnong W, Bangs M, Arunyawat U (2021) Genetic variation of circadian clock genes in a cavernicolous Anopheles dirus (Diptera: Culicidae) in western Thailand. Agric Nat Resour 55:968–975. https://doi.org/10.34044/j.anres.2021.55.6.07

Christophers SR (1933) The Fauna of British India, including Ceylon and Burma. Diptera: family Culicidae; tribe Anophelini, vol 4. Taylor & Francis, London, p 271

Clements AN (1992) *Biology of mosquitoes*, Volume 1: development, nutrition and reproduction. CABI Publishing; Revised ed. Edition. pp. 552

Clements AN (1999) *Biology of mosquitoes*, volume 2: sensory reception and behaviour. CABI Publishing, p 752

Clements AN (2012) *The biology of mosquitoes*. Volume 3: transmission of viruses and interactions with bacteria. CABI Publishing, p 571

Clements AN, Harbach RE (2018) Controversies over the scientific name of the principal mosquito vector of yellow fever virus—expediency versus validity. J Vector Ecol 43(1):1–14

Downes JA (1978) Feeding and mating in the insectivorous Ceratopogoninae (Diptera). Mem Ent Soc Can 110(S104):1–62. https://doi.org/10.4039/entm110104fv

Gajanana A, Rajendran R, Samuel PP, Thenmozhi V, Tsai TF, Kimura-Kuroda J et al (1997) Japanese encephalitis in South Arcot district, TamilNadu, India: a three-year longitudinal study of vector abundance andinfection frequency. J Med Entomol 34:651–659

Harbach RE (2018) Culicipedia: species-group, genus-group and family-group names in Culicidae (Diptera). CABI, p 378

Harrison G (1978) Mosquitoes, malaria, and man: a history of the hostilities since 1880. E.P. Dutton, New York, p 314

Jambulingam PK, Gunasekharan K, Sahoo SS, Hota PK, Tyagi BK, Kalyanasundaram M (1989) Effect of permethrin-impregnated bednets in reducing population of malaria vector, *anopheles culicifacies* in a tribal village of Orissa state (India). Indian J Med Res 89:48–51

Kalyanasundaram M (1982) A preliminary report on the synthesis and testing of mosquito repellents. Indian J Med Res 76:190–194

Kandasamy C (2021) DEPA (N, N-diethyl Phenylacetamide): commercial journey of India's first indigenous mosquito repellent with novel properties. In: Tyagi BK (ed) Genetically modified and other innovative vector control technologies: eco-bio-social considerations for safe application. Springer, Singapore. 449, pp 369–377. https://doi.org/10.1007/978-981-16-2964-8

Linnaeus C [as Caroli Linnæi]. (1758) *Systema naturæ per regna tria naturæ, secundum classes, ordines, genera, species, cum characteribus, differentiis, synonymis, locis*. Tomus I. Editio Decima, Reformata. Impensis Direct. Laurentii Salvii, Holmiæ [Stockholm]. [10th edition; *Culex* at pp. 602–603]

Linnaeus C (1762). Zweyter Theil, enthalt Beschreibungen verschiedenerwichtiger Naturalien, pp. 267-606. In: Gadebusch, T.H., Reise nach Palastina in den Jahren von 1749 bis 1752. viii + 606 pp. J. C.Koppe, Rostock. [1762.07.??]

Lippi CA, Mundis SJ, Sippy R et al (2023) Trends in mosquito species distribution modeling: insights for vector surveillance and disease control. Parasit Vectors 16:302. https://doi.org/10.1186/s13071-023-05912-z

Mains JW, Brelsfoard CL, Dobson SL (2015) Male mosquitoes as vehicles for insecticide. PLoS Negl Trop Dis 9(1):e0003406. https://doi.org/10.1371/journal.pntd.0003406

Mains JW, Brelsfoard CL, Rose RI, Dobson SL (2016) Female adult *Aedes albopictus* suppression by Wolbachia-infected male mosquitoes. Sci Rep 6:33846. https://doi.org/10.1038/srep33846

Manson P (1878) On the development of *Filaria sanguis hominis* and on the mosquito considered as a nurse. J Linn Soc (Zool) 14:304–311

Patil PB, Dasgupta SK, Gorman K, Pickl-Herk A, Puinean M, McKemey A, et al. (2022) Elimination of a closed population of the yellow fever mosquito, *Aedes aegypti*, through releases of self-limiting male mosquitoes. PLoS Negl Trop Dis 16(5): e0010315. https://doi.org/10.1371/journal.pntd.0010315

Rao TR (1984) The Anopheline of India (revised edition.). Malaria Research Centre, Indian Council of Medical Research, New Delhi, p 518

Ross R (1897) On some peculiar pigmented cells found in two mosquitoes fed on malaria blood. Br Med J 2:1786–1788

Ross R (1898a) Further observations on the transformation of crescents. Indian Medical Gazette

Ross R (1898b) Preliminary report on the infection of birds with Proteosomaby the bites of mosquitoes. Government Press Calcutta (11 Oct, 1898); Indian Medical Gazette (Jan. 1899. *British Medical Journal*

Suárez MF, Nelson MJ (2024) Registro de altitud del *Aedes aegypti* en Colombia. biomedica [Internet]. 1 de diciembre de 1981 [citado 26 de febrero de 1(4):225. https://revistabiomedica.org/index.php/biomedica/article/view/1809

Thenmozhi V, Thiruppathi B, Kasiviswanathan V, Dhananjeyan KJ, Alagarsamy S, Rajamannar V, Tyagi BK (2016) Role of Anopheles subpictus Grassi inJapanese encephalitis virus transmission in Tirunelveli, South India. Indian J Med Res 144(3):477–481

Theobald FV (1901a) A monograph of the Culicidae or mosquitoes. Volume 1. British Museum (Natural History), London

Theobald FV (1901b) A monograph of the Culicidae or mosquitoes. Volume 2. British Museum (Natural History), London

Theobald FV (1903) A monograph of the Culicidae or mosquitoes. Volume 3. British Museum (Natural History), London

Theobald FV (1907) A monograph of the Culicidae or mosquitoes. Volume 4. British Museum (Natural History), London

Theobald FV (1908) First report on the collection of Culicidae and Corethridae in the Indian Museum, Calcutta, with descriptions of new genera and species. Rec Indian Mus. 2: 287–302

Theobald FV (1910) A monograph of the Culicidae or mosquitoes. Volume 5. British Museum (Natural History, London

Tyagi BK (1993) An indigenously developed mechanical device for sampling the resting adults of mosquitoes. Journal of Applied Zoological Researches 4: 180–184

Tyagi BK (2003) Medical entomology: a handbook of insects and other arthropods of medical importance. Scientific Publishers (India), Jodhpur, p 262

Tyagi BK (2004) The invincible deadly mosquitoes. Scientific Publishers (India), Jodhpur, p 265

Tyagi BK (2008) Vector-borne diseases: epidemiology and control. Scientific Publishers (India), p 298

Tyagi BK (2020) Mosquito hunters: a history of hostilities against man's deadliest foe—the mosquito—since 1881. Scientific Publishers (India), p 474

Tyagi BK (2023a) Male mosquito: a zaddy and subaltern sex in the disease epidemiology and control. Scientific Publishers (India), p 232

Tyagi BK (2023b) Desert malaria: an emerging malaria paradigm and its global impact on disease elimination. Springer-Nature, Singapore, p 416

Tyagi BK, Hiriyan J, Samuel PP, Tewari SC, Paramasivan R (2006) Dengue in Kerala: a critical review. ICMR Bull 36(4-5):13–22

Tyagi BK, Munirathinam A and Venkatesh A (2015) A catalogue of Indian mosquitoes. International Journal of Mosquito Research 2 (2): 50–97

Tyagi BK, Reddya NB, Sajal B, Pushpalatha E (2024) *Mosquito: ugly, bold and beautiful*. Bisen Singh Mahendra Pal Singh, Dehra Dun (India).(*in press*)

Wahid I, Sunahara T, an d Mogi M (2002) Mouthparts of male Aedes (Stegomyia) mosquitoes. J. Am. Mosq. Control Assoc. 18(2):73–80

Wilkerson, R.C., Linton, Y-M. and Strickman, D (2021) Mosquitoes of the World. Johns Hopkins University Press, Baltimore, Vol. 1 and Vol. 2, 1332 pp

Positive Aspects of Mosquitoes

17

Shriza Rai, Omkar, and Geetanjali Mishra

Abstract

Mosquitoes are among the most misunderstood creatures on the planet. Contrary to the common belief that mosquitoes serve no purpose except to suck blood, out of estimated 3500 species of mosquitoes, majority do not suck blood. Looking beyond their vampiric tendency, mosquitoes are pollinators of crop plants, decomposers, and provide other ecological services. Huge biomass contributed by mosquitoes makes their existence essential, and their eradication may destabilize the ecosystem. These bizarre creatures hold great potential for innovation as well. Their peculiar painless blood-sucking has inspired the development of microneedles, apart from other mosquito-inspired technologies. This chapter uncovers the positive aspects of the mosquitoes, highlighting their relevance and underestimated potential.

Keywords

Mosquito benefits · Bioinspiration · Elephant mosquitoes · Biomimetics · Microneedles · Pollinators

"A mountain is not nearly as much a wonder as a mosquito".

—Mokokoma Mokhonoana

S. Rai · G. Mishra (✉)
Ladybird Research Laboratory, Department of Zoology, University of Lucknow, Lucknow, Uttar Pradesh, India

Omkar
Ladybird Research Laboratory, Department of Zoology, University of Lucknow, Lucknow, Uttar Pradesh, India

© The Author(s), under exclusive license to Springer Nature Singapore Pte Ltd. 2024

Omkar (ed.), *Mosquitoes*, https://doi.org/10.1007/978-981-97-4163-2_17

Learning Objectives

1. Contrary to the belief that all mosquitoes are harmful, mosquitoes do share some ecological importance. Acknowledging their importance will help us break the myth that "good mosquito is a dead mosquito".
2. Mosquitoes have inhabited nearly all habitats and have survived since ancient times. Understanding the behaviour and ecology of mosquitoes helps us acknowledge their goodness as well as helps us control the harmful effects of mosquitoes.
3. All living creatures have a purpose to serve in the ecosystem; thus, reckless eradication of mosquitoes may create imbalance. There is yet a lot to learn from mosquitoes.
4. Mosquitoes have inspired the development of various technology, and further studying mosquitoes with this approach may further contribute to research and development.

17.1 Introduction

No matter how pointless and irksome mosquitoes seem to humans, they are of considerable importance in the ecosystem. From being pollinators to being an important source of biomass in the food chain, mosquitoes perform many important ecological functions. They also serve as food for fish, bats, frogs, etc. Larvae of certain species of mosquitoes are decomposers, feeding on bacteria, algae, and other microbes in water. While larvae of certain species exhibit fierce cannibalism, i.e., feeding on the larvae of other mosquito species, their biological control potential has been overlooked.

There are more than 3500 species of mosquitoes around the world (Wilkerson et al. 2021), and out of these only a few cause diseases, like malaria, dengue, chikungunya, West Nile virus, etc. In spite of being the deadliest animal on earth causing maximum number of deaths than any other animal on the planet as illustrated in an infographic by Gatesnotes (https://www.gatesnotes.com/Most-Lethal-Animal-Mosquito-Week), majority of mosquitoes do not feed on humans at all and are in fact beneficial to us.

Research on biological structures and behaviour of mosquitoes have bio-inspired designs, paving path for new technologies and developments (Dixon and Vondra 2022). Painless microneedles, blood monitoring devices, drones, etc., are a few such mosquito-inspired technologies. People often assume we have figured mosquitoes out by now, but we are far from done. This chapter uncovers the secret life of mosquitoes which is both bizarre and ecologically important. Also, exploring these tiny creatures from positive aspects may reveal more interesting innovations and help us understand them better.

17.2 Ecological Importance

Mosquitoes are among the most hated creatures in the world, causing more than one million deaths each year as per US Centers for Disease Control and Prevention (2019). However, to live in a world without mosquitoes may not be as great as it sounds. Paul Ehrlich in his "rivet popper hypothesis" has evoked the importance of all living beings in the ecosystem (Ehrlich and Ehrlich 1981). Like thousands of rivets hold various components of an aeroplane together, numerous species maintain the balance in the ecosystem. If every passenger travelling in the aeroplane starts pulling out a rivet to take home, it may eventually compromise the safety of the aeroplane. Likewise, loss or extinction of a mosquito species may initially seem trivial, but their removal can eventually disrupt other species in the ecosystem. No matter how insignificant and pestering their (mosquitoes) existence appear to humans, they are of significant ecological importance.

17.2.1 As Pollinators

There is a popular misconception that all mosquitoes bite, while in reality out of estimated 3500 species of mosquito, only a few hundred bite humans and livestock and are vectors of diseases; even then, only the females bite at the time of egg laying, in order to draw certain proteins from the blood essential for egg laying. We often overlook that the prime food source for mosquitoes is not blood but flower nectar.

Owing to their tiny in appearance and being active around dusk, pollination by mosquitoes is hard to see. During the short growing season in arctic, swarm of nectar-hungry mosquitoes aid in pollination of plants. Being active at dusk and after that, mosquitoes are thus crucial night pollinators, yet our hatred towards them have overshadowed their benefits.

The importance of night pollinators has been demonstrated by flower-blocking experiments. In these experiments, when the night pollinators were excluded, flowers produced significantly fewer viable seeds compared to the flowers whose daytime pollinators were blocked out. One could easily witness huge swarm of mosquitoes on flowers at night, suggesting the importance of mosquitoes as night pollinators contributing to ecosystem functioning as pollinators.

Pollination by mosquitoes has been de-emphasized and is far more common than we realise. Mosquitoes as pollinators go back to 1913, when John Smith Dexter first reported mosquitoes pollinating orchids (Dexter 1913). He discovered pollen from *Platanthera obtusata*, the blunt-leaved bog orchid, stuck to their head and eyes, while feeding on orchid blooms. Later in the 1970s, researchers described the snowpool mosquitoes, *Aedes communis*, as an important pollinator of *P. obtusata* (Gorham 1976). While feeding on the nectar from the flower spur of these orchids, their eyes naturally come into contact with a cluster of pollen also known as the pollinium and get stuck to the eyes of mosquitoes. The pollinium stuck onto the

Fig. 17.1 Mosquitoes feeding on flower nectar and pollinating flowers

eyes of a mosquito touches the stigma of another flower the mosquito visits, and this is how the flower gets pollinated.

The role of mosquitoes as pollinators has been underestimated and not studied as much as their role as blood suckers and diseases vectors. There are multiple evidences of mosquitoes taking up the role of generalist pollinators of some plant families and helping in the pollination of goldenrod, orchids, and other small flowers of wetter region (Hansen 2016). For instance, various species of elephant mosquitoes, like *Toxorhynchites* spp., do not feed on animal blood but on flower nectar and plant sap. While looking for nectar, they pollinate multiple flowers they visit (Fig. 17.1). Thus, elephant mosquitoes are one of many species that might be doing more good for humanity than bad.

While mosquito proboscis has hardly left any ecological pie untasted, their intricately twisted relationship with plants, animals, microorganisms, and pathogens is overshadowed by the itchy bumpy bite and diseases they spread.

With limited ecological niche and only a few pollinators available, the rare arctic orchids are more dependent on mosquitoes for pollination. Thus, if mosquitoes are eradicated completely, the beautiful rare arctic orchids would be among the worst of sufferers.

Blinded with the hatred towards mosquitoes, we are unable to admit anything in favour of our enemy number one. This is probably why they are often excluded from pollinator studies. Thus, which plants they visit and how successfully they pollinate in comparison with more famous pollinators, like bees, butterflies, and beetles, are yet to be explored.

17.2.2 Sourcing Honey dew

The primary source of food for mosquitoes is flower nectar, but they also consume plant sugar in the form of honeydew. Honeydew is excreted as waste product by plant-sucking insects, like aphids. Smell emitted by the microbes present in the honeydew helps mosquitoes to locate honeydew which is otherwise hard to find in the environment (Peach et al. 2019a, b).

Ants are the prime consumers of honeydew, and farm aphids to extract honeydew from the plants. Mosquitoes have learned an interesting way to exploit honey dew from ants. A mosquito tricks ants into regurgitating and sharing its freshly ingested meal by inserting its mouthparts into the ant's mouth and stroking the ant's head with its antennae. However, this interesting mechanism is understudied and deserves more attention.

17.2.3 Contribution in Food Web and in Biomass Transfer

Mosquito larvae feed on microorganisms, like bacteria and algae in water, and larvae are eaten by small and large fish, turtles, and other predators. Thus, mosquitoes contribute an important share in the aquatic food chain and also make up a large biomass of aquatic ecosystem. In Alaska alone, mosquito biomass accounts for up to 96 million pounds. Hence, from where this huge biomass will be contributed if entire mosquito species are wiped out remains a mystery.

Mosquito larvae surviving to adulthood become terrestrial from aquatic. Thus, there is a transformation of biomass from aquatic to terrestrial ecosystems. Larvae and adult mosquitoes are eaten by dragonflies, fish, turtles, birds, bats, and other predators. Therefore, mosquitoes are important to the terrestrial food chain as well. Adult mosquitoes, dead or eaten and excreted, then undergo decomposition and are turned into nutrients for plants, thus contributing to another ecological function. However, their role in nutrient-cycling remains understudied.

17.2.4 Disease Reduction and Ecological Balance

Mosquitoes, being the world's deadliest animal, cause immense suffering worldwide. However, not all species of mosquitoes is responsible for the spread of diseases. Thus, targeting all the species of mosquitoes would aid in already collapsing ecosystem and declining pollinator populations. By acknowledging the sophisticated lifestyle of mosquitoes along with developing better understanding of their behaviour will be helpful in coming up with sustainable strategies to control the spread of mosquito and mosquito-borne diseases.

Biologists and entomologists come face to face when it comes to thinking of the consequences of mosquito extinction. Biologist Janet McAllister at the Centre for Disease Control and Prevention in Colorado (Fang, 2010) shares a negative opinion for mosquitoes and finds little use for the mosquito as either a pollinator or significant player in the insect food chain. Sharing his opinion, entomologist Daniel Strickman (Fang, 2010) belittles mosquito by stating that by eradicating all mosquitoes more human lives could be saved and the government will be freed of their high malaria burden which could be used in recovering country's gross domestic product, potentially accelerating the development.

The largest concentration of mosquitoes is observed in arctic tundra of Canada, Alaska, and Russia; the effects of extinction would be most observable in these places. As per Bruce Harrison (Brothers 2022), an expert and entomologist at North Carolina Department of Environment and Natural Resources, mosquito extinction may lead to 50% drop in the number of migratory birds that nest in the arctic tundra. Agreeing with this, entomologist Rich Merritt, University of Michigan (Brothers 2022), anticipated that with so much biological mass that mosquitoes contribute to the food chain (as discussed before), their extinction would affect many aquatic animals to switch their diet in order to survive. Without mosquito larvae, animals like fishes that rely on mosquito larvae would have to adapt to changes in their diet in order to survive. This is not as simple as it sounds, as traits like feeding behaviour are deeply imprinted genetically. For instance, *Gambusia affinis,* commonly known as mosquito fish, is specialised in killing mosquitoes and is so effective that *Gambusia* spp. are implied to control mosquitoes' breeding in swimming pools. Thus, the extinction of these specialised predators could have a long-lasting effect throughout the food chain.

17.3 Other Ecological Services

Indicator of water quality: Transparent larvae of tufted mosquitoes belonging to the genus *Chaoborus* (Corethra) are indicator of very good quality of water (Teilen 2018).

Decomposers: Mosquitoes in larval stage act as ecosystem's clean-up crew by feeding on dead organic matter, decaying leaves, and other microbes (Donald et al. 2020). Larvae filter bacteria and algae from water and keeping it clean. Though many biologists believe that the complete eradication of mosquito species will hardly be an issue as similar organisms will soon step in to fill the roles of mosquitoes. As yet, this is an understatement in view of our limited understanding of these bold, beautiful, and bizarre creatures, and mosquitoes do serve a purpose on earth.

Biological control: Certain species of mosquitoes have fiercely cannibalistic larvae. For instance, elephant mosquitoes *(Toxorhynchites)* are known as "mosquito-eaters", and the larvae prey on the larvae of other mosquitoes as well as on each other. Being harmless in itself, elephant mosquitoes can be used as biological control agents; thus, further studies need to be conducted in this direction.

17.4 Inspiring Technology

17.4.1 Drones

Night owls of the insect world and how mosquitoes avoid running into the walls in complete darkness have inspired researchers to develop drones with improved night vision. Scientists are working on building a sensor that may prevent the crashing of helicopters in low vision (Young and Garratt 2020).

The team of researchers have filmed *Culex quinquefasciatus*, a pesky mosquito species known to transmit Zika and West Nile viruses. An array of around 12,000 cells is arranged in circle at and around the base of each antenna of the mosquito. These resemble upside down umbrella and is responsible for the wobbling of antenna. Multiple images taken at multiple angles have increased the understanding of how air moves off the insect's long, slender wings and how the flow alters as the mosquito flies closer to the surface without colliding, greatly affecting the air flowing around the antennae. Making use of the way wings generate a downward draft when flying near the surface, scientists have developed computer simulations of this flow.

17.4.2 Medical Discoveries

Mosquitoes have inspired researchers that have led to the advancement of technologies (Fig. 17.2). Bitten by a mosquito, the pain is often not felt because the mosquito essentially anesthetizes the biting surface with its mouth. These anaesthetic properties of mosquito saliva are being studied and being incorporated in synthetic forms into local and topical anaesthetics. In addition, according to "Mosquito Info", prototype products have been developed to allow diabetics to test their blood sugar more painlessly with the use of a device resembling the tentacles of the mouth of a mosquito.

For implantation of intracortical microelectrodes into the brain, next-generation microelectrode has been designed biomimicking mosquito insertion of its proboscis into the host's skin (Shoffstall et al. 2018). This prototype provides a reproducible method to augment the insertion of small, flexible devices into the brain with great precision.

Apart from mosquito biting, movement, vision, and olfaction are other behaviours that hold the potential for inspiring technologies (Dixon and Vondra 2022). For

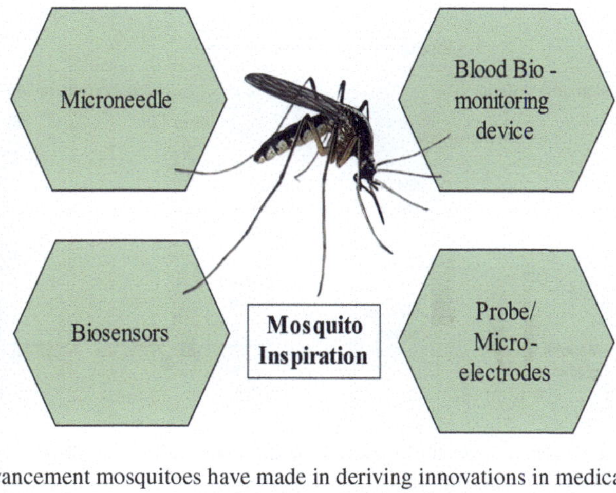

Fig. 17.2 Advancement mosquitoes have made in deriving innovations in medical field

example, research is being conducted on viscoelastic proteins used in mosquito movement for use in tissue engineering and drug delivery applications. Superhydrophobic surface coating of the mosquito eye has inspired the creation of a robust and water-resistant biomaterial for drug delivery use. Biomimicking olfactory receptors of mosquito, biosensors have been developed that will aid in the detection of disease-specific volatile organic compounds. Mosquitoes thus hold a lot of scope for the exploration and by further dissecting mosquito features; various other prospects could be uncovered for the continued development of novel innovations.

17.4.2.1 Microneedles

Researchers Bharat Bhushan and Navin Kumar from the Indian Institute of Technology (IIT) Ropar, in collaboration with scholars at Ohio State University, have developed a technology for painless microneedles (Gurera et al. 2018). Inspired by the painless bite of mosquitoes, the team devised a method where individuals do not feel discomfort when pierced (Fig. 17.3). This breakthrough offers potential applications in various medical procedures.

Mosquito proboscis-inspired (MPI) insertion technology is used in biopsy needles in order to reduce tissue deformation and organ displacement (Li et al. 2020). With the current available needle delivery systems, the soft tissue and organs surrounding the area of piercing get deformed, resulting in a non-diagnostic biopsy or under-sampling of the target.

MPI insertion has adopted two features inspired by mosquitoes in prostate biopsy: (1) the harpoon-shape notches at tip of the needle and (2) reciprocating needle-cannula motions for incremental insertion (Li et al. 2020). It has been found that the MPI needle insertion has reduced both local tissue deformation and global prostate displacement. Thus, MPI insertion in the clinical biopsy procedures as well as insights of needle–tissue interaction offer less painful and advanced microneedles.

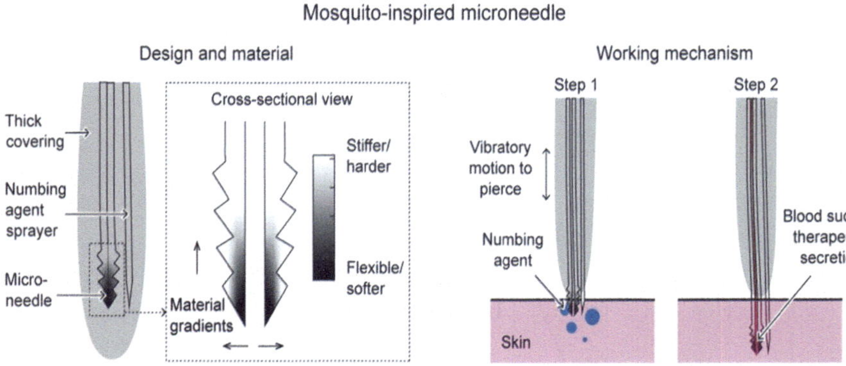

Fig. 17.3 Researchers developed the proposed needle using principles they learned from how mosquitoes pierce skin without pain. Image courtesy of Bhushan, Kumar, Guerra

17.4.2.2 Biomonitoring of Blood: E-Mosquito

Exploiting the mechanism by which mosquito sucks blood, scientists of the University of Calgary, Canada, have come up with a blood monitoring device (Wang et al. 2017). This device pierces the skin in a similar manner to that of mosquitoes' mouthparts and extracts a small amount of blood to use for glucose testing.

The present prototype is the hard work of Mintchev and his team (Matchar 2017) who have been working for a decade on e-mosquito device and have been successful in miniaturizing the device to the size of a watch, making it more convenient to wear. This device can penetrate the skin with greater force and controllability, and the prick is as painless as that of mosquitoes. The e-mosquito watch can be worn on the wrist like any smartwatch but could also be strapped anywhere on the body. However, there are a few challenges to be overcome before the introduction of this device in the market. While the e-mosquito can accurately hit a capillary, at times the blood it brings to the surface is not enough for testing. In this sense, the device is truly similar to a mosquito, which rarely leaves behind a pool of blood on the surface of the skin. To resolve this issue, Mintchev and his team is working on developing a needle that doubles as a sensor. As the needle penetrates the skin, the sensor checks the blood while still embedded and transmits the results.

John L. Smith, a former diabetes industry consultant, has written a book on the failures of various continuous blood glucose-monitoring technologies (Smith and Rice 2015). In his book, he describes tiny needle technologies, like e-mosquito as "[a] recurrent technological theme" that has been tried many times over the years but has yet to bear fruit.

Developed in a watch-like band, the e-mosquito can be programmed to automatically prick the skin multiple times a day. Since a person with type-1 diabetes need to prick fingers up to eight times a day, e-mosquito can make the process less tiring and less injuring, and the results can be analysed and alert a diabetic person of their blood glucose level. Thus, an e-mosquito device inspired from the mosquito's blood-sucking jab could relieve people, children, the elderly, and the blind, in particular, of the traditional way of testing sugar by pricking a finger and wiping the blood on the test strips. The latest version is a "watch-like wearable microsystem for Minimally Invasive, Pseudo-Continuous Blood Glucose Monitoring" that "bites"; a person wearing it every few hours draws blood and tests glucose levels (Matchar 2017). The device can also be used for genetic testing or any sort of blood analysis that needs to be done regularly. The improved consumer-ready e-mosquito will soon be available in the market, as little as three years, depending on FDA approval.

17.5 Busting the Myths

Mosquitoes are among the most misrepresented lifeforms on earth. There is a popular belief that "a good mosquito is a dead mosquito". However, this myth about mosquitoes needs to be dispelled. Only a few mosquitoes are harmful out of 3500 species, but the annoyance these few species have created globally gives all the mosquitoes a bad rap. According to the National Wildlife Federation, mosquitoes have more to serve than being just an everyday pollinator. "Mosquitoes are part of the food web, serving as important prey in both winged adult and aquatic larval form for a lot of other wildlife from dragonflies and turtles to bats and birds—including hummingbirds, which rely on small flying insects and spiders as a primary food source".

Not all mosquitoes are thirsty for blood; rather both males and females are vegetarian, preferring plant matter, like nectar, honeydew, etc. Females blood-feed only before oviposition, drawing nutrients from blood to lay eggs. Also, it's not likely that any mosquito can carry any of the diseases. For instance, malaria is carried by only one genus of mosquito, i.e., female *Anopheles*.

Contrary to what many believe that mosquitoes like hot, moist places and the tropics, this is a bit misleading. Mosquitoes rather prefer cool, moist areas, which is why they proliferate in cool and damp conditions. In tropical areas where there is a lot of canopy cover, it can be warm and humid, but it is actually colder under the canopy. Probably, this is the reason why it is unlikely to encounter mosquitoes in the middle of a field in hot sunny conditions with no trees, unless it's dawn or dusk or when things start to cool down.

17.6 A Lot to Learn

As humans, we tend to be a little self-absorbed when it comes to nature pestering us. Anyways, mosquitoes as a group are generous beasts. Blind to the benefits each organism in nature renders, we have a lot to learn from mosquitoes. Talking of their feeding patterns, does the mosquito opportunistically feed on humans? Why does mosquitoes feed at dusk, during night or in the daytime? How do they transfer diseases without themselves getting infected? Exploring such questions and understanding their behaviour is essential in effectively controlling mosquitoes.

Our hatred towards mosquitoes has outsmarted their excellent survival instincts. Each organism has evolved to exploit its niche. Interestingly, mosquitoes have found the means of tapping into a nutrient-rich depository to make offspring without having to expend the energy to overcome an animal, as a predator normally would. Their remarkable sensory systems help them in locating their hosts by detecting small quantities of carbon dioxide from human breath up to 50 metres away. After relocating the host, they pick body heat and odours prior to any use of vision for landing on suitable spot.

Though mosquito larva can be found from temporary snow-melted pools to lakes and a few likes of *Deinocerites cancer* can even be found in a few types of saltwater habitats, such as crab burrows (Adames 1971). Another interesting habitat for mosquito larvae is the phytotelma (small water cavity) of the carnivorous plant, *Sarracenia purpurea* (Heard 1994). The mosquito larva of *Wyeomyia smithii* resides in the phytotelma (Fig. 17.4) till it goes through the fifth instar, feeding on decomposing insects. Larvae belonging to other species of mosquito, like *Toxorhynchites* spp., are also reported from pitcher plants (Maquart et al. 2023). Researchers have busted the secret of surviving in this bizarre habitat and have found that digestive

Fig. 17.4 Mosquito larvae of *Wyeomyia smithii* residing in phytotelma (small water cavity) of the carnivorous plant *Sarracenia purpurea*

enzymes in this plant are too weak to dissolve the mosquito larvae. Other such bizarre habitats include the placement of eggs by some mosquito species (*Culex* spp.) in the water that get collected between the leaves of tropical plants in the Brazilian Atlantic Forest (Cardoso et al. 2015), and the larvae of some other mosquito species (*Coquillettidia perturbans* (Walker), also known as cattail mosquito) attach themselves to the roots of aquatic plants to breathe (Morris et al. 1990).

Ancient association of mosquitoes and flowers is evident from the mosquito scales found in flower fossils from the mid-Cretaceous era. Owing to the ancient connection between mosquitoes and flowers, flowers have strongly influenced the evolution of mosquitoes. Probably this is a reason for rapid increase in mosquito diversity corresponding with the appearance of flowering plants. Recent studies have found that some of the odour constituents of certain flowers that mosquitoes feed on (and pollinate) are shared with humans (Peach et al. 2019a, b). More study in this interaction may possibly indicate the evolutionary origins of why some mosquitoes take blood.

While enjoying a blood meal, chances of female mosquito picking up a pathogen or a parasite from their host is high. However, the mosquitoes do not get sick while carrying these pathogens and transmitting them to another host. The midgut lining of mosquito has properties that can prevent the pathogen or parasite from getting out of the gut as it has to make its way through the mosquito's haemolymph, containing circulating immune cells, or haemocytes, which are similar to our white blood cells. Thus, more studies are required to be conducted on the immune system of mosquitoes, which is not possible if these species are eradicated. Also, not all mosquitoes are vectors, and what makes some mosquitoes carriers of some of the deadliest diseases is a subject of research. Another problem of getting rid of these most hated creatures is that we still do not know everything about them; thus, there is a possibility that mosquitoes can be more useful in ways we cannot imagine.

The more we know about these fascinating creatures, the better we can defend ourselves against the diseases mosquitoes spread and appreciate their goodness.

17.7 A World without Mosquitoes

Mosquitoes top the list of the deadliest animals on the planet. However, it is not the mosquitoes that cause the deaths, but the pathogens they transmit cause the illness. Looking at the bigger picture, mosquitoes are not as useless as they facilitate the ecosystem and can drive innovation and research. Mosquitoes have also inspired new technologies that have reduced our discomfort such as microneedles based on the specifications of a mosquito's proboscis, which have reduced the pain from injections.

Mosquitoes just offer uber-like transportation services for many parasites and pathogens, and not all of these passengers are killers. If we get rid of all mosquito species, what will happen to them is unclear, but the risk of it being catastrophic is too great to actually attempt eradication. Owing to the large biomass mosquitoes create, their complete eradication would be a huge loss in terms of the other fauna that rely on them for food. Also, with honeybees and other pollinators declining, mosquitoes could stabilise the pace of pollination and protect the world from food shortages (Peach and Gries 2016).

Eradication of one species might affect the whole food web, thus producing large alterations over time. Hence, we cannot be sure if these pesky creatures deserve so much of hatred, and who knows more studies on their behaviour and ecology may someday uncover their values in the future that we are not currently aware of.

17.8 Conclusion

The debate on the value of mosquitoes on earth is not new in the scientific community. Though we may not care to acknowledge the purpose mosquitoes serve, mosquitoes contribute their share in nature and have ecological significance. Although often overlooked, mosquitoes play crucial roles in the ecosystem. They contribute to pollination and add significantly to the overall biomass.

From the deadliest creature on earth to the most misunderstood animal on the planet, mosquitoes have come far away. While they are among the most disliked creatures, particularly by humans, they have the potential to revolutionise science and technology. Looking beyond their vampiric tendencies, understanding mosquitoes in the context of their environment may lead to discovery of more such positive aspects.

With limited knowledge on the behaviour and ecology of mosquitoes, there is so much to explore about them- from fancy colours, peculiar sex lives, and so many hosts. All these reflect a rich diversity of mosquitoes that is hard to ignore once you take a close look into their bizarre lives. With so much to learn about mosquitoes, we should give away the thought that we have figured them out, while the reality is we are far from done.

Points to Remember

1. There are approximately 3500 species of mosquitoes known around the world.
2. Majority of the mosquitoes do not bite mammals. Out of those that do, only the females feed on blood.
3. Mosquitoes serve various ecological purposes, viz., pollination, decomposition, biomass production, nutrient cycling, etc.
4. Mosquitoes have inspired development and advancements in science and technology.
5. Making use of the way wings generate a downward draft when a mosquito is flying near the surface, scientists have developed computer simulations of this flow. This is being used in preventing drones from crashing.
6. Microneedles, blood monitoring devices, micro-electrodes, etc., are inspired by mosquitoes and the way they pierce their proboscis into the skin of the host without causing pain.
7. The more we know about these fascinating creatures, the better we can defend ourselves against the diseases mosquitoes spread and appreciate their goodness.

References

Adames AJ (1971) Mosquito studies (Diptera, Culicidae). XXIV. A revision of the crabhole mosquitoes of the genus Deinocerites. Contrib Am Entomol Inst 7:1–154

Brothers BB (2022) What are the benefits of mosquitoes? Best Bee Brothers. https://bestbeebrothers.com/blogs/blog/what-are-mosquitoes-good-for

Cardoso CAA, Lourenço-de-Oliveira R, Codeço CT, Motta MA (2015) Mosquitoes in bromeliads at ground level of the Brazilian Atlantic Forest: the relationship between mosquito Fauna, water volume, and plant type. Ann Entomol Soc Am 108(4):449–458. https://doi.org/10.1093/aesa/sav040

Dexter JS (1913) Mosquitoes pollinating orchids. Science 37(962):867–867

Dixon AR, Vondra I (2022) Biting innovations of mosquito-based biomaterials and medical devices. Materials 15(13):4587

Donald CL, Siriyasatien P, Kohl A (2020) Toxorhynchites species: a review of current knowledge. Insects 11(11):747

Ehrlich P, Ehrlich A (1981) Extinction: the causes and consequences of the disappearance of species. Random House. http://hdl.handle.net/10822/788604

Fang, J (2010) Ecology: A world without mosquitoes. Nature, 466(7305)

Gorham JR (1976) Orchid pollination by Aedes mosquitoes in Alaska. American Midland Naturalist, 208–210

Gurera D, Bhushan B, Kumar N (2018) Lessons from mosquitoes' painless piercing. J Mech Behav Biomed Mater 84:178–187

Hansen E (2016) Orchid fever: a horticultural tale of love, lust, and lunacy. Vintage

Heard SB (1994) Pitcher-plant midges and mosquitoes: a processing chain commensalism. Ecology 75(6):1647–1660. https://www.gatesnotes.com/Most-Lethal-Animal-Mosquito-Week

Li AD, Putra KB, Chen L, Montgomery JS, Shih A (2020) Mosquito proboscis-inspired needle insertion to reduce tissue deformation and organ displacement. Sci Rep 10(1):1–14

Maquart PO, Rahola N, Chhuoy K, Heng K, Chhum M, Suor K, Boyer S (2023) Description of a new species of Toxorhynchites (Diptera: Culicidae) from Nepenthes pitchers in Cambodia. J Asia Pac Entomol 26(2):102064

Matchar E (2017) A blood-monitoring device inspired by mosquitoes . Smithsonian Magazine, July 14, 2017. https://www.smithsonianmag.com/innovation/blood-monitoring-device-inspired-mosquitoes-180963989/

Morris CD, Callahan JL, Lewis RH (1990) Distribution and abundance of larval Coquillettidia perturbans in a Florida freshwater marsh. J Am Mosq Control Assoc 6:452–460

Peach DA, Gries G (2016) Nectar thieves or invited pollinators? A case study of tansy flowers and common house mosquitoes. Arthropod Plant Interact 10:497–506

Peach DA, Gries R, Young N, Lakes R, Galloway E, Alamsetti SK et al (2019a) Attraction of female Aedes aegypti (L.) to aphid honeydew. Insects 10(2):43

Peach DA, Gries R, Zhai H, Young N, Gries G (2019b) Multimodal floral cues guide mosquitoes to tansy inflorescences. Sci Rep 9(1):1–10

Shoffstall AJ, Srinivasan S, Willis M et al (2018) A mosquito inspired strategy to implant microprobes into the brain. Sci Rep 8:122. https://doi.org/10.1038/s41598-017-18522-4

Smith JL, Rice MJ (2015) Why have so many intravascular glucose monitoring devices failed? (2015). J Diabetes Sci Technol 9(4):782–791. https://doi.org/10.1177/1932296815587013

Teilen B (2018) Insect larvae as live food—mosquito larvae & co. Garnelio, 13 July 2018. https://www.garnelio.de/en/blog/nutrition/insect-larvae-as-live-food-mosquito-larvae-co.

U.S. Centers for Disease Control and Prevention (2019) “Fighting the world’s deadliest animal.” https://www.cdc.gov/globalhealth/stories/2019/world-deadliest-animal.html. August 5, 2019

Wang G, Poscente MD, Park SS, Andrews CN, Yadid-Pecht O, Mintchev MP (2017) Wearable microsystem for minimally invasive, pseudo-continuous blood glucose monitoring: the e-mosquito. IEEE Trans Biomed Circuits Syst 11(5):979–987. https://doi.org/10.1109/TBCAS.2017.2669440

Wilkerson RC, Linton Y-M, Strickman D (2021) Mosquitoes of the world. Johns Hopkins University Press, Baltimore

Young J, Garratt M (2020) Drones become even more insect-like. Science 368(6491):586–587

SPRINGER NATURE

GPSR Compliance

The European Union's (EU) General Product Safety Regulation (GPSR) is a set of rules that requires consumer products to be safe and our obligations to ensure this.

If you have any concerns about our products, you can contact us on ProductSafety@springernature.com

In case Publisher is established outside the EU, the EU authorized representative is:

Springer Nature Customer Service Center GmbH
Europaplatz 3
69115 Heidelberg, Germany

The manufacturer's authorised representative in the EU is Springer Nature Customer Service Centre GmbH, Europaplatz 3, 69115 Heidelberg, Germany. If you have any concerns regarding our products, please contact ProductSafety@springernature.com

Printed and bound by CPI Group (UK) Ltd, Croydon, CR0 4YY
01/10/2026
02225862-0012